Excel

Get the Results You Want!

SUCCESS ONE® HSC TOPIC-BY-TOPIC

MATHEMATICS EXTENSION 1

Past HSC questions arranged into topics with worked answers **1992–2022**

BRAND NEW SERIES

PASCAL PRESS

ISBN 978 1 74125 731 1

Pascal Press
PO Box 250
Glebe NSW 2037
(02) 8585 4050
www.pascalpress.com.au

Publisher: Vivienne Joannou
Commissioning and series editor: Mark Dixon
Project editor: Rosemary Peers
Edited by Rosemary Peers
Answers checked by Peter Little
Cover and page design by Sonia Woo
Page layout and typesetting by lj Design (Julianne Billington)
Printed by Vivar Printing/Green Giant Press

CONTENTS

ABOUT THIS BOOK

Revise the smart way and use this guide during the year to complete HSC questions whenever you have completed a topic in class and need to study for a topic test, an assessment task or lastly for the HSC Exam.

This book contains over 700 questions from 31 years of HSC past papers arranged into syllabus topics.

Each topic:

- has questions arranged in order with the most recent HSC papers first
- has questions marked with one of three levels of difficulty: Easy, Medium and Hard
- has questions relevant only to the current course (the current syllabus was first examined in 2020).

Extra information on the questions

- **Bonus question** Some questions have been written to give extra practice on a topic. These are usually new syllabus topics where there are only past HSC questions from 2020 onwards.
- The exam from which each question is drawn is indicated at the end of each question. **HSC** indicates the exam is Mathematics Extension 1 or the course that preceded it: Mathematics 3 Unit (1992–2000). Due to the changes in the syllabus, questions from HSC Mathematics Advanced or the versions of the course that preceded it are also included: Mathematics (2001–2019) or Mathematics 2 Unit (1992–2000). These are indicated by **MA HSC.**
- Some questions do not have marks included because these were not indicated in the HSC Exam paper from which they came. The answer section will then include the note: (**No mark allocation in exam**).
- Some questions only have total marks included because marks for the individual parts were not indicated in the HSC Exam paper from which they came. The answer section will then include the note: (**Total mark allocation only in exam**).
- In some instances if there is a multipart question and one part is not on the chapter's topic or one part is no longer part of the syllabus for this course, it will be omitted. However, if the answer to the omitted part is needed to complete the remaining part it will be **provided in square brackets**.

Self-assessment summary

Students can complete the self-assessment summary on page vi to assess their progress.

Past HSC Examination papers

- The past HSC Examination questions contained in this publication have been reproduced under licence from the NSW Education Standards Authority (NESA) in whom copyright is vested.
- NESA takes no responsibility for errors in the reproduction of past HSC Examination questions contained in this publication.
- NESA was the first publisher of the HSC Examination papers from which the questions come in the years 1992–2022.
- HSC Examination papers in Mathematics 3 Unit 1992–2000 © NSW Education Standards Authority for and on behalf of the Crown in right of the State of New South Wales 1992–2000
- HSC Examination papers in Mathematics Extension 1 2001–2022 © NSW Education Standards Authority for and on behalf of the Crown in right of the State of New South Wales 2001–2021
- HSC Examination papers in Mathematics 2 Unit 1992–2000 © NSW Education Standards Authority for and on behalf of the Crown in right of the State of New South Wales 1992–2000
- HSC Examination papers in Mathematics 2001–2019 © NSW Education Standards Authority for and on behalf of the Crown in right of the State of New South Wales 2001–2019
- HSC Examination papers in Mathematics Advanced 2020–2022 © NSW Education Standards Authority for and on behalf of the Crown in right of the State of New South Wales 2020–2022

Worked answers

- The worked answers contained in this publication are examples of answers which the authors believe would score full marks.
- They are not necessarily ideal or model answers, nor are they the only answers which would score full marks.
- They are not endorsed by NESA.
- For Mathematics Extension 1 (and the previous course Mathematics 3 Unit) worked answers for the 1992–2003 HSC Examination papers were written by José Carreno, for the 2004–2008 HSC Examination papers by Barbara D'Angelo and Megan McKeown, for 2009 by Barbara D'Angelo, for 2010–2020 by Lyn Baker and for 2021–2022 by Mitchell Jones.
- For Mathematics Advanced (and the previous courses Mathematics and Mathematics 2 Unit) worked answers for the 1992–2003 HSC Examination papers were written by Keith Saines, for the 2004–2005 HSC Examination papers by Lynne Knapman and Berra Mossemenear, for 2006–2009 by Berra Mossemenear, Graeme Downward, Bronwyn Opferkuch and Robert Russell, and for 2010–2022 by Allyn Jones.
- Bonus questions and answers were written by Allyn Jones and Lyn Baker.
- The ticks in the solutions are provided as a guide only to the division of marks and are not endorsed by NESA.

SELF-ASSESSMENT SUMMARY

Note: For the purposes of the functionality of this summary, all parts of questions which have been assigned a level are counted as a whole question.

YEAR 11	EASY	MEDIUM	HARD	MASTERY
FUNCTIONS				
Graphical relationships	/29	/5	/2	1 2 3 4 5
Inequalities	/33	/12	/2	1 2 3 4 5
Inverse functions	/33	/33	/4	1 2 3 4 5
Parametric form of functions or relations	/9	/5	/1	1 2 3 4 5
Remainder and factor theorems	/26	/9		1 2 3 4 5
Sums and products of roots of polynomials	/18	/18		1 2 3 4 5
TRIGONOMETRIC FUNCTIONS				
Inverse trigonometric functions	/15	/5	/1	1 2 3 4 5
Further trigonometric identities	/8	/11	/1	1 2 3 4 5
CALCULUS				
Rates of change with respect to time	/16	/5		1 2 3 4 5
Exponential growth and decay	/62	/18	/2	1 2 3 4 5
Related rates of change	/17	/14	/6	1 2 3 4 5
COMBINATORICS				
Permutations and combinations	/35	/20	/6	1 2 3 4 5
Binomial expansions and Pascal's triangle	/12	/4	/6	1 2 3 4 5

YEAR 12	EASY	MEDIUM	HARD	MASTERY
PROOF				
Proof by mathematical induction	/11	/15		1 2 3 4 5
VECTORS				
Introduction to vectors	/19	/8	/3	1 2 3 4 5
Further operations with vectors	/29	/29	/12	1 2 3 4 5
Projectile motion	/28	/42	/37	1 2 3 4 5
TRIGONOMETRIC FUNCTIONS				
Trigonometric equations	/18	/24	/3	1 2 3 4 5
CALCULUS				
Further calculus skills	/84	/38	/6	1 2 3 4 5
Further area and volume of solids	/31	/38	/4	1 2 3 4 5
Differential equations	/13	/13	/9	1 2 3 4 5
STATISTICAL ANALYSIS				
Bernoulli and binomial distributions	/43	/26	/7	1 2 3 4 5
Normal approximation for the sample proportion	/9	/8	/4	1 2 3 4 5

1 The graph of $f(x) = \dfrac{3}{x-1}$ is shown.

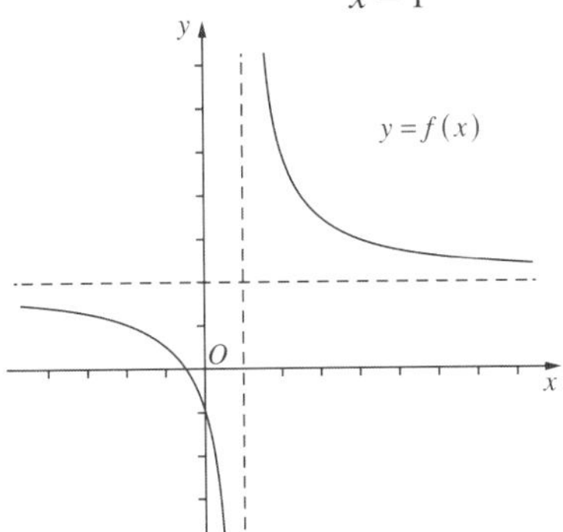

The graph of $f(x)$ was transformed to get the graph of $g(x)$ as shown.

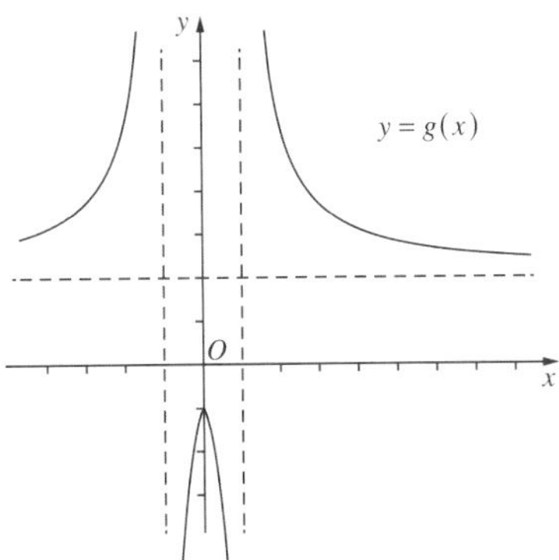

What transformation was applied?

A $g(x) = f(|x|)$ **B** $g(x) = \sqrt{f(x)}$

C $g(x) = f(-x)$ **D** $g(x) = \dfrac{1}{f(x)}$ *(1 mark)*

(Q2, **2022 HSC**) Easy

2 The diagram shows the graph of the sum of the functions $f(x)$ and $g(x)$.

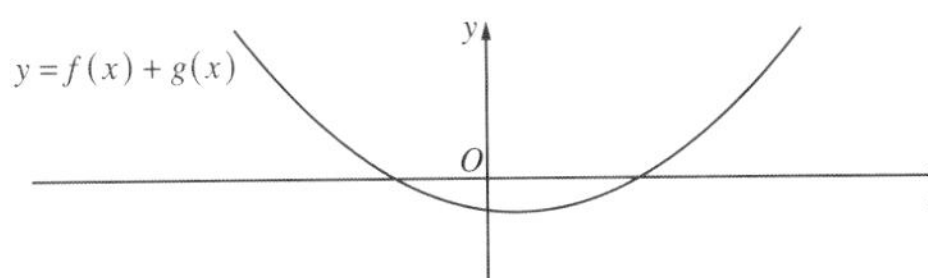

Which of the following best represents the graphs of both $f(x)$ and $g(x)$?

A

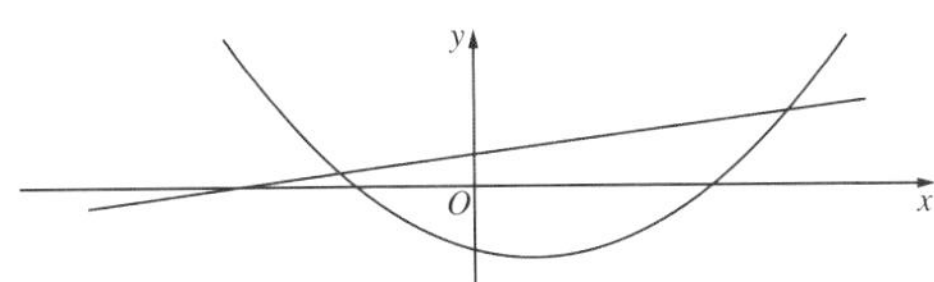

B

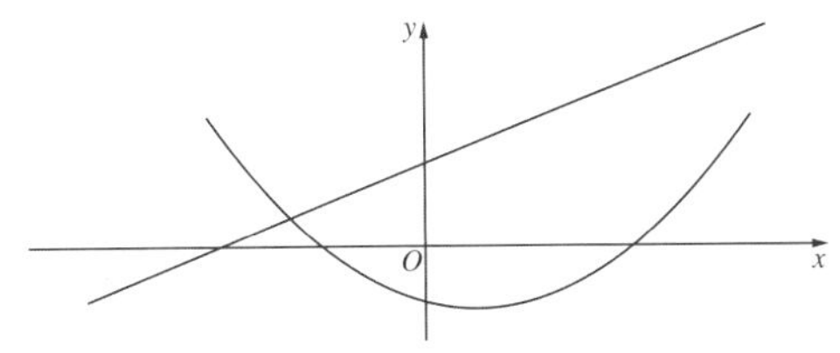

C

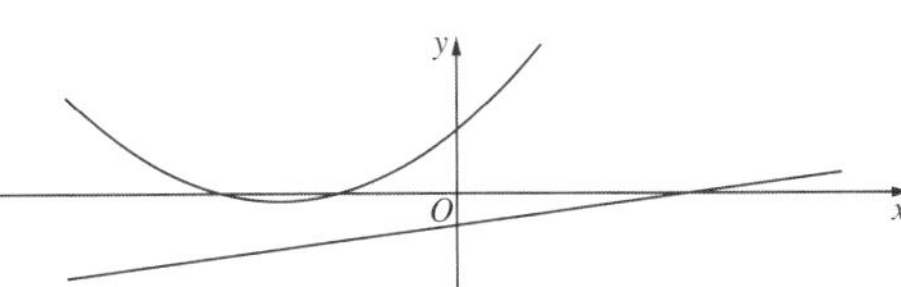

D

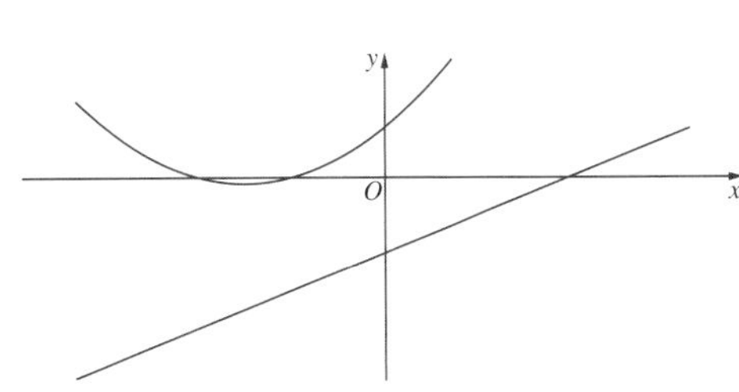

(1 mark)

(Q4, **2022 HSC**) Easy

3 The diagram shows the graph of $y = f(x)$.

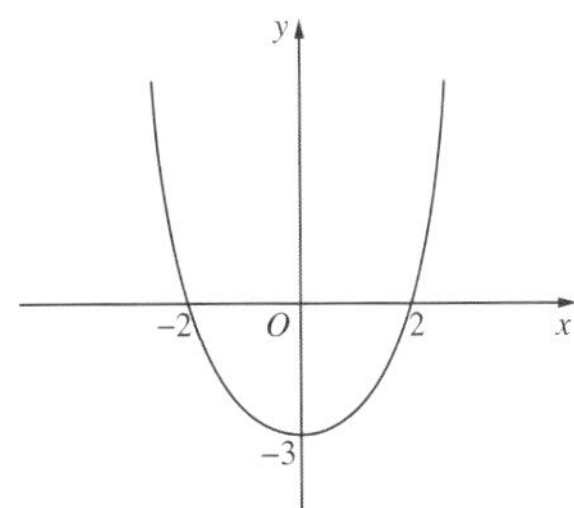

Sketch the graph of $y = \dfrac{1}{f(x)}$. *(3 marks)*

(Q11c, **2020 HSC**) Easy

4 Each of the diagrams below shows the graph of $y = f(x)$ which has x-intercepts of $-4, 0, 2$ and 4 passes through the point $(1, 1)$.

On each diagram draw the graphs of the following functions:

i $y = |f(x)|$ *(1 mark)* Easy

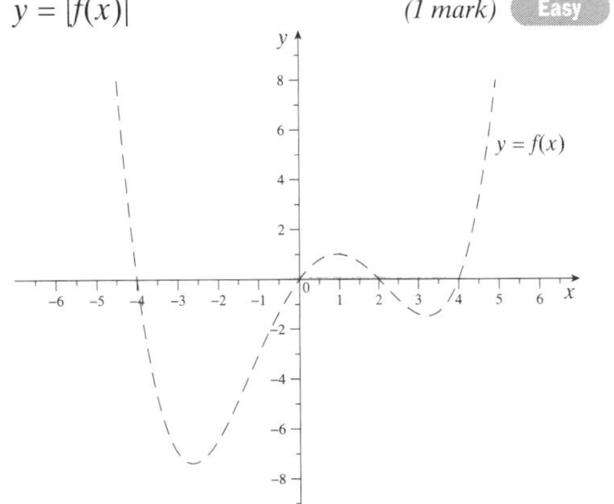

ii $y = \sqrt{f(x)}$ *(2 marks)* Easy

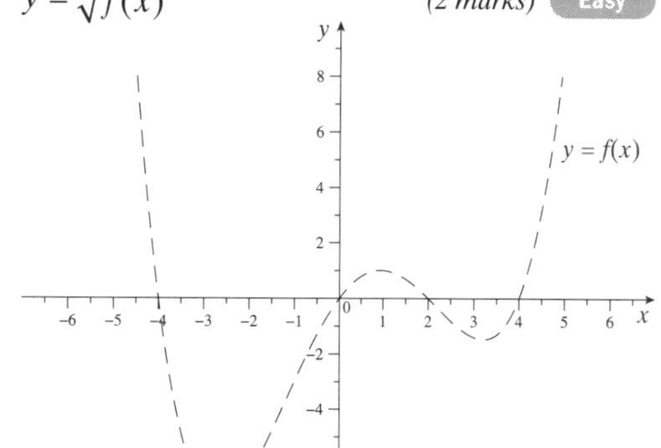

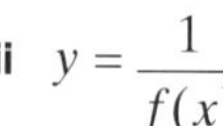

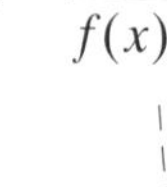

iii $y = \dfrac{1}{f(x)}$ *(2 marks)* Easy

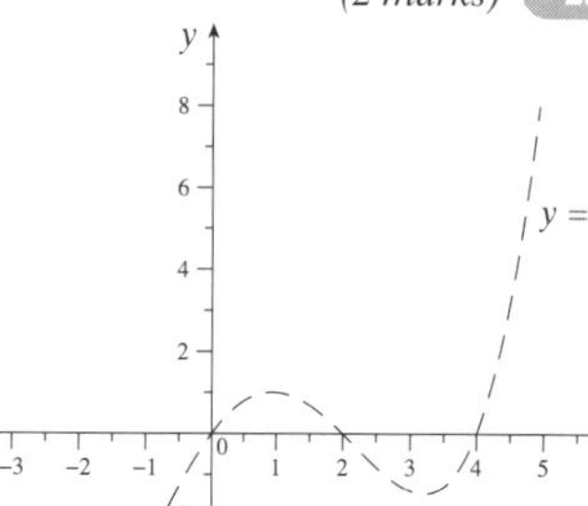

Bonus question (see page iv)

5 The diagram shows the graph of $y = f(x)$.

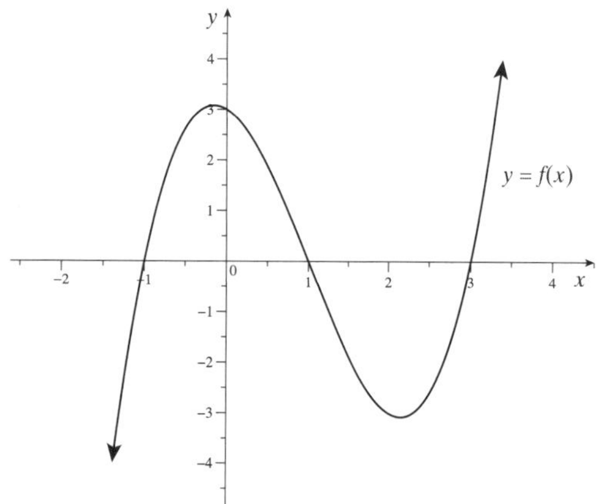

On the number plane below draw the graph of $y = f(\sqrt{x+1})$. *(2 marks)*

Bonus question Easy

6 The graph of $y = f(x)$ is shown below.

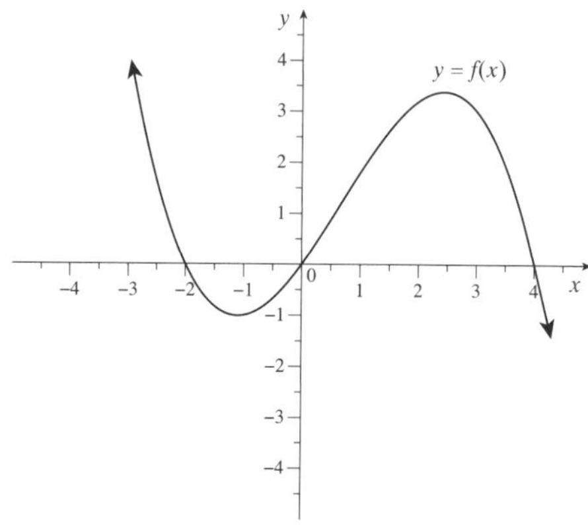

The graph of $y = \dfrac{1}{f(x)}$ is best represented by

A

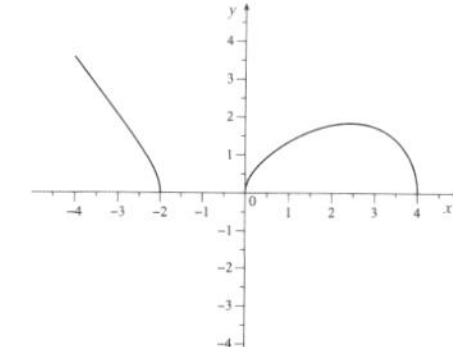

B

C

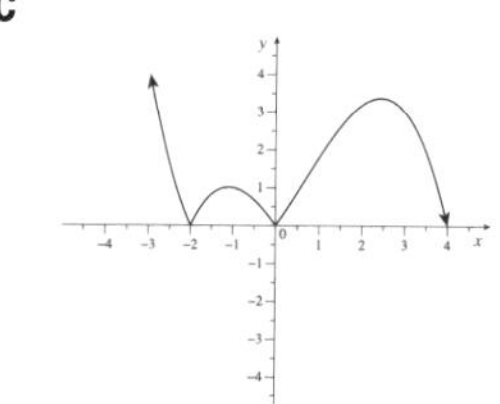

D

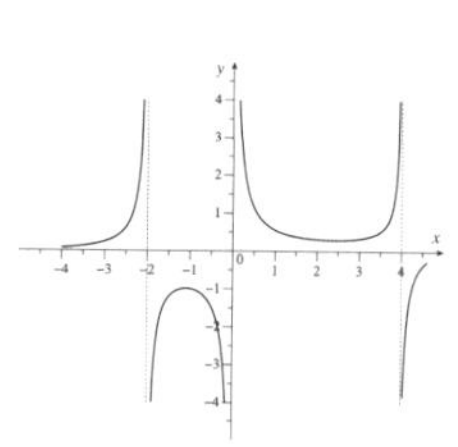

(1 mark)

Bonus question Easy

7 The number plane shows the graph of $y = f(x)$.

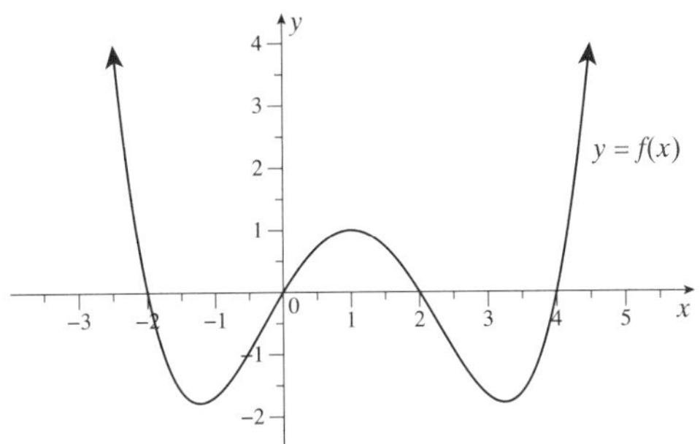

A second graph is drawn based on the graph of $y = f(x)$.

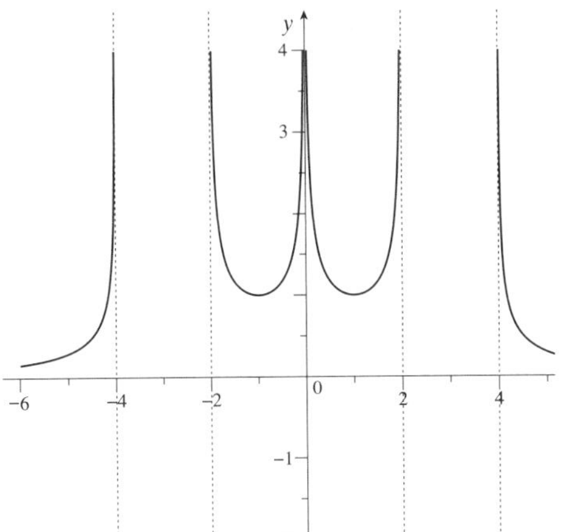

Which of these is the equation for the second graph?

A $y = \sqrt{f(x)}$ **B** $y = \dfrac{1}{\sqrt{f(x)}}$

C $y = \sqrt{f(|x|)}$ **D** $y = \dfrac{1}{\sqrt{f(|x|)}}$ *(1 mark)*

Bonus question 

8 The diagram shows the graph of a function $y = f(x)$.

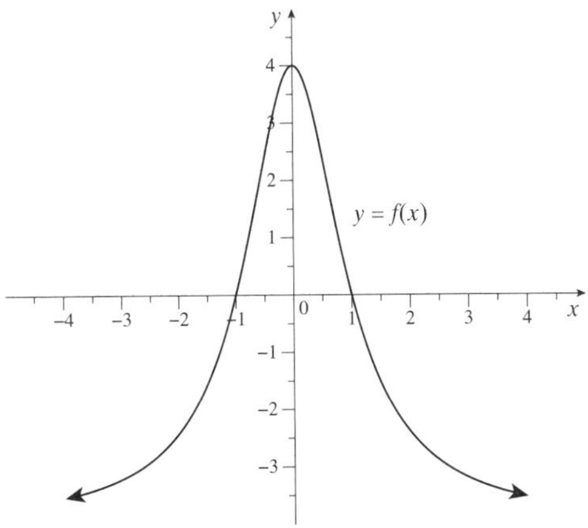

On the diagrams below draw the graphs of the functions:

i $y = \sqrt{f(x)}$ *(2 marks)* Easy

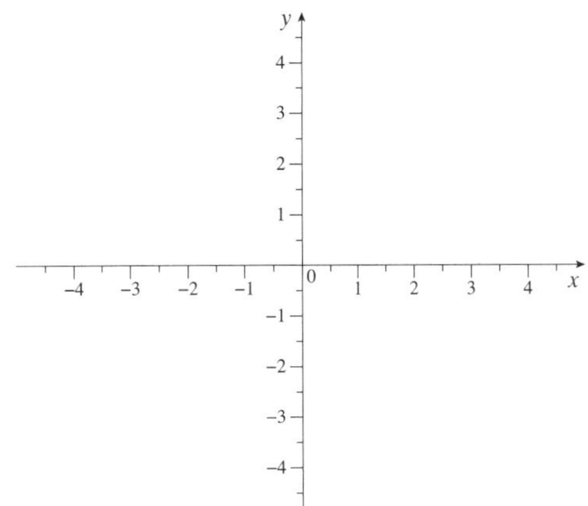

ii $y = \dfrac{1}{f(x)}$ *(2 marks)* Easy

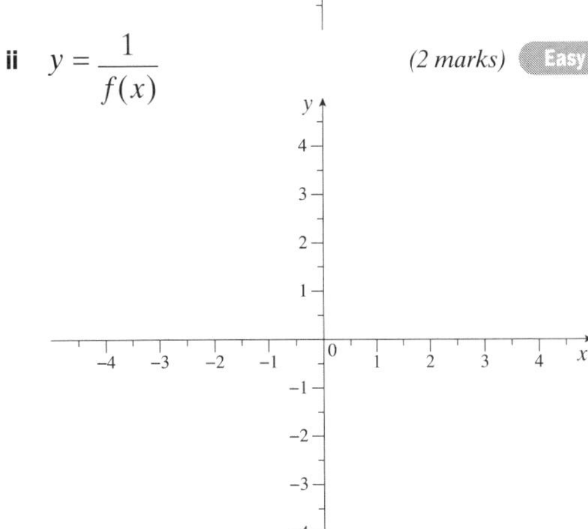

Bonus question

9 The graph of the function $y = a|x + b| + c$ is drawn on the diagram below.

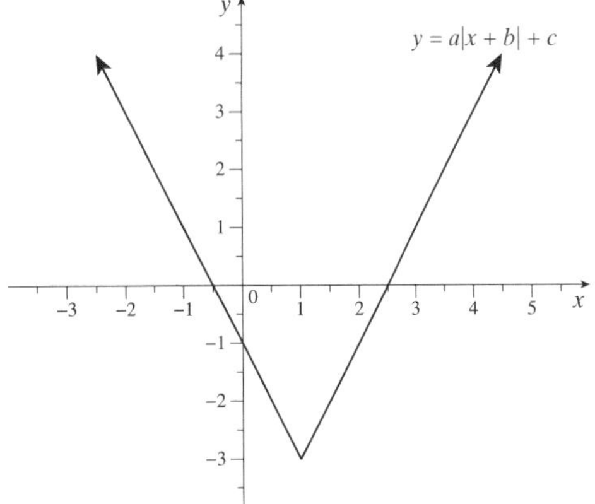

What are the values of a, b and c ? *(2 marks)*

Bonus question Easy

10 The diagram shows the graph of a function $y = f(x)$.

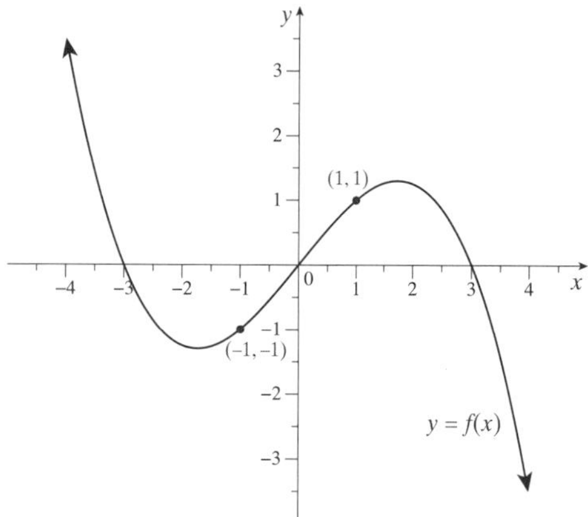

On the diagrams below draw the graphs of the functions:

i $y = \dfrac{1}{f(x)}$ *(2 marks)* Easy

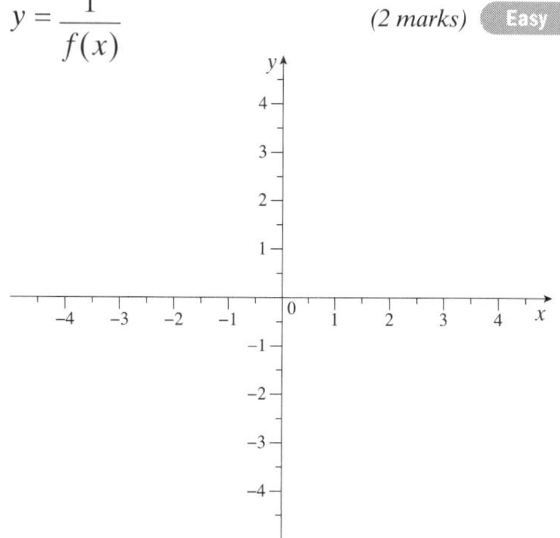

ii $y = \sqrt{f(x)}$ *(2 marks)* Easy

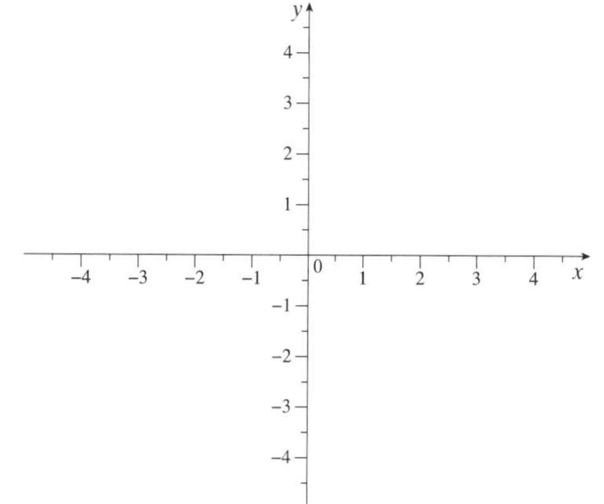

Bonus question

11 The diagram shows the graph of a function $y = f(x)$.

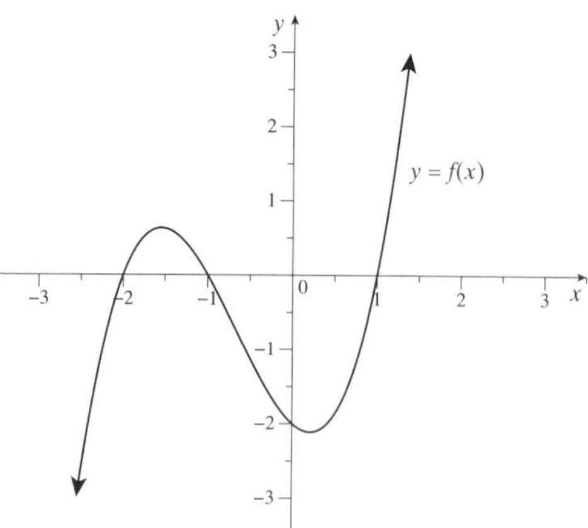

On the diagram below draw the graph of the function $y = |f(x - 1)|$. *(2 marks)*

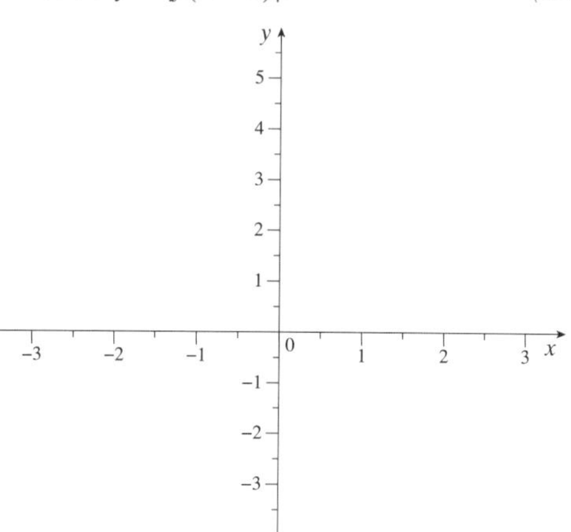

Bonus question Easy

12 The diagram shows the graph of $y = f(x)$ represented as a dotted curve.

The points (1, 1) and (–1, 1) lie on $y = f(x)$.

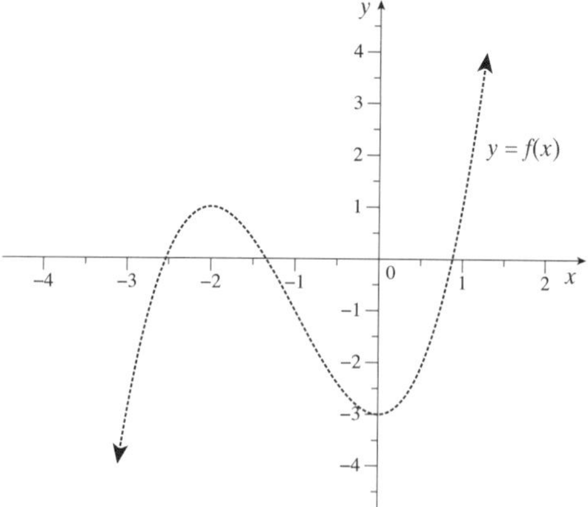

On the same diagram draw the graph of $y = \dfrac{1}{f(x)}$. *(2 marks)*

Bonus question 

13 The diagram shows the graph of $y = f(x)$ represented as a dotted line.

a On the same diagram sketch the graph of $y^2 = f(x)$. *(2 marks)* Easy

b State the domain of $y^2 = f(x)$. *(1 mark)* Easy

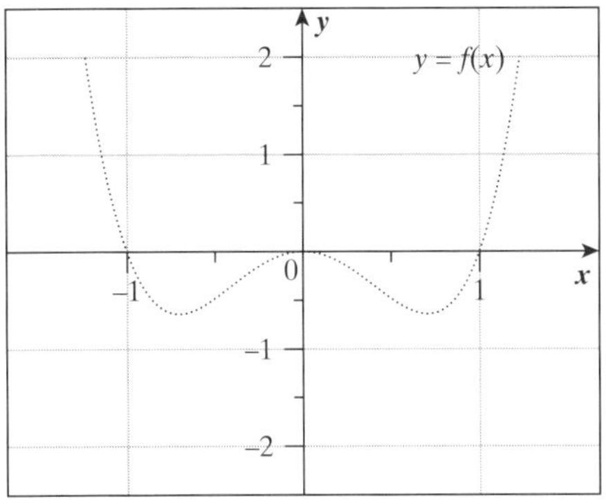

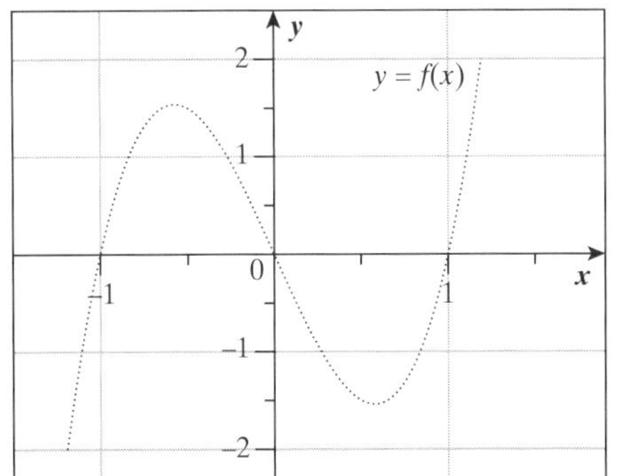

Bonus question

14 The diagram shows the graph of $y = f(x)$ represented as a dotted line.

On the same diagram sketch the graph of $y^2 = f(x)$. *(2 marks)*

Bonus question Easy

15 The diagram shows the graph of $y = f(x)$ over the domain $-2\pi \le x \le 2\pi$ represented as a dotted line.

On the same diagram sketch the graph of $y^2 = f(x)$. *(2 marks)*

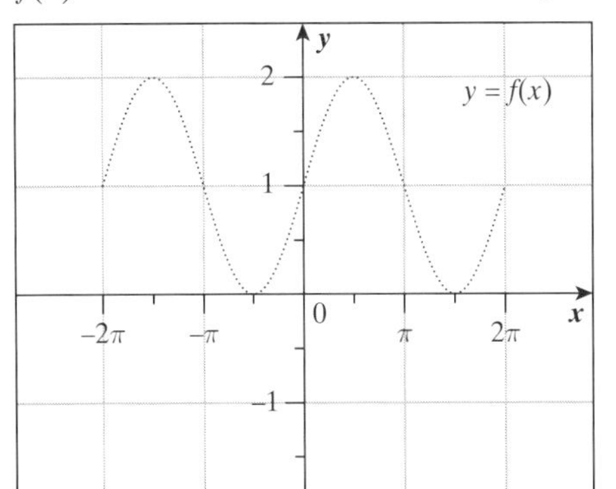

Bonus question Easy

16 The diagram shows the graph of $y = f(x)$ represented as a dashed line and $y = g(x)$ as a dotted line.

On the same diagram sketch the graph of $y = f(x) + g(x)$. *(2 marks)*

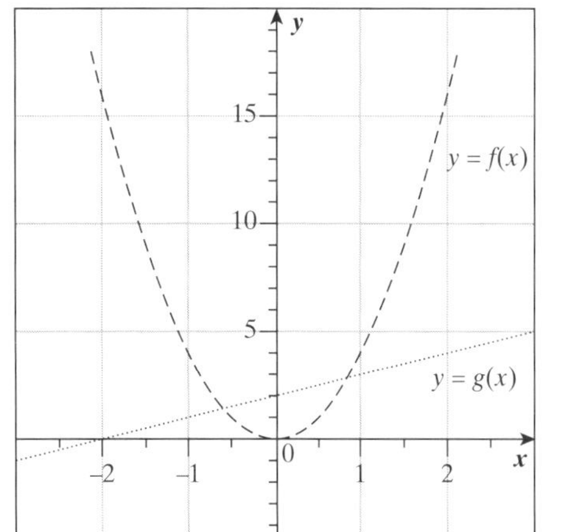

Bonus question Easy

17 The diagram shows the graph of $y = f(x)$ represented as a dashed line and $y = g(x)$ as a dotted line.

On the same diagram sketch the graph of $y = f(x) + g(x)$. *(2 marks)*

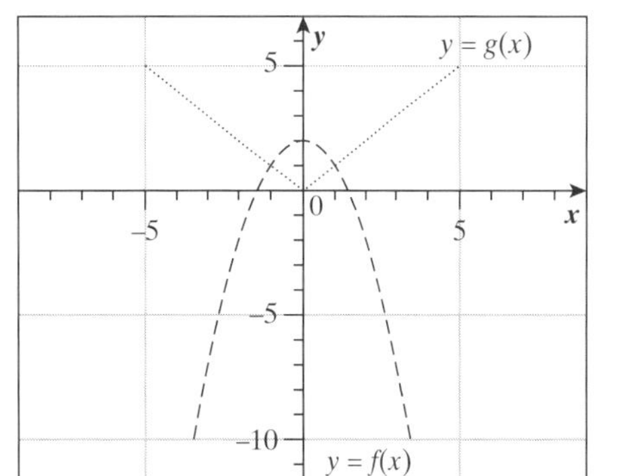

Bonus question Easy

18 The diagram shows the graph of $y = f(x)$ represented as a dashed line and $y = g(x)$ as a dotted line.

On the same diagram sketch the graph of $y = f(x) - g(x)$. *(2 marks)*

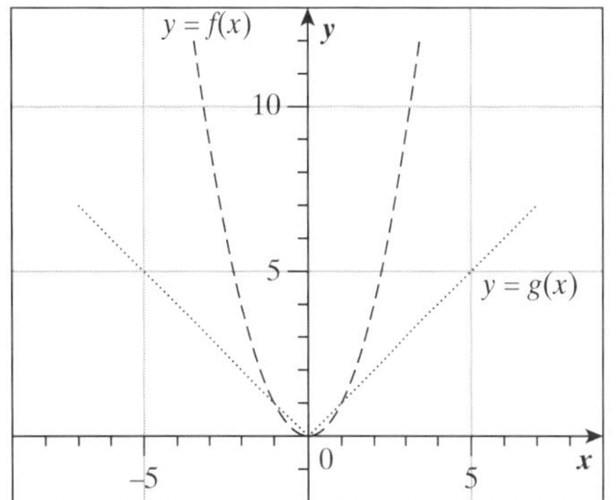

Bonus question Medium

19 The diagram shows the graph of $y = f(x)$ represented as a dashed line and $y = g(x)$ as a dotted line.
On the same diagram sketch the graph of $y = f(x) - g(x)$. *(2 marks)*

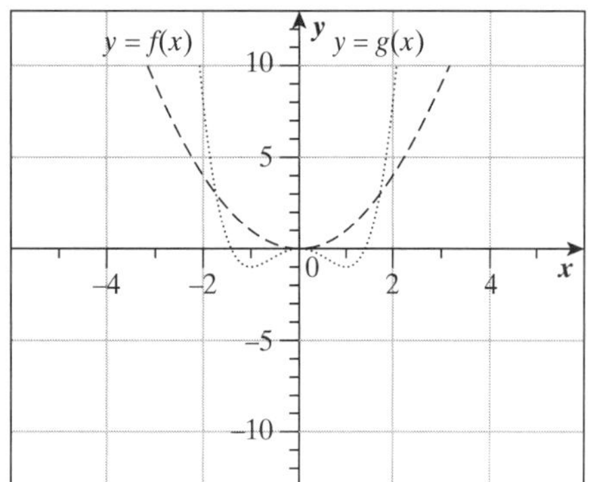

Bonus question Medium

20 The diagram shows the graph of $y = f(x)$ represented as a dashed line and $y = f(x) \times g(x)$ as a dotted line.
On the same diagram sketch the graph of $y = g(x)$ given $g(0) = 0$. *(3 marks)*

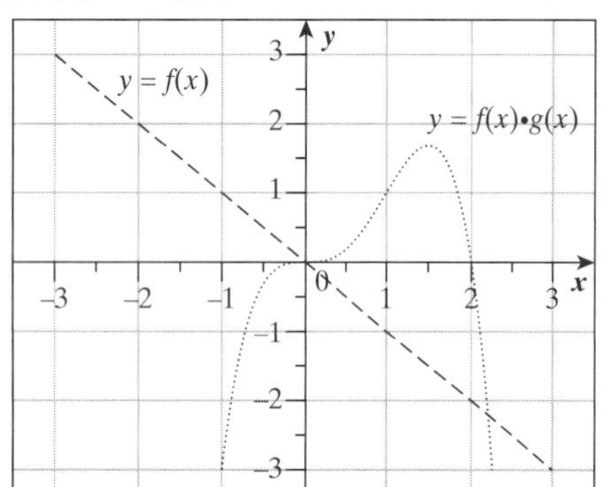

Bonus question Hard

21 The diagram shows the graph of $y = f(x)$ represented as a dashed line and $y = g(x)$ as a dotted line.
On the same diagram sketch the graph of $y = f(x) \times g(x)$. *(3 marks)*

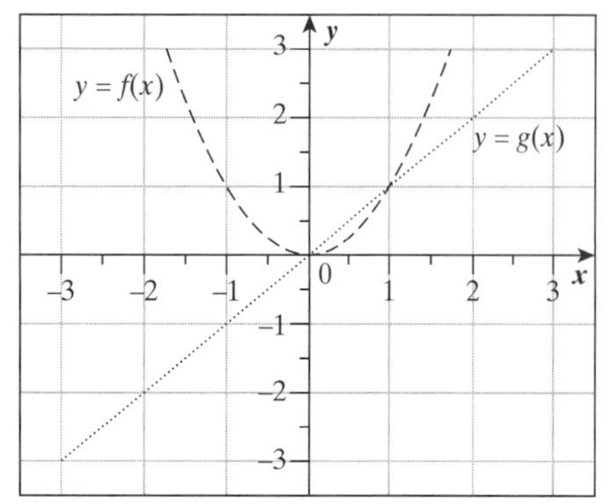

Bonus question Medium

22 The diagram shows the graph of $y = f(x)$ represented as a dashed line and $y = g(x)$ as a dotted line.
On the same diagram sketch the graph of $y = f(x) \times g(x)$. *(2 marks)*

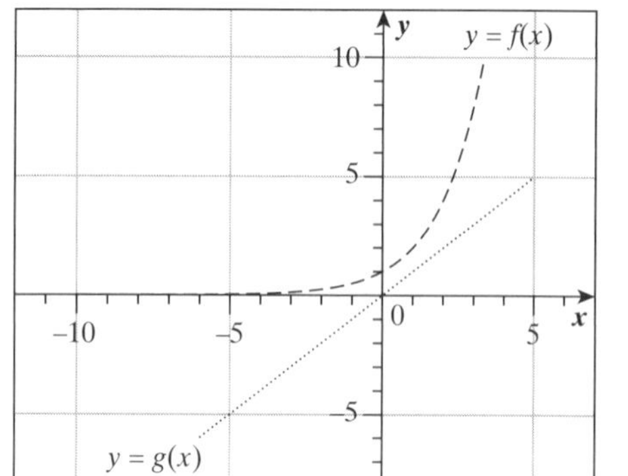

Bonus question Medium

23 The diagram shows the graph of $y = f(x)$ represented as a dashed line and $y = f(x) + g(x)$ as a dotted line.
On the same diagram sketch the graph of $y = g(x)$. *(2 marks)*

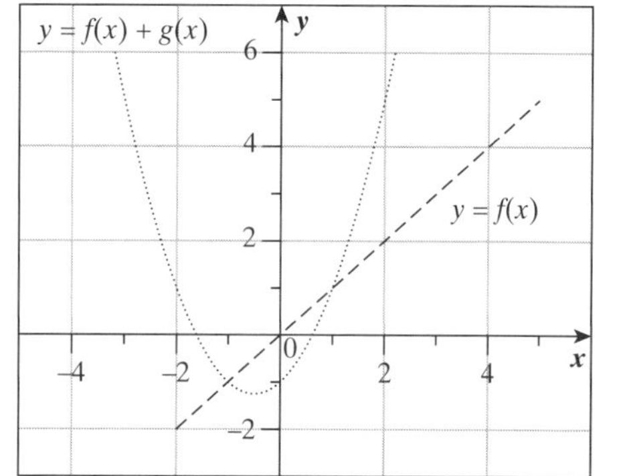

Bonus question Hard

24 The diagram shows the graph of $f(x) = x + 1$ represented as a dashed line and $g(x) = x - 1$ as a dotted line.

- **i** On the same diagram sketch the graph of $y = |f(x)|$ and $y = |g(x)|$. *(2 marks)* Easy
- **ii** Hence sketch the graph of $y = |f(x)| + |g(x)|$. *(2 marks)* Easy
- **iii** Use your sketch to solve $|x + 1| + |x - 1| < 4$. *(2 marks)* Easy

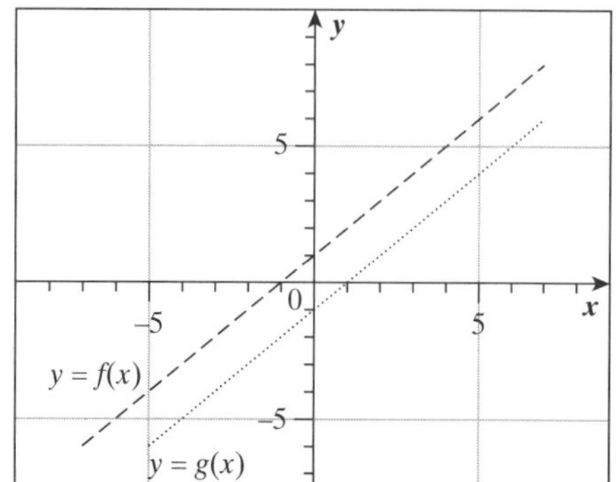

Bonus question

25 The diagram shows the graph of $y = f(x)$ represented as a dotted line.
On the same diagram sketch the graph of $y = f(|x|)$. *(2 marks)*

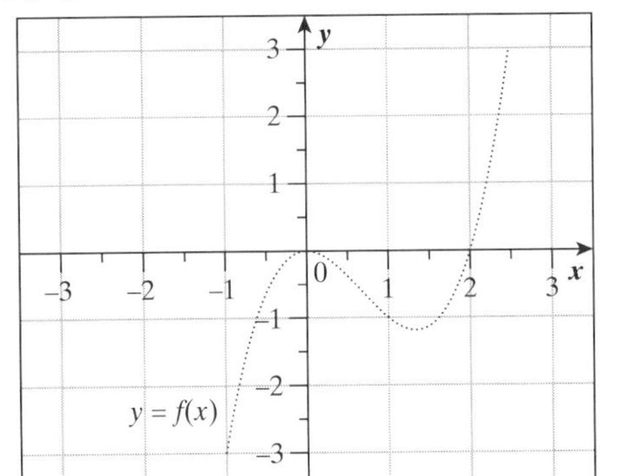

Bonus question Easy

26 The diagram shows the graph of $y = f(x)$.

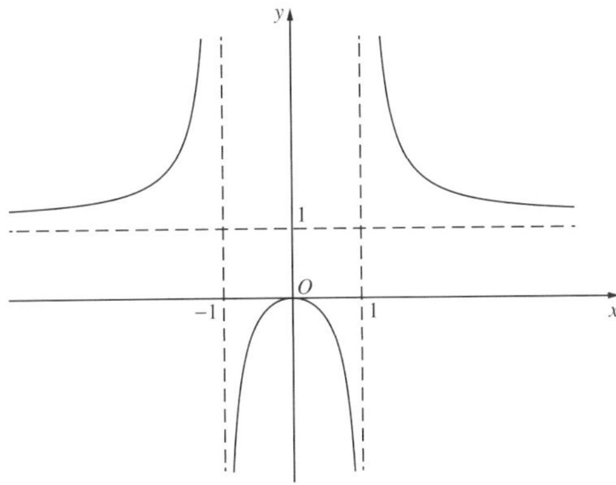

Which equation best describes the graph?

A $y = \dfrac{x}{x^2 - 1}$ **B** $y = \dfrac{x^2}{x^2 - 1}$

C $y = \dfrac{x}{1 - x^2}$ **D** $y = \dfrac{x^2}{1 - x^2}$ *(1 mark)*

(Q4, **2019 HSC**) Easy

27 The diagram shows the number of penguins, $P(t)$, on an island at time t.

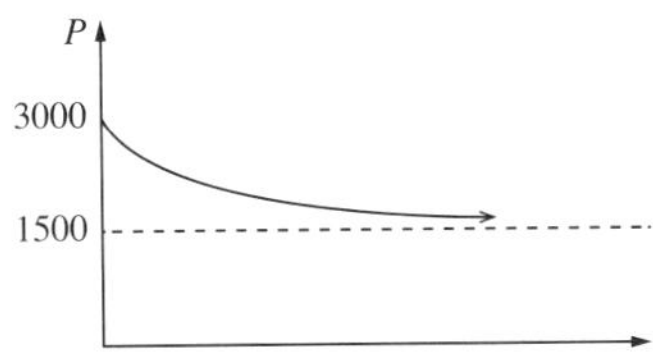

Which equation best represents this graph?

A $P(t) = 1500 + 1500e^{-kt}$
B $P(t) = 3000 - 1500e^{-kt}$
C $P(t) = 3000 + 1500e^{-kt}$
D $P(t) = 4500 - 1500e^{-kt}$

(Q5, **2018 HSC**) Easy

28 **i** Find the horizontal asymptote of the graph $y = \dfrac{2x^2}{x^2 + 9}$. *(1 mark)* Easy

ii Without the use of calculus, sketch the graph $y = \dfrac{2x^2}{x^2 + 9}$, showing the asymptote found in part **i**. *(2 marks)* Medium

(Q13b, **2012 HSC**)

Year 11 Graphical relationships—Worked Answers

1 Since the graph of $y = f(x)$ has been reflected about the y-axis to form $g(x)$, then $g(x) = f(|x|)$.
Answer A

2 The roots of the function $f(x) + g(x)$ occur when $f(x) = a$ and $g(x) = -a$ such that $f(x) + g(x) = 0$.
Answer A

3 $f(2) = 0$ and $f(-2) = 0$ so

$y = \dfrac{1}{f(x)}$ will be undefined when $x = \pm 2$.

$\dfrac{1}{f(x)} \neq 0$ for any value of x.

So there are asymptotes at $x = \pm 2$ and $y = 0$. ✓

$f(x) > 0$, and so $\dfrac{1}{f(x)} > 0$, when $x < -2$ or $x > 2$.

$f(x) < 0$, and so $\dfrac{1}{f(x)} < 0$, when $-2 < x < 2$.

$y = f(x)$ has a stationary point at $x = 0$, so $y = \dfrac{1}{f(x)}$ will also have a stationary point at $x = 0$.

$f(0) = -3$ so $\dfrac{1}{f(0)} = -\dfrac{1}{3}$. ✓

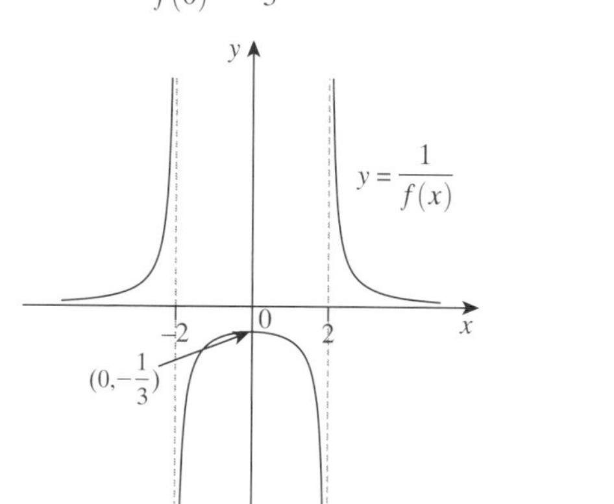

✓
(3 marks)

4 **i**

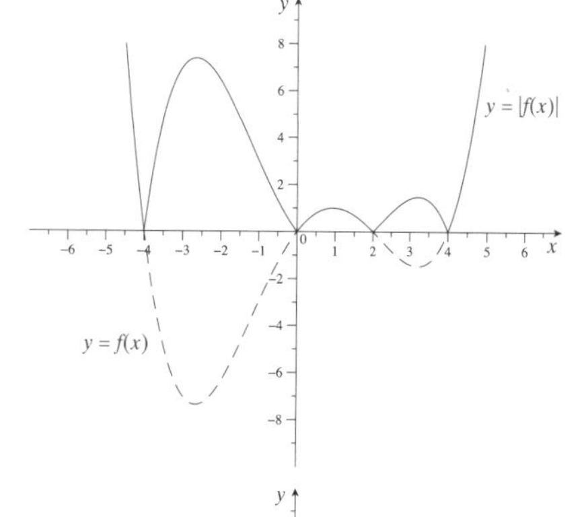

✓
(1 mark)

ii

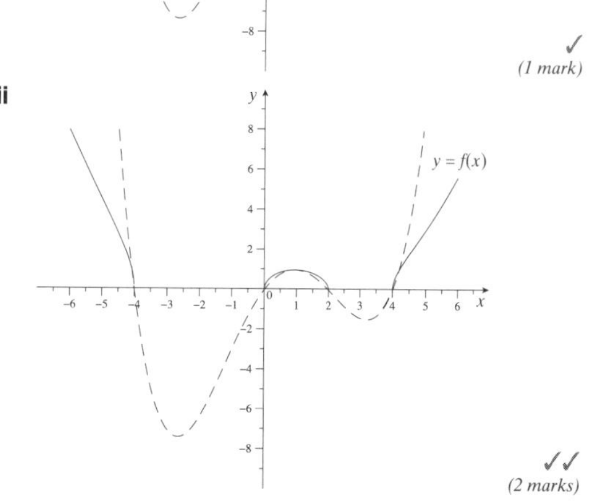

✓✓
(2 marks)

iii

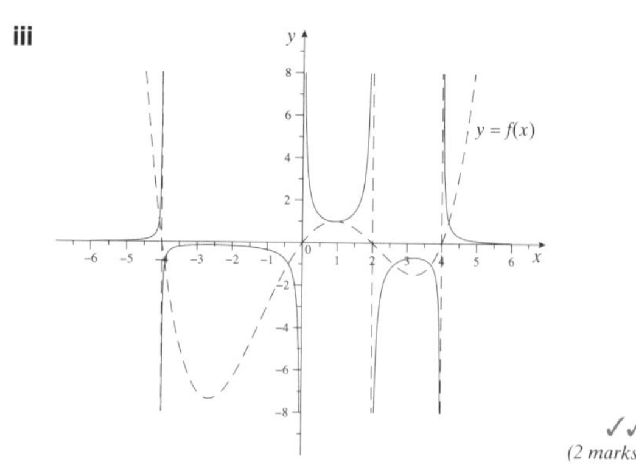

✓✓
(2 marks)

5

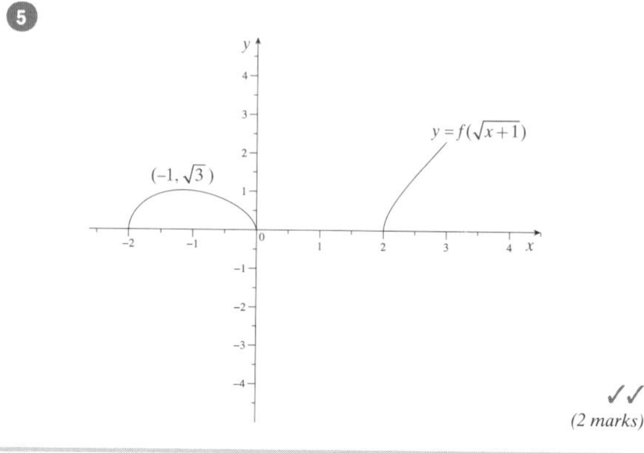

✓✓
(2 marks)

6

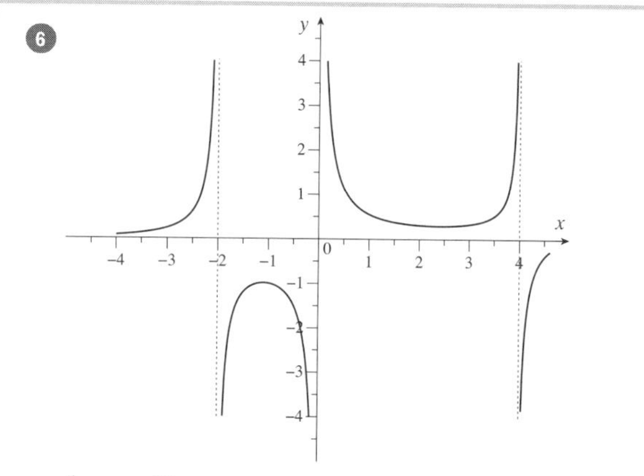

Answer D

7 The second graph is reflected about the y-axis. Also, there are discontinuities at the values of x where $f(x) = 0$ in the first graph. $\therefore$ the second graph is

$y = \dfrac{1}{\sqrt{f(|x|)}}$.

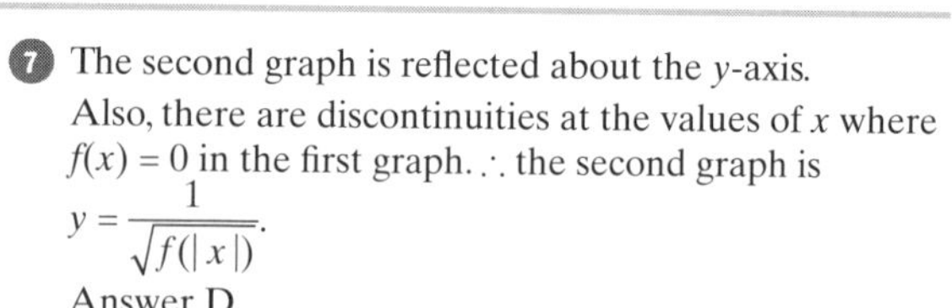

Answer D

8 **i**

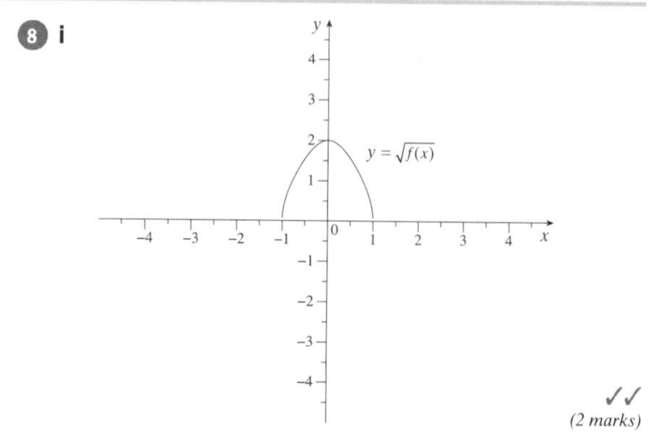

✓✓
(2 marks)

ii

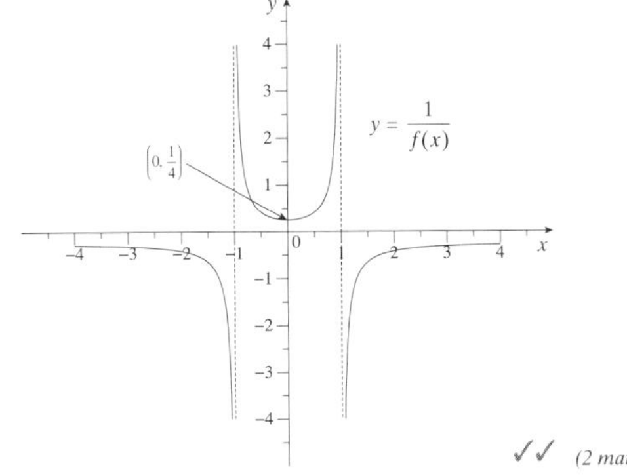

✓✓ *(2 marks)*

9 The minimum value occurs at $x = 1$.
This means $b = -1$, and so the equation is
$y = a|x - 1| + c$. The minimum value is $y = -3$.
This means $c = -3$, and so the equation is
$y = a|x - 1| - 3$. ✓
The graph has y-intercept of -1, so substitute $(0, -1)$:
$-1 = a|0 - 1| - 3$
$-1 = a - 3$
$a = 2$
$\therefore a = 2, b = -1, c = -3$. ✓ *(2 marks)*

10 i

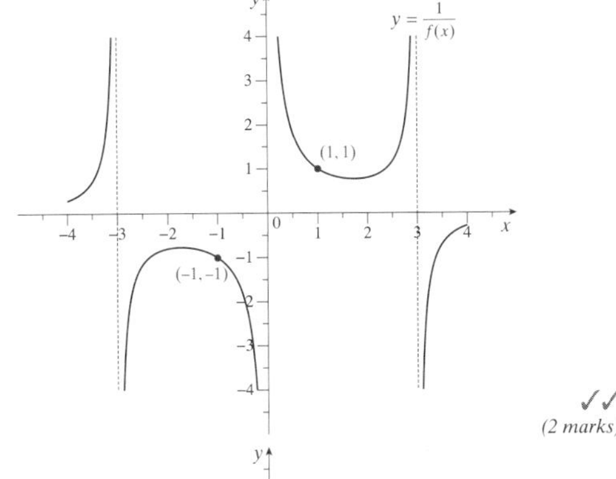

✓✓ *(2 marks)*

ii

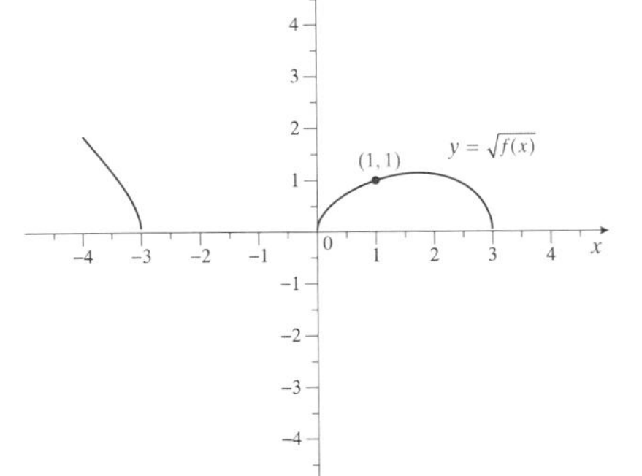

✓✓ *(2 marks)*

11

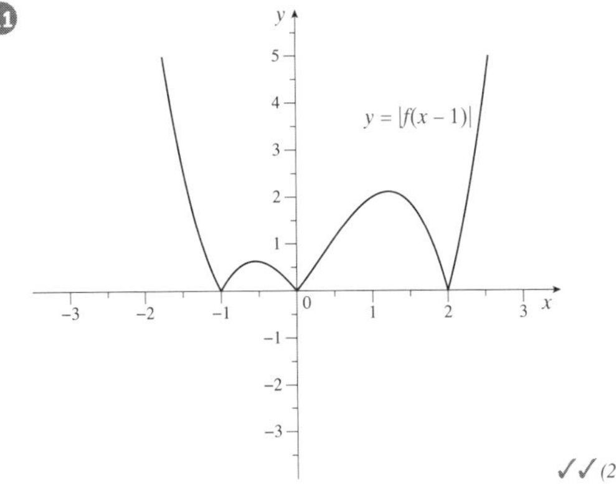

✓✓ *(2 marks)*

12

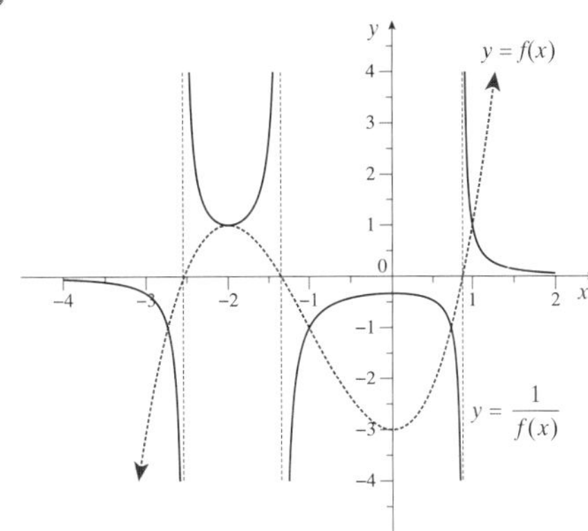

✓✓ *(2 marks)*

13 a

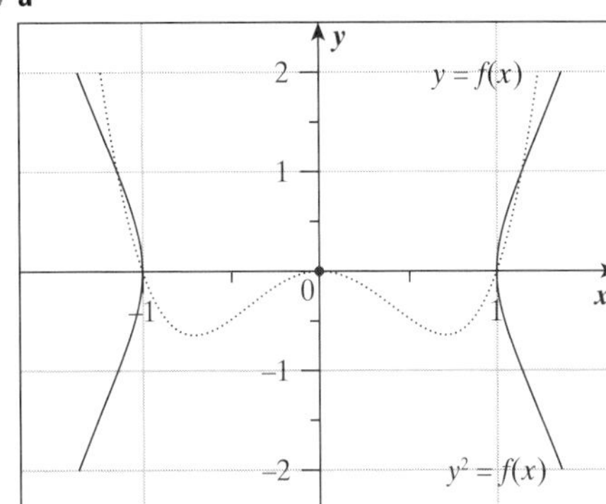

✓

Your solution should show curves intersecting at $y = 0$ and $y = 1$. ✓
Don't forget there is a point at the origin. *(2 marks)*

b *Domain*: $x \in (-\infty, -1] \cup [0] \cup [1, \infty)$ ✓ *(1 mark)*

14

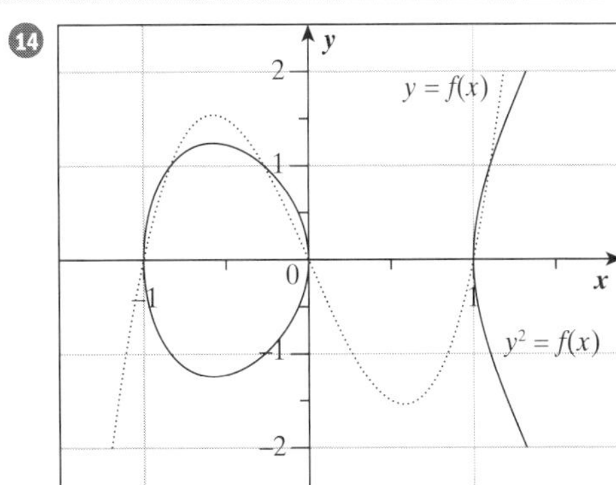

✓

Your solution should clearly show the three points of intersection at $y = 1$ and the x-axis as an axis of symmetry. ✓ *(2 marks)*

15

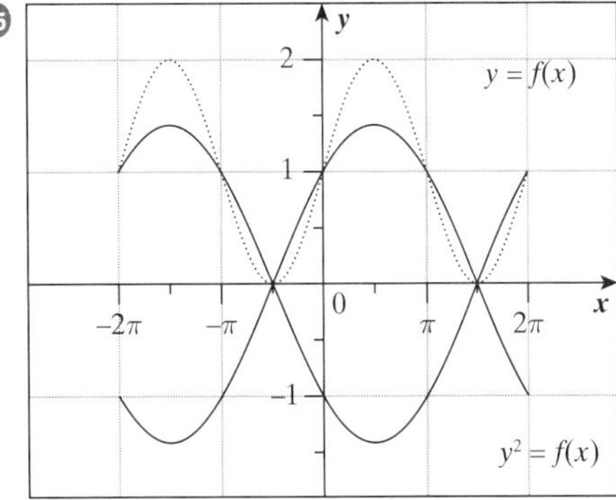

✓

The x intercepts are $-\frac{\pi}{2}$ and $\frac{3\pi}{2}$. ✓ *(2 marks)*

16

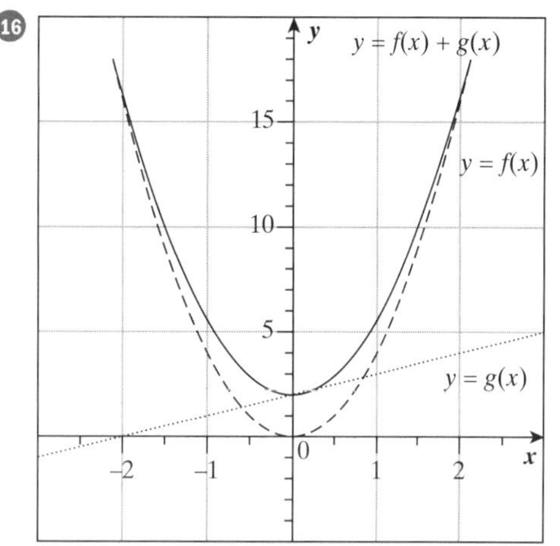

✓✓ *(2 marks)*

17

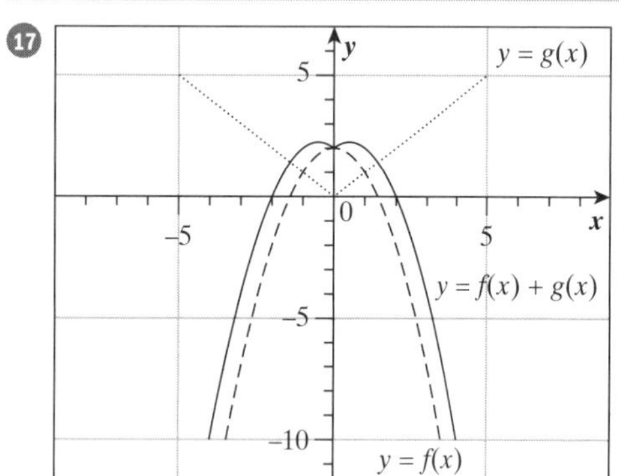

✓✓ *(2 marks)*

18

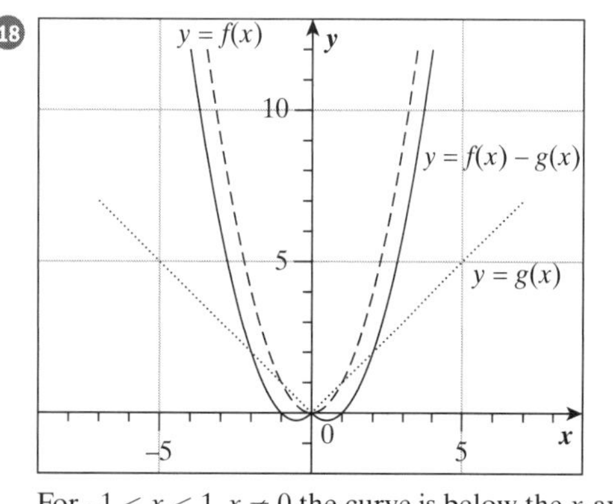

✓

For $-1 < x < 1, x \neq 0$ the curve is below the x-axis. ✓ *(2 marks)*

19

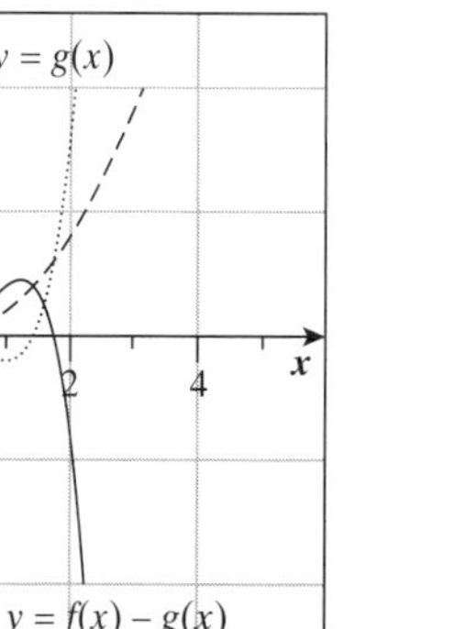

✓

The other two points at $y = 2$ should be approximately shown. ✓ *(2 marks)*

20

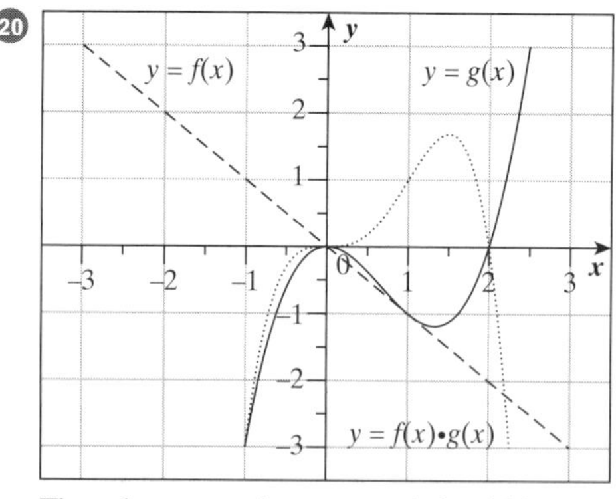

✓✓

The other two points at $y = -1$ should be approximately shown. ✓ *(3 marks)*

21

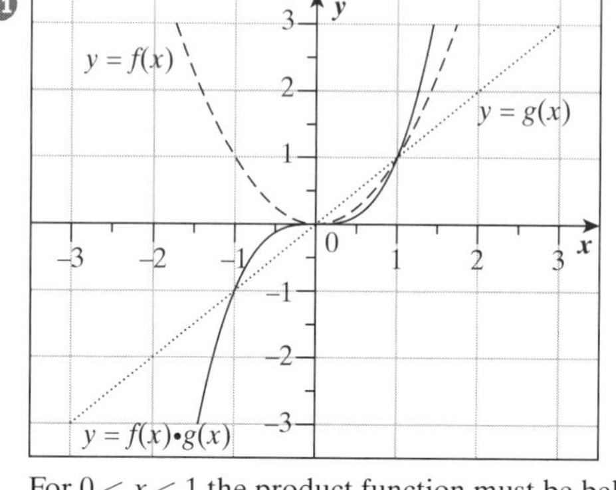

✓✓

For $0 < x < 1$ the product function must be below $y = f(x)$, but above the x-axis. ✓ *(3 marks)*

22

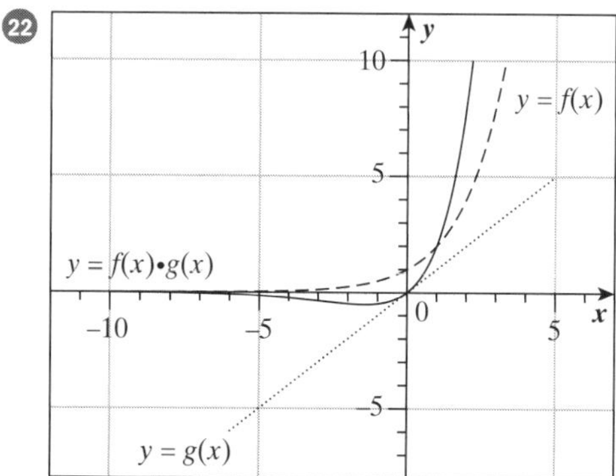

✓

As $x \to -\infty$ the product function approaches zero from below the x-axis, i.e. $f(x) \times g(x) \to 0^-$ or the product function is asymptotic to the x-axis from below. ✓ *(2 marks)*

23

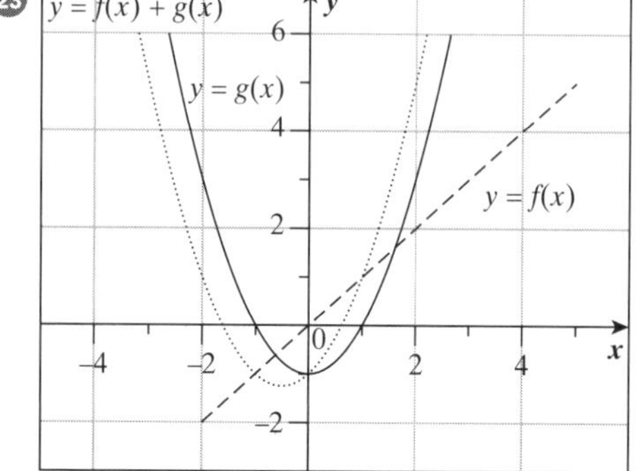

✓

It should be obvious that the y-axis is the axis of symmetry. ✓ *(2 marks)*

24 **i**

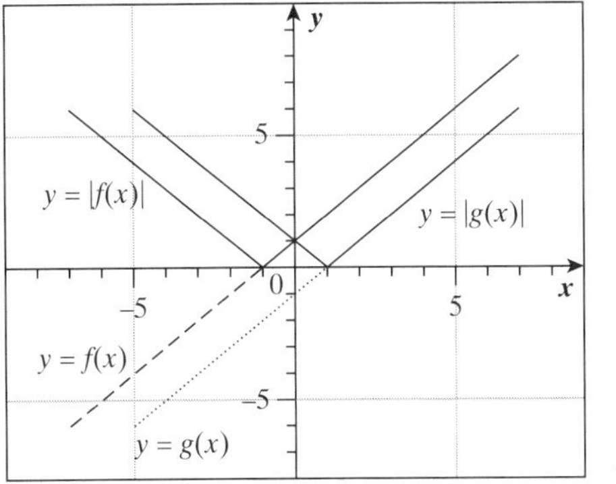

✓✓
(2 marks)

ii

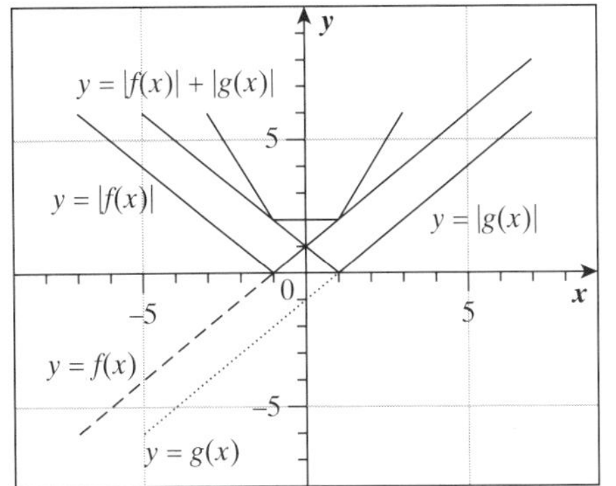

✓

Your solution should intersect with $y = |g(x)|$ at $(1, 2)$ and $y = |f(x)|$ at $(-1, 2)$. ✓
(2 marks)

iii

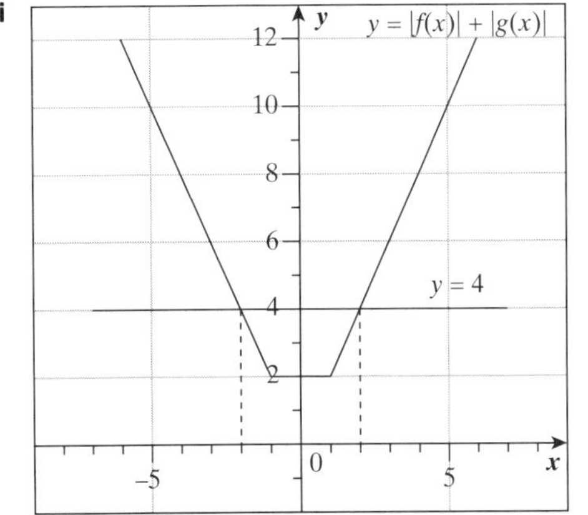

✓

Therefore, graphically the solution to solve $|x + 1| + |x - 1| < 4$ is $-2 < x < 2$. ✓
(2 marks)

25

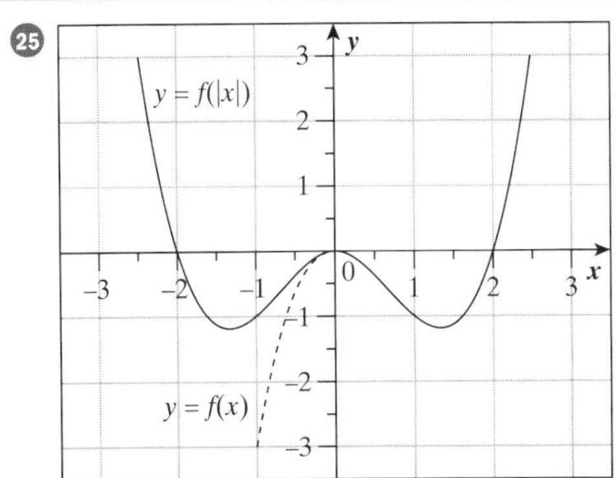

✓

It should be obvious that the y-axis is the axis of symmetry. ✓
(2 marks)

26 From the diagram, as $x \to \infty, f(x) \to 1^+$

Of the options, this is only true for $y = \dfrac{x^2}{x^2 - 1}$.

Answer B

27 $P(0) = 3000$
Now $1500e^0 = 1500$
So possible equations are A and D.
As $t \to \infty$, $1500e^{-kt} \to 0$
So the equation that best represents the graph is $P(t) = 1500 + 1500e^{-kt}$.
Answer A

28 **i**

$$\begin{aligned} y &= \frac{2x^2}{x^2 + 9} \\ y(x^2 + 9) &= 2x^2 \\ x^2y + 9y &= 2x^2 \\ 2x^2 - x^2y &= 9y \\ x^2(2 - y) &= 9y \\ x^2 &= \frac{9y}{2 - y} \end{aligned}$$

Now $2 - y \neq 0$
$y \neq 2$

The horizontal asymptote is $y = 2$. ✓
(1 mark)

ii $x^2 \geq 0$ for all values of x

$\therefore \dfrac{2x^2}{x^2 + 9} \geq 0$ for all values of x

$$\begin{aligned} f(x) &= \frac{2x^2}{x^2 + 9} \\ f(0) &= 0 \\ f(-x) &= \frac{2(-x)^2}{(-x)^2 + 9} \\ &= \frac{2x^2}{x^2 + 9} \\ &= f(x) \end{aligned}$$

$\therefore$ the function is even

$$f(x) = \frac{2}{1 + \frac{9}{x^2}}$$

$\therefore$ as $x \to \infty, f(x) \to 2^-$ ✓

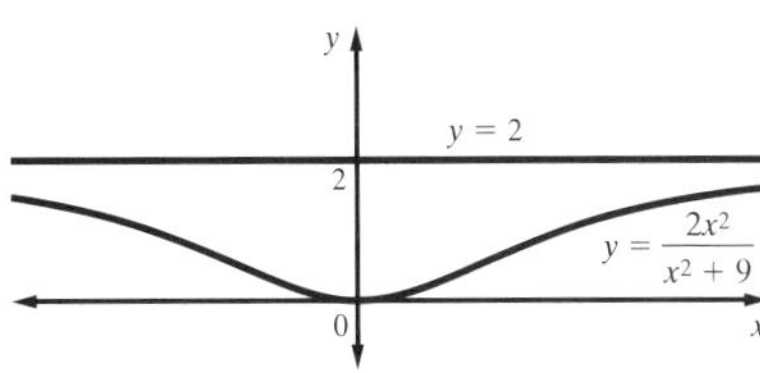

✓
(2 marks)

1 Solve $\dfrac{x}{2-x} \geq 5$. *(3 marks)*

(Q11f, **2022 HSC**) Easy

2 Which diagram best represents the solution set of $x^2 - 2x - 3 \geq 0$?

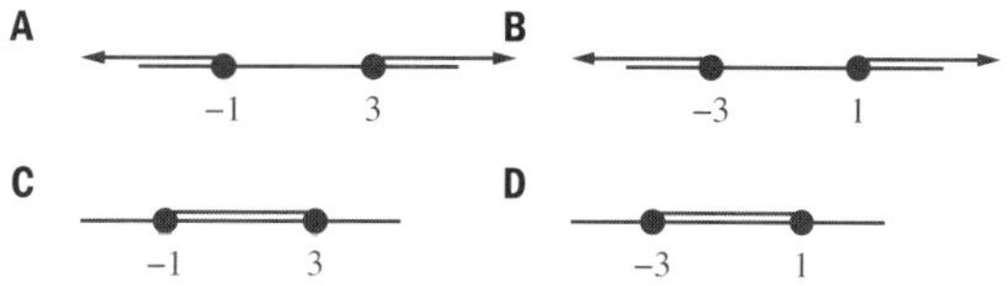

(1 mark)

(Q1, **2020 HSC**) Easy

3 Which inequality is the solution of $\dfrac{|x+3|}{2} \leq 1$?

A $-5 \leq x \leq -1$ **B** $-5 \leq x \leq 1$

C $-1 \leq x \leq 5$ **D** $1 \leq x \leq 5$ *(1 mark)*

Bonus question (see page iv) Easy

4 For what values of k is the polynomial $x^2 - (k+2)x + 2(k+2)$ positive for all values of x? *(2 marks)*

Bonus question Easy

5 Solve $|x^2 - 4x| \leq x - 4$. *(3 marks)*

Bonus question Medium

6 Find the range of values of k for which the expression $(1 - 2k) + 2x - x^2$ is always negative. *(1 mark)*

Bonus question Easy

7 Solve $\dfrac{|1-4x|}{2} \leq x$. *(2 marks)*

Bonus question Easy

8 Solve $\dfrac{|5-2x|}{3} \leq 1$. *(2 marks)*

Bonus question Medium

9 For what values of x is $\dfrac{x}{x+1} < 2$? *(3 marks)*

(Q11b, **2019 HSC**) Medium

10 Consider the function $f(x) = \dfrac{1}{4x-1}$.

i Find the domain of $f(x)$. *(1 mark)* Easy

ii For what values of x is $f(x) < 1$? *(2 marks)* Medium

(Q11e, **2018 HSC**)

11 Solve $\dfrac{2x}{x+1} > 1$. *(3 marks)*

(Q11c, **2017 HSC**) Easy

12 Solve $\dfrac{3}{2x+5} - x > 0$. *(3 marks)*

(Q11e, **2016 HSC**) Medium

13 Solve $|x - 2| \leq 3$. *(2 marks)*

(Q11c, **2016 MA HSC**) Easy

14 Solve the inequality $\dfrac{4}{x+3} \geq 1$. *(3 marks)*

(Q11c, **2015 HSC**) Easy

15 **i** Find the domain and range for the function $f(x) = \sqrt{9 - x^2}$. *(2 marks)* Easy

ii On a number plane, shade the region where the points (x, y) satisfy both of the inequalities $y \leq \sqrt{9 - x^2}$ and $y \geq x$. *(2 marks)* Easy

(Q13b, **2015 MA HSC**)

16 Solve $\dfrac{x^2+5}{x} > 6$. *(3 marks)*

(Q11e, **2014 HSC**) Medium

17 Which inequality has the same solution as $|x + 2| + |x - 3| = 5$?

A $\dfrac{5}{3-x} \geq 1$ **B** $\dfrac{1}{x-3} - \dfrac{1}{x+2} \leq 0$

C $x^2 - x - 6 \leq 0$ **D** $|2x - 1| \geq 5$ *(1 mark)*

(Q10, **2013 HSC**) Hard

18 Which inequality defines the domain of the function $f(x) = \dfrac{1}{\sqrt{x+3}}$?

A $x > -3$ **B** $x \geq -3$

C $x < -3$ **D** $x \leq -3$ *(1 mark)*

(Q3, **2013 MA HSC**) Easy

19 Solve $\dfrac{x}{x-3} < 2$. *(3 marks)*

(Q11c, **2012 HSC**) Medium

20 Solve $\dfrac{4-x}{x} < 1$. *(3 marks)*

(Q1c, **2011 HSC**) Medium

21 Solve $\dfrac{3}{x+2} < 4$. *(3 marks)*

(Q1d, **2010 HSC**) Easy

22 Solve the inequality $x^2 - x - 12 < 0$. *(2 marks)*

(Q2b, **2010 MA HSC**) Easy

23 Solve the inequality $\frac{x+3}{2x} > 1$. *(3 marks)*

(Q1d, **2009 HSC**) Medium

24 **i** Find the vertical and horizontal asymptotes of the hyperbola $y = \frac{x-2}{x-4}$ and hence sketch the graph of $y = \frac{x-2}{x-4}$.

(3 marks) Medium

ii Hence, or otherwise, find the values of x for which $\frac{x-2}{x-4} \leq 3$. *(2 marks)* Easy

(Q3b, **2007 HSC**)

25 **i** Write down the discriminant of $2x^2 + (k-2)x + 8$, where k is a constant.

(1 mark) Easy

ii Hence, or otherwise, find the values of k for which the parabola $y = 2x^2 + kx + 9$ does not intersect the line $y = 2x + 1$.

(2 marks) Easy

(Q7c, **2006 MA HSC**)

26 Find the values of x for which $|x - 3| \leq 1$.

(2 marks)

(Q1e, **2005 MA HSC**) Easy

27 Solve $\frac{4}{x+1} < 3$. *(3 marks)*

(Q1b, **2004 HSC**) Easy

28 Solve $\frac{3}{x-2} \leq 1$. *(3 marks)*

(Q1b, **2003 HSC**) Easy

29 Solve $|x - 1| \geq 3$ and graph your solution on the number line. *(2 marks)*

(Q4a, **2002 MA HSC**) Easy

30 Solve $\frac{3x}{x-2} \leq 1$. *(3 marks)*

(Q4a, **2001 HSC**) Medium

31 Find the values of k for which the quadratic equation $3x^2 + 2x + k = 0$ has no real roots.

(2 marks)

(Q4a, **2001 MA HSC**) Easy

32 Solve $\frac{5}{x+2} \leq 1$. *(3 marks)*

(Q1e, **2000 HSC**) Easy

33 Find all real x such that $|4x - 1| > 2\sqrt{x(1-x)}$.

(3 marks)

(Q7c, **1998 HSC**) Hard

34 **i** Sketch the graph of $y = x^2 - 6$, and label all intercepts with the axes.

ii On the same set of axes, carefully sketch the graph of $y = |x|$.

iii Find the x coordinates of the two points where the graphs intersect.

iv Hence solve the inequality $x^2 - 6 \leq |x|$.

(6 marks)

(Q4b, **1997 MA HSC**) Easy

35 Solve the inequality $\frac{2}{x-1} \leq 1$. *(3 marks)*

(Q1e, **1996 HSC**) Easy

36 A layer of plastic cuts out 15% of the light and lets through the remaining 85%.

i Show that two layers of the plastic let through 72·25% of the light.

Easy

ii How many layers of the plastic are required to cut out at least 90% of the light? Medium

(3 marks) (Q3b, **1996 MA HSC**)

37 **i** Verify that $x = \frac{1}{3}$ and $x = 2$ satisfy the equation $7 - 3x = \frac{2}{x}$. Easy

ii On the same set of axes, sketch the graphs of $y = 7 - 3x$ and $y = \frac{2}{x}$. Easy

iii Using part **ii**, or otherwise, write down all values of x for which $7 - 3x < \frac{2}{x}$.

Easy

(Q2c, **1994 HSC**)

38 Find the values of m for which $12 + 4m - m^2 > 0$.

(Q5a, **1994 MA HSC**) Easy

39 Solve the inequality $x^2 - 3x < 4$.

(Q1a, **1993 HSC**) Easy

40 Solve $x^2 - x - 2 > 0$.

(Q1a, **1992 HSC**) Easy

Year 11 Inequalities—Worked Answers

1
$$\frac{x}{2-x} \geq 5$$
$x \neq 2$ ✓
$$\frac{x}{2-x} = 5$$
$$x = 5(2-x) = 10 - 5x$$
$$6x = 10$$
$$x = \frac{5}{3}$$ ✓

Testing regions:

When $x = \frac{2}{3}$
$$\frac{\frac{2}{3}}{2-\frac{2}{3}} = \frac{1}{2} < 5$$

When $x = \frac{9}{5}$
$$\frac{\frac{9}{5}}{2-\frac{9}{5}} = 9 > 5$$

When $x = \frac{11}{5}$
$$\frac{\frac{11}{5}}{2-\frac{11}{5}} = -11 < 5$$

$\therefore \frac{5}{3} \leq x < 2$ ✓

$\frac{2}{3}$ $\frac{5}{3}$ $\frac{9}{5}$ 2 2.2

(3 marks)

2
$$x^2 - 2x - 3 \geq 0$$
$$(x-3)(x+1) \geq 0$$

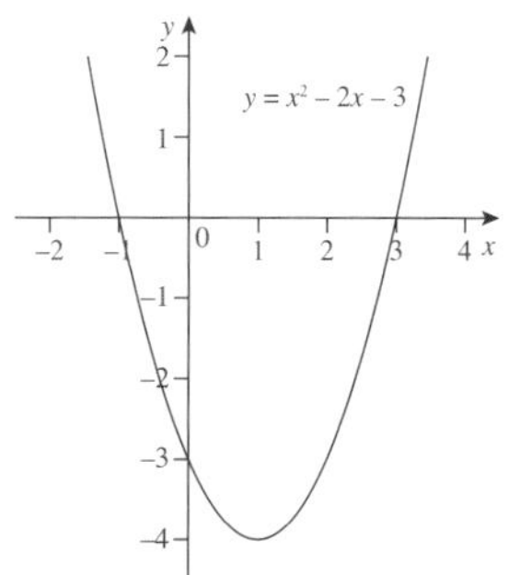

From the graph $x \leq -1$ or $x \geq 3$

−1 3

Answer A

3
$$\frac{|x+3|}{2} \leq 1$$
$$-1 \leq \frac{x+3}{4} \leq 1$$
$$-2 \leq x+3 \leq 2$$
$$-5 \leq x \leq -1$$
Answer A

4 $\Delta = (-(k+2))^2 - 4(1)(2k+4) < 0$
$$k^2 + 4k + 4 - 8k - 16 < 0$$
$$k^2 - 4k - 12 < 0$$ ✓
$$(k-6)(k+2) < 0$$
Consider $(k-6)(k+2) = 0$
$$k = 6, -2$$
$\therefore -2 < k < 6$ ✓

(2 marks)

5 Solve graphically $y = |x^2 - 4x|$ and $y = x - 4$:

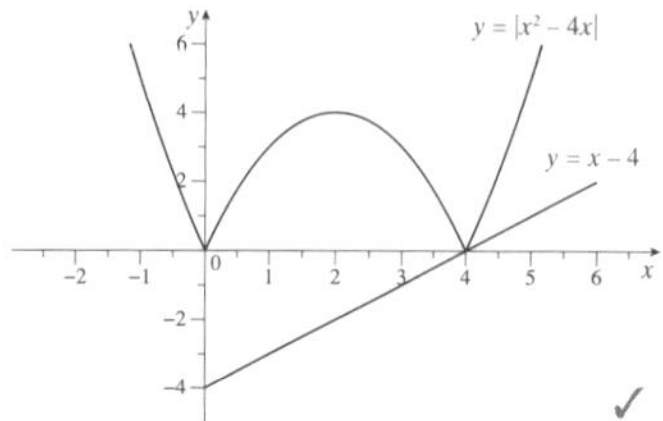

✓

The solution to $|x^2 - 4x| \leq x - 4$ is $x = 4$. ✓

Note: Solving $|x^2 - 4x| \leq x - 4$ algebraically would provide two cases:

$x^2 - 4x \leq x - 4$	$-x^2 + 4x \leq x - 4$
$x^2 - 5x + 4 \leq 0$	$x^2 - 3x - 4 \geq 0$
$(x-4)(x-1)$	$(x-4)(x+1) \geq 0$
$\therefore 1 \leq x \leq 4$	$\therefore x \leq -1$ or $x \geq 4$

However, testing of solutions would yield $x = 4$ as the only solution. ✓

(3 marks)

6 Consider $a = -1 < 0$ and $\Delta < 0$:
$$\Delta = 2^2 - 4(-1)(1-2k) < 0$$
$$4 + 4 - 8k < 0$$
$$8k > 8$$
$$k > 1$$ ✓

(1 mark)

7
$$\frac{|1-4x|}{2} \leq x$$
$$-x \leq \frac{1-4x}{2} \leq x$$
Two cases:

$-2x \leq 1 - 4x$	$1 - 4x \leq 2x$
$2x \leq 1$	$6x \geq 1$
$x \leq \frac{1}{2}$ ✓	$x \geq \frac{1}{6}$ ✓

$\therefore \frac{1}{6} \leq x \leq \frac{1}{2}$

(2 marks)

8
$$\frac{|5-2x|}{3} \leq 1$$
$$-1 \leq \frac{5-2x}{3} \leq 1$$
$-3 \leq 5 - 2x \leq 3$ ✓
$$-8 \leq -2x \leq -2$$
$$4 \geq x \geq 1$$
$\therefore 1 \leq x \leq 4$ ✓

(2 marks)

9
$$\frac{x}{x+1} < 2$$
If $x + 1 > 0$
i.e. $x > -1$,
then $x < 2(x+1)$
$$x < 2x + 2$$
$$-x < 2$$
$x > -2$ ✓
But if $x > -1$, it is greater than −2.
If $x < -1$,
then $x > 2x + 2$
$$-x > 2$$
$x < -2$ ✓
So $x < -2$ or $x > -1$. ✓

(3 marks)

10 $f(x) = \frac{1}{4x-1}$

i Now $4x - 1 \neq 0$
$$4x \neq 1$$
$$x \neq \frac{1}{4}$$
Domain is all real x except $x = \frac{1}{4}$. ✓

(1 mark)

ii $f(x) < 1$
If $x < \frac{1}{4}$, the denominator is negative and so $f(x)$ is also negative and hence less than 1. ✓
If $x > \frac{1}{4}$,
$$\frac{1}{4x-1} < 1$$
$$1 < 4x - 1$$
$$2 < 4x$$
$\frac{1}{2} < x$ ✓
So $f(x) < 1$ when $x < \frac{1}{4}$ and when $x > \frac{1}{2}$.

(2 marks)

11
$$\frac{2x}{x+1} > 1 \quad (x \neq -1)$$
Multiply both sides by $(x+1)^2$
$2x(x+1) > (x+1)^2$ ✓
$$2x^2 + 2x > x^2 + 2x + 1$$
$x^2 > 1$ ✓
$\therefore x < -1$ or $x > 1$ ✓

(3 marks)

12
$$\frac{3}{2x+5} - x > 0$$
$$\frac{3}{2x+5} > x$$
If $2x + 5 > 0$
$$2x > -5$$
$$x > -2\frac{1}{2}$$
Then $3 > x(2x+5)$
$$3 > 2x^2 + 5x$$

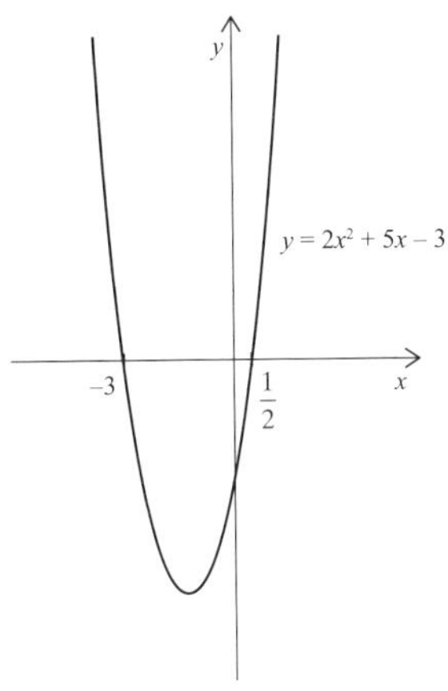

$2x^2 + 5x - 3 < 0$

$(2x - 1)(x + 3) < 0$

$-3 < x < \frac{1}{2}$ ✓

But $x > -2\frac{1}{2}$

So $-2\frac{1}{2} < x < \frac{1}{2}$ ✓

If $2x + 5 < 0$

$x < -2\frac{1}{2}$

and $2x^2 + 5x - 3 > 0$

So $x < -3$ ✓

$\therefore x < -3$ or $-2\frac{1}{2} < x < \frac{1}{2}$ *(3 marks)*

13 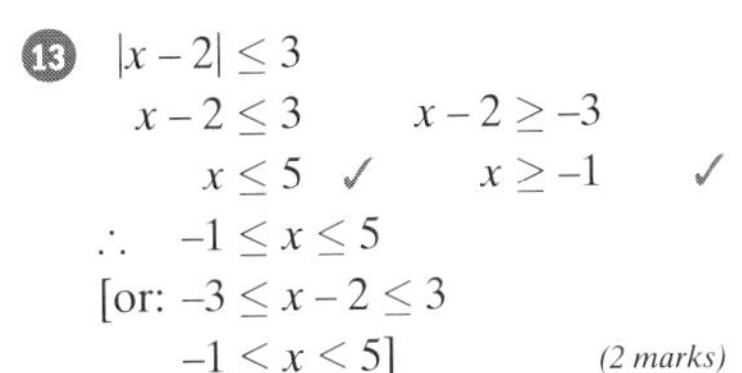

$|x - 2| \leq 3$

$x - 2 \leq 3$ $x - 2 \geq -3$

$x \leq 5$ ✓ $x \geq -1$ ✓

$\therefore -1 \leq x \leq 5$

[or: $-3 \leq x - 2 \leq 3$

$-1 \leq x \leq 5$] *(2 marks)*

14 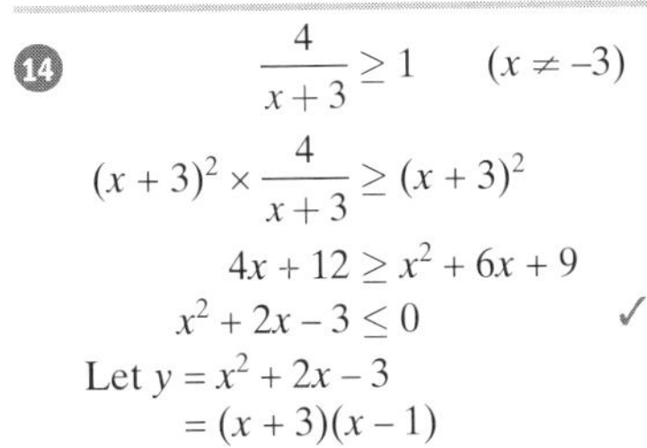

$\frac{4}{x+3} \geq 1 \quad (x \neq -3)$

$(x + 3)^2 \times \frac{4}{x+3} \geq (x + 3)^2$

$4x + 12 \geq x^2 + 6x + 9$

$x^2 + 2x - 3 \leq 0$ ✓

Let $y = x^2 + 2x - 3$

$= (x + 3)(x - 1)$

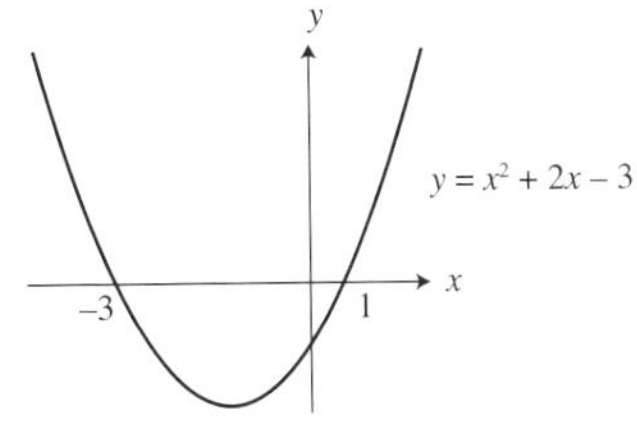

From the graph $x^2 + 2x - 3 \leq 0$ when $-3 \leq x \leq 1$ ✓

But $x \neq -3$

So $-3 < x \leq 1$ ✓ *(3 marks)*

15 **i** Domain: $-3 \leq x \leq 3$ ✓

Range: $0 \leq f(x) \leq 3$ ✓ *(2 marks)*

ii

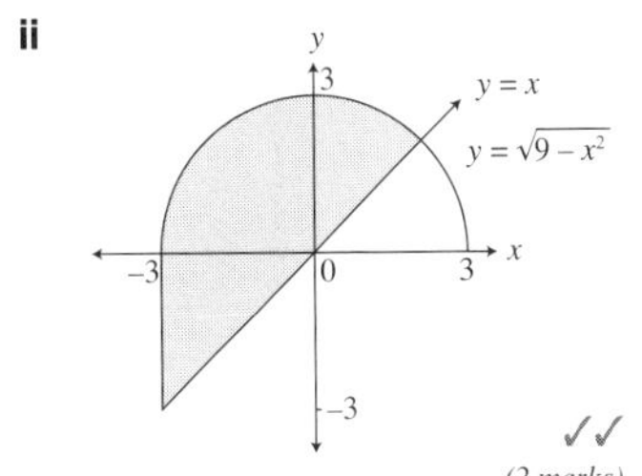

✓✓ *(2 marks)*

16 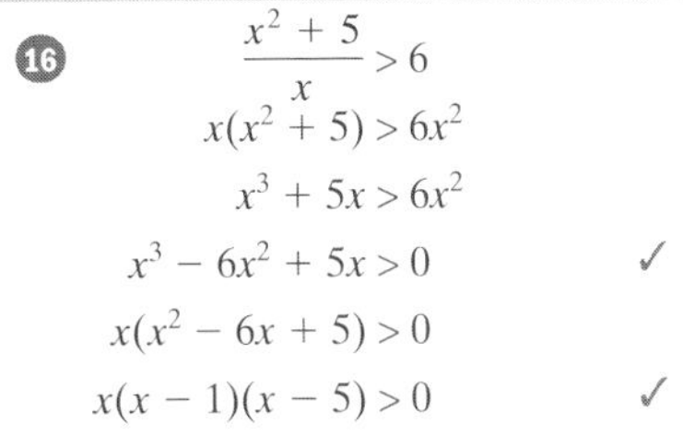

$\frac{x^2 + 5}{x} > 6$

$x(x^2 + 5) > 6x^2$

$x^3 + 5x > 6x^2$

$x^3 - 6x^2 + 5x > 0$ ✓

$x(x^2 - 6x + 5) > 0$

$x(x - 1)(x - 5) > 0$ ✓

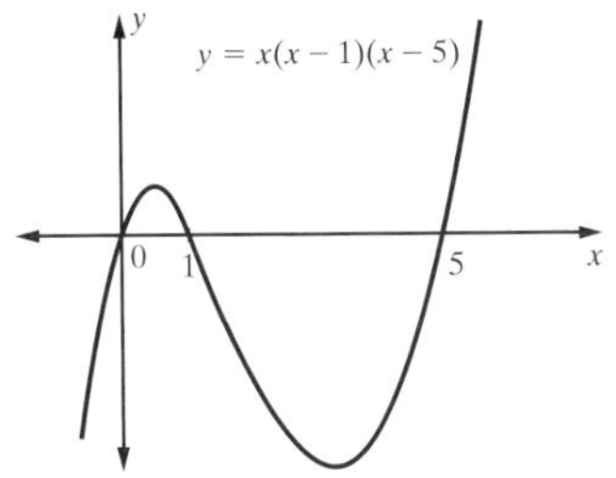

From the graph $x(x - 1)(x - 5) > 0$ when $0 < x < 1$ or $x > 5$.

$\therefore \frac{x^2 + 5}{x} > 6$ when $0 < x < 1$ or $x > 5$ ✓ *(3 marks)*

17 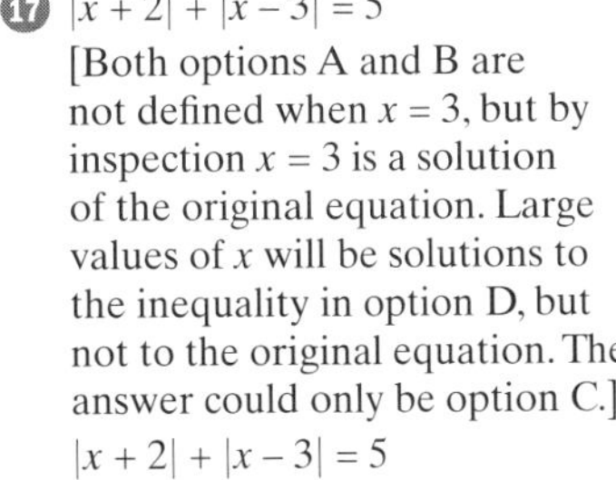

$|x + 2| + |x - 3| = 5$

[Both options A and B are not defined when $x = 3$, but by inspection $x = 3$ is a solution of the original equation. Large values of x will be solutions to the inequality in option D, but not to the original equation. The answer could only be option C.]

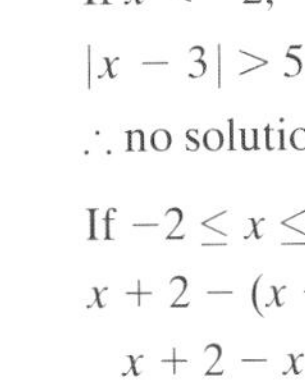

$|x + 2| + |x - 3| = 5$

If $x < -2$,

$|x - 3| > 5$

$\therefore$ no solution for $x < -2$

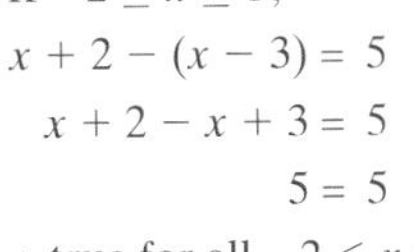

If $-2 \leq x \leq 3$,

$x + 2 - (x - 3) = 5$

$x + 2 - x + 3 = 5$

$5 = 5$

$\therefore$ true for all $-2 \leq x \leq 3$

If $x > 3$,

$|x + 2| > 5$

$\therefore$ no solution for $x > 3$

$x^2 - x - 6 \leq 0$

$(x + 2)(x - 3) \leq 0$

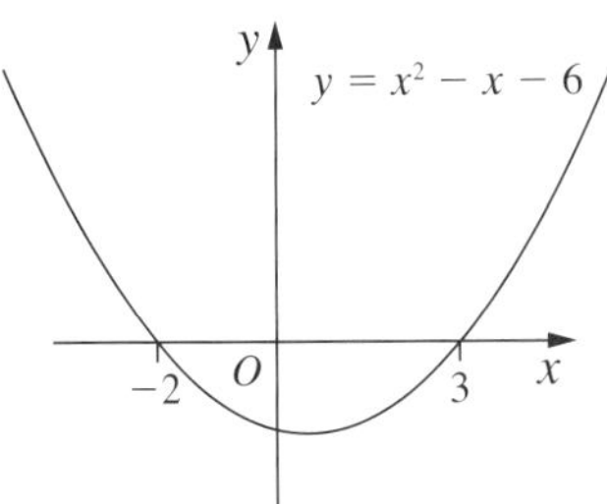

From the graph, $x^2 - x - 6 \leq 0$ for $-2 \leq x \leq 3$

Answer C

18 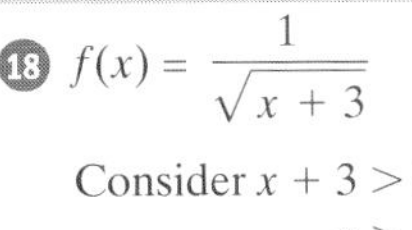

$f(x) = \frac{1}{\sqrt{x + 3}}$

Consider $x + 3 > 0$

$x > -3$

Answer A

19 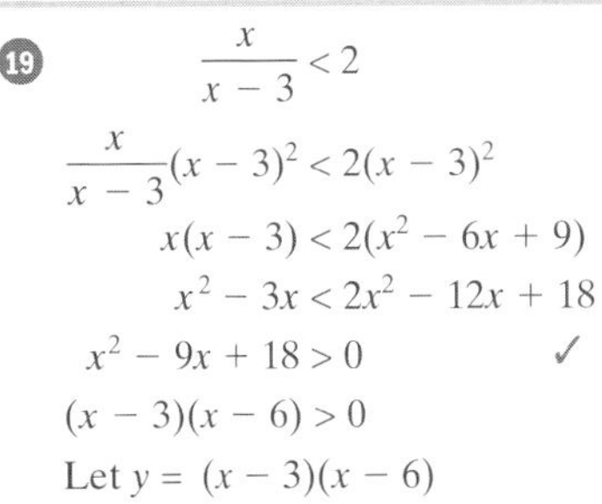

$\frac{x}{x - 3} < 2$

$\frac{x}{x - 3}(x - 3)^2 < 2(x - 3)^2$

$x(x - 3) < 2(x^2 - 6x + 9)$

$x^2 - 3x < 2x^2 - 12x + 18$

$x^2 - 9x + 18 > 0$ ✓

$(x - 3)(x - 6) > 0$

Let $y = (x - 3)(x - 6)$

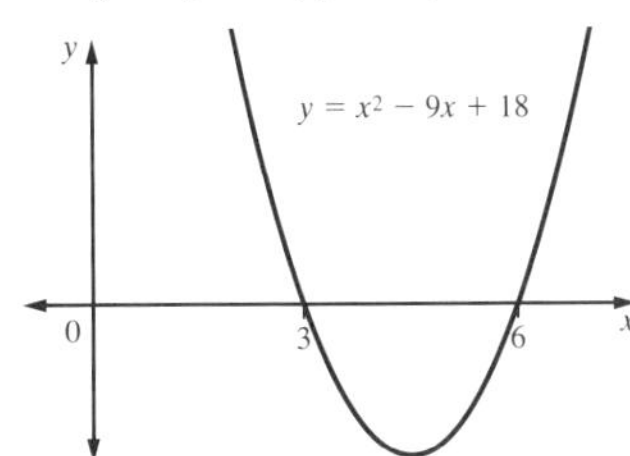

From the diagram, $y > 0$ when $x < 3$ or $x > 6$. ✓

$\therefore \frac{x}{x - 3} < 2$ when $x < 3$ or $x > 6$ ✓ *(3 marks)*

20 Method 1

$\frac{4 - x}{x} < 1, x \neq 0$

If $x < 0$, $4 - x > 0$

and so $\frac{4 - x}{x} < 0$

$\therefore x < 0$ is a solution. ✓

If $x > 0$,

$4 - x < x$

$4 < 2x$

$x > 2$ ✓

$\therefore x < 0$ or $x > 2$

Method 2

$\frac{4-x}{x} < 1, x \neq 0$

$x^2 \times \frac{4-x}{x} < x^2$

$4x - x^2 < x^2$

$2x^2 - 4x > 0$

$2x(x-2) > 0$

Let $y = 2x(x-2)$

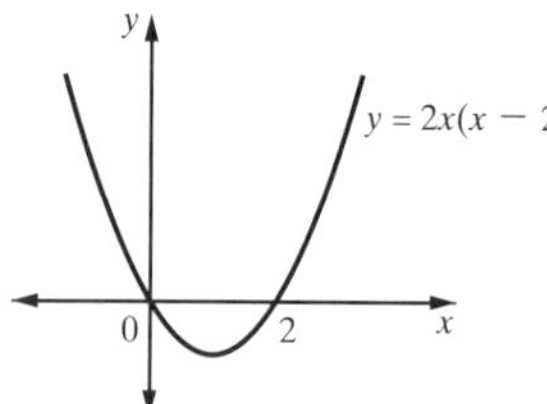

From the diagram,

$y > 0$ when $x < 0$ and $x > 2$

$\therefore \frac{4-x}{x} < 1$ when $x < 0$ or $x > 2$ ✓ *(3 marks)*

21 *Method 1*

$\frac{3}{x+2} < 4$

Now $x + 2 \neq 0$

$\therefore x \neq -2$ ✓

If $x < -2$,

$\frac{3}{x+2}$ is negative and hence less than 4.

So $x < -2$ is a solution. ✓

If $x > -2$,

$\frac{3}{x+2} < 4$

$3 < 4(x+2)$

$3 < 4x + 8$

$-5 < 4x$

$-1\frac{1}{4} < x$

$\therefore x < -2$ or $x > -1\frac{1}{4}$ ✓

Method 2

$\frac{3}{x+2} < 4$

$\frac{3}{x+2}(x+2)^2 < 4(x+2)^2$

$3(x+2) < 4(x^2 + 4x + 4)$

$3x + 6 < 4x^2 + 16x + 16$

$4x^2 + 13x + 10 > 0$

$(4x+5)(x+2) > 0$

Let $y = (4x+5)(x+2)$

When $y = 0$,

$4x + 5 = 0$ or $x + 2 = 0$

$x = -1\frac{1}{4}$ or $x = -2$

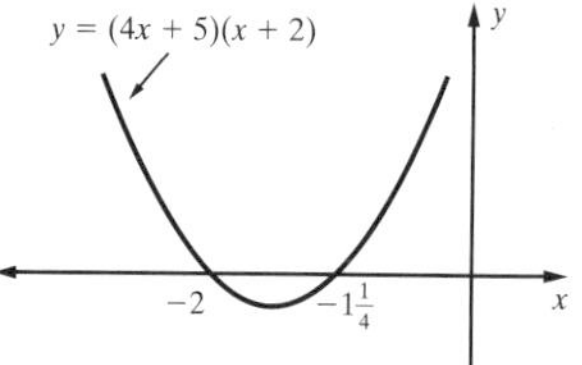

From the graph $y > 0$ when

$x < -2$ or

$x > -1\frac{1}{4}$

$\therefore \frac{3}{x+2} < 4$ when $x < -2$ or

$x > -1\frac{1}{4}$

(3 marks)

22 $x^2 - x - 12 < 0$

$(x-4)(x+3) < 0$ ✓

Method 1:

Let $y = x^2 - x - 12$,
so we need to consider values of x for which $y < 0$:

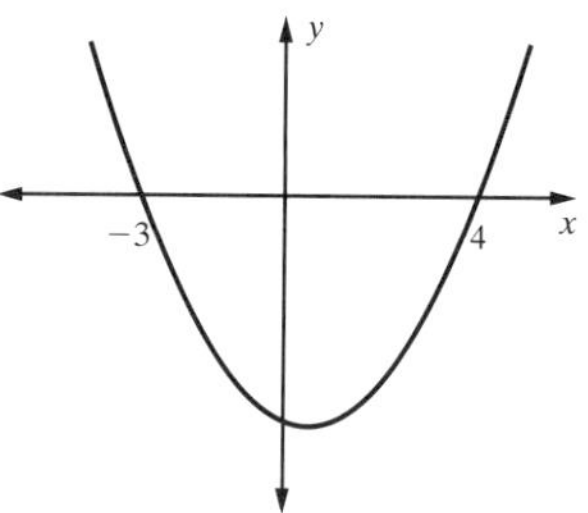

$\therefore -3 < x < 4$ ✓

Method 2:

Let $(x-4)(x+3) = 0$ gives
$x = 4, -3$:

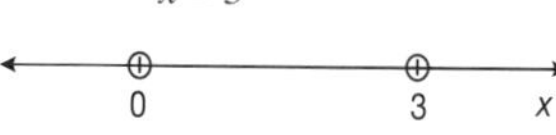

Choose 0: subs in

$(x-4)(x+3) < 0$

LHS $= (0-4)(0+3)$

$= -12 < 0$, $\therefore$ true

$\therefore -3 < x < 4$ *(2 marks)*

23 $\frac{x+3}{2x} > 1$

Method 1:

Let $\frac{x+3}{2x} = 1, x \neq 0$

$x + 3 = 2x$

$x = 3$ ✓

$$\longleftarrow \overset{\circ}{0} \quad\quad \overset{\circ}{3} \longrightarrow x$$

By checking intervals:

When $x > 3$, e.g. $x = 4$,

$\frac{4+3}{2(4)} = \frac{7}{8} \ngtr 1$

$\therefore x \ngtr 3$ ✓

When $0 < x < 3$, e.g. $x = 2$,

$\frac{2+3}{2(2)} = \frac{5}{4} > 1$

$\therefore$ True

$\therefore$ Solution is $0 < x < 3$. ✓

Method 2:

$\frac{x+3}{2x} > 1$

Multiplying both sides by $(2x)^2$:

$2x(x+3) > (2x)^2$

$2x^2 + 6x > 4x^2$

$0 > 2x^2 - 6x$

i.e. $2x^2 - 6x < 0$

$2x(x-3) < 0$

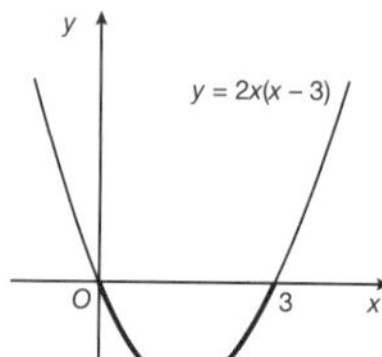

$\therefore 0 < x < 3$. *(3 marks)*

24 **i** Vertical asymptote is $x = 4$. ✓

Horizontal asymptote:

$y = \lim_{x \to \infty} \frac{x-2}{x-4}$

$= \lim_{x \to \infty} \frac{1 - \frac{2}{x}}{1 - \frac{4}{x}}$

$= 1$. ✓

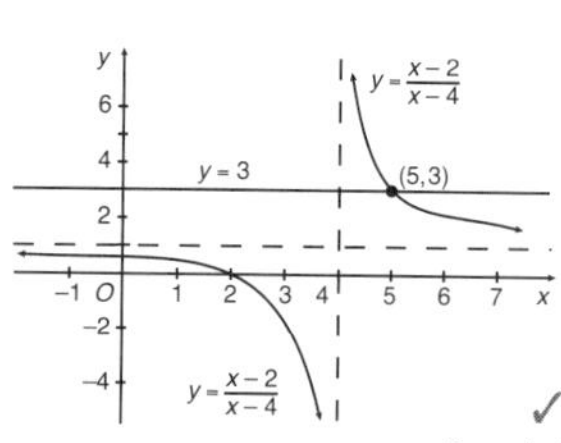

✓ *(3 marks)*

ii On the graph of $y = \frac{x-2}{x-4}$, sketch $y = 3$.

Now $\frac{x-2}{x-4} = 3$

$x - 2 = 3x - 12$

$-2x = -10$

$x = 5$

$\therefore$ The graph intersects $y = 3$ at $(5, 3)$. ✓

The hyperbola is touching or below the line,

$y = 3$ when $x < 4$ and $x \geqslant 5$.

i.e. $\frac{x-2}{x-4} \leqslant 3$ when $x < 4$ and $x \geqslant 5$. ✓

(2 marks)

25 **i** $\Delta = b^2 - 4ac$
$= (k-2)^2 - 4(2)(8)$
$= (k-2)^2 - 64.$ ✓ *(1 mark)*

ii $y = 2x^2 + kx + 9$
$y = 2x + 1$

Solving simultaneously:

$2x^2 + kx + 9 = 2x + 1$
$2x^2 + kx - 2x + 8 = 0$
$2x^2 + (k-2)x + 8 = 0$

If the parabola does not intersect the line, then $2x^2 + (k-2)x + 8 = 0$ has no real roots, ✓

i.e. $\Delta < 0$
$(k-2)^2 - 4(2)(8) < 0$
$k^2 - 4k + 4 - 64 < 0$
$k^2 - 4k - 60 < 0$
$(k-10)(k+6) < 0$

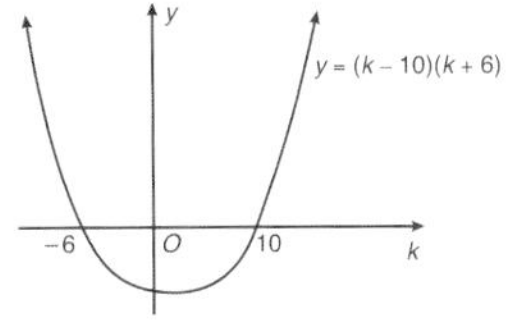

$\therefore -6 \le k \le 10$ ✓
(2 marks)

26 $|x-3| \leqslant 1$
$\therefore -1 \leqslant x - 3 \leqslant 1$ ✓
$\therefore 2 \leqslant x \leqslant 4.$ ✓
(2 marks)

27 $\dfrac{4}{x+1} < 3$

$\therefore (x+1)^2 \dfrac{4}{x+1} < 3(x+1)^2$

(on multiplying both sides by $(x+1)^2$)

$4(x+1) < 3(x^2 + 2x + 1)$ ✓
$4x + 4 < 3x^2 + 6x + 3$
$0 < 3x^2 + 2x - 1$

i.e. $3x^2 + 2x - 1 > 0$
$(3x-1)(x+1) > 0$ ✓

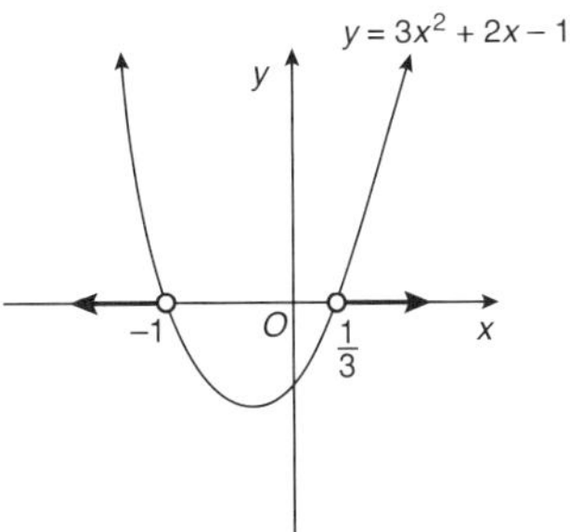

$\therefore x < -1$ or $x > \dfrac{1}{3}$ ✓

(N.B. $x \neq -1$).
(3 marks)

28 $\dfrac{3}{x-2} \leqslant 1$

$\therefore 3(x-2) \leqslant (x-2)^2$
(On multiplying both sides by the positive number, $(x-2)^2, x \neq 2$.) ✓

$\therefore 3(x-2) - (x-2)^2 \leqslant 0, \quad x \neq 2$
$\therefore (x-2)(3-(x-2)) \leqslant 0$
$\therefore (x-2)(5-x) \leqslant 0$
$\therefore x - 2 \leqslant 0$ or $5 - x \leqslant 0$ ✓
$\therefore x \leqslant 2$ or $5 \leqslant x$
$\therefore x < 2$ or $x \geqslant 5$ ✓
(n.b. $x \neq 2$)
(3 marks)

29 $|x-1| \ge 3$
$\therefore x - 1 \ge 3$ or $-(x-1) \ge 3$
$\therefore x \ge 4$ or $-x + 1 \ge 3$
$\therefore x \ge 4$ or $x \le -2$ ✓

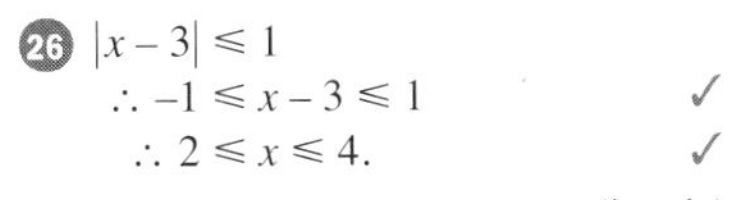

✓ *(2 marks)*

30 $\dfrac{3x}{x-2} \le 1$

$\therefore 3x(x-2) \le (x-2)^2$
(on multiplying both sides by the positive number $(x-2)^2$ for $x \neq 2$) ✓
$\therefore 3x(x-2) - (x-2)^2 \le 0$
$(x-2)(3x-(x-2)) \le 0$
$(x-2)(2x+2) \le 0$
$2(x-2)(x+1) \le 0$ ✓

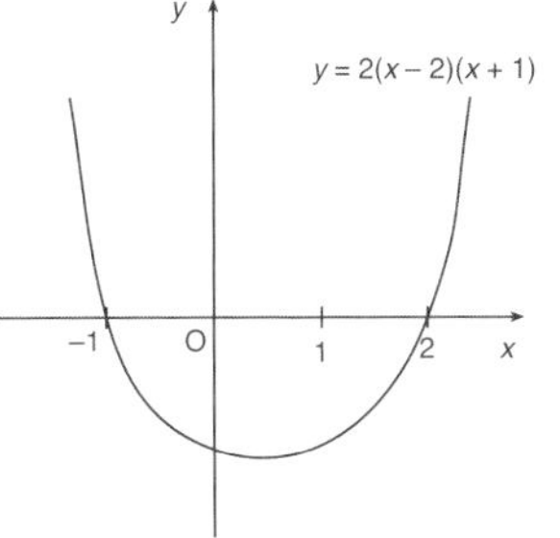

$\therefore$ From graph, and the condition $x \neq 2, -1 \le x < 2.$ ✓
(3 marks)

31 No real roots occur when
$\Delta < 0$
$b^2 - 4ac < 0$
$(2)^2 - 4(3)(k) < 0$
$4 - 12k < 0$ ✓
$-12k < -4$
$k > \dfrac{4}{12}$
$\therefore k > \dfrac{1}{3}.$ ✓
(2 marks)

32 $\dfrac{5}{x+2} \leqslant 1$

$\therefore 5(x+2) \leqslant (x+2)^2 \quad x \neq -2$ ✓

(on multiplying both sides by the positive number $(x+2)^2$)
$\therefore 5x + 10 \leqslant x^2 + 4x + 4$
$\therefore x^2 - x - 6 \geqslant 0$
$\therefore (x-3)(x+2) \geqslant 0$ ✓

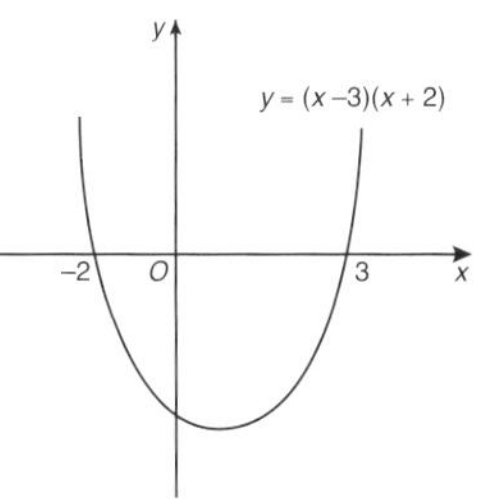

$\therefore x \geqslant 3$ or $x < -2$ (since $x \neq -2$).
✓ *(3 marks)*

33 Now $\sqrt{x(1-x)}$ is only defined for $0 \leqslant x \leqslant 1$.

$|4x-1| > 2\sqrt{x(1-x)}$
$(4x-1)^2 > 4x(1-x)$
$16x^2 - 8x + 1 > 4x - 4x^2$
$20x^2 - 12x + 1 > 0$ ✓
$(10x-1)(2x-1) > 0$
$\therefore x < \dfrac{1}{10}$ or $x > \dfrac{1}{2}$ ✓

But since $0 \leqslant x \leqslant 1$,
$\therefore 0 \leqslant x < \dfrac{1}{10}$ or $\dfrac{1}{2} < x \leqslant 1.$ ✓

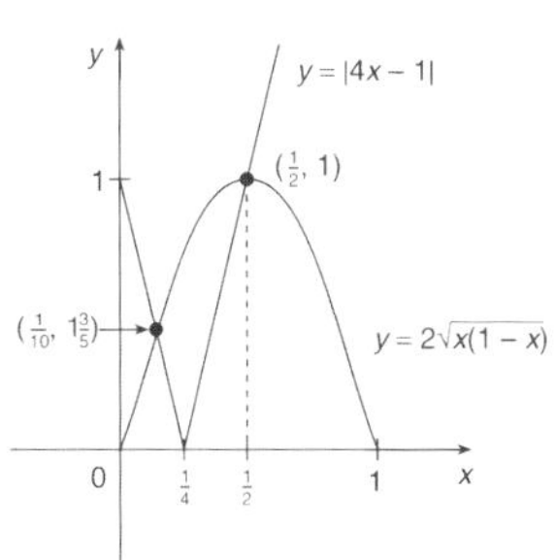

(3 marks)

34 **i and ii**

$y = x^2 - 6$
$= (x + \sqrt{6})(x - \sqrt{6})$
$\therefore x$ intercepts are $\pm\sqrt{6}$
y intercept occurs when $x = 0$
$\therefore y = 0 - 6$
$= -6$ ✓

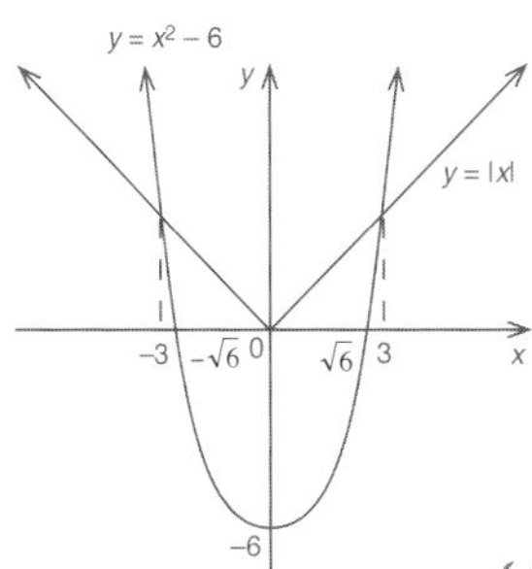

✓✓

iii Graphs intersect when
$x^2 - 6 = |x|$
$(x^2 - 6)^2 = x^2$
$x^4 - 12x^2 + 36 = x^2$
$x^4 - 13x^2 + 36 = 0$
$(x^2 - 9)(x^2 - 4) = 0$ ✓
$(x+3)(x-3)$
$(x+2)(x-2) = 0$
$\therefore x = \pm 3$ or ± 2
From the graph $x = \pm 3.$ ✓

iv $x^2 - 6 \leqslant |x|$
when $-3 \leqslant x \leqslant 3$
✓(from graph).
(6 marks)
(Total mark allocation only in exam)

35

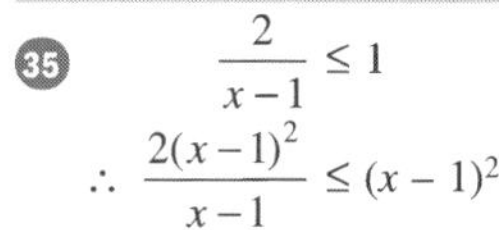

$$\frac{2}{x-1} \leq 1$$

$$\therefore \frac{2(x-1)^2}{x-1} \leq (x-1)^2$$

(On multiplying both sides by the positive number $(x-1)^2$) ✓

$\therefore 2(x-1) \leq (x-1)^2 \quad x \neq 1$

$(x-1)^2 - 2(x-1) \geq 0$

$(x-1)(x-1-2) \geq 0$

$(x-1)(x-3) \geq 0 \qquad x \neq 1$ ✓

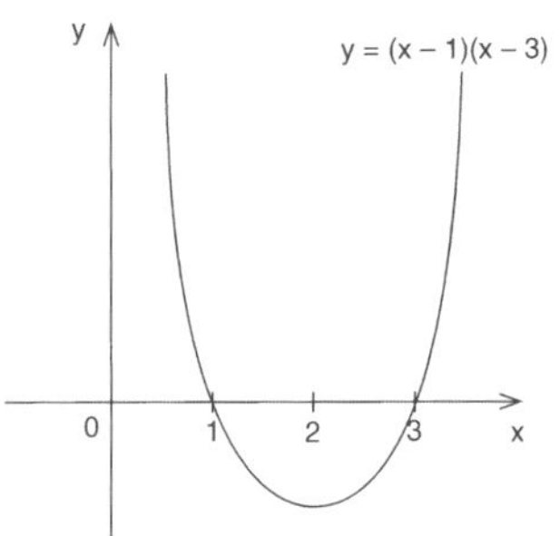

$\therefore x < 1$ or $x \geq 3$ (from graph) ✓
(3 marks)

36 **i** Two layers let through
$85\% \times 85\%$
$= \frac{85}{100} \times \frac{85}{100}$
$= 0.7225$
$= 72.25\%.$ ✓

ii Cut out at least 90% means let through 10% or less

$\therefore \left(\frac{85}{100}\right)^n \leqslant 10\%$

$\left(\frac{85}{100}\right)^n \leqslant 0.1$ ✓

$n \log(0.85) \leqslant \log(0.1)$

$n \geqslant \frac{\log(0.1)}{\log(0.85)}$

$n \geqslant 14.1681\ldots$

$\therefore n = 15$

$\therefore$ 15 layers are required. ✓
(3 marks)
(Total mark allocation only in exam)

37 **i** For $x = \frac{1}{3}$:

$\text{LHS} = 7 - 3\left(\frac{1}{3}\right)$
$= 7 - 1$
$= 6$

$\text{RHS} = \frac{2}{\frac{1}{3}}$
$= 2 \times 3$
$= 6$

$\therefore x = \frac{1}{3}$ satisfies the equation.

For $x = 2$:

$\text{LHS} = 7 - 3(2) \qquad \text{RHS} = \frac{2}{2}$
$= 7 - 6 \qquad\qquad = 1$
$= 1$

$\therefore x = 2$ satisfies the equation.

ii

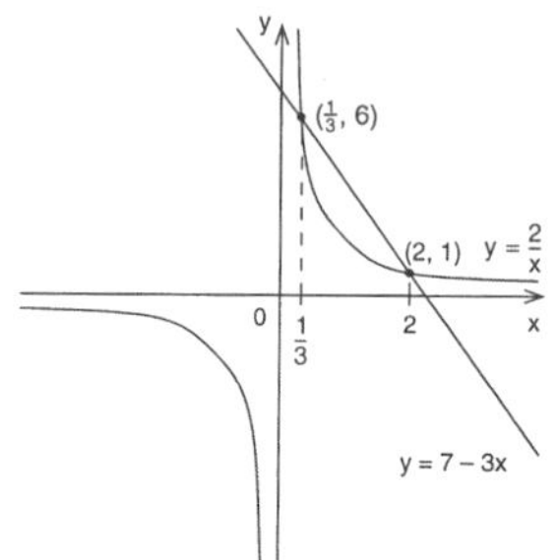

iii From the diagram:

$7 - 3x < \frac{2}{x}$

for $0 < x < \frac{1}{3}$ and $x > 2$.
(No mark allocation in exam)

38 $12 + 4m - m^2 > 0$
$(6 - m)(2 + m) > 0$

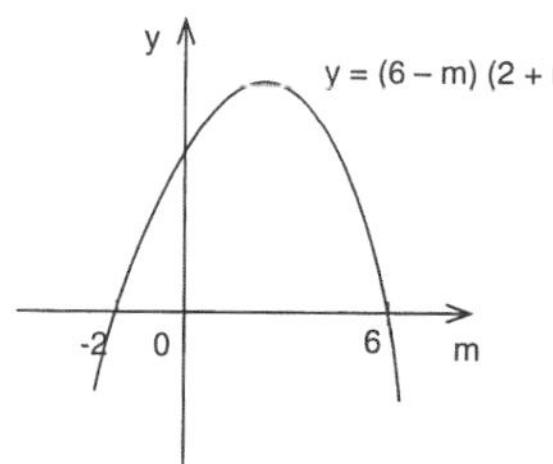

$\therefore -2 < m < 6$
(No mark allocation in exam)

39 $x^2 - 3x < 4$
$x^2 - 3x - 4 < 0$
$(x - 4)(x + 1) < 0$

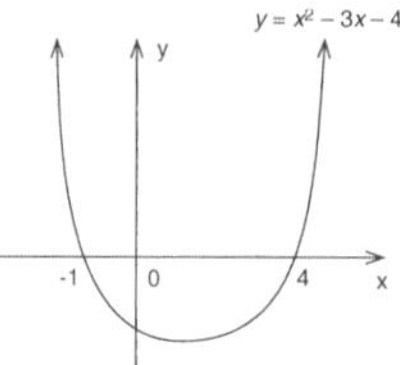

$\therefore -1 < x < 4.$
(No mark allocation in exam)

40 $x^2 - x - 2 > 0$
$(x + 1)(x - 2) > 0$

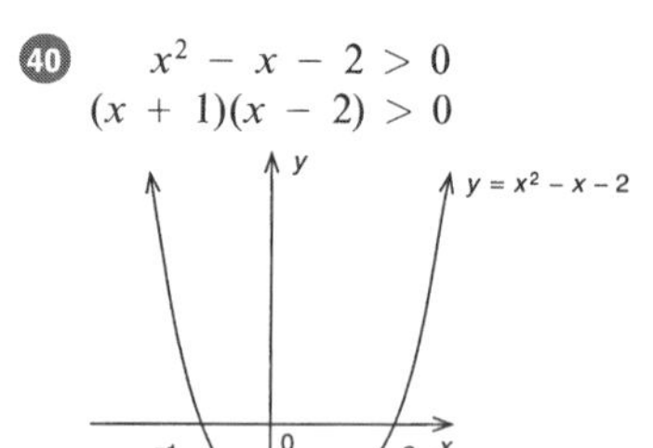

$\therefore x < -1$ and $x > 2$.
(No mark allocation in exam)

1 A given function $f(x)$ has an inverse $f^{-1}(x)$.

The derivatives of $f(x)$ and $f^{-1}(x)$ exist for all real numbers x.

The graphs $y = f(x)$ and $y = f^{-1}(x)$ have at least one point of intersection.

Which statement is true for all points of intersection of these graphs?

A All points of intersection lie on the line $y = x$.

B None of the points of intersection lie on the line $y = x$.

C At no point of intersection are the tangents to the graphs parallel.

D At no point of intersection are the tangents to the graphs perpendicular. *(1 mark)*

(Q9, **2022 HSC**) Easy

2 A function is defined by $f(x) = 4 - \left(1 - \dfrac{x}{2}\right)^2$ for x in the domain $(-\infty, 2]$.

i Sketch the graph of $y = f(x)$ showing the x- and y-intercepts. *(2 marks)* Medium

ii Find the equation of the inverse function, $f^{-1}(x)$, and state its domain. *(3 marks)* Medium

iii Sketch the graph of $y = f^{-1}(x)$. *(1 mark)* Medium

(Q12d, **2021 HSC**)

3 Given $f(x) = 1 + \sqrt{x}$, what are the domain and range of $f^{-1}(x)$?

A $x \geq 0, y \geq 0$ **B** $x \geq 0, y \geq 1$

C $x \geq 1, y \geq 0$ **D** $x \geq 1, y \geq 1$ *(1 mark)*

(Q2, **2020 HSC**) Medium

4 Functions f and g are defined by $f(x) = 2x - 3$ and $g(x) = 5 - 4x$.

Find the inverse of the composite function $f \circ g$, written in terms of x. *(2 marks)*

Bonus question (see page iv) Medium

5 A function is defined as $f(x) = 1 - \sqrt{x-3}$. The graph of $y = f(x)$ is shown below.

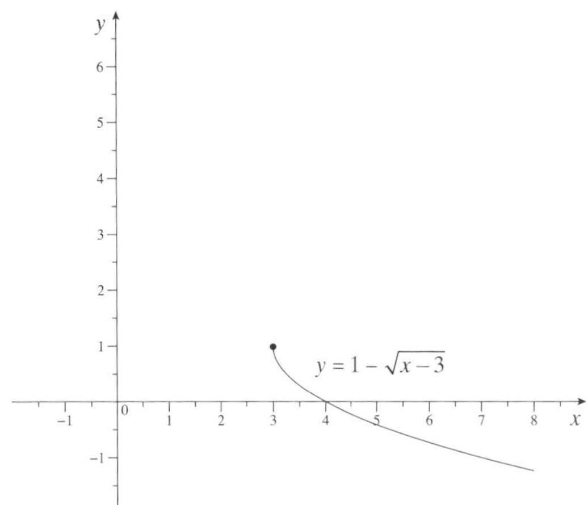

Sketch the graph of $y = f^{-1}(x)$ on the diagram above. *(2 marks)*

Bonus question Easy

6 The function $f(x) = -\sqrt{1 + \sqrt{1+x}}$ has inverse $f^{-1}(x)$.

The graph of $y = f^{-1}(x)$ forms part of the curve $y = x^4 - 2x^2$.

The diagram shows the curve $y = x^4 - 2x^2$.

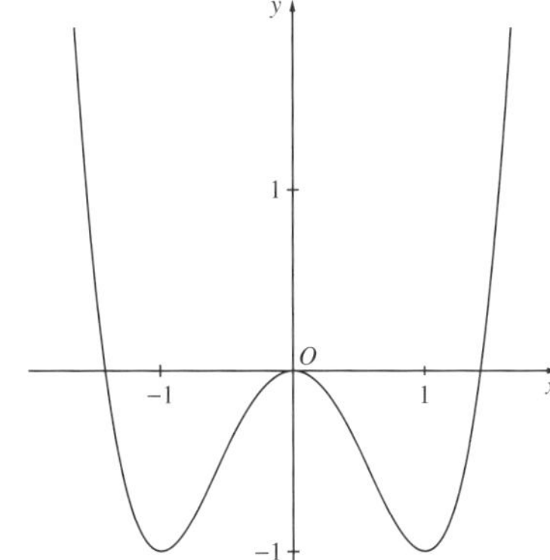

How many points do the graphs of $y = f(x)$ and $y = f^{-1}(x)$ have in common?

A 1 **B** 2

C 3 **D** 4 *(1 mark)*

(Q10, **2019 HSC**) Hard

7 The diagram shows the graph $y = \dfrac{x}{x^2+1}$, for all real x.

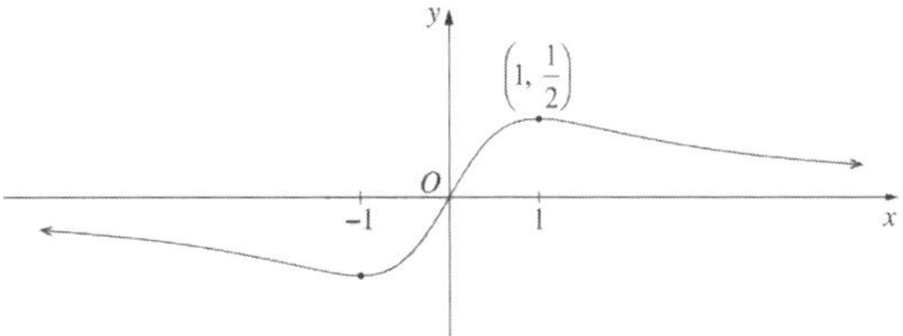

Consider the function $f(x) = \dfrac{x}{x^2+1}$, for $x \geq 1$.

The function $f(x)$ has an inverse. (Do NOT prove this.)

i State the domain and range of $f^{-1}(x)$. *(2 marks)* Medium

ii Sketch the graph $y = f^{-1}(x)$. *(1 mark)* Medium

iii Find an expression for $f^{-1}(x)$. *(3 marks)* Medium

(Q13b, **2018 HSC**)

8 Find the inverse of the function $y = x^3 - 2$. *(2 marks)*

(Q11a, **2016 HSC**) Easy

9 The diagram shows the graph $y = f(x)$.

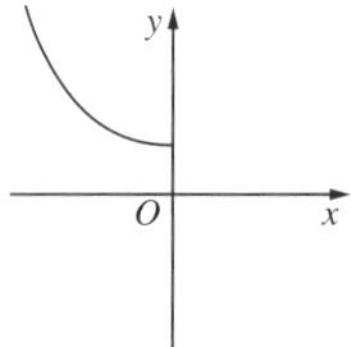

Which diagram shows the graph $y = f^{-1}(x)$?

A

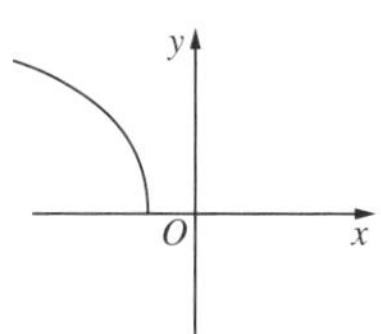

B

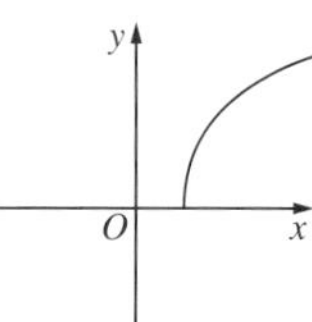

C

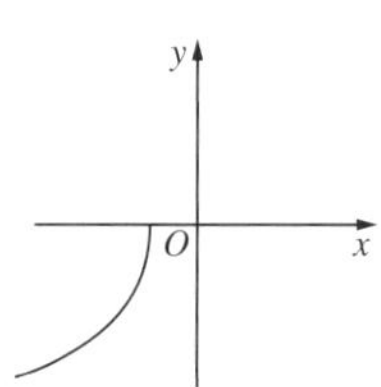

D

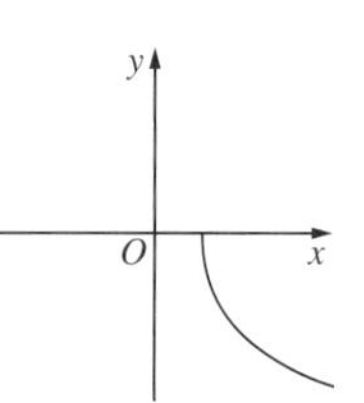

(1 mark)

(Q2, **2013 HSC**) Easy

10 Let $f(x) = \sqrt{4x - 3}$.

i Find the domain of $f(x)$. *(1 mark)* Easy

ii Find an expression for the inverse function $f^{-1}(x)$. *(2 marks)* Easy

iii Find the points where the graphs $y = f(x)$ and $y = x$ intersect. *(1 mark)* Easy

iv On the same set of axes, sketch the graphs $y = f(x)$ and $y = f^{-1}(x)$ showing the information found in part **iii**. *(2 marks)* Medium

(Q12b, **2012 HSC**)

11 Let $f(x) = e^{-x^2}$. The diagram shows the graph $y = f(x)$.

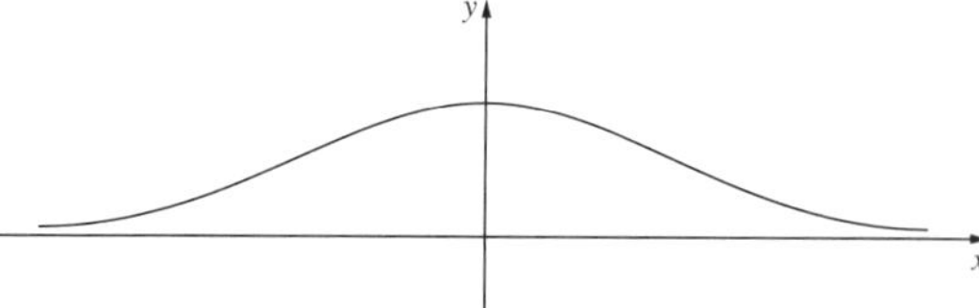

i The graph has two points of inflexion. Find the x coordinates of these points. *(3 marks)* Medium

ii Explain why the domain of $f(x)$ must be restricted if $f(x)$ is to have an inverse function. *(1 mark)* Easy

iii Find a formula for $f^{-1}(x)$ if the domain of $f(x)$ is restricted to $x \geq 0$. *(2 marks)* Medium

iv State the domain of $f^{-1}(x)$. *(1 mark)* Medium

v Sketch the curve $f^{-1}(x)$. *(1 mark)* Medium

(Q3b, **2010 HSC**)

12 Let $f(x) = \dfrac{3 + e^{2x}}{4}$

i Find the range of $f(x)$. *(1 mark)* Easy

ii Find the inverse function $f^{-1}(x)$. *(2 marks)* Medium

(Q3a, **2009 HSC**)

13 Let $f(x) = x - \frac{1}{2}x^2$ for $x \leq 1$. This function has an inverse, $f^{-1}(x)$.

i Sketch the graphs of $y = f(x)$ and $y = f^{-1}(x)$ on the same set of axes. (Use the same scale on both axes.) *(2 marks)* Medium

ii Find an expression for $f^{-1}(x)$. *(3 marks)* Medium

iii Evaluate $f^{-1}\left(\dfrac{3}{8}\right)$. *(1 mark)* Easy

(Q5a, **2008 HSC**)

14 Consider the function $f(x) = e^x - e^{-x}$.

i Show that $f(x)$ is increasing for all values of x. *(1 mark)* Easy

ii Show that the inverse function is given by

$$f^{-1}(x) = \log_e\left(\frac{x + \sqrt{x^2 + 4}}{2}\right).$$

(3 marks) Hard

iii Hence, or otherwise, solve $e^x - e^{-x} = 5$. Give your answer correct to two decimal places. *(1 mark)* Medium

(Q6b, **2007 HSC**)

15 Let $f(x) = \log_e(1 + e^x)$ for all x. Show that $f(x)$ has an inverse. *(2 marks)*

(Q5b, **2006 HSC**) Medium

16 The diagram below shows a sketch of the graph of $y = f(x)$, where $f(x) = \dfrac{1}{1 + x^2}$ for $x \geq 0$.

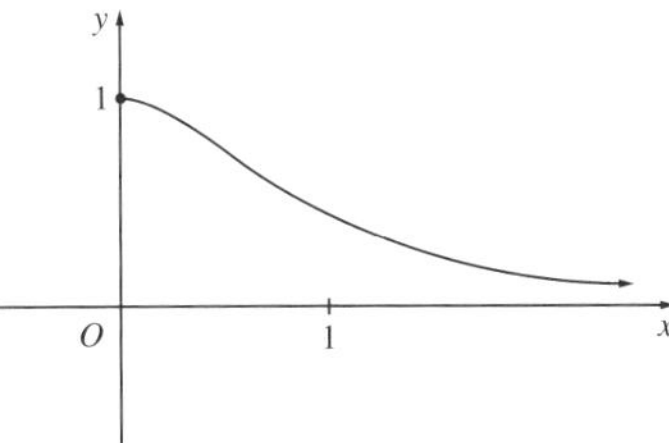

- **i** Copy or trace this diagram into your writing booklet. On the same set of axes, sketch the graph of the inverse function, $y = f^{-1}(x)$. *(1 mark)* Easy
- **ii** State the domain of $f^{-1}(x)$. *(1 mark)* Easy
- **iii** Find an expression for $y = f^{-1}(x)$ in terms of x. *(2 marks)* Medium
- **iv** The graphs of $y = f(x)$ and $y = f^{-1}(x)$ meet at exactly one point P.
  Let α be the x-coordinate of P. Explain why α is a root of the equation $x^3 + x - 1 = 0$. *(1 mark)* Medium

(Q5b, **2004 HSC**)

17 The graph of $f(x) = x^2 - 4x + 5$ is shown in the diagram.

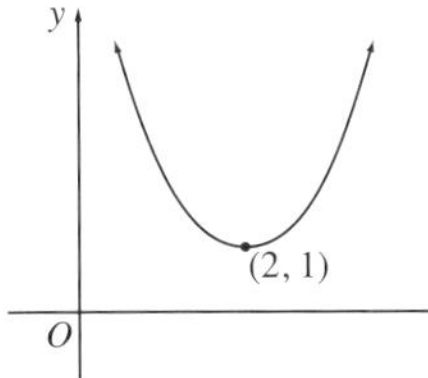

- **i** Explain why $f(x)$ does not have an inverse function. *(1 mark)* Easy
- **ii** Sketch the graph of the inverse function, $g^{-1}(x)$, of $g(x)$, where $g(x) = x^2 - 4x + 5$, $x \leq 2$. *(1 mark)* Easy
- **iii** State the domain of $g^{-1}(x)$. *(1 mark)* Easy
- **iv** Find an expression for $y = g^{-1}(x)$ in terms of x. *(2 marks)* Medium

(Q5b, **2003 HSC**)

18 Let $g(x) = e^x + \dfrac{1}{e^x}$ for all real values of x and let $f(x) = e^x + \dfrac{1}{e^x}$ for $x \leq 0$.

- **i** Sketch the graph $y = g(x)$ and explain why $g(x)$ does not have an inverse function. *(2 marks)* Easy
- **ii** On a separate diagram, sketch the graph of the inverse function $y = f^{-1}(x)$. *(1 mark)* Easy
- **iii** Find an expression for $y = f^{-1}(x)$ in terms of x. *(3 marks)* Hard

(Q7a, **2002 HSC**)

19 Consider the function $f(x) = \dfrac{x}{x + 2}$.

- **i** Show that $f'(x) > 0$ for all x in the domain. Easy
- **ii** State the equation of the horizontal asymptote of $y = f(x)$. Easy
- **iii** Without using any further calculus, sketch the graph of $y = f(x)$. Easy
- **iv** Explain why $f(x)$ has an inverse function $f^{-1}(x)$. Easy
- **v** Find an expression for $f^{-1}(x)$. Medium
- **vi** Write down the domain of $f^{-1}(x)$. Easy

(8 marks) (Q5b, **2000 HSC**)

20 Consider the function $f(x) = e^x - 1 - x$.

- **i** Show that the minimum of $f(x)$ occurs at $x = 0$. Easy
- **ii** Deduce that $f(x) \geq 0$ for all x. Easy
- **iii** On the same set of axes, sketch $y = e^x - 1$ and $y = x$. Easy
- **iv** Find the inverse function of $g(x) = e^x - 1$. Easy
- **v** State the domain of $g^{-1}(x)$. Easy
- **vi** For what values of x is $\log_e(1 + x) \leq x$? Justify your answer. Medium

(9 marks) (Q5b, **1999 HSC**)

21 Consider the function $f(x) = 1 + \dfrac{3}{(x - 2)}$ for $x > 2$.

- **i** Give the equations of the horizontal and vertical asymptotes for $y = f(x)$. Easy
- **ii** Find the inverse function $f^{-1}(x)$. Medium
- **iii** State the domain of $f^{-1}(x)$. Easy

(5 marks) (Q4b, **1998 HSC**)

22 Consider the function $f(x) = \dfrac{1}{4}[(x - 1)^2 + 7]$.

- **i** Sketch the parabola $y = f(x)$, showing clearly any intercepts with the axes, and the coordinates of its vertex. Use the same scale on both axes. Medium
- **ii** What is the largest domain containing the value $x = 3$, for which the function has an inverse function $f^{-1}(x)$? Medium
- **iii** Sketch the graph of $y = f^{-1}(x)$ on the same set of axes as your graph in part **i**. Label the two graphs clearly. Easy
- **iv** What is the domain of the inverse function? Easy
- **v** Let a be a real number not in the domain found in part **ii**. Find $f^{-1}(f(a))$. Hard
- **vi** Find the coordinates of any points of intersection of the two curves $y = f(x)$ and $y = f^{-1}(x)$. Easy

(10 marks) (Q7b, **1996 HSC**)

23 Consider the function $f(x) = \dfrac{e^x}{3+e^x}$.

Note that e^x is always positive, and that $f(x)$ is defined for all real x.

a Show that $f(x)$ has no stationary points. *(2 marks)* **Medium**

b Find the coordinates of the point of inflexion, given that $f''(x) = \dfrac{3e^x\left(3-e^x\right)}{\left(3+e^x\right)^3}$. *(1 mark)* **Medium**

c Show that $0 < f(x) < 1$ for all x. *(2 marks)* **Medium**

d Describe the behaviour of $f(x)$ for very large positive and very large negative values of x, i.e. as $x \to \infty$ and $x \to -\infty$. *(2 marks)* **Medium**

e Sketch the curve $y = f(x)$. *(2 marks)* **Medium**

f Explain why $f(x)$ has an inverse function. *(1 mark)* **Medium**

g Find the inverse function $y = f^{-1}(x)$. *(2 marks)* **Medium**

(Q4, **1995 HSC**)

Year 11 Inverse functions—Worked Answers

1 Points of intersection of function $f(x)$ and its $f^{-1}(x)$ inverse may have points of intersection on and off the line $y = x$.
There are an infinite number of points with parallel tangents. The graph of $f(x) = \frac{1}{x}$ can be used as a counter example for A, B and C.
Answer D

2 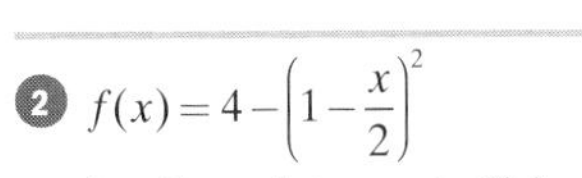

$f(x) = 4 - \left(1 - \frac{x}{2}\right)^2$

i For x-intercepts $f(x) = 0$.

$$4 - \left(1 - \frac{x}{2}\right)^2 = 0$$
$$\left(1 - \frac{x}{2}\right)^2 = 4$$
$$1 - \frac{x}{2} = \pm 2$$
$$-\frac{x}{2} = \pm 2 - 1$$

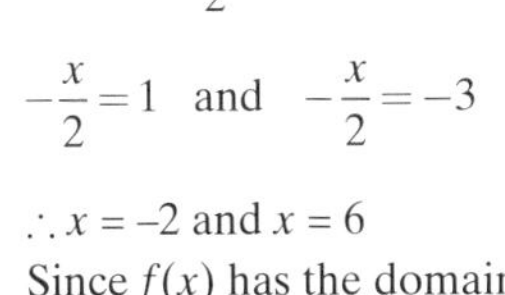

$-\frac{x}{2} = 1$ and $-\frac{x}{2} = -3$

$\therefore x = -2$ and $x = 6$

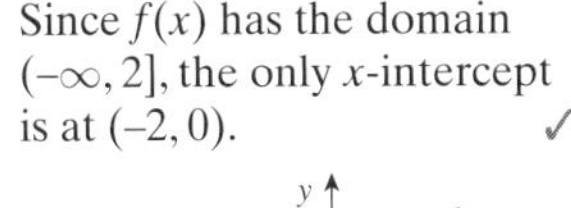

Since $f(x)$ has the domain $(-\infty, 2]$, the only x-intercept is at $(-2, 0)$. ✓

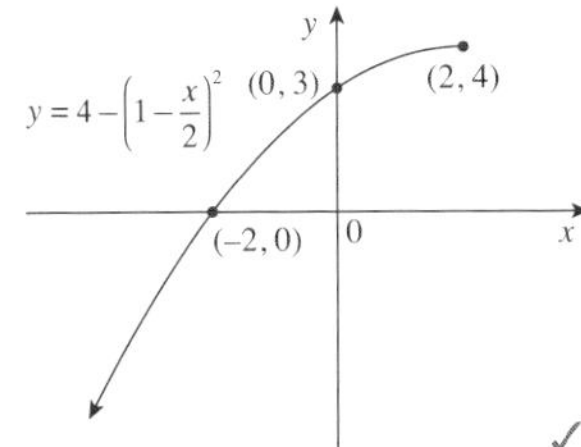

✓ *(2 marks)*

ii $f(x) = 4 - \left(1 - \frac{x}{2}\right)^2$

$f^{-1}(x)$: $x = 4 - \left(1 - \frac{y}{2}\right)^2$ ✓

$$\left(1 - \frac{y}{2}\right)^2 = 4 - x$$
$$1 - \frac{y}{2} = \pm\sqrt{4 - x}$$
$$-\frac{y}{2} = -1 \pm \sqrt{4 - x}$$
$$y = 2 \pm 2\sqrt{4 - x}$$ ✓

Since $f(x)$ has the domain $(-\infty, 2]$, then $f^{-1}(x)$ has the range of $(-\infty, 2]$.

$2 = 2 \pm 2\sqrt{4 - x}$

For $f^{-1}(x) = 2$, then $f^{-1}(x)$ must be the negative case.

$\therefore f^{-1}(x) = 2 - 2\sqrt{4 - x}$ ✓

(3 marks)

iii Sketch $f^{-1}(x) = 2 - 2\sqrt{4 - x}$

y, (4, 2), (3, 0), 0, x, (0, −2), $f^{-1}(x) = 2 - 2\sqrt{4 - x}$

✓ *(1 mark)*

3 $f(x) = 1 + \sqrt{x}$
Domain: $x \geq 0$
Range: $y \geq 1$
So $f^{-1}(x)$ has domain $x \geq 1$ and range $y \geq 0$.
Answer C

4 $f \circ g = f(g(x))$

$f \circ g = 2(5 - 4x) - 3$
$= 10 - 8x - 3$
$= 7 - 8x$ ✓

Let $y = 7 - 8x$

$(f \circ g)^{-1}$: $x = 7 - 8y$
$8y = 7 - x$
$y = \frac{7 - x}{8}$

$\therefore (f \circ g)^{-1} = \frac{7 - x}{8}$ ✓ *(2 marks)*

5

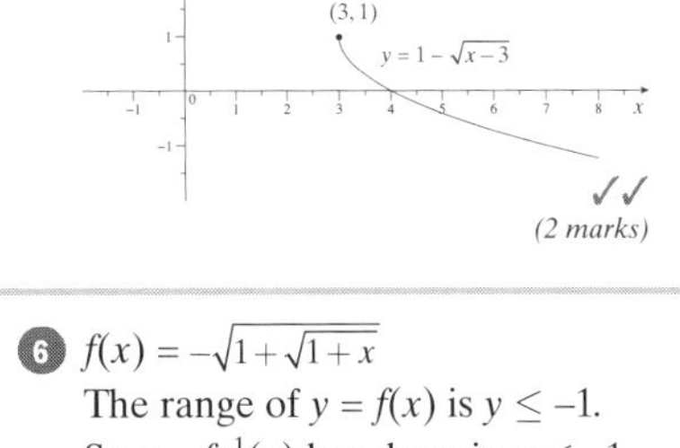

✓✓ *(2 marks)*

6 $f(x) = -\sqrt{1 + \sqrt{1 + x}}$
The range of $y = f(x)$ is $y \leq -1$.
So $y = f^{-1}(x)$ has domain $x \leq -1$.
The graphs have one point in common $(-1, -1)$.

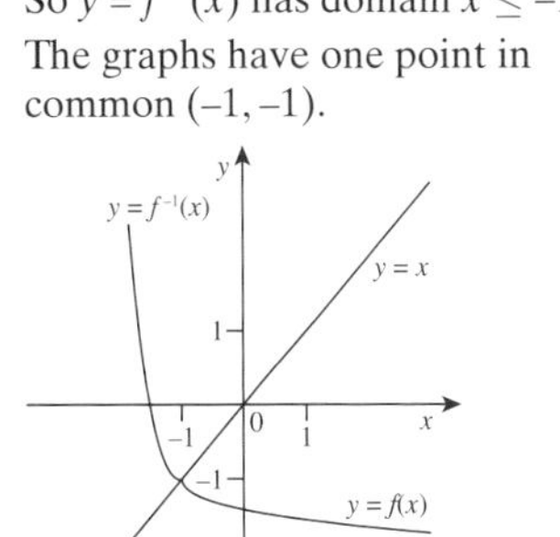

Answer A

7 **i** $f(x) = \frac{x}{x^2 + 1}$, $\quad x \geq 1$

Range of $f(x)$ is $0 < y \leq \frac{1}{2}$.

So the domain of $f^{-1}(x)$ is $0 < x \leq \frac{1}{2}$. ✓

Range of $f^{-1}(x)$ is $y \geq 1$. ✓ *(2 marks)*

ii

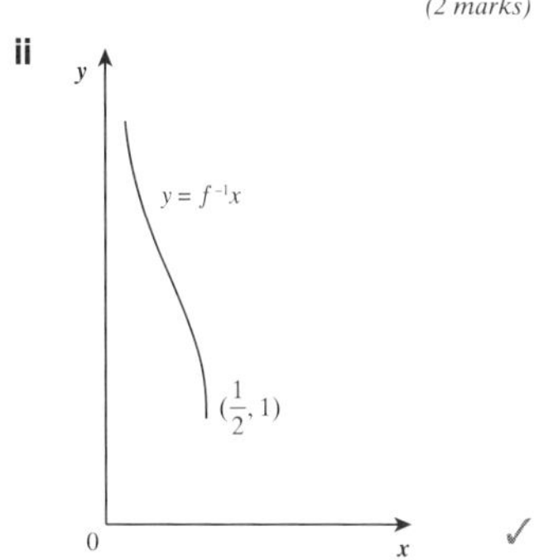

✓ *(1 mark)*

iii 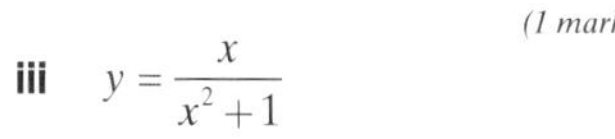

$y = \frac{x}{x^2 + 1}$

Inverse function:

$$x = \frac{y}{y^2 + 1}$$ ✓

$xy^2 + x = y$

$xy^2 - y + x = 0$

$$y = \frac{-b \pm \sqrt{b^2 - 4ac}}{2a}$$
$$= \frac{-(-1) \pm \sqrt{(-1)^2 - 4 \times x \times x}}{2x}$$
$$= \frac{1 \pm \sqrt{1 - 4x^2}}{2x}$$ ✓

But, $y \geq 1$

$\therefore y = \frac{1 + \sqrt{1 - 4x^2}}{2x}$ ✓ *(3 marks)*

8 $y = x^3 - 2$
Inverse function is $x = y^3 - 2$ ✓
$y^3 = x + 2$
$y = \sqrt[3]{x + 2}$ ✓ *(2 marks)*

9 The graph of $y = f^{-1}(x)$ is the reflection of the graph of $y = f(x)$ in the line $y = x$.

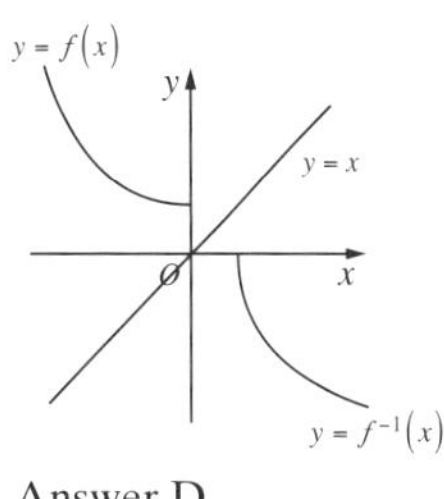

Answer D

10 $f(x) = \sqrt{4x - 3}$

i $4x - 3 \geq 0$

$4x \geq 3$

$x \geq \frac{3}{4}$

The domain is all real numbers $x \geq \frac{3}{4}$. ✓

(1 mark)

ii $y = \sqrt{4x - 3}$

Range is $y \geq 0$

Inverse function:

$x = \sqrt{4y - 3}$ ✓

$x^2 = 4y - 3$

$4y = x^2 + 3$

$y = \frac{x^2 + 3}{4}$

So $f^{-1}(x) = \frac{x^2 + 3}{4}, x \geq 0$ ✓

(2 marks)

iii Graphs intersect when $f(x) = x$

$\sqrt{4x - 3} = x$

$4x - 3 = x^2$

$x^2 - 4x + 3 = 0$

$(x - 1)(x - 3) = 0$

$x = 1$ or $x = 3$

So the points of intersection are (1, 1) and (3, 3). ✓

(1 mark)

iv

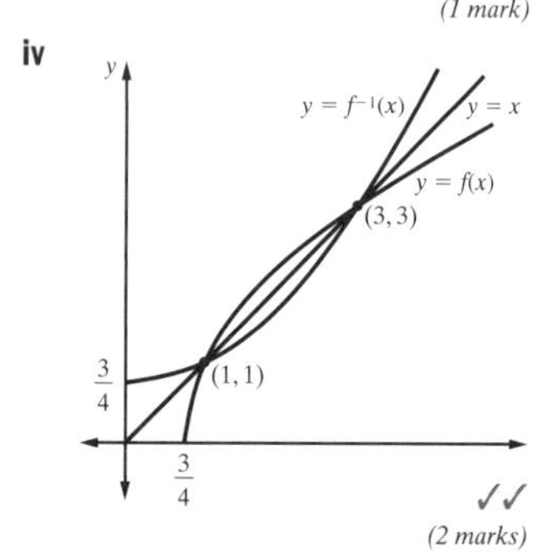

✓✓ *(2 marks)*

11 **i** $f(x) = e^{-x^2}$

$f'(x) = -2xe^{-x^2}$

$f''(x) = -2x \times (-2xe^{-x^2}) + e^{-x^2} \times (-2)$

$= 4x^2e^{-x^2} - 2e^{-x^2}$

$= e^{-x^2}(4x^2 - 2)$ ✓

At points of inflexion

$f''(x) = 0$

i.e. $e^{-x^2}(4x^2 - 2) = 0$

Now $e^{-x^2} \neq 0$

$\therefore 4x^2 - 2 = 0$

$4x^2 = 2$

$x^2 = \frac{1}{2}$

$x = \pm\frac{1}{\sqrt{2}}$ ✓

If $x < \frac{1}{\sqrt{2}}, f''(x) < 0$

If $x > \frac{1}{\sqrt{2}}, f''(x) > 0$

$\therefore$ There is a point of inflexion at $x = \frac{1}{\sqrt{2}}$.

The function is an even function, so there is also a point of inflexion at $x = -\frac{1}{\sqrt{2}}$.

$\therefore$ The points of inflexion occur at $x = \pm\frac{1}{\sqrt{2}}$. ✓

(3 marks)

ii For most values of y, there are two values of x. The domain must be restricted so that there is only one value of x for each value of y in order for an inverse function to exist. ✓

(1 mark)

iii $y = e^{-x^2}$

Inverse function:

$x = e^{-y^2}$

$\ln x = -y^2$

$y^2 = -\ln x$

$= \ln x^{-1}$

$= \ln\left(\frac{1}{x}\right)$ ✓

$y = \pm\sqrt{\ln\left(\frac{1}{x}\right)}$

But if the domain of $f(x)$ is restricted to $x \geq 0$, then the range of the inverse function is $y \geq 0$.

$\therefore y = \sqrt{\ln\left(\frac{1}{x}\right)}$

$f^{-1}(x) = \sqrt{\ln\left(\frac{1}{x}\right)}$ ✓

(2 marks)

iv $f(x) = e^{-x^2}$

$f(0) = e^{-(0)^2}$

$= 1$

The range of $f(x)$ is $0 < y \leq 1$.

So the domain of $f^{-1}(x)$ is $0 < x \leq 1$. ✓

(1 mark)

v

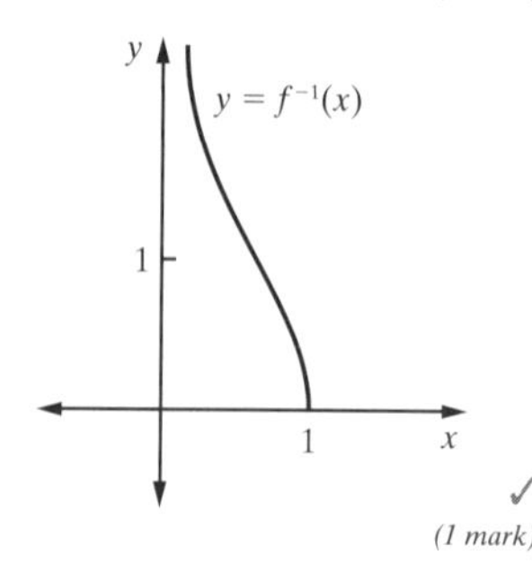

✓ *(1 mark)*

12 **i** $f(x) = \frac{3 + e^{2x}}{4}$

Since $e^{2x} > 0$ for all x, then $f(x) > \frac{3 + 0}{4}$.

$\therefore$ Range of $f(x)$ is $y > \frac{3}{4}$. ✓

(1 mark)

ii The inverse function is given by

$x = \frac{3 + e^{2y}}{4}$ ✓

$\therefore 4x = 3 + e^{2y}$

$\therefore e^{2y} = 4x - 3$

$\therefore 2y = \ln(4x - 3)$

$\therefore y = \frac{1}{2}\ln(4x - 3)$ for $x > \frac{3}{4}$. ✓

(2 marks)

13 **i** $f(x) = x - \frac{1}{2}x^2$ for $x \leqslant 1$

$= x\left(1 - \frac{1}{2}x\right)$

$f(x)$ is a parabola with x-intercepts at $x = 0$ and $x = 2$. However, $f(x)$ is only defined for $x \leq 1$.

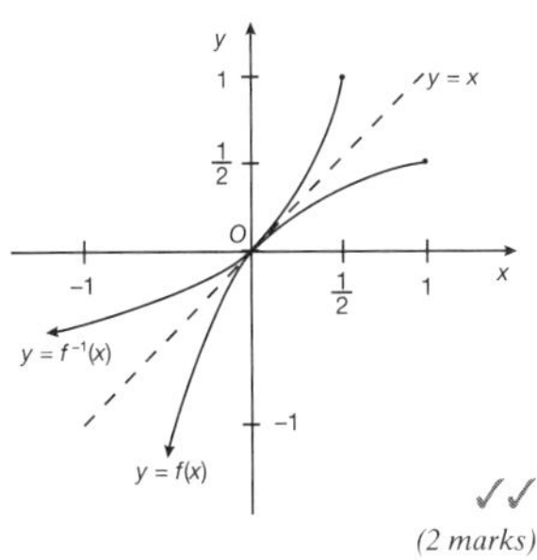

✓✓ *(2 marks)*

ii $y = x - \frac{1}{2}x^2$

Inverse function, $f^{-1}(x)$:

$x = y - \frac{1}{2}y^2$ for $y \leq 1$ ✓

$\frac{1}{2}y^2 - y + x = 0$

Using the quadratic formula to solve for y:

$$y = \frac{1 \pm \sqrt{1 - 4\left(\frac{1}{2}\right)x}}{2\left(\frac{1}{2}\right)}$$

$= 1 \pm \sqrt{1 - 2x}$ ✓

but $y \leq 1$,

$\therefore y = 1 - \sqrt{1 - 2x}$

i.e. $f^{-1}(x) = 1 - \sqrt{1 - 2x}$. ✓

(3 marks)

iii $f^{-1}\left(\frac{3}{8}\right) = 1 - \sqrt{1 - 2\left(\frac{3}{8}\right)}$

$= 1 - \sqrt{1 - \frac{3}{4}}$

$= 1 - \frac{1}{2}$

$= \frac{1}{2}$. ✓

(1 mark)

14 i $f(x) = e^x - e^{-x}$

$f'(x) = e^x + e^{-x} > 0$ for all x, since e^x and $e^{-x} > 0$ for all x.

$\therefore f(x)$ is increasing for all values of x. ✓

(1 mark)

ii $f(x) = e^x - e^{-x}$

Inverse function, $f^{-1}(x)$:

$x = e^y - e^{-y}$

$x = e^y - \frac{1}{e^y}$

$e^y x = e^{2y} - 1$

$e^{2y} - e^y x - 1 = 0$ ✓

Let $m = e^y$,

$\therefore m^2 - mx - 1 = 0$

$$m = \frac{x \pm \sqrt{x^2 - 4(1)(-1)}}{2}$$

$$= \frac{x \pm \sqrt{x^2 + 4}}{2}$$

$$\therefore e^y = \frac{x \pm \sqrt{x^2 + 4}}{2}$$ ✓

Since $e^y > 0$ for all y, then

$$e^y = \frac{x + \sqrt{x^2 + 4}}{2}$$

$$\therefore y = \log_e \left\{\frac{x + \sqrt{x^2 + 4}}{2}\right\}.$$

i.e.

$$f^{-1}(x) = \log_e \left\{\frac{x + \sqrt{x^2 + 4}}{2}\right\}.$$ ✓

(3 marks)

iii Now $e^x - e^{-x} = 5$

If $f(x) = e^x - e^{-x}$

$\therefore f(x) = 5$

$\therefore f^{-1}(5) = x$

and

$f^{-1}(5) = \log_e \left(\frac{5 + \sqrt{25 + 4}}{2}\right)$ (from **ii**)

$= \log_e \left(\frac{5 + \sqrt{29}}{2}\right)$

$= 1.6472\ldots$

$= 1.65$ to 2 decimal places

i.e.
$x = 1.65$ to 2 decimal places. ✓

(1 mark)

15 $f(x) = \log_e(1 + e^x)$

$$f'(x) = \frac{e^x}{1 + e^x}$$ ✓

Since $e^x > 0$ for all x,

$\frac{e^x}{1 + e^x} > 0$ for all x

and $f(x)$ is monotonic increasing for all x.

$\therefore f(x)$ has an inverse function. ✓

(2 marks)

16 i

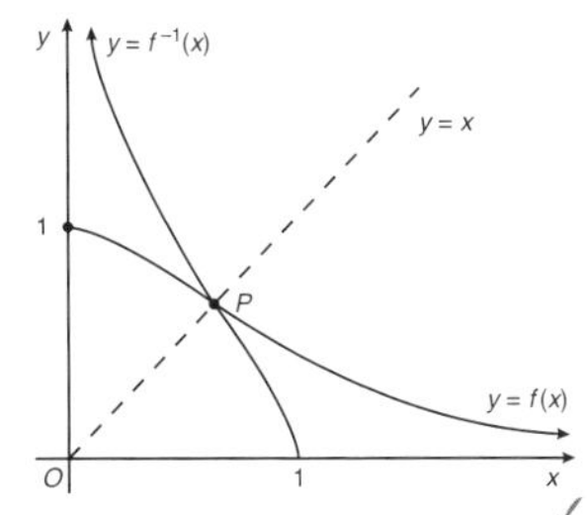

✓

The graph of the inverse function is a reflection of $y = f(x)$ in the line $y = x$.

(1 mark)

ii From the graph in **i**, the domain of $f^{-1}(x)$ is $0 < x \leq 1$.

Note: The range of $y = f(x)$ is $0 < y \leq 1$. Therefore the domain of $y = f^{-1}(x)$ will be $0 < x \leq 1$. ✓

(1 mark)

iii $y = f(x) = \frac{1}{1 + x^2}$

Interchanging x and y (reflection through $y = x$) gives:

$x = \frac{1}{1 + y^2}$

$x(1 + y^2) = 1$

$x + xy^2 = 1$

$xy^2 = 1 - x$

$y^2 = \frac{1 - x}{x}$ ✓

$y = \pm\sqrt{\frac{1 - x}{x}}$

Since $y \geq 0$,

$\therefore f^{-1}(x) = \sqrt{\frac{1 - x}{x}}$. ✓

(2 marks)

iv The graphs of $y = f(x)$ and $y = f^{-1}(x)$ also meet with the graph of $y = x$ at P. (See diagram.)

Solving $y = x$ and $y = \frac{1}{1 + x^2}$ simultaneously:

$x = \frac{1}{1 + x^2}$

$x(1 + x^2) = 1$

$x + x^3 = 1$

$x^3 + x - 1 = 0$

Since $x = \alpha$ at P, then α is a root of $x^3 + x - 1 = 0$. ✓

(1 mark)

17 i For every value of $y > 1$, there are 2 different values of x, therefore the inverse of $f(x)$ is not a function. ✓

(1 mark)

ii The graph of the inverse function is a reflection of $g(x)$ in the line $y = x$. Note that (2, 1) becomes (1, 2).

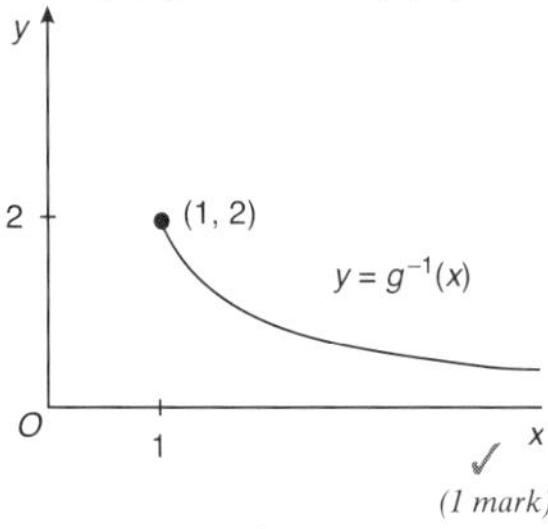

✓ *(1 mark)*

iii Domain of $g^{-1}(x)$ is $x \geq 1$. ✓

(1 mark)

iv $x = y^2 - 4y + 5$

$y^2 - 4y = x - 5$ $\quad y \leq 2$

$\therefore y^2 - 4y + 4 = x - 5 + 4$

(on completing the square) ✓

$\therefore (y-2)^2 = x-1$

$\therefore y-2 = -\sqrt{x-1}$

(Since $y \leqslant 2, y-2 \leqslant 0$ $\therefore$ the negative sign is taken.)

$\therefore y = 2-\sqrt{x-1}$.

$g^{-1}(x) = 2-\sqrt{x-1}$ ✓

(2 marks)

18 i

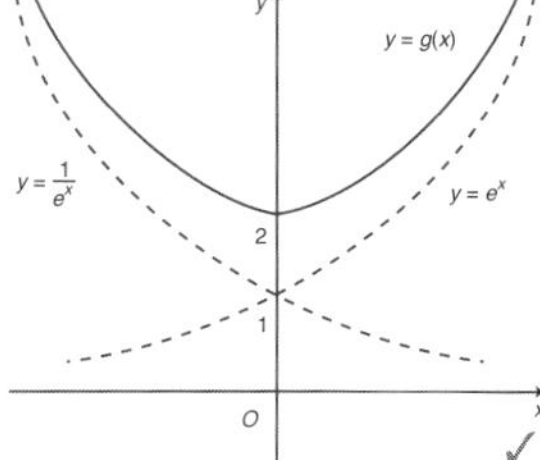

The graph of $y = g(x)$ is obtained by adding the graphs of $y = e^x$ and $y = \frac{1}{e^x}$.

$g(x)$ does not have an inverse function because for every value of $y > 2$, there are 2 values of x. ✓

(2 marks)

ii

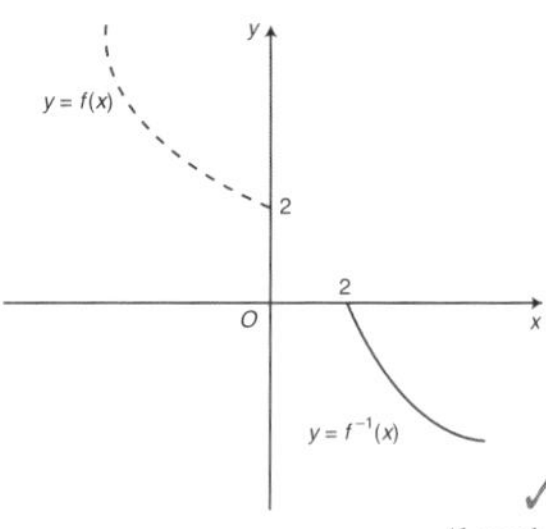

(1 mark)

iii $x = e^y + \frac{1}{e^y}$ for $y \leqslant 0$ and $x \geqslant 2$

$xe^y = e^{2y} + 1$

$\therefore e^{2y} - xe^y + 1 = 0$ ✓

Let $u = e^y$

$u^2 - xu + 1 = 0$

$$u = \frac{-(-x) \pm \sqrt{(-x)^2 - 4(1)(1)}}{2(1)}$$

$$= \frac{x \pm \sqrt{x^2-4}}{2}$$ ✓

i.e. $e^y = \frac{x \pm \sqrt{x^2-4}}{2}$ (where $x \geqslant 2$)

$$\therefore f^{-1}(x) = \ln\left(\frac{x-\sqrt{x^2-4}}{2}\right)$$

where $x \geqslant 2$ ✓

(The negative sign before $\sqrt{x^2-4}$ is chosen because $\frac{x-\sqrt{x^2-4}}{2} \leq 1$ for all $x \geq 2$ and hence $\ln\left(\frac{x-\sqrt{x^2-4}}{2}\right) \leq 0$, and we know from the graph that $y \leq 0$.)

(3 marks)

19 i $f(x) = \frac{x}{x+2}$

The domain is the set of all real numbers, $x \neq -2$.

$$f'(x) = \frac{(x+2).1 - x.(1)}{(x+2)^2}$$

$$= \frac{x+2-x}{(x+2)^2}$$

$$= \frac{2}{(x+2)^2}$$ ✓

> 0 for all $x \neq -2$.

(since $(x+2)^2$ is always positive) ✓

ii Let $y = f(x)$

$\therefore y = \frac{x}{x+2}$

$\therefore xy + 2y = x$

$x - xy = 2y$

$x(1-y) = 2y$

$x = \frac{2y}{1-y}$ ✓

$\therefore$ The horizontal asymptote has the equation

$1 - y = 0$

i.e. $y = 1$. ✓

iii

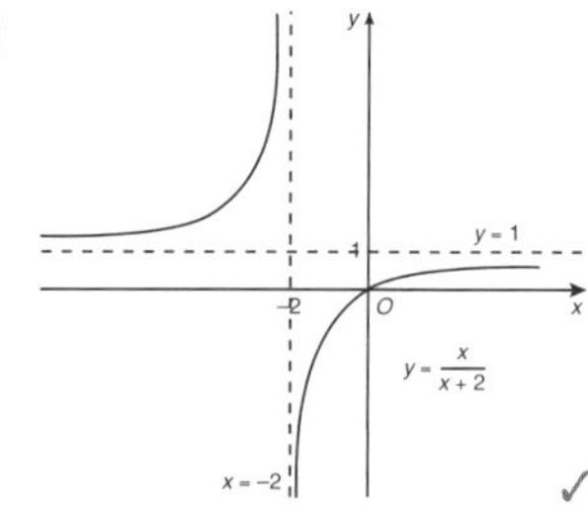

Note $f(0) = 0$ and the function is increasing for all values in the domain because $f'(x) > 0$ for all x in the domain.

iv $f(x)$ has an inverse function because for every value of the range, there is a corresponding unique value in the domain. ✓

Graphically, any line parallel to the x-axis cuts the graph of the function once (except at the asymptote $y = 1$.)

v From **ii**, $x = \frac{2y}{1-y}$

$\therefore f^{-1}(x) = \frac{2x}{1-x}$. ✓

vi All real numbers except $x = 1$. ✓

(8 marks)

(Total mark allocation only in exam)

20 i $f(x) = e^x - 1 - x$

$f'(x) = e^x - 1$

$= 0$ when $x = 0$

$f''(x) = e^x$

$f''(0) = e^0$

$= 1 > 0$

$\therefore$ The minimum of $f(x)$ occurs at $x = 0$. ✓

ii Now $f(0) = e^0 - 1 - 0$

$= 0$ ✓

Since the minimum of $f(x)$ occurs at $(0, 0)$ and this is the only turning point of $f(x)$, $\therefore f(x) \geq 0$ for all x. ✓

iii

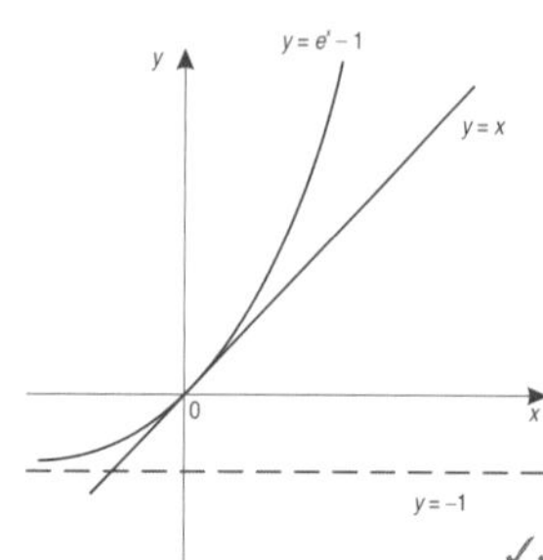

✓✓

iv The inverse function of $g(x) = e^x - 1$ is

$x = e^y - 1$

i.e. $e^y = x + 1$

i.e. $y = \log_e(x+1)$

$\therefore g^{-1}(x) = \log_e(x+1)$ ✓

v The domain of $g^{-1}(x)$ is the set of all real numbers x, such that

$x + 1 > 0$

$\therefore x > -1$ ✓

vi Now $e^x - 1 - x \geqslant 0$ (from **ii**)

$\therefore e^x - 1 \geqslant x$

$\therefore e^x \geqslant x + 1$

$\therefore \log_e e^x \geqslant \log_e(x+1)$ ✓

(since the logarithmic function is a monotonic increasing function)

$\therefore x \geq \log_e(x+1)$ for all $x > -1$ (the domain of $y = \log_e(x+1)$ from **v**) ✓

OR

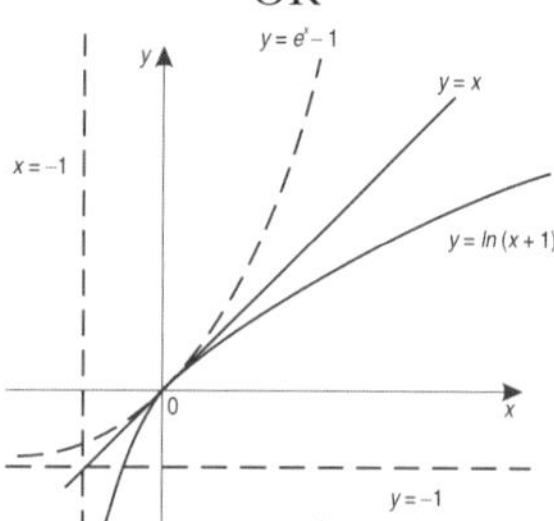

From the graph, $\log_e(x+1) \leq x$ for all $x > -1$.

(9 marks)

(Total mark allocation only in exam)

21 **i** Vertical asymptote is $x = 2$. ✓

Now $y = 1 + \frac{3}{x-2}$

$\therefore y - 1 = \frac{3}{x-2}$

$\therefore x - 2 = \frac{3}{y-1}$

$x = 2 + \frac{3}{y-1}$

$\therefore$ Horizontal asymptote is $y = 1$. ✓

ii Since

$x = 2 + \frac{3}{y-1}$ (from **i**) ✓

$f^{-1}(x) = 2 + \frac{3}{x-1}$. ✓

iii The domain of $f^{-1}(x)$ is $x > 1$. (This is the range of $f(x)$.) ✓

(5 marks)

(Total mark allocation only in exam)

22 **i** $y = \frac{1}{4}[(x-1)^2 + 7]$

Axis of symmetry is $x = 1$. ✓

When $x = 1$,

$y = \frac{1}{4}[(1-1)^2 + 7]$

$= \frac{7}{4}$

So, the vertex is at $(1, \frac{7}{4})$. ✓

At $x = 0$, $y = \frac{1}{4}[1 + 7]$

$= \frac{8}{4}$

$= 2$

$\therefore$ y intercept is (0, 2) ✓

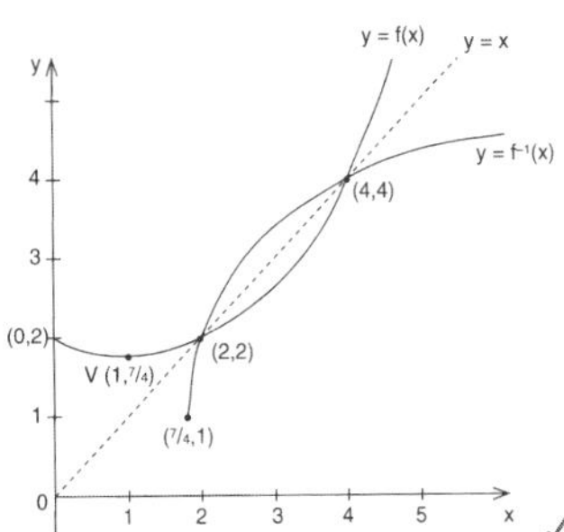

✓

ii Since the vertex is at $(1, \frac{7}{4})$, the greatest domain containing the value 3 for which the function has an inverse is $x \geq 1$. ✓

iii See graph in **i**. ✓

iv The domain of $f^{-1}(x)$ is $x \geq \frac{7}{4}$ ✓

v If $a < 1$, then

$f(a) = f(1 + (1 - a)) = f(2 - a)$

since $x = 1$ is the axis of symmetry of the parabola.

Now $f(a) = f(2 - a)$

$\therefore f^{-1}(f(a)) = f^{-1}(f(2 - a))$

$= 2 - a$ ✓

N.B. Since $a < 1, 2 - a > 1$ and hence $(2 - a)$ is a value in the range of $f^{-1}(x)$.

vi The points of the intersection of $y = f(x)$ and $y = f^{-1}(x)$ must be on the line $y = x$.

$\therefore$ Solving simultaneously $y = x$ and $y = f(x)$:

$x = \frac{1}{4}[(x-1)^2 + 7]$

$4x = x^2 - 2x + 1 + 7$

$x^2 - 6x + 8 = 0$ ✓

$(x - 2)(x - 4) = 0$

$\therefore x = 2$ or $x = 4$

$\therefore y = 2$ or $y = 4$

$\therefore$ The coordinates are (2, 2) and (4, 4). ✓

(10 marks)

(Total mark allocation only in exam)

23 **a** $f(x) = \frac{e^x}{3 + e^x}$

$f'(x) = \frac{(3 + e^x).e^x - e^x.e^x}{(3 + e^x)^2}$

$= \frac{3e^x + e^{2x} - e^{2x}}{(3 + e^x)^2}$

$= \frac{3e^x}{(3 + e^x)^2}$ ✓

$\therefore f'(x) > 0$ for all values of x since $e^x > 0$.

$\therefore f'(x)$ has no stationary points.

✓ *(2 marks)*

b $f''(x) = \frac{3e^x(3 - e^x)}{(3 + e^x)^3}$

Since $f'(x) > 0$, a point of inflexion occurs when $f''(x) = 0$

$\frac{3e^x(3 - e^x)}{(3 + e^x)^3} = 0$

$\therefore 3 - e^x = 0$

$e^x = 3$

$x = \ln 3$.

At $x = \ln 3$, $y = \frac{3}{3 + 3}$

$= \frac{1}{2}$

$\therefore$ The coordinates of the point of inflexion are $\left(\ln 3, \frac{1}{2}\right)$. ✓

(1 mark)

c $f(x) > 0$ for all x since $e^x > 0$ ✓

$f(x) = \frac{e^x}{3 + e^x} < \frac{3 + e^x}{3 + e^x} = 1$

$\therefore f(x) < 1$ ✓

$\therefore 0 < f(x) < 1$ for all x.

(2 marks)

d $\lim_{x \to \infty} \frac{e^x}{3 + e^x} = \lim_{x \to \infty} \frac{1}{\frac{3}{e^x} + 1}$

(on dividing numerator and denominator by e^x)

$= \frac{1}{0 + 1}$

$= 1$

$\lim_{x \to -\infty} \frac{e^x}{3 + e^x} = \frac{0}{3 + 0}$

$= 0$.

$\therefore f(x)$ approaches the x-axis as $x \to -\infty$ ✓

and it approaches $y = 1$ as $x \to \infty$. ✓

(2 marks)

e

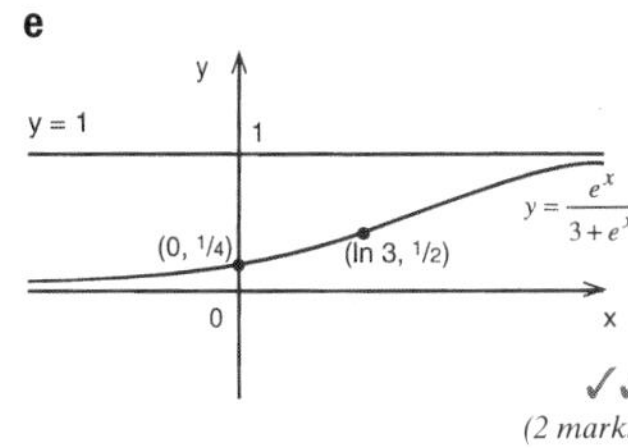

✓✓

(2 marks)

f $f(x)$ has an inverse function as there is only one value of x for each value of y, i.e. Lines parallel to the x-axis cannot cut $y = f(x)$ at more than one point. ✓

(1 mark)

g The inverse function is

$x = \frac{e^y}{3 + e^y}$ ✓

$x(3 + e^y) = e^y$

$3x + xe^y = e^y$

$3x = e^y - xe^y$

$e^y(1 - x) = 3x$

$e^y = \frac{3x}{1 - x}$

$y = \ln\left(\frac{3x}{1 - x}\right)$. ✓

(2 marks)

1 A curve is defined in parametric form by $x = 2 + t$ and $y = 3 - 2t^2$ for $-1 \le t \le 0$.

Which diagram best represents this curve?

A

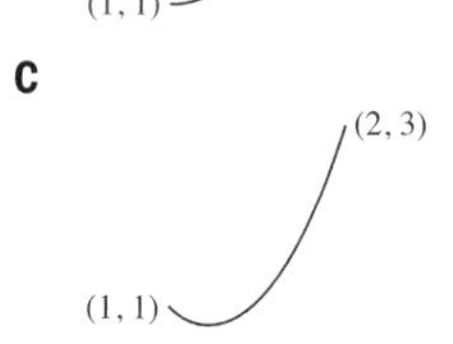

B

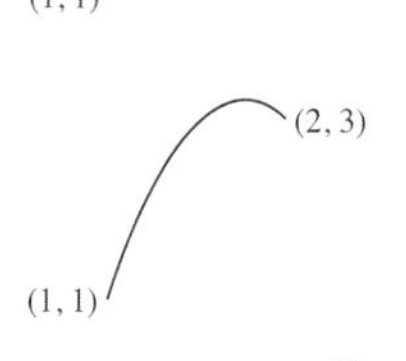

C

(2, 3)
(1, 1)

D

(2, 3)
(1, 1)

(1 mark)

(Q5, **2022 HSC**) Easy

2 The diagram shows a semicircle.

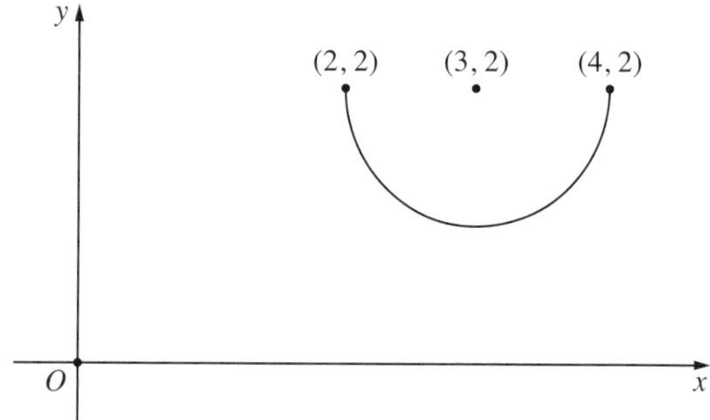

Which pair of parametric equations represents the semicircle shown?

A $\begin{cases} x = 3 + \sin t \\ y = 2 + \cos t \end{cases}$ for $-\frac{\pi}{2} \le t \le \frac{\pi}{2}$

B $\begin{cases} x = 3 + \cos t \\ y = 2 + \sin t \end{cases}$ for $-\frac{\pi}{2} \le t \le \frac{\pi}{2}$

C $\begin{cases} x = 3 - \sin t \\ y = 2 - \cos t \end{cases}$ for $-\frac{\pi}{2} \le t \le \frac{\pi}{2}$

D $\begin{cases} x = 3 - \cos t \\ y = 2 - \sin t \end{cases}$ for $-\frac{\pi}{2} \le t \le \frac{\pi}{2}$ *(1 mark)*

(Q8, **2021 HSC**) Hard

3 Consider the parametric equations $x = 4\cos\theta - 3$ and $y = 4\sin\theta + 2$.

Which of these is the corresponding cartesian equation?

A $x^2 - 6x + y^2 + 4y = 3$
B $x^2 - 2x + y^2 + 4y = 1$
C $x^2 + 6x + y^2 - 4y = 3$
D $x^2 + 6x + y^2 - 4y = 16$ *(1 mark)*

Bonus question (see page iv) Easy

4 A circle with centre $(3, -4)$ passes through the origin.

Which of these is the equation of the circle?

A $x = 3 - 5\cos\theta$ $y = 4 + 5\sin\theta$
B $x = 3 - 5\cos\theta$ $y = 4 - 5\sin\theta$
C $x = 5\cos\theta - 3$ $y = 5\sin\theta + 4$
D $x = 5\cos\theta + 3$ $y = 5\sin\theta - 4$ *(1 mark)*

Bonus question Easy

5 Find the Cartesian equation of the curve represented by the equations $x = \frac{t}{2} + 1$ and $y = \frac{t^2}{4} - 1$. *(2 marks)*

Bonus question Easy

6 The parametric equations of a curve are $x = 4\cos t$ and $y = 4\sin t$, $0 \le t \le 2\pi$. What is the value of t at the point $(-2, 2\sqrt{3})$? *(3 marks)*

Bonus question Medium

7 If $x = \sin t + \cos t$ and $y = \frac{1}{2}\sin 2t$, find the Cartesian equation of the curve. *(2 marks)*

Bonus question Medium

8 Sketch the curve given by $x = 2t$, $y = t^2$. *(2 marks)*

Bonus question Easy

9 Find the Cartesian equation of the curve represented by the equations $x = -3 + \cos 2\theta$ and $y = 1 + \sin 2\theta$ and hence find the equation of the tangent to the curve at the point $(-3, 0)$. *(3 marks)*

Bonus question Medium

10 Using the parameter θ, find the pair of parametric equations of the circle with Cartesian equation

$x^2 - 8x + y^2 + 4y = 5$. *(2 marks)*

Bonus question Medium

11 Find the Cartesian equation for the parametric equations $x = t - \frac{1}{t}$ and $y = 3(t + \frac{1}{t})$. *(2 marks)*

Bonus question Easy

12 Let $A(0, -k)$ be a fixed point on the y-axis with $k > 0$. The point $C(t, 0)$ is on the x-axis. The point $B(0, y)$ is on the y-axis so that ΔABC is right-angled with the right angle at C. The point P is chosen so that $OBPC$ is a rectangle as shown in the diagram.

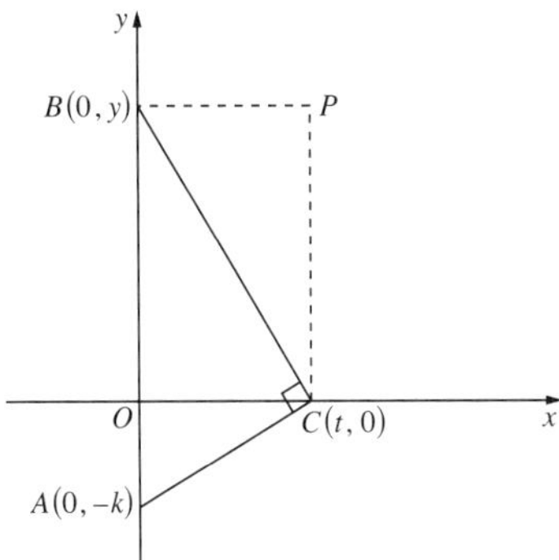

Show that P lies on the parabola given parametrically by

$x = t$ and $y = \dfrac{t^2}{k}$. *(2 marks)*

(Q12d i, **2012 HSC**) **Medium**

---

13 A curve has parametric equations $x = \dfrac{t}{2}$, $y = 3t^2$. Find the Cartesian equation for this curve. *(2 marks)*

(Q1d, **2003 HSC**) **Easy**

---

14 The variable point $(3t, 2t^2)$ lies on a parabola. Find the Cartesian equation for this parabola. *(2 marks)*

(Q1e, **2002 HSC**) **Easy**

Year 11 Parametric form of functions or relations—Worked Answers

1 If $x = 2 + t$ and $y = 3 - 2t^2$, then
$x - 2 = t$
$\therefore y = 3 - 2(x - 2)^2$
The Cartesian equation shows a concave down parabola with vertex at $(2, 3)$.
Answer B

2 In the domain $-\frac{\pi}{2} \leq t \leq \frac{\pi}{2}$,
$-1 \leq \sin t \leq 1$ and $0 \leq \cos t \leq 1$.
In the diagram, y is always less than the centre of the semicircle.
$\therefore y = 2 - \cos t$
Answer C

3 $x = 4\cos\theta - 3 \quad y = 4\sin\theta + 2$
$\frac{x+3}{4} = \cos\theta \quad \frac{y-2}{4} = \sin\theta$
As $\cos^2\theta + \sin^2\theta = 1$, then
$$\left(\frac{x+3}{4}\right)^2 + \left(\frac{y-2}{4}\right)^2 = 1$$
$$(x+3)^2 + (y-2)^2 = 16$$
$$x^2 + 6x + 9 + y^2 - 4y + 4 = 16$$
$$x^2 + 6x + y^2 - 4y = 16 - 13$$
$$x^2 + 6x + y^2 - 4y = 3$$
Answer C

4

From the diagram, the radius is 5 units.
$\therefore$ the equation of the circle is
$(x-3)^2 + (y+4)^2 = 25$
$$\left(\frac{x-3}{5}\right)^2 + \left(\frac{y+4}{5}\right)^2 = 1$$
As $\sin^2\theta + \cos^2\theta = 1$, then consider:
$$\cos\theta = \frac{x-3}{5}$$
$$\sin\theta = \frac{y+4}{5}$$
$5\cos\theta = x - 3$
$5\sin\theta = y + 4$
$x = 5\cos\theta + 3$
$y = 5\sin\theta - 4$
Answer D

5 From $x = \frac{t}{2} + 1$,
$t = 2(x - 1)$ ✓
Substitute into y:
$$y = \frac{[2(x-1)]^2}{4} - 1$$
$$= (x-1)^2 - 1$$
$$= x^2 - 2x + 1 - 1$$
$$= x^2 - 2x$$
$\therefore \quad y = x^2 - 2x$ ✓
(2 marks)

6 $x = 4\cos t$
At $(-2, 2\sqrt{3})$,
$4\cos t = -2$
$$\cos t = -\frac{1}{2}$$
$$t = \frac{2\pi}{3} \text{ or } \frac{4\pi}{3}, \ 0 \leq t \leq 2\pi$$ ✓
$y = 4\sin t$
At $(-2, 2\sqrt{3})$,
$4\sin t = 2\sqrt{3}$
$$\sin t = \frac{\sqrt{3}}{2}$$
$$t = \frac{\pi}{3} \text{ or } \frac{2\pi}{3}, \ 0 \leq t \leq 2\pi$$ ✓
$\therefore$ at $(-2, 2\sqrt{3})$, $t = \frac{2\pi}{3}$. ✓
(3 marks)

7 $x = \sin t + \cos t$
$x^2 = (\sin t + \cos t)^2$
$= \sin^2 t + 2\sin t\cos t + \cos^2 t$
$= 1 + 2\sin t\cos t$
$= 1 + \sin 2t$ ✓
$\sin 2t = x^2 - 1$
But $y = \frac{1}{2}\sin 2t$
$$\therefore y = \frac{1}{2}(x^2 - 1)$$
$$= \frac{x^2}{2} - \frac{1}{2}$$ ✓
(2 marks)

8 $x = 2t$
$$\therefore t = \frac{x}{2}$$
$$y = t^2$$
$$= \left(\frac{x}{2}\right)^2$$
$$= \frac{x^2}{4}$$ ✓

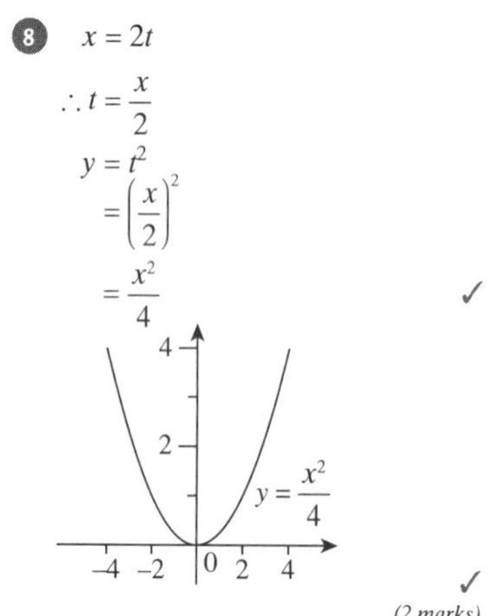

✓
(2 marks)

9
$$x = -3 + \cos 2\theta$$
$$x + 3 = \cos 2\theta$$
$$(x+3)^2 = \cos^2 2\theta \ \ldots ①$$
$$y = 1 + \sin 2\theta$$
$$y - 1 = \sin 2\theta$$
$$(y-1)^2 = \sin^2 2\theta \ \ldots ②$$ ✓
① + ②:
$(x+3)^2 + (y-1)^2$
$= \cos^2 2\theta + \sin^2 2\theta$
$(x+3)^2 + (y-1)^2 = 1$
$\therefore$ circle with centre $(-3, 1)$ and radius 1. ✓

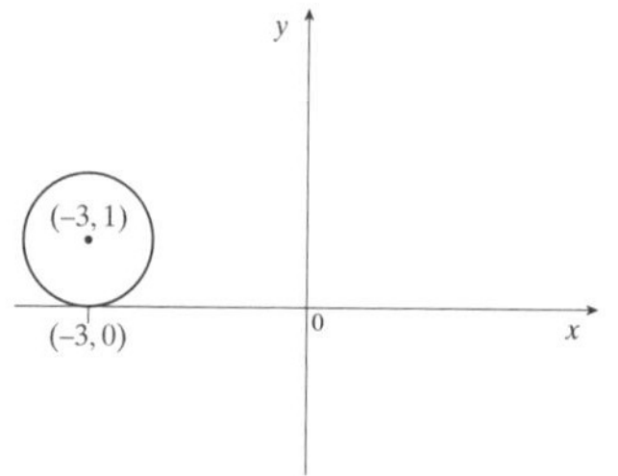

Hence, tangent at $(-3, 0)$ is x-axis.
$\therefore$ equation of tangent is $y = 0$. ✓
(3 marks)

10 $x^2 - 8x + y^2 + 4y = 5$
$x^2 - 8x + 16 + y^2 + 4y + 4$
$= 5 + 16 + 4$
$(x-4)^2 + (y+2)^2 - 25$
Circle, centre $(4, -2)$ and radius 5 units. ✓
Parametric equations in the form $x = 4 + 5\cos\theta$ and $y = -2 + 5\sin\theta$.
✓ *(2 marks)*

11
$$x = t - \frac{1}{t}$$
$$x^2 = \left(t - \frac{1}{t}\right)^2$$
$$= t^2 - 2 + \frac{1}{t^2}$$
$$\therefore t^2 + \frac{1}{t^2} = x^2 + 2$$
Also, $y = 3\left(t + \frac{1}{t}\right)$
$$\frac{y}{3} = 3\left(t + \frac{1}{t}\right)$$
$$\frac{y^2}{9} = \left(t + \frac{1}{t}\right)^2$$
$$= t^2 + 2 + \frac{1}{t^2}$$ ✓
$$= x^2 + 2 + 2$$
$$= x^2 + 4$$
$\therefore \quad y^2 = 9x^2 + 36$
$\therefore 9x^2 - y^2 + 36 = 0$ ✓
(2 marks)

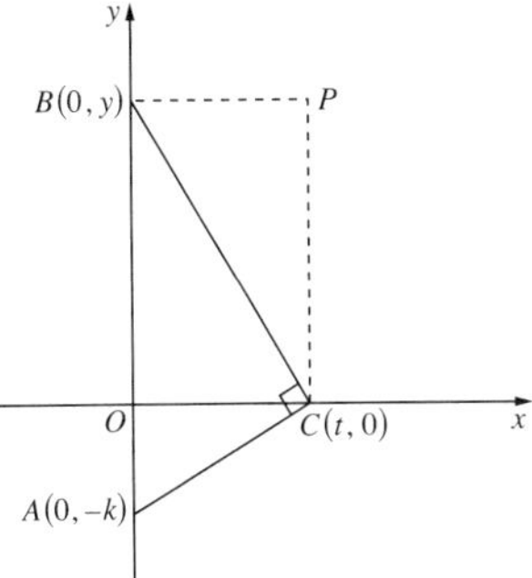

$OBPC$ is a rectangle.
So PC is parallel to the y-axis.
So at P, $x = t$

$$m = \frac{y_2 - y_1}{x_2 - x_1}$$

$$m_{AC} = \frac{0 - (-k)}{t - 0} = \frac{k}{t}$$

$$m_{BC} = \frac{y - 0}{0 - t} = -\frac{y}{t} \quad ✓$$

Now AC is perpendicular to BC.

So $\frac{k}{t} \times -\frac{y}{t} = -1$

$$ky = t^2$$

$$y = \frac{t^2}{k}$$

BP is parallel to the x-axis so, at P,

$y = \frac{t^2}{k}$.

$\therefore$ P lies on the parabola given parametrically by

$x = t$ and $y = \frac{t^2}{k}$. ✓

(2 marks)

13 $x = \frac{t}{2}$, $y = 3t^2$

$\therefore t = 2x$, ✓

$\therefore y = 3(2x)^2$
$= 3(4x^2)$
$= 12x^2$

$\therefore$ Cartesian equation is $y = 12x^2$. ✓

(2 marks)

14 $x = 3t$ and $y = 2t^2$

$\therefore t = \frac{x}{3}$ ✓

$\therefore y = 2\left(\frac{x}{3}\right)^2$

$\therefore y = \frac{2x^2}{9}$

or $x^2 = \frac{9}{2}y$. ✓

(2 marks)

1 Let $P(x)$ be a polynomial of degree 5. When $P(x)$ is divided by the polynomial $Q(x)$, the remainder is $2x + 5$.

Which of the following is true about the degree of Q?

A The degree must be 1.
B The degree could be 1.
C The degree must be 2.
D The degree could be 2. *(1 mark)*

(Q3, **2022 HSC**) Easy

2 What is the remainder when $P(x) = -x^3 - 2x^2 - 3x + 8$ is divided by $x + 2$?

A -14 **B** -2
C 2 **D** 14 *(1 mark)*

(Q3, **2021 HSC**) Easy

3 Let $P(x) = x^3 + 3x^2 - 13x + 6$.

i Show that $P(2) = 0$. *(1 mark)* Easy

ii Hence, factor the polynomial $P(x)$ as $A(x)B(x)$, where $B(x)$ is a quadratic polynomial. *(2 marks)* Easy

(Q11a, **2020 HSC**)

4 Let $P(x) = qx^3 + rx^2 + rx + q$ where q and r are constants, $q \neq 0$. One of the zeros of $P(x)$ is -1.

Given that α is a zero of $P(x)$, $\alpha \neq -1$, which of the following is also a zero?

A $-\dfrac{1}{\alpha}$ **B** $-\dfrac{q}{\alpha}$
C $\dfrac{1}{\alpha}$ **D** $\dfrac{q}{\alpha}$ *(1 mark)*

(Q7, **2019 HSC**) Medium

5 Find the polynomial $Q(x)$ that satisfies $x^3 + 2x^2 - 3x - 7 = (x - 2)Q(x) + 3$. *(2 marks)*

(Q11d, **2019 HSC**) Easy

6 Consider the polynomial $P(x) = x^3 - 2x^2 - 5x + 6$.

i Show that $x = 1$ is a zero of $P(x)$. *(1 mark)* Easy

ii Find the other zeros. *(2 marks)* Easy

(Q11a, **2018 HSC**)

7 Which polynomial is a factor of $x^3 - 5x^2 + 11x - 10$?

A $x - 2$ **B** $x + 2$
C $11x - 10$ **D** $x^2 - 5x + 11$ *(1 mark)*

(Q1, **2017 HSC**) Easy

8 What is the remainder when $2x^3 - 10x^2 + 6x + 2$ is divided by $x - 2$?

A -66 **B** -10
C $-x^3 + 5x^2 - 3x - 1$ **D** $x^3 - 5x^2 + 3x + 1$ *(1 mark)*

(Q2, **2016 HSC**) Easy

9 What is the remainder when $x^3 - 6x$ is divided by $x + 3$?

A -9 **B** 9
C $x^2 - 2x$ **D** $x^2 - 3x + 3$ *(1 mark)*

(Q1, **2015 HSC**) Easy

10 Consider the polynomials $P(x) = x^3 - kx^2 + 5x + 12$ and $A(x) = x - 3$.

i Given that $P(x)$ is divisible by $A(x)$, show that $k = 6$. *(1 mark)* Easy

ii Find all the zeros of $P(x)$ when $k = 6$. *(2 marks)* Easy

(Q11f, **2015 HSC**)

11 The remainder when the polynomial $P(x) = x^4 - 8x^3 - 7x^2 + 3$ is divided by $x^2 + x$ is $ax + 3$.

What is the value of a?

A -14 **B** -11
C -2 **D** 5 *(1 mark)*

(Q9, **2014 HSC**) Easy

12 The polynomial $P(x) = x^3 - 4x^2 - 6x + k$ has a factor $x - 2$.

What is the value of k?

A 2 **B** 12
C 20 **D** 36 *(1 mark)*

(Q1, **2013 HSC**) Easy

13 When the polynomial $P(x)$ is divided by $(x + 1)(x - 3)$, the remainder is $2x + 7$.

What is the remainder when $P(x)$ is divided by $x - 3$?

A 1 **B** 7
C 9 **D** 13 *(1 mark)*

(Q8, **2012 HSC**) Medium

14 Let $P(x) = x^3 - ax^2 + x$ be a polynomial, where a is a real number.

When $P(x)$ is divided by $x - 3$ the remainder is 12.

Find the remainder when $P(x)$ is divided by $x + 1$. *(3 marks)*

(Q2a, **2011 HSC**) Easy

15 Let

$$P(x) = (x + 1)(x - 3)\,Q(x) + ax + b,$$

where $Q(x)$ is a polynomial and a and b are real numbers.
The polynomial $P(x)$ has a factor of $x - 3$.
When $P(x)$ is divided by $x + 1$ the remainder is 8.

i Find the values of a and b. *(2 marks)* Easy

ii Find the remainder when $P(x)$ is divided by $(x + 1)(x - 3)$. *(1 mark)* Easy

(Q2c, **2010 HSC**)

16 The polynomial $p(x) = x^3 - ax + b$ has a remainder of 2 when divided by $(x - 1)$ and a remainder of 5 when divided by $(x + 2)$.

Find the values of a and b. *(3 marks)*

(Q2a, **2009 HSC**) Easy

17 The polynomial x^3 is divided by $x + 3$. Calculate the remainder. *(2 marks)*

(Q1a, **2008 HSC**) Easy

18 The polynomial $P(x) = x^2 + ax + b$ has a zero at $x = 2$. When $P(x)$ is divided by $x + 1$, the remainder is 18.

Find the values of a and b. *(3 marks)*

(Q2c, **2007 HSC**) Medium

19 Let $P(x) = (x + 1)(x - 3)Q(x) + a(x + 1) + b$, where $Q(x)$ is a polynomial and a and b are real numbers.

When $P(x)$ is divided by $(x + 1)$ the remainder is -11.

When $P(x)$ is divided by $(x - 3)$ the remainder is 1.

i What is the value of b? *(1 mark)* Easy

ii What is the remainder when $P(x)$ is divided by $(x + 1)(x - 3)$? *(2 marks)* Medium

(Q3b, **2004 HSC**)

20 Suppose $x^3 - 2x^2 + a \equiv (x + 2)Q(x) + 3$ where $Q(x)$ is a polynomial.

Find the value of a. *(2 marks)*

(Q2c, **2002 HSC**) Medium

21 Is $x + 3$ a factor of $x^3 - 5x + 12$? Give reasons for your answer. *(2 marks)*

(Q1e, **2001 HSC**) Easy

22 Find the value of k if $x - 3$ is a factor of $P(x) = x^3 - 3kx + 6$. *(2 marks)*

(Q1c, **2000 HSC**) Easy

23 Find the remainder when the polynominal $P(x) = x^3 - 4x$ is divided by $x + 3$. *(2 marks)*

(Q1e, **1999 HSC**) Easy

24 Find the quotient, $Q(x)$, and the remainder, $R(x)$, when the polynomial $P(x) = x^4 - x^2 + 1$ is divided by $x^2 + 1$. *(3 marks)*

(Q2a, **1998 HSC**) Easy

25 $(x - 2)$ is a factor of the polynomial $P(x) = 2x^3 + x + a$. Find the value of a. *(1 mark)*

(Q1a, **1996 HSC**) Easy

26 When the polynomial $P(x)$ is divided by $(x + 1)(x - 4)$, the quotient is $Q(x)$ and the remainder is $R(x)$.

i Why is the most general form of $R(x)$ given by $R(x) = ax + b$? Easy

ii Given that $P(4) = -5$, show that $R(4) = -5$. Easy

iii Further, when $P(x)$ is divided by $(x + 1)$, the remainder is 5. Find $R(x)$. Medium

(Q4a, **1994 HSC**)

27 When the polynomial $P(x)$ is divided by $x^2 - 1$ the remainder is $3x - 1$.

What is the remainder when $P(x)$ is divided by $x - 1$?

(Q3b, **1993 HSC**) Medium

28 Show that $(x - 1)(x - 2)$ is a factor of

$$P(x) = x^n(2^m - 1) + x^m(1 - 2^n) + (2^n - 2^m)$$

where m and n are positive integers.

(Q6a, **1992 HSC**) Medium

Year 11 Remainder and factor theorems—Worked Answers

1 If when $P(x)$ is divided by $Q(x)$ the remainder is $2x + 5$, which is degree 1, then $deg(Q)$ must be more than 1 but less than 5, such that $2 \le deg(Q) \le 4$.
Answer D

2 If $P(x)$ is divided by $x - k$, then the remainder is $P(k)$.

$P(x) = -x^3 - 2x^2 - 3x + 8$

$P(-2) = -(-2)^3 - 2(-2)^2 - 3(-2) + 8$

$P(-2) = 14$

Answer D

3 $P(x) = x^3 + 3x^2 - 13x + 6$

i $P(2) = 2^3 + 3 \times 2^2 - 13 \times 2 + 6$
$= 0$ ✓ *(1 mark)*

ii As $P(2) = 0$, $(x - 2)$ is a factor of $P(x)$.

$$\begin{array}{r} x^2 + 5x - 3 \\ x - 2 \overline{)\, x^3 + 3x^2 - 13x + 6} \\ \underline{x^3 - 2x^2} \\ 5x^2 - 13x \\ \underline{5x^2 - 10x} \\ -3x + 6 \\ \underline{-3x + 6} \\ 0 \end{array}$$
✓

$\therefore P(x) = (x - 2)(x^2 + 5x - 3)$ ✓ *(2 marks)*

4 $P(x) = qx^3 + rx^2 + rx + q$

$\alpha\beta\gamma = -\dfrac{d}{a}$

$= \dfrac{q}{q}$

$= -1$

But one root, γ say, $= -1$

So $\alpha\beta = 1$

$\beta = \dfrac{1}{\alpha}$

Answer C

5 $x^3 + 2x^2 - 3x - 7$
$= (x - 2)\,Q(x) + 3$

$$\begin{array}{r} x^2 + 4x + 5 \\ x - 2 \overline{)\, x^3 + 2x^2 - 3x - 7} \\ \underline{x^3 - 2x^2} \\ 4x^2 - 3x \\ \underline{4x^2 - 8x} \\ 5x - 7 \\ \underline{5x - 10} \\ 3 \end{array}$$
✓

$\therefore Q(x) = x^2 + 4x + 5$ ✓ *(2 marks)*

6 $P(x) = x^3 - 2x^2 - 5x + 6$

i $P(1) = 1^3 - 2 \times 1^2 - 5 \times 1 + 6$
$= 0$
$\therefore x = 1$ is a zero of $P(x)$. ✓ *(1 mark)*

ii
$$\begin{array}{r} x^2 - x - 6 \\ x - 1 \overline{)\, x^3 - 2x^2 - 5x + 6} \\ \underline{x^3 - x^2} \\ -x^2 - 5x \\ \underline{-x^2 + x} \\ -6x + 6 \\ \underline{-6x + 6} \\ 0 \end{array}$$
✓

So $P(x) = (x - 1)(x^2 - x - 6)$
$= (x - 1)(x + 2)(x - 3)$

$\therefore$ the other zeros are $x = -2$ and $x = 3$. ✓ *(2 marks)*

7 $P(x) = x^3 - 5x^2 + 11x - 10$
$P(2) = 2^3 - 5 \times 2^2 + 11 \times 2 - 10$
$= 0$
$\therefore x - 2$ is a factor of $P(x)$.
Answer A

8 $P(x) = 2x^3 - 10x^2 + 6x + 2$
$P(2) = 2 \times 2^3 - 10 \times 2^2 + 6 \times 2 + 2$
$= -10$
Answer B

9 $P(x) = x^3 - 6x$
$P(-3) = (-3)^3 - 6 \times -3$
$= -9$
Answer A

10 **i** $P(x) = x^3 - kx^2 + 5x + 12$
$P(x)$ is divisible by $(x - 3)$ so $P(3) = 0$

$\therefore 3^3 - k \times 3^2 + 5 \times 3 + 12 = 0$
$54 - 9k = 0$
$9k = 54$
$k = 6$ ✓ *(1 mark)*

ii $P(x) = x^3 - 6x^2 + 5x + 12$

$$\begin{array}{r} x^2 - 3x - 4 \\ x - 3 \overline{)\, x^3 - 6x^2 + 5x + 12} \\ \underline{x^3 - 3x^2} \\ -3x^2 + 5x \\ \underline{-3x^2 + 9x} \\ -4x + 12 \\ \underline{-4x + 12} \\ 0 \end{array}$$

$\therefore \ P(x) = (x - 3)(x^2 - 3x - 4)$ ✓
$= (x - 3)(x - 4)(x + 1)$

So the zeros of $P(x)$ are -1, 3 and 4. ✓ *(2 marks)*

11
$$\begin{array}{r} x^2 - 9x + 2 \\ x^2 + x \overline{)\, x^4 - 8x^3 - 7x^2 + 0x + 3} \\ \underline{x^4 + x^3} \\ -9x^3 - 7x^2 \\ \underline{-9x^3 - 9x^2} \\ 2x^2 + 0x \\ \underline{2x^2 + 2x} \\ -2x + 3 \end{array}$$

The remainder is $-2x + 3$.
So $a = -2$.
Answer C

12 $P(x) = x^3 - 4x^2 - 6x + k$
$P(2) = 2^3 - 4 \times 2^2 - 6 \times 2 + k$
$= -20 + k$
But $P(2) = 0$
$\therefore k = 20$
Answer C

13 $P(x) = (x + 1)(x - 3)Q(x) + 2x + 7$

$P(3) = (3 + 1)(3 - 3)Q(3) + 2(3) + 7$
$= 13$

$\therefore$ the remainder is 13.

Answer D

14 $P(x) = x^3 - ax^2 + x$

Now $P(3) = 12$
$12 = 3^3 - a(3^2) + 3$
$12 = 30 - 9a$
$9a = 18$
$a = 2$ ✓

$\therefore P(x) = x^3 - 2x^2 + x$ ✓

$P(-1) = (-1)^3 - 2(-1)^2 + (-1)$
$= -4$

$\therefore$ the remainder when $P(x)$ is divided by $x + 1$ is -4. ✓ *(3 marks)*

15 **i** $P(x) = (x + 1)(x - 3)Q(x) + ax + b$

$x - 3$ is a factor of $P(x)$
$\therefore P(3) = 0$
$\therefore (3 + 1)(3 - 3)Q(3) + a(3) + b = 0$
$3a + b = 0$ [1]

When $P(x)$ is divided by $(x + 1)$ the remainder is 8.
$\therefore P(-1) = 8$
$\therefore (-1 + 1)(-1 - 3)Q(-1) + a(-1) + b = 8$
$-a + b = 8$ [2]

[1] − [2] $\quad 4a = -8$
$a = -2$ ✓

Substitute in [1],
$3(-2) + b = 0$
$-6 + b = 0$
$b = 6$

$\therefore a = -2$ and $b = 6$ ✓ *(2 marks)*

ii $P(x) = (x+1)(x-3)Q(x) - 2x + 6$

The remainder is $-2x + 6$. ✓

[The degree of $(-2x + 6)$ is less than the degree of $(x + 1)(x - 3)$.] *(1 mark)*

16 $p(x) = x^3 - ax + b$

$p(1) = 2$

$\therefore 1 - a + b = 2$

$-a + b = 1 \quad \ldots(1)$

$p(-2) = 5$

$\therefore -8 + 2a + b = 5$

$2a + b = 13 \quad \ldots(2)$ ✓

Subtracting (1) from (2):

$3a = 12$

$\therefore a = 4$ ✓

Substituting $a = 4$ into (1)

$-4 + b = 1$

$\therefore b = 5$ ✓

(3 marks)

17 $P(x) = x^3$

Using the remainder theorem,

$P(-3) = (-3)^3$ ✓

$= -27$

$\therefore$ The remainder is -27. ✓

(2 marks)

18 $P(x) = x^2 + ax + b$

Now $P(2) = 0$

$\therefore 2^2 + 2a + b = 0$

$4 + 2a + b = 0$

$2a + b = -4 \quad \ldots(1)$ ✓

Also $P(-1) = 18$

$\therefore (-1)^2 + a(-1) + b = 18$

$1 - a + b = 18$

$-a + b = 17$

$b = 17 + a \ldots(2)$ ✓

Substituting $b = 17 + a$ into (1):

$2a + 17 + a = -4$

$3a = -21$

$a = -7$

$\therefore b = 17 - 7$

$= 10$

i.e. $a = -7$ and $b = 10$. ✓

(3 marks)

19 i $P(x) = (x+1)(x-3)Q(x) + a(x+1) + b$

Now $P(-1) = -11$

$\therefore (-1+1)(-1-3)Q(x) + a(-1+1) + b = -11$

$0 + 0 + b = -11$

$\therefore b = -11$ ✓

(1 mark)

ii Now $P(3) = 1$

$\therefore (3+1)(3-3)Q(x) + a(3+1) + b = 1$

$\therefore 0 + a(3+1) + b = 1$

$\therefore 3a + a + b = 1$ ✓

$\therefore 4a - 11 = 1$

(substituting for b)

$\therefore 4a = 12$

$\therefore a = 3.$ ✓

(2 marks)

20 $x^3 - 2x^2 + a \equiv (x+2)Q(x) + 3$

For $x = -2$,

$(-2)^3 - 2(-2)^2 + a = 0 \times Q(-2) + 3$ ✓

$-8 - 8 + a = 3$

$\therefore a = 19.$ ✓

(2 marks)

21 Let $P(x) = x^3 - 5x + 12$

$P(-3) = (-3)^3 - 5(-3) + 12$

$= -27 + 15 + 12$

$= 0$ ✓

$\therefore x + 3$ is a factor of $x^3 - 5x + 12$. ✓

(2 marks)

22 If $x - 3$ is a factor of $P(x)$, then $P(3) = 0$

i.e. $(3)^3 - 3k(3) + 6 = 0$

$27 - 9k + 6 = 0$ ✓

$33 = 9k$

$k = \frac{33}{9}$

$= \frac{11}{3}.$ ✓

(2 marks)

23 $P(x) = x^3 - 4x$

$P(-3) = (-3)^3 - 4(-3)$ ✓

$= -27 + 12$

$= -15$

$\therefore$ The remainder is -15. ✓

(2 marks)

24

$$\begin{array}{r} x^2 - 2 \\ x^2 + 1 \overline{) x^4 - x^2 + 1} \\ \underline{x^4 + x^2} \\ -2x^2 + 1 \\ \underline{-2x^2 - 2} \\ 3 \end{array}$$ ✓

$\therefore R(x) = 3$ ✓

$\therefore Q(x) = x^2 - 2.$ ✓

(3 marks)

25 $P(x) = 2x^3 + x + a$

If $(x - 2)$ is a factor of $P(x)$,

$P(2) = 2(2)^3 + 2 + a = 0$

$16 + 2 + a = 0$

$a = -18$ ✓

(1 mark)

26 i The divisor $(x+1)(x-4)$ is of degree 2, hence the remainder will be at most of degree 1, i.e. $R(x)$ is of the form $ax + b$.

ii $P(x) = (x+1)(x-4)Q(x) + R(x)$

If $P(4) = -5$

$-5 = (4+1) \times (4-4)Q(4) + R(4)$

$-5 = 0 + R(4)$

$\therefore R(4) = -5.$

iii If $P(-1) = 5$

$5 = (-1+1)(-1-4)Q(-1) + R(-1)$

$5 = 0 + R(-1)$

$\therefore R(-1) = 5$

If $R(x) = ax + b$

$R(-1) = -a + b = 5 \quad \ldots(1)$

$R(4) = 4a + b = -5 \ldots(2)$

(from **ii**)

Solving simultaneously:

(1) − (2) $\quad -5a = 10$

$a = -2$

Substituting $a = -2$ into (1):

$-(-2) + b = 5$

$2 + b = 5$

$b = 3$

$\therefore R(x) = -2x + 3.$

(No mark allocation in exam)

27 $\frac{P(x)}{x^2 - 1} = Q(x) + \frac{3x - 1}{x^2 - 1}$

$\frac{P(x)}{(x+1)(x-1)} = Q(x) + \frac{3x - 1}{(x+1)(x-1)}$

$\therefore \frac{P(x)}{x - 1} = Q(x)(x+1) + \frac{3x - 1}{x - 1}$

Now $\frac{3x - 1}{x - 1} = 3 + \frac{2}{x - 1}$

$\therefore P(x) = [Q(x)(x+1) + 3] + \frac{2}{x - 1}$

$\therefore$ The remainder is 2.

(No mark allocation in exam)

28 $P(x) = x^n(2^m - 1) + x^m(1 - 2^n) + (2^n - 2^m)$
$P(1) = 1(2^m - 1) + 1(1 - 2^n) + (2^n - 2^m)$
$= 2^m - 1 + 1 - 2^n + 2^n - 2^m$
$= 0$
$\therefore (x - 1)$ is a factor of $P(x)$.

$P(2) = 2^n(2^m - 1) + 2^m(1 - 2^n) + 2^n - 2^m$
$= 2^{n+m} - 2^n + 2^m - 2^{n+m} + 2^n - 2^m$
$= 0$
$\therefore (x - 2)$ is a factor of $P(x)$.

Since $(x - 1)$ and $(x - 2)$ are prime factors of $P(x)$, $(x - 1)(x - 2)$ is a factor of $P(x)$.

(No mark allocation in exam)

1 The monic polynomial, P, has degree 3 and roots α, β, γ.

It is given that

$\alpha^2 + \beta^2 + \gamma^2 = 85$ and
$P'(\alpha) + P'(\beta) + P'(\gamma) = 87$.

Find $\alpha\beta + \beta\gamma + \gamma\alpha$. *(3 marks)*

(Q13d, **2022 HSC**) Medium

2 The roots of $x^4 - 3x + 6 = 0$ are α, β, γ and δ.

What is the value of $\frac{1}{\alpha} + \frac{1}{\beta} + \frac{1}{\gamma} + \frac{1}{\delta}$? *(2 marks)*

(Q11h, **2021 HSC**) Medium

3 A monic polynomial $p(x)$ of degree 4 has one repeated zero of multiplicity 2 and is divisible by $x^2 + x + 1$.

Which of the following could be the graph of $p(x)$?

A

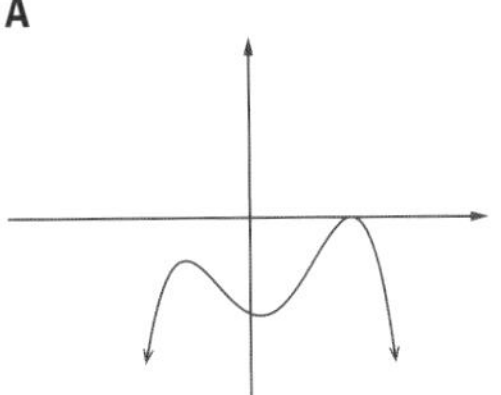

B

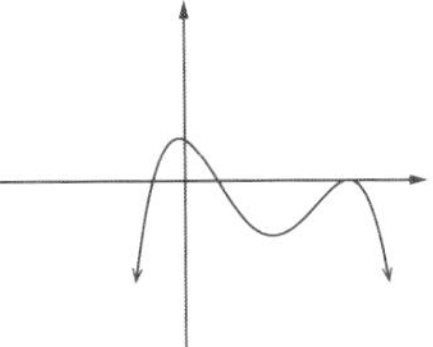

C

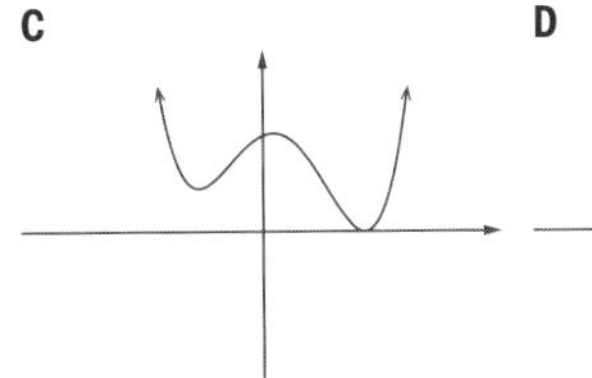

D

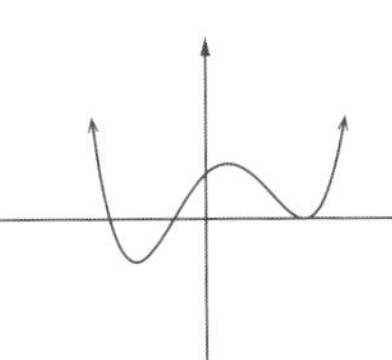

(1 mark)

(Q5, **2020 HSC**) Medium

4 The polynomial $P(x) = x^4 + 4x^3 + 3x^2 - 4x - 4$ has a double root at $x = \alpha$.

What is the value of α?

A 1 **B** –1
C 2 **D** –2 *(1 mark)*

Bonus question (see page iv) Easy

5 $P(x) = x^3 + ax^2 + bx - 4$.
$(x + 1)^2$ is a factor of $P(x)$.

Find the values of a and b. *(3 marks)*

Bonus question Medium

6 A polynomial $P(x) = px^3 - qx^2 + r$ has a multiple zero at $x = 1$ and, when divided by $x - 2$, the remainder is 5.

Find the values of p, q and r. *(3 marks)*

Bonus question Medium

7 $(x - m)^2$ is a factor of a polynomial $P(x)$.

Show that $P'(m) = 0$. *(2 marks)*

Bonus question Medium

8 The diagram shows the graph of $y = \frac{1}{x - k}$, where k is a positive real number.

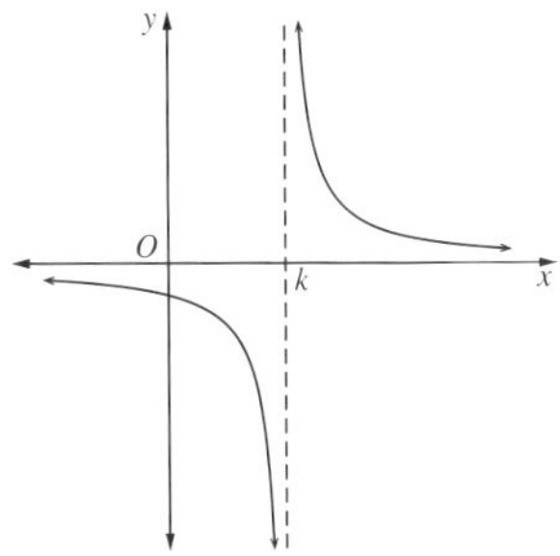

By considering the graphs of $y = x^2$ and $y = \frac{1}{x - k}$, explain why the function $f(x) = x^3 - kx^2 - 1$ has exactly one real zero. *(2 marks)* Medium

(Q14b i, **2019 HSC**)

9 The polynomial $2x^3 + 6x^2 - 7x - 10$ has zeros α, β and γ.

What is the value of $\alpha\beta\gamma(\alpha + \beta + \gamma)$?

A –60 **B** –15
C 15 **C** 60 *(1 mark)*

(Q1, **2018 HSC**) Easy

10 The diagram shows the graph of $y = a(x + b)(x + c)(x + d)^2$.

What are possible values of a, b, c and d?

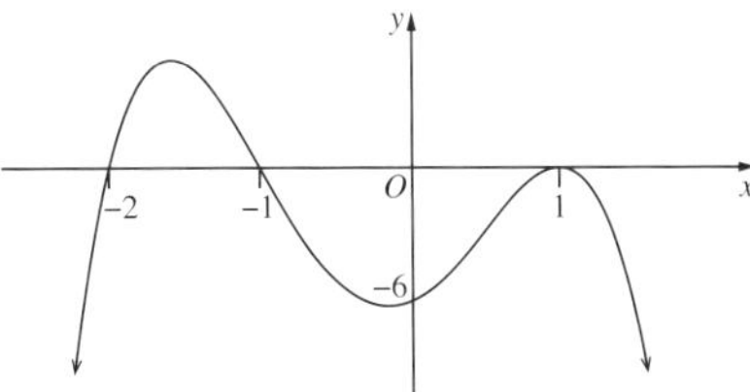

A $a = -6, b = -2, c = -1, d = 1$
B $a = -6, b = 2, c = 1, d = -1$
C $a = -3, b = -2, c = -1, d = 1$
D $a = -3, b = 2, c = 1, d = -1$ *(1 mark)*

(Q4, **2018 HSC**) Easy

11 Consider the polynomial $p(x) = ax^3 + bx^2 + cx - 6$ with a and b positive.

Which graph could represent $p(x)$?

A

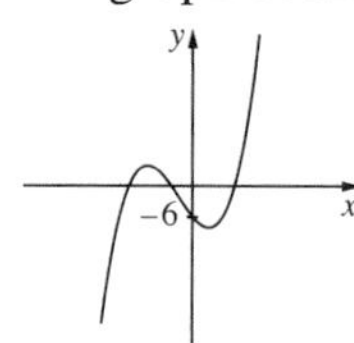

B

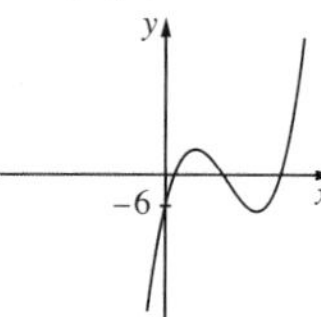

C

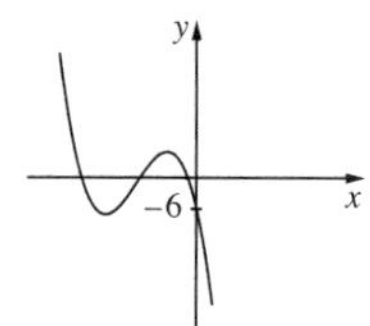

D

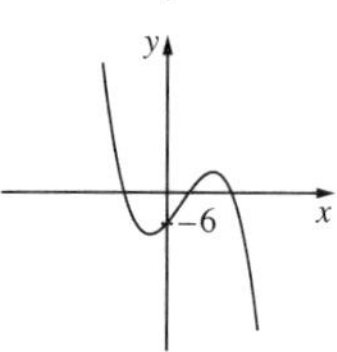

(1 mark)

(Q10, **2016 HSC**) Medium

12 Which group of three numbers could be the roots of the polynomial equation $x^3 + ax^2 - 41x + 42 = 0$?

A $2, 3, 7$ **B** $1, -6, 7$

C $-1, -2, 21$ **D** $-1, -3, -14$ *(1 mark)*

(Q5, **2014 HSC**) Medium

13 The roots of the quadratic equation $2x^2 + 8x + k = 0$ are α and β.

i Find the value of $\alpha + \beta$. *(1 mark)* Easy

ii Given that $\alpha^2\beta + \alpha\beta^2 = 6$, find the value of k. *(2 marks)* Easy

(Q14b, **2014 MA HSC**)

14 Which diagram best represents the graph $y = x(1 - x)^3(3 - x)^2$?

A

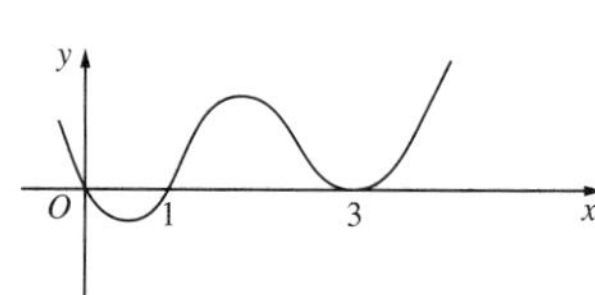

B

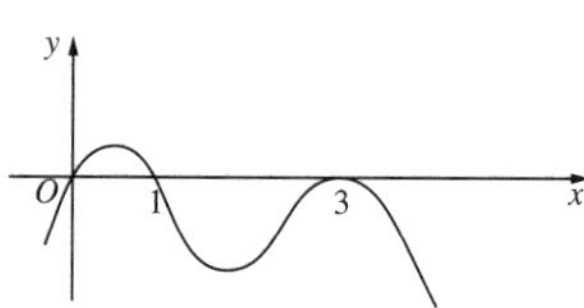

C

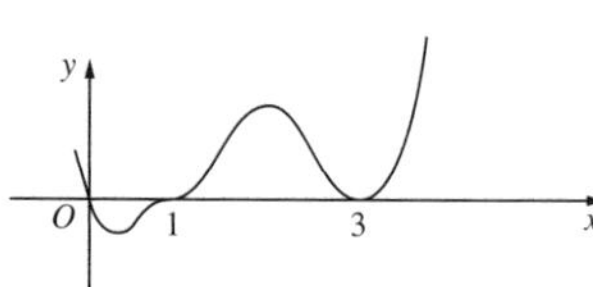

D

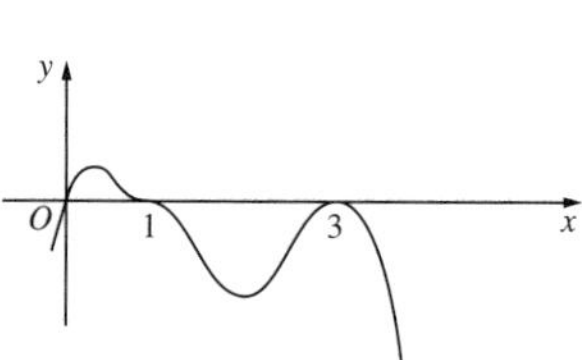

(1 mark)

(Q4, **2013 HSC**) Easy

15 The polynomial equation $2x^3 - 3x^2 - 11x + 7 = 0$ has roots α, β and γ.

Find $\alpha\beta\gamma$. *(1 mark)*

(Q11a, **2013 HSC**) Easy

16 A polynomial equation has roots α, β and γ where

$\alpha + \beta + \gamma = -2$, $\alpha\beta + \alpha\gamma + \beta\gamma = 3$ and $\alpha\beta\gamma = 1$.

Which polynomial equation has the roots α, β and γ?

A $x^3 + 2x^2 + 3x + 1 = 0$

B $x^3 + 2x^2 + 3x - 1 = 0$

C $x^3 - 2x^2 + 3x + 1 = 0$

D $x^3 - 2x^2 + 3x - 1 = 0$ *(1 mark)*

(Q3, **2012 HSC**) Medium

17 The quadratic equation $x^2 - 6x + 2 = 0$ has roots α and β.

i Find $\alpha + \beta$. *(1 mark)* Easy

ii Find $\alpha\beta$. *(1 mark)* Easy

iii Find $\frac{1}{\alpha} + \frac{1}{\beta}$. *(1 mark)* Easy

(Q2a, **2011 MA HSC**)

18 The polynomial $p(x)$ is given by $p(x) = ax^3 + 16x^2 + cx - 120$, where a and c are constants.

The three zeros of $p(x)$ are -2, 3 and α.

Find the value of α. *(3 marks)*

(Q2c, **2008 HSC**) Medium

19 The cubic polynomial $P(x) = x^3 + rx^2 + sx + t$, where r, s and t are real numbers, has three real zeros, 1, α and $-\alpha$.

i Find the value of r. *(1 mark)* Easy

ii Find the value of $s + t$. *(2 marks)* Medium

(Q4a, **2006 HSC**)

20 Let α and β be the solutions of $x^2 - 3x + 1 = 0$.

i Find $\alpha\beta$. *(1 mark)* Easy

ii Hence find $\alpha + \frac{1}{\alpha}$. *(1 mark)* Easy

(Q7a, **2006 MA HSC**)

21 It is known that two of the roots of the equation $2x^3 + x^2 - kx + 6 = 0$ are reciprocals of each other. Find the value of k. *(2 marks)*

(Q4c, **2003 HSC**) Medium

22 The polynomial $P(x) = x^3 - 2x^2 + kx + 24$ has roots α, β, γ.

i Find the value of $\alpha + \beta + \gamma$. *(1 mark)* Easy

ii Find the value of $\alpha\beta\gamma$. *(1 mark)* Easy

iii It is known that two of the roots are equal in magnitude but opposite in sign.
Find the third root and hence find the value of k. *(2 marks)* Medium

(Q4b, **2002 HSC**)

23 The polynomial $P(x) = x^3 + px^2 + qx + r$ has real roots $\sqrt{k}, -\sqrt{k}$ and α.

i Explain why $\alpha + p = 0$. Easy

ii Show that $k\alpha = r$. Easy

iii Show that $pq = r$. Medium

(4 marks) (Q4c, **2000 HSC**)

24 Let α, β and γ be the roots of the polynomial $2x^3 - 14x - 1 = 0$. Find $\alpha\beta\gamma$. *(2 marks)*

(Q1e, **1998 HSC**) Easy

25 The polynomial $P(x) = x^3 + bx^2 + cx + d$ has roots 0, 3, and –3.

i Find b, c, and d. Medium

ii Without using calculus, sketch the graph of $y = P(x)$. Easy

iii Hence, or otherwise, solve the inequality $\dfrac{x^2 - 9}{x} > 0$. Medium

(5 marks) (Q2c, **1997 HSC**)

26 Consider the equation

$$x^3 + 6x^2 - x - 30 = 0.$$

One of the roots of this equation is equal to the sum of the other two roots.

Find the values of the three roots. *(4 marks)*

(Q2c, **1995 HSC**) Medium

Year 11 Sums and products of roots of polynomials—Worked Answers

1 $P(x) = x^3 + bx^2 + cx + d$

$\alpha + \beta + \gamma = -b$

$\alpha\beta + \beta\gamma + \alpha\gamma = c$

$P'(x) = 3x^2 + 2bx + c$ ✓

Substituting α, β and γ into $P'(x)$ gives:

$P'(\alpha) = 3\alpha^2 + 2b\alpha + c$

$P'(\beta) = 3\beta^2 + 2b\beta + c$

$P'(\gamma) = 3\gamma^2 + 2b\gamma + c$ ✓

$P'(\alpha) + P'(\beta) + P'(\gamma) = 87$

$$\begin{aligned}87 &= 3\alpha^2 + 2b\alpha + c + 3\beta^2 + 2b\beta + c + 3\gamma^2 + 2b\gamma + c\\ &= 3(\alpha^2 + \beta^2 + \gamma^2) + 2b(\alpha + \beta + \gamma) + 3c\\ &= 3(\alpha^2 + \beta^2 + \gamma^2) - 2 \times -b(\alpha + \beta + \gamma) + 3(\alpha\beta + \beta\gamma + \alpha\gamma)\\ &= 3(\alpha^2 + \beta^2 + \gamma^2) - 2(\alpha + \beta + \gamma)^2 + 3(\alpha\beta + \beta\gamma + \alpha\gamma)\\ &= 3(\alpha^2 + \beta^2 + \gamma^2) - 2[\alpha^2 + \beta^2 + \gamma^2 + (2(\alpha\beta + \beta\gamma + \alpha\gamma))] + 3(\alpha\beta + \beta\gamma + \alpha\gamma)\\ &= \alpha^2 + \beta^2 + \gamma^2 - (\alpha\beta + \beta\gamma + \alpha\gamma)\\ &= 85 - (\alpha\beta + \beta\gamma + \alpha\gamma)\end{aligned}$$

$\therefore \alpha\beta + \beta\gamma + \alpha\gamma = -2$ ✓

(3 marks)

2 For $x^4 - 3x + 6 = 0$, $a = 1$, $d = -3$ and $e = 6$.

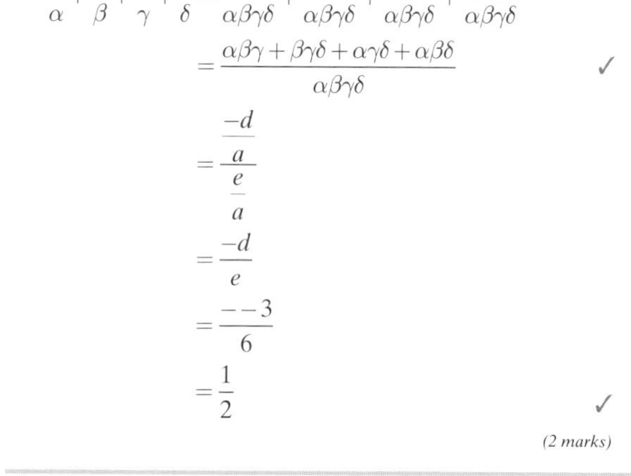

$$\begin{aligned}\frac{1}{\alpha} + \frac{1}{\beta} + \frac{1}{\gamma} + \frac{1}{\delta} &= \frac{\beta\gamma\delta}{\alpha\beta\gamma\delta} + \frac{\alpha\gamma\delta}{\alpha\beta\gamma\delta} + \frac{\alpha\beta\delta}{\alpha\beta\gamma\delta} + \frac{\alpha\beta\gamma}{\alpha\beta\gamma\delta}\\ &= \frac{\alpha\beta\gamma + \beta\gamma\delta + \alpha\gamma\delta + \alpha\beta\delta}{\alpha\beta\gamma\delta} \quad ✓\\ &= \frac{\frac{-d}{a}}{\frac{e}{a}}\\ &= \frac{-d}{e}\\ &= \frac{--3}{6}\\ &= \frac{1}{2}\end{aligned}$$ ✓

(2 marks)

3 $x^2 + x + 1$ is not divisible by 2.

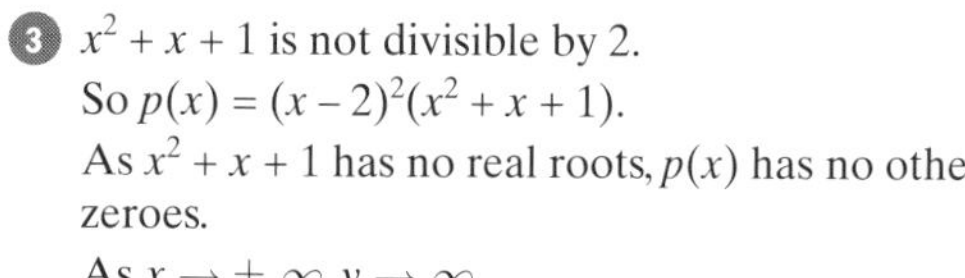

So $p(x) = (x - 2)^2(x^2 + x + 1)$.

As $x^2 + x + 1$ has no real roots, $p(x)$ has no other zeroes.

As $x \to \pm\infty, y \to \infty$

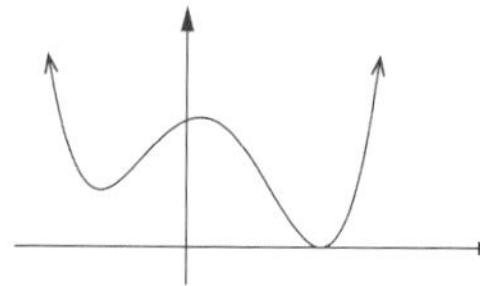

Answer C

4 $P(x) = x^4 + 4x^3 + 3x^2 - 4x - 4$

$P'(x) = 4x^3 + 12x^2 + 6x - 4$

$= 2(2x^3 + 6x^2 + 3x - 2)$

Now $P(x)$ has a double root at $x = \alpha$ so $P'(\alpha) = 0$.

Try each option:

$P'(1) \neq 0$

$P'(-1) \neq 0$

$P'(2) \neq 0$

$P'(-2) = 0$

So, $\alpha = -2$.

Answer D

5 $P(x) = x^3 + ax^2 + bx - 4$

$P'(x) = 3x^2 + 2ax + b$

$(x + 1)^2$ is a factor of $P(x)$.

So $P(-1) = 0$ and $P'(-1) = 0$. ✓

$(-1)^3 + a(-1)^2 + b(-1) - 4 = 0$

$-1 + a - b - 4 = 0$

$a - b = 5$ .... (1)

$3(-1)^2 + 2a(-1) + b = 0$

$3 - 2a + b = 0$

$2a - b = 3$ .... (2) ✓

(2) – (1) $a = -2$

Sub. in (1) $-2 - b = 5$

$b = -7$

$\therefore a = -2$ and $b = -7$ ✓

(3 marks)

6 $P(x) = px^3 - qx^2 + r$

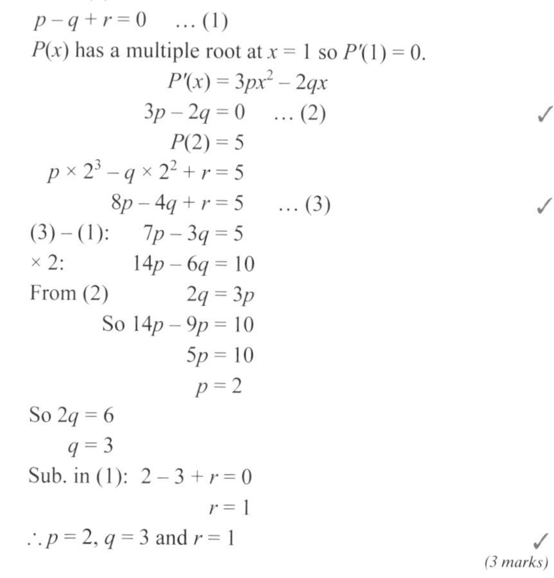

$P(1) = 0$

$p - q + r = 0$ ... (1)

$P(x)$ has a multiple root at $x = 1$ so $P'(1) = 0$.

$P'(x) = 3px^2 - 2qx$

$3p - 2q = 0$ ... (2) ✓

$P(2) = 5$

$p \times 2^3 - q \times 2^2 + r = 5$

$8p - 4q + r = 5$ ... (3) ✓

(3) – (1): $7p - 3q = 5$

× 2: $14p - 6q = 10$

From (2) $2q = 3p$

So $14p - 9p = 10$

$5p = 10$

$p = 2$

So $2q = 6$

$q = 3$

Sub. in (1): $2 - 3 + r = 0$

$r = 1$

$\therefore p = 2$, $q = 3$ and $r = 1$ ✓

(3 marks)

7 $(x - m)^2$ is a factor of a polynomial $P(x)$.

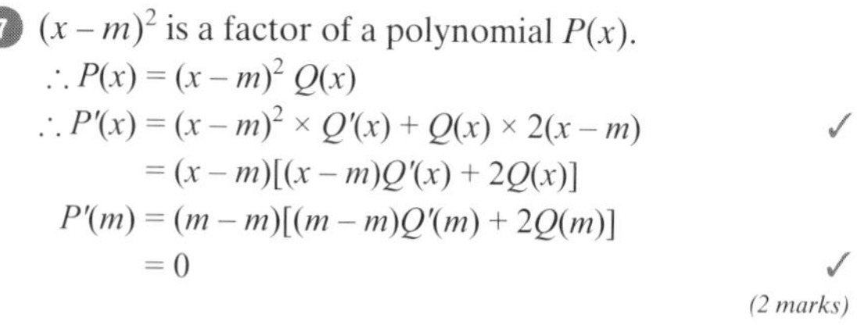

$\therefore P(x) = (x - m)^2\, Q(x)$

$\therefore P'(x) = (x - m)^2 \times Q'(x) + Q(x) \times 2(x - m)$ ✓

$= (x - m)[(x - m)Q'(x) + 2Q(x)]$

$P'(m) = (m - m)[(m - m)Q'(m) + 2Q(m)]$

$= 0$ ✓

(2 marks)

8 $y = x^2$ is a parabola, concave up and with vertex the origin, so it will only intersect $y = \dfrac{1}{x-k}$ in one place.

So there is only one real solution of $x^2 = \dfrac{1}{x-k}$.

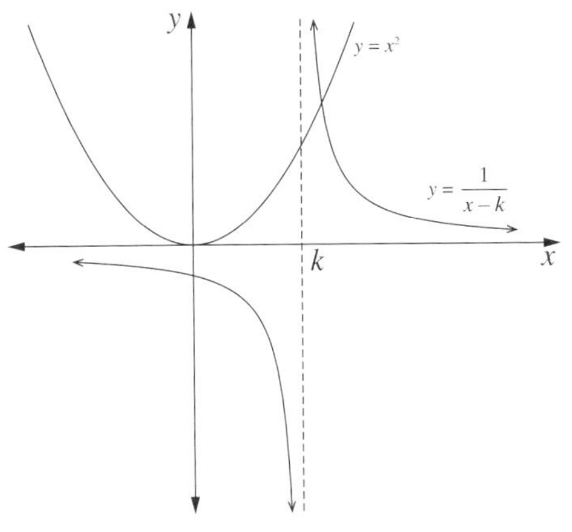

Now zeroes of

$f(x) = x^3 - kx^2 - 1$ occur when $f(x) = 0$ ✓

i.e. $x^3 - kx^2 - 1 = 0$

$x^3 - kx^2 = 1$

$x^2(x - k) = 1$

$x^2 = \dfrac{1}{x-k} \quad (x \neq k)$

$f(k) \neq 0$

So $f(x)$ has exactly one real zero. ✓

(2 marks)

9 $P(x) = 2x^3 + 6x^2 - 7x - 10$

$$\alpha\beta\gamma = -\frac{d}{a} = -\frac{-10}{2} = 5$$

$$\alpha + \beta + \gamma = -\frac{b}{a} = -\frac{6}{2} = -3$$

So $\alpha\beta\gamma\,(\alpha + \beta + \gamma) = 5 \times -3$

$= -15$

Answer B

10 There are single roots at –2 and –1 and a double root at 1.

So possible values of b and c are 2 and 1 and a possible value of d is –1.

So $y = a(x + 2)(x + 1)(x - 1)^2$

When $x = 0$, $y = -6$

$-6 = a(0 + 2)(0 + 1)(0 - 1)^2$

$-6 = 2a$

$a = -3$

Answer D

11 $p(x) = ax^3 + bx^2 + cx - 6$

$a > 0$ so as $x \to \infty$, $p(x) \to \infty$

So options C and D are not correct.

Sum of roots $= -\dfrac{b}{a}$

So the sum of the roots is negative as both a and b are positive.

So option B is not correct.

Answer A

12 $x^3 + ax^2 - 41x + 42 = 0$

$\alpha\beta\gamma = -42$

[So options A and C are eliminated.]

$\alpha\beta + \alpha\gamma + \beta\gamma = -41$

[So option D is eliminated.]

Of the options only 1, −6 and 7 could be the roots of the equation.

Answer B

13 **i**
$$\alpha + \beta = \frac{-b}{a} = \frac{-8}{2} = -4$$
✓

(1 mark)

ii
$$\alpha\beta = \frac{c}{a} = \frac{k}{2}$$

$$\alpha^2\beta + \alpha\beta^2 = \alpha\beta(\alpha + \beta)$$
✓
$$= \frac{k}{2}(-4) = -2k$$

$-2k = 6$

$k = -3$ ✓

(2 marks)

14 $y = x(1 - x)^3(3 - x)^2$

As $x \to \infty$, $y \to -\infty$

The polynomial $y = P(x)$ has a triple root at $x = 1$ and a double root at $x = 3$.

So the curve will have stationary points at $x = 1$ and $x = 3$.

Only graph D will satisfy these requirements.

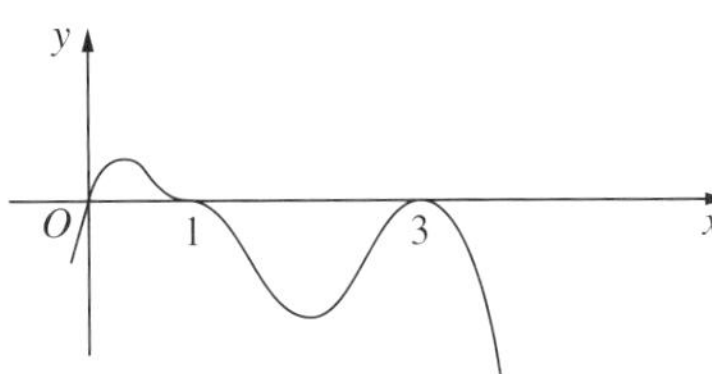

Answer D

15 $P(x) = 2x^3 - 3x^2 - 11x + 7$

$$\alpha\beta\gamma = -\frac{d}{a} = -\frac{7}{2}$$
✓

(1 mark)

16 $ax^3 + bx^2 + cx + d = 0$

[Note: all the choices have $a = 1$.]

Now $\alpha + \beta + \gamma = \dfrac{-b}{a}$

$\therefore \alpha + \beta + \gamma = -b$ when $a = 1$

But $\alpha + \beta + \gamma = -2$

$\therefore -b = -2$

$b = 2$

$\alpha\beta + \alpha\gamma + \beta\gamma = \dfrac{c}{a}$

$\therefore \alpha\beta + \alpha\gamma + \beta\gamma = c$ when $a = 1$

But $\alpha\beta + \alpha\gamma + \beta\gamma = 3$

$\therefore c = 3$

$\alpha\beta\gamma = \frac{-d}{a}$

$\therefore \alpha\beta\gamma = -d$ when $a = 1$

But $\alpha\beta\gamma = 1$

$\therefore -d = 1$

$d = -1$

So $b = 2, c = 3$ and $d = -1$

$\therefore x^3 + 2x^2 + 3x - 1 = 0$

has roots α, β and γ.

Answer B

17 **i** $\alpha + \beta = -\frac{b}{a}$

$= -\frac{(-6)}{1}$

$= 6$ ✓ *(1 mark)*

ii $\alpha\beta = \frac{c}{a}$

$= \frac{2}{1}$

$= 2$ ✓ *(1 mark)*

iii $\frac{1}{\alpha} + \frac{1}{\beta} = \frac{\beta + \alpha}{\alpha\beta}$

$= \frac{6}{2}$

$= 3$ ✓ *(1 mark)*

18 $p(x) = ax^3 + 16x^2 + cx - 120$

Sum of the roots $= -\frac{B}{A}$

$= -\frac{16}{a}$

$\therefore -2 + 3 + \alpha = -\frac{16}{a}$

$\alpha = -\frac{16}{a} - 1 \quad \ldots(1)$ ✓

Product of the roots $= -\frac{D}{A}$

$= \frac{120}{a}$

$\therefore -2 \times 3 \times \alpha = \frac{120}{a}$

$-6\alpha = \frac{120}{a}$

$\alpha = -\frac{20}{a} \quad \ldots(2)$ ✓

Equating (1) and (2) above:

$-\frac{16}{a} - 1 = -\frac{20}{a}$

$-\frac{16}{a} + \frac{20}{a} = 1$

$\frac{4}{a} = 1$

$a = 4$

Substituting $a = 4$ into (2):

$\therefore \alpha = -\frac{20}{4}$

$= -5.$ ✓ *(3 marks)*

19 **i** $P(x) = x^3 + rx^2 + sx + t$

Now $1 + \alpha - \alpha = -\frac{B}{A}$ (sum of roots)

$\therefore 1 = -r$

$\therefore r = -1.$ ✓ *(1 mark)*

ii If 1 is a zero, and $r = -1$, then

$P(1) = 1 + r + s + t = 0$ ✓

$\therefore 1 - 1 + s + t = 0$

$s + t = 0.$ ✓ *(2 marks)*

20 **i** Product of the roots $= \alpha\beta$

$= \frac{1}{1}$

$= 1.$ ✓ *(1 mark)*

ii Sum of the roots $= \alpha + \beta$

$= \alpha + \frac{1}{\alpha}$ (using **i**)

$= \frac{3}{1}$

$= 3.$ ✓ *(1 mark)*

21 $2x^3 + x^2 - kx + 6 = 0 \quad \ldots(1)$

Now $\alpha + \frac{1}{\alpha} + \beta = -\frac{1}{2}$

and $\alpha\left(\frac{1}{\alpha}\right) + \alpha(\beta) + \frac{1}{\alpha}(\beta) = -\frac{k}{2}$

i.e. $1 + \alpha\beta + \frac{\beta}{\alpha} = -\frac{k}{2}$

and $\alpha\left(\frac{1}{\alpha}\right)(\beta) = -3$

i.e. $\beta = -3$ ✓

Substituting $\beta = -3$ for x into (1):

$2(-3)^3 + (-3)^2 - k(-3) + 6 = 0$

$-54 + 9 + 3k + 6 = 0$

$3k = 39$

$\therefore k = 13$ ✓ *(2 marks)*

22 $P(x) = x^3 - 2x^2 + kx + 24$

i $\alpha + \beta + \gamma = 2$ ✓ *(1 mark)*

ii $\alpha\beta\gamma = -24$ ✓ *(1 mark)*

iii Let $\alpha = -\beta$, then the roots are $-\beta, \beta$ and γ.

$\therefore -\beta + \beta + \gamma = 2$ (from **i**)

$\therefore \gamma = 2$

Now $k = \alpha\beta + \alpha\gamma + \beta\gamma$

$= (-\beta)\beta + (-\beta)\gamma + \beta\gamma$

$= -\beta^2 - \beta\gamma + \beta\gamma$

$= -\beta^2$ ✓

But $\alpha\beta\gamma = -24$ (from **ii**)
$\therefore (-\beta)\beta\gamma = -24$
$\therefore -\beta^2\gamma = -24$
$\therefore -2\beta^2 = -24$ (since $\gamma = 2$)

$\therefore -\beta^2 = -12$
$\therefore k = -12.$ (since $k = -\beta^2$) ✓

(2 marks)

---

23 **i** The polynomial
$P(x) = x^3 + px^2 + qx + r$ has roots:
$x_1 = \sqrt{k}$, $x_2 = -\sqrt{k}$ and $x_3 = \alpha$

Now $x_1 + x_2 + x_3 = -p$
i.e. $\sqrt{k} - \sqrt{k} + \alpha = -p$
$\therefore \alpha + p = 0$ ✓

ii Now $x_1x_2x_3 = -r$
i.e. $\sqrt{k}(-\sqrt{k})\alpha = -r$
$\therefore -k\alpha = -r$
$\therefore k\alpha = r$ ✓

iii Now $x_1x_2 + x_1x_3 + x_2x_3 = q$
$\therefore \sqrt{k}(-\sqrt{k}) + \sqrt{k}\,\alpha + (-\sqrt{k})\alpha = q$
$\therefore -k + \sqrt{k}\,\alpha - \sqrt{k}\,\alpha = q$
$\therefore -k = q$ ✓

Also, $p = -\alpha$ (from **i**)
$\therefore pq = -\alpha(-k)$
$= \alpha k$
$= r$ (from **ii**) ✓ *(4 marks)*

(Total mark allocation only in exam)

---

24 $2x^3 - 14x - 1$
$= 2x^3 + 0x^2 + (-14)x + (-1) = 0$ ✓
$\therefore \alpha\beta\gamma = -\dfrac{-1}{2}$
$= \dfrac{1}{2}.$ ✓

(2 marks)

---

25 **i** $P(x) = x^3 + bx^2 + cx + d$

Roots: $\alpha = 0, \beta = 3, \gamma = -3$

$\alpha + \beta + \gamma = -b$
$0 + 3 + (-3) = -b$
$\therefore b = 0$

$\alpha\beta + \alpha\gamma + \beta\gamma = c$
$-3(3) = c$
$\therefore c = -9$ ✓

$\alpha\beta\gamma = -d$
$0 = -d$
$\therefore d = 0$ ✓

ii $y = P(x)$
$= x^3 - 9x$
$= x(x^2 - 9)$
$= x(x - 3)(x + 3)$

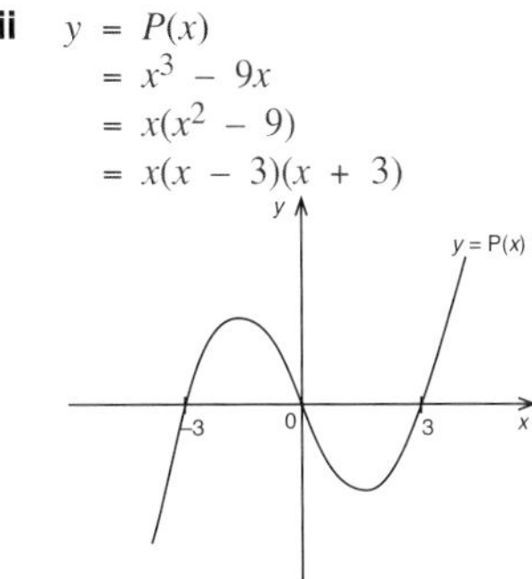

✓

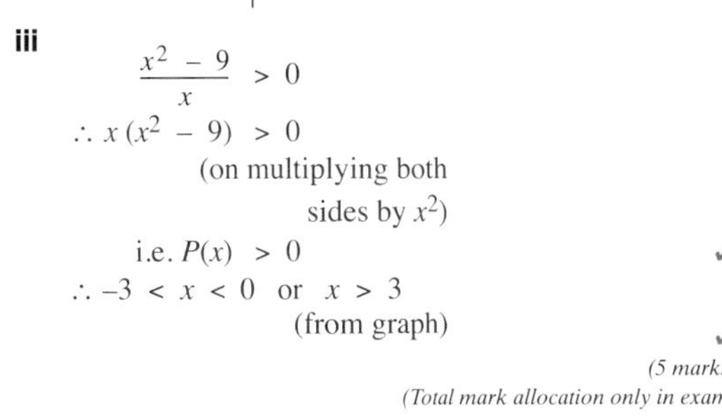

iii $\dfrac{x^2 - 9}{x} > 0$
$\therefore x(x^2 - 9) > 0$
(on multiplying both sides by x^2)
i.e. $P(x) > 0$ ✓
$\therefore -3 < x < 0$ or $x > 3$
(from graph) ✓

(5 marks)
(Total mark allocation only in exam)

---

26 Let the roots of $x^3 + 6x^2 - x - 30 = 0$ be α, β and γ where $\alpha = \beta + \gamma$.

Now $\alpha + \beta + \gamma = -6$
$\therefore \alpha + \alpha = -6$
$2\alpha = -6$
$\alpha = -3$
$\therefore \beta + \gamma = -3$ . . . (1) ✓

and $\alpha\beta\gamma = 30$
$\therefore -3\beta\gamma = 30$
$\beta\gamma = -10$ . . . (2) ✓

Solving equations (1) and (2) simultaneously:

From (1) $\gamma = -3 - \beta$ . . . (3) ✓
Substituting (3) into (2):
$\beta(-3 - \beta) = -10$
$-3\beta - \beta^2 = -10$
$\beta^2 + 3\beta - 10 = 0$
$(\beta + 5)(\beta - 2) = 0$
$\therefore \beta = -5$ or 2

$\therefore$ The three roots are -3, -5 and 2. ✓

(4 marks)
(Total mark allocation only in exam)

1 It is given that $\cos\left(\dfrac{23\pi}{12}\right) = \dfrac{\sqrt{6}+\sqrt{2}}{4}$.

Which of the following is the value of $\cos^{-1}\left(\dfrac{\sqrt{6}+\sqrt{2}}{4}\right)$?

A $\dfrac{23\pi}{12}$ **B** $\dfrac{11\pi}{12}$

C $\dfrac{\pi}{12}$ **D** $-\dfrac{11\pi}{12}$ *(1 mark)*

(Q1, **2022 HSC**) Easy

2 Find the equation of the tangent to the curve $y = x \arctan(x)$ at the point with coordinates $\left(1, \dfrac{\pi}{4}\right)$. Give your answer in the form $y = mx + c$. *(3 marks)*

(Q12c, **2022 HSC**) Easy

3 The function f is defined by $f(x) = \sin(x)$ for all real numbers x. Let g be the function defined on $[-1, 1]$ by $g(x) = \arcsin(x)$.

Is g the inverse of f? Justify your answer. *(2 marks)*

(Q13c, **2022 HSC**) Easy

4 Which graph represents the function $y = \sin^{-1}(\sin x)$?

A

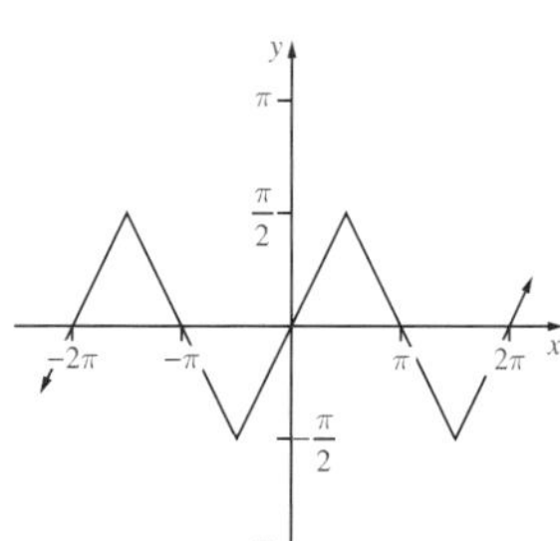

B

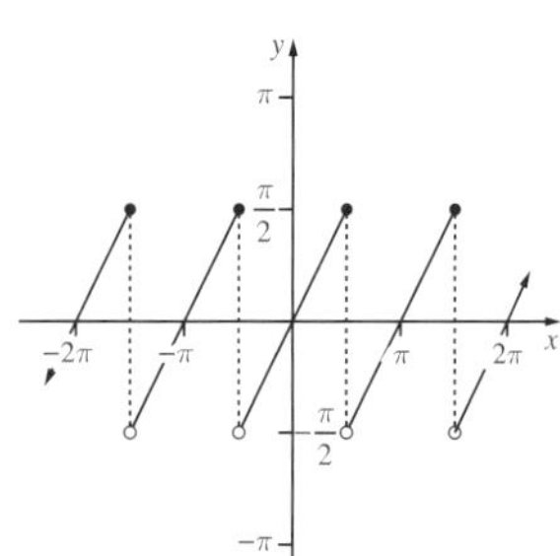

C

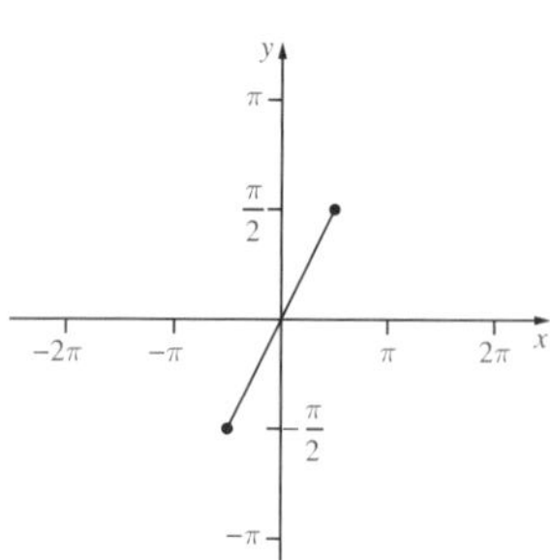

D

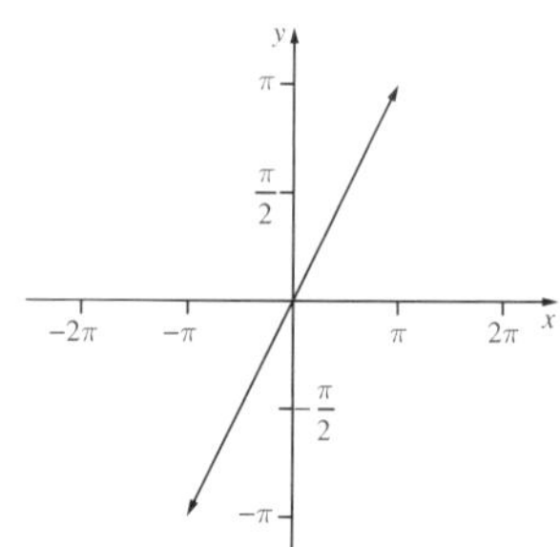

(1 mark)

(Q9, **2021 HSC**) Medium

5 Which graph best represents $y = \cos^{-1}(-\sin x)$, for $-\dfrac{\pi}{2} \le x \le \dfrac{\pi}{2}$?

A **B**

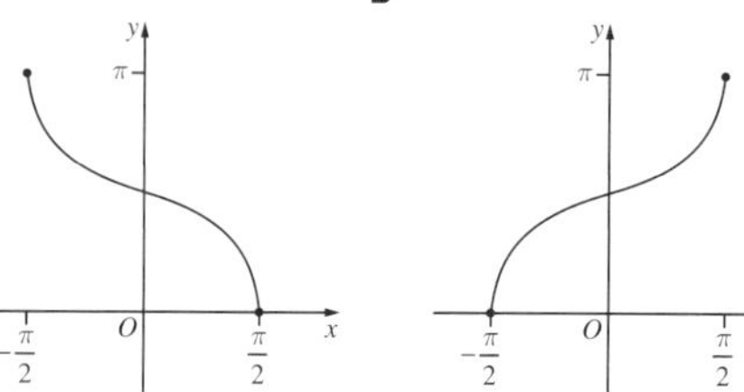

C

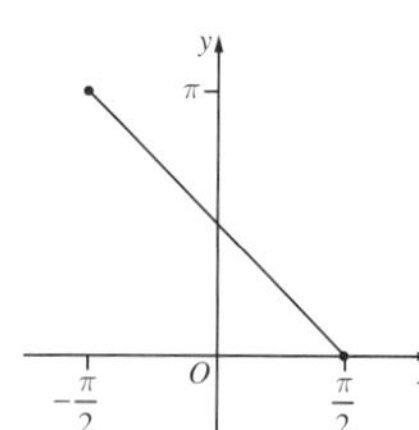

D

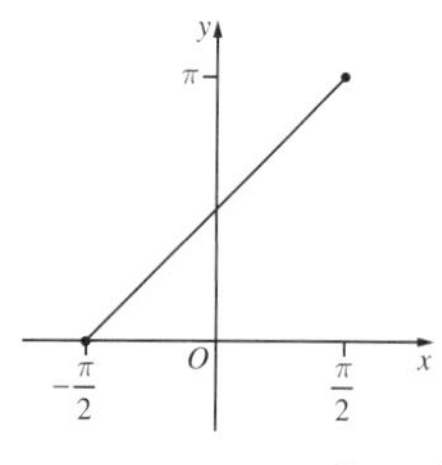

(1 mark)

(Q9, **2019 HSC**) Medium

6 Which diagram represents the domain of the function $f(x) = \sin^{-1}\left(\dfrac{3}{x}\right)$?

A

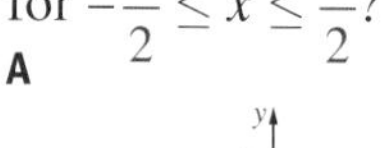

B

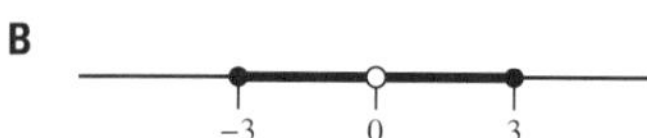

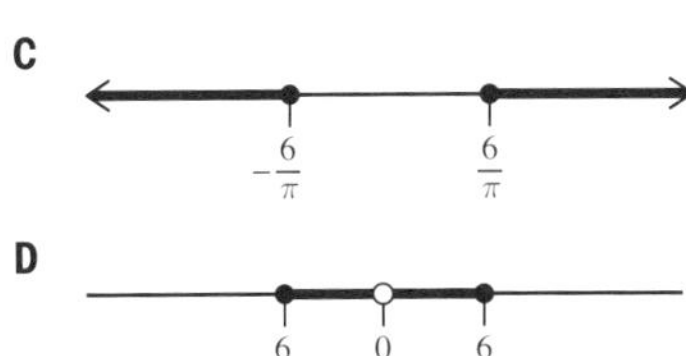

(1 mark)

(Q7, **2017 HSC**) Medium

7 Sketch the graph of the function $y = 2\cos^{-1}x$. *(2 marks)*

(Q11d, **2017 HSC**) Easy

8 What is the domain of the function $f(x) = \sin^{-1}(2x)$?

A $-\pi \le x \le \pi$ **B** $-2 \le x \le 2$

C $-\frac{\pi}{4} \le x \le \frac{\pi}{4}$ **D** $-\frac{1}{2} \le x \le \frac{1}{2}$ *(1 mark)*

(Q6, **2015 HSC**) Easy

9 Sketch the graph $y = 6\tan^{-1}x$, clearly indicating the range. *(2 marks)*

(Q11c, **2014 HSC**) Easy

10 The diagram shows the graph of a function.

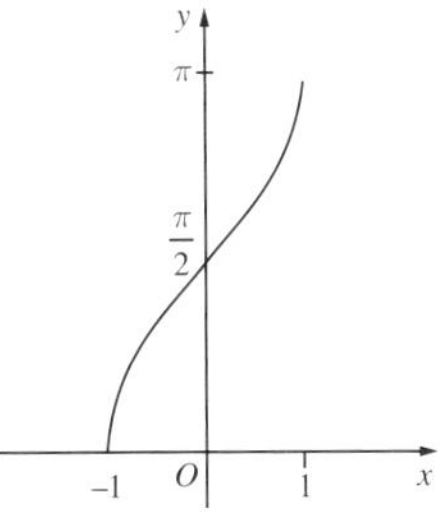

Which function does the graph represent?

A $y = \cos^{-1}x$ **B** $y = \frac{\pi}{2} + \sin^{-1}x$

C $y = -\cos^{-1}x$ **D** $y = -\frac{\pi}{2} - \sin^{-1}x$

(1 mark)

(Q9, **2013 HSC**) Medium

11 Which function best describes the following graph?

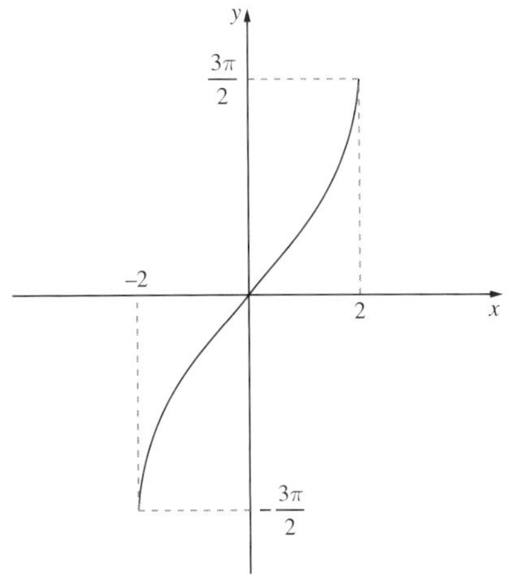

A $y = 3\sin^{-1}2x$ **B** $y = \frac{3}{2}\sin^{-1}2x$

C $y = 3\sin^{-1}\frac{x}{2}$ **D** $y = \frac{3}{2}\sin^{-1}\frac{x}{2}$ *(1 mark)*

(Q4, **2012 HSC**) Easy

12 Write $\sin\left(2\cos^{-1}\left(\frac{2}{3}\right)\right)$ in the form $a\sqrt{b}$, where a and b are rational. *(2 marks)*

(Q13a, **2012 HSC**) Medium

13 Find the exact value of $\cos^{-1}\left(-\frac{1}{2}\right)$. *(1 mark)*

(Q1e, **2011 HSC**) Easy

14 Sketch the graph of the function $f(x) = 2\cos^{-1}x$. Clearly indicate the domain and range of the function. *(2 marks)*

(Q2d, **2011 HSC**) Easy

15 Let $f(x) = \cos^{-1}\left(\frac{x}{2}\right)$. What is the domain of $f(x)$? *(1 mark)*

(Q1b, **2010 HSC**) Easy

16 Let $f(x) = 2\cos^{-1}x$.

i Sketch the graph of $y = f(x)$, indicating clearly the coordinates of the endpoints of the graph. *(2 marks)* Easy

ii State the range of $f(x)$. *(1 mark)* Easy

(Q2b, **2007 HSC**)

17 Find the exact values of x and y which satisfy the simultaneous equations

$\sin^{-1}x + \frac{1}{2}\cos^{-1}y = \frac{\pi}{3}$ and

$3\sin^{-1}x - \frac{1}{2}\cos^{-1}y = \frac{2\pi}{3}$. *(3 marks)* Hard

(Q5c, **2007 HSC**)

18 State the domain and range of $y = \cos^{-1}\left(\frac{x}{4}\right)$. *(2 marks)*

(Q1c, **2005 HSC**) Easy

19 Sketch the graph of $y = 3\cos^{-1}2x$. Your graph must clearly indicate the domain and the range. *(2 marks)*

(Q2a, **2003 HSC**) Easy

20 State the domain and range of the function $f(x) = 3\sin^{-1}\left(\frac{x}{2}\right)$. *(2 marks)*

(Q1d, **2002 HSC**) Easy

Year 11 Inverse trigonometric functions—Worked Answers

1 Since $\cos\left(\frac{23\pi}{12}\right)=\frac{\sqrt{6}+\sqrt{2}}{4}$ then

$\cos\left(2\pi-\frac{23\pi}{12}\right)=\frac{\sqrt{6}+\sqrt{2}}{4}.$

As $y=\cos^{-1}(\theta)$ has a range of $[0,\pi]$, then

$\cos^{-1}\left(\frac{\sqrt{6}+\sqrt{2}}{4}\right)=\frac{\pi}{12}.$

Answer C

2 $y = x\arctan(x)$

Let $u = x$ and $v = \arctan(x)$

$\frac{dy}{dx}=u\frac{dv}{dx}+v\frac{du}{dx}$

$=x\left(\frac{1}{1+x^2}\right)+\arctan(x)$ ✓

At $\left(1,\frac{\pi}{4}\right)$,

$m=(1)\left(\frac{1}{1+1^2}\right)+\arctan(1)$

$\therefore m=\frac{1}{2}+\frac{\pi}{4}$ ✓

$y-y_1=m(x-x_1)$

$y-\frac{\pi}{4}=\left(\frac{1}{2}+\frac{\pi}{4}\right)(x-1)$

$y-\frac{\pi}{4}=\frac{1}{2}x-\frac{1}{2}+\frac{\pi}{4}x-\frac{\pi}{4}$

$y=\frac{1}{2}x-\frac{1}{2}+\frac{\pi}{4}x$

$\therefore y=\left(\frac{1}{2}+\frac{\pi}{4}\right)x-\frac{1}{2}$ ✓

(3 marks)

3 $f(x)$ is a many-to-one function.

$\therefore$ the inverse of $f(x)$ is not a function. ✓

$g(x)=\sin^{-1}(x)$ is a function so it cannot be the inverse of f. ✓

(2 marks)

4 By substituting $x=\pi$ we see that options A or B are possible graphs of $y=\sin^{-1}(\sin x)$.

Consider the derivative function:

$\frac{dy}{dx}=\frac{1}{\sqrt{1-(\sin x)^2}}\times\cos x$

$\frac{dy}{dx}=\frac{\cos x}{\sqrt{1-\sin^2 x}}$

$\frac{dy}{dx}=\frac{\cos x}{|\cos x|}$

$\frac{dy}{dx}=\pm 1$

Neither $\sin^{-1}x$ nor $\sin x$ are discontinuous.

Answer A

5 $y=\cos^{-1}(-\sin x)$

$\frac{dy}{dx}=\frac{-1}{\sqrt{1-(-\sin x)^2}}\times -\cos x$

$=\frac{\cos x}{\sqrt{1-\sin^2 x}}$

$=1 \quad \left(-\frac{\pi}{2}\le x\le\frac{\pi}{2}\right)$

So y has a constant gradient of 1.

[Or: Using $\cos^{-1}(-\sin x)=\pi-\cos^{-1}(\sin x)$ it follows that $y=x+\frac{\pi}{2}$]

Answer D

6 $f(x)=\sin^{-1}\left(\frac{3}{x}\right)$

Domain: $-1\le\frac{3}{x}\le 1$

If $\frac{3}{x}\ge -1$

$\frac{x}{3}\le -1$ or $x>0$

$x\le -3$ or $x>0$

If $\frac{3}{x}\le 1$

$\frac{x}{3}\ge 1$ or $x<0$

$x\ge 3$ or $x<0$

But $\frac{3}{x}\ge -1$ and $\frac{3}{x}\le 1$ must both hold at the same time.

$\therefore x\le -3$ or $x\ge 3$

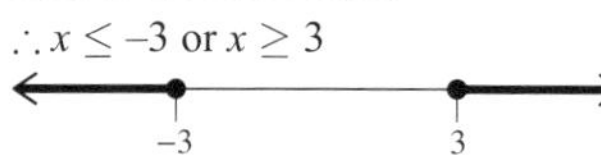

Answer A

7

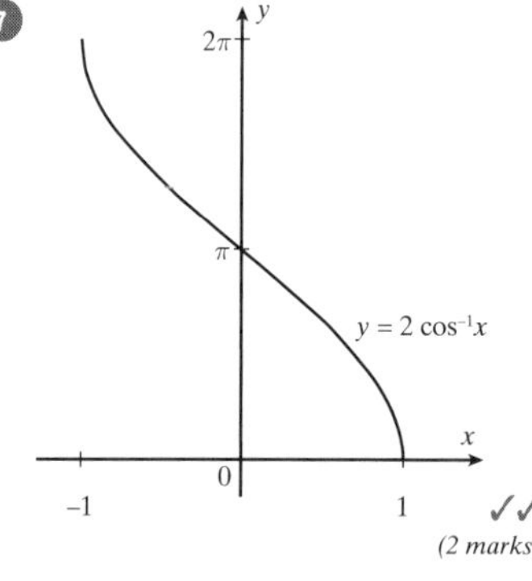

✓✓ *(2 marks)*

8 $f(x)=\sin^{-1}(2x)$

Now $f(u)=\sin^{-1}u$ has domain $-1\le u\le 1$.

So $\sin^{-1}(2x)$ has domain $-1\le 2x\le 1$.

$-\frac{1}{2}\le x\le\frac{1}{2}$

Answer D

9 $y = 6\tan^{-1}x$

Now $-\frac{\pi}{2}<\tan^{-1}x<\frac{\pi}{2}$

$\therefore -3\pi<6\tan^{-1}x<3\pi$ ✓

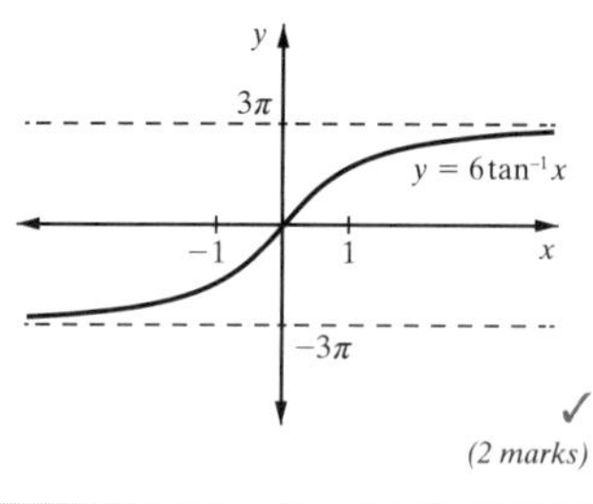

✓ *(2 marks)*

10 Consider the graphs of $y=\sin^{-1}x$ and $y=\cos^{-1}x$.

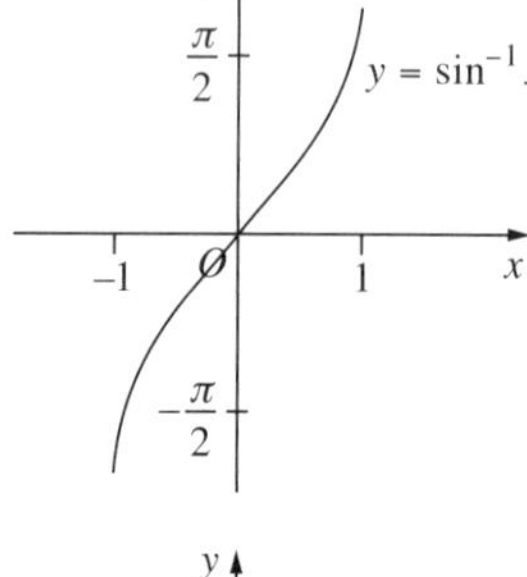

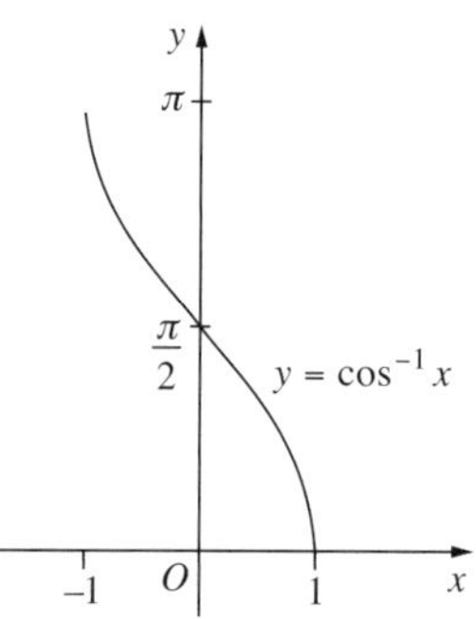

The graph in the question is that of $y=\sin^{-1}x$ moved up by $\frac{\pi}{2}$ units.

So it is the graph of

$y=\frac{\pi}{2}+\sin^{-1}x.$

Answer B

11

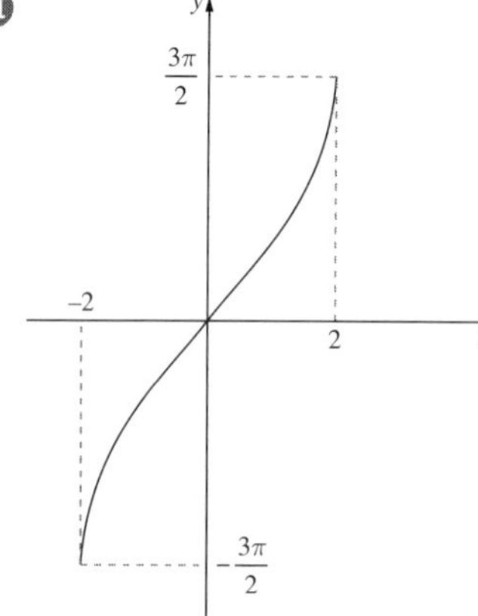

$y = \sin^{-1}u$ has domain $-1 \le u \le 1$
Here, $-2 \le x \le 2$

$\therefore -1 \le \frac{x}{2} \le 1$

So the equation is of the form
$y = a\sin^{-1}\frac{x}{2}$.

When $x = 2, y = \frac{3\pi}{2}$

$\frac{3\pi}{2} = a\sin^{-1}1$

$= a \times \frac{\pi}{2}$

$\therefore a = 3$
The equation is $y = 3\sin^{-1}\frac{x}{2}$.
Answer C

12 $\sin\left(2\cos^{-1}\left(\frac{2}{3}\right)\right)$

Let $x = \cos^{-1}\left(\frac{2}{3}\right)$

$\therefore \cos x = \frac{2}{3}$

$\cos^2 x = \frac{4}{9}$

$\sin^2 x = 1 - \cos^2 x$

$= 1 - \frac{4}{9}$

$= \frac{5}{9}$

$\sin x = \pm\frac{\sqrt{5}}{3}$

But $0 \le \cos^{-1}u \le \pi$ for all u
$\therefore 0 \le x \le \pi$
$\therefore \sin x \ge 0$

So $\sin x = \frac{\sqrt{5}}{3}$ ✓

$\sin\left(2\cos^{-1}\left(\frac{2}{3}\right)\right) = \sin 2x$

$= 2\sin x\cos x$

$= 2 \times \frac{\sqrt{5}}{3} \times \frac{2}{3}$

$= \frac{4\sqrt{5}}{9}$ ✓

(2 marks)

13 $\cos^{-1}\left(-\frac{1}{2}\right) = \pi - \frac{\pi}{3}$

$= \frac{2\pi}{3}$ ✓

(1 mark)

14 $f(x) = 2\cos^{-1}x$

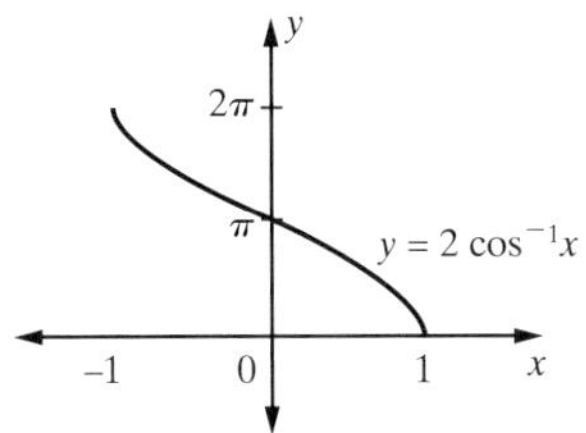

✓

Domain: $-1 \le x \le 1$

Range: $0 \le y \le 2\pi$ ✓

(2 marks)

15 $f(x) = \cos^{-1}\left(\frac{x}{2}\right)$

Now $-1 \le \frac{x}{2} \le 1$

$\therefore -2 \le x \le 2$

Domain: $-2 \le x \le 2$ ✓

(1 mark)

16 **i**

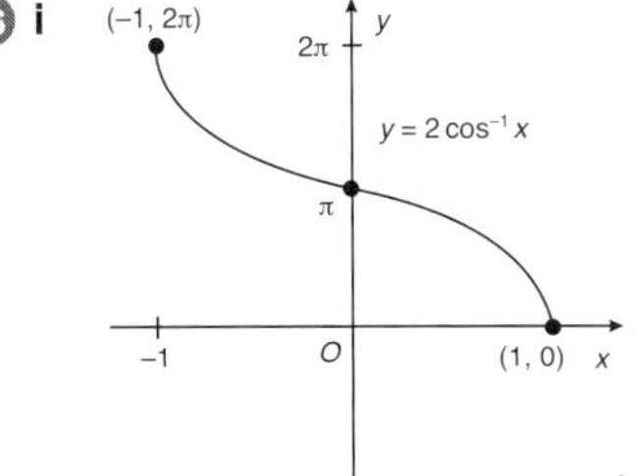

✓✓

(2 marks)

ii Range: $0 \le y \le 2\pi$. ✓

(1 mark)

17 $\sin^{-1}x + \frac{1}{2}\cos^{-1}y = \frac{\pi}{3}$ . . . (1)

$3\sin^{-1}x - \frac{1}{2}\cos^{-1}y = \frac{2\pi}{3}$ . . . (2)

Adding (1) and (2):

$4\sin^{-1}x = \frac{3\pi}{3}$

$4\sin^{-1}x = \pi$

$\sin^{-1}x = \frac{\pi}{4}$

$\therefore x = \frac{1}{\sqrt{2}}$. ✓

Substituting $x = \frac{1}{\sqrt{2}}$ into (1):

$\sin^{-1}\left(\frac{1}{\sqrt{2}}\right) + \frac{1}{2}\cos^{-1}y = \frac{\pi}{3}$

$\frac{\pi}{4} + \frac{1}{2}\cos^{-1}y = \frac{\pi}{3}$ ✓

$\frac{1}{2}\cos^{-1}y = \frac{\pi}{3} - \frac{\pi}{4}$

$\frac{1}{2}\cos^{-1}y = \frac{\pi}{12}$

$\cos^{-1}y = \frac{\pi}{6}$

$\therefore y = \frac{\sqrt{3}}{2}$. ✓

(3 marks)

18 The domain of $y = \cos^{-1}x$ is
$-1 \leqslant x \leqslant 1$

$\therefore$ The domain of $y = \cos^{-1}\left(\frac{x}{4}\right)$ is

$-1 \leqslant \frac{x}{4} \leqslant 1$

i.e. $-4 \leqslant x \leqslant 4$. ✓

The range of $y = \cos^{-1}x$ is
$0 \leqslant \cos^{-1}x \leqslant \pi$

$\therefore$ The range of $y = \cos^{-1}\left(\frac{x}{4}\right)$ is

$0 \leqslant \cos^{-1}\left(\frac{x}{4}\right) \leqslant \pi$. ✓

(2 marks)

19 $y = 3\cos^{-1}2x$

$\therefore \frac{y}{3} = \cos^{-1}2x$

domain: $-1 \leqslant 2x \leqslant 1$

$\therefore -\frac{1}{2} \leqslant x \leqslant \frac{1}{2}$

range: $0 \leqslant \frac{y}{3} \leqslant \pi$

$\therefore 0 \leqslant y \leqslant 3\pi$. ✓

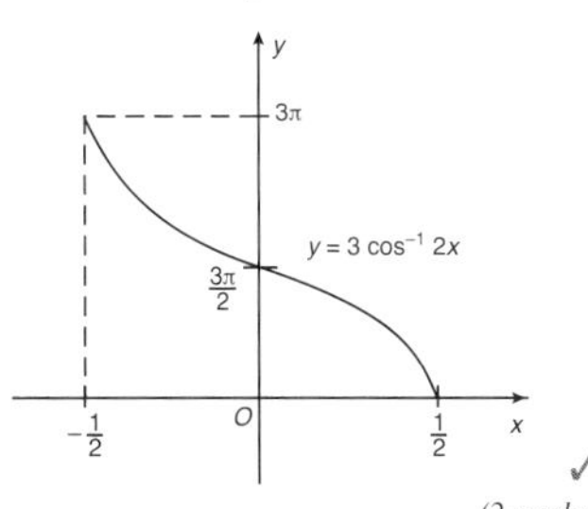

✓

(2 marks)

20 The domain of $f(u) = \sin^{-1}u$ is
$-1 \le u \le 1$ and its range is
$-\frac{\pi}{2} \leqslant f(u) \leqslant \frac{\pi}{2}$.

Hence the domain of
$f(x) = 3\sin^{-1}\left(\frac{x}{2}\right)$

is $-1 \leqslant \frac{x}{2} \leqslant 1$
i.e. $-2 \leqslant x \leqslant 2$ ✓
and the range is
$-\frac{3\pi}{2} \leqslant f(x) \leqslant \frac{3\pi}{2}$.

✓

(2 marks)

1 Which of these is equal to $\cos 3\alpha \sin 2\alpha$?

A $\frac{1}{2}(\sin 5\alpha + \sin \alpha)$ **B** $\frac{1}{2}(\sin 5\alpha - \sin \alpha)$

C $\frac{1}{2}\left(\sin\frac{5\alpha}{2} + \sin\frac{\alpha}{2}\right)$ **D** $\frac{1}{2}\left(\sin\frac{5\alpha}{2} - \sin\frac{\alpha}{2}\right)$

(1 mark)

Bonus question (see page iv) Easy

2 Prove $\sin 255° \cos 15° = \frac{-2-\sqrt{3}}{4}$ *(2 marks)*

Bonus question Easy

3 Use the double angle result to show that $\tan\frac{\pi}{8} = \sqrt{2} - 1$. *(4 marks)*

Bonus question Hard

4 Which expression is equivalent to $\frac{\tan 2x - \tan x}{1 + \tan 2x \tan x}$?

A $\tan x$ **B** $\tan 3x$

C $\frac{\tan 2x - 1}{1 + \tan 2x}$ **D** $\frac{\tan x}{1 + \tan 2x \tan x}$

(1 mark)

(Q3, **2016 HSC**) Easy

5 The angle θ satisfies $\sin\theta = \frac{5}{13}$ and $\frac{\pi}{2} < \theta < \pi$.
What is the value of $\sin 2\theta$?

A $\frac{10}{13}$ **B** $-\frac{10}{13}$

C $\frac{120}{169}$ **D** $-\frac{120}{169}$ *(1 mark)*

(Q8, **2013 HSC**) Medium

6 **i** Prove that $\tan^2\theta = \frac{1 - \cos 2\theta}{1 + \cos 2\theta}$ provided that $\cos 2\theta \neq -1$. *(2 marks)* Easy

ii Hence find the exact value of $\tan\frac{\pi}{8}$. *(1 mark)* Medium

(Q3c, **2009 HSC**)

7 By making the substitution $t = \tan\frac{\theta}{2}$, or otherwise, show that

$\operatorname{cosec}\theta + \cot\theta = \cot\frac{\theta}{2}$. *(2 marks)*

(Q4b, **2005 HSC**) Medium

8 Prove that $\frac{\sin 3\theta}{\sin\theta} - \frac{\cos 3\theta}{\cos\theta} = 2$
(for $\sin\theta \neq 0, \cos\theta \neq 0$). *(3 marks)*

(Q4a, **1996 HSC**) Medium

9 Prove the following identity:

$$\frac{2\tan A}{1 + \tan^2 A} = \sin 2A.$$

(Q2a, **1994 HSC**) Medium

10 Prove the following identity:

$$\frac{\sin A}{\cos A + \sin A} + \frac{\sin A}{\cos A - \sin A} = \tan 2A.$$

(Q2a, **1993 HSC**) Medium

11 Show that $\sin\left(x + \frac{\pi}{2}\right) = \cos x$. *(1 mark)*

Bonus question Easy

12 Given that $\cos x = \frac{\sqrt{5}}{3}$, find the exact value of $\cos 2x$. *(2 marks)*

Bonus question Medium

13 Prove that $\sin 2x \cot x - \cos 2x \equiv 1$ (provided $\sin x \neq 0$). *(2 marks)*

Bonus question Easy

14 Find the values of a and b $(a > b)$ given that $\sin 75° = \frac{1}{4}(\sqrt{a} + \sqrt{b})$. *(2 marks)*

Bonus question Medium

15 **i** Show that $\tan\frac{\theta}{2} = \operatorname{cosec}\theta - \cot\theta$. *(2 marks)* Easy

ii Hence find the exact value of $\tan\frac{\pi}{12}$. *(1 mark)* Easy

Bonus question

16 Given that $0 < x < \frac{\pi}{3}$, show that

$\tan\left(x + \frac{\pi}{6}\right) = \frac{\sqrt{3}\sin x + \cos x}{\sqrt{3}\cos x - \sin x}$. *(2 marks)*

Bonus question Medium

17 **i** Use the t-formulae to show that

$\frac{\sin x - \cos x + 1}{\sin x + \cos x + 1} = \tan\left(\frac{x}{2}\right)$. *(2 marks)* Medium

ii Hence find the exact value of $\tan\frac{5\pi}{12}$. *(3 marks)* Medium

Bonus question

Year 11 Further trigonometric identities—Worked Answers

1 $\cos A \sin B$

$= \frac{1}{2}[\sin(A+B) - (\sin(A-B)]$

$\therefore \cos 3\alpha \sin 2\alpha$

$= \sin(3\alpha + 2\alpha) - (\sin 3\alpha - 2\alpha)$

$= \sin 5\alpha - \sin \alpha$

Answer B

2 $\sin 255° \cos 15°$

$= \frac{1}{2}(\sin(255° + 15°) + \sin(255° - 15°))$

$= \frac{1}{2}(\sin 270° + \sin 240°)$ ✓

$= \frac{1}{2}\left(-1 + \left(-\frac{\sqrt{3}}{2}\right)\right)$

$= -\frac{1}{2}\left(1 + \frac{\sqrt{3}}{2}\right)$

$= \frac{-2-\sqrt{3}}{4}$ ✓ *(2 marks)*

3 Let $A = \frac{\pi}{8}$.

As $\tan 2A = \frac{2\tan A}{1-\tan^2 A}$

$\therefore \tan 2\left(\frac{\pi}{8}\right) = \frac{2\tan\frac{\pi}{8}}{1-\tan^2\frac{\pi}{8}}$

$\tan\frac{\pi}{4} = \frac{2\tan\frac{\pi}{8}}{1-\tan^2\frac{\pi}{8}}$

$1 = \frac{2\tan\frac{\pi}{8}}{1-\tan^2\frac{\pi}{8}}$ ✓

$1 - \tan^2\frac{\pi}{8} = 2\tan\frac{\pi}{8}$

$\tan^2\frac{\pi}{8} + 2\tan\frac{\pi}{8} - 1 = 0$ ✓

$\tan\frac{\pi}{8} = \frac{-2 \pm \sqrt{2^2 - 4(1)(-1)}}{2(1)}$

$= \frac{-2 \pm \sqrt{8}}{2}$

$= \frac{-2 \pm 2\sqrt{2}}{2}$

$= \frac{2(-1 \pm \sqrt{2})}{2}$

$= -1 \pm \sqrt{2}$ ✓

But $\tan\frac{\pi}{8} > 0$, hence

$\tan\frac{\pi}{8} = -1 + \sqrt{2}$. ✓ *(4 marks)*

4 $\frac{\tan 2x - \tan x}{1 + \tan 2x \tan x} = \tan(2x - x)$

$= \tan x$

Answer A

5 $\sin\theta = \frac{5}{13}$

Now $\sin^2\theta + \cos^2\theta = 1$

So $\cos^2\theta = 1 - \sin^2\theta$

$= 1 - \left(\frac{5}{13}\right)^2$

$= \frac{144}{169}$

$\therefore \cos\theta = -\frac{12}{13} \quad \left(\frac{\pi}{2} < \theta < \pi\right)$

$\sin 2\theta = 2\sin\theta\cos\theta$

$= 2 \times \frac{5}{13} \times -\frac{12}{13}$

$= -\frac{120}{169}$

[Or simply use your calculator to find the value of θ and hence 2θ and $\sin 2\theta$.]

Answer D

6 i

$\tan^2\theta = \frac{1-\cos 2\theta}{1+\cos 2\theta} \quad (\cos 2\theta \neq -1)$

$\text{RHS} = \frac{1-(\cos^2\theta - \sin^2\theta)}{1+(\cos^2\theta - \sin^2\theta)}$

$= \frac{1-\cos^2\theta + \sin^2\theta}{1+\cos^2\theta - \sin^2\theta}$ ✓

Dividing every term by $\cos^2\theta$:

$\text{RHS} = \frac{\sec^2\theta - 1 + \tan^2\theta}{\sec^2\theta + 1 - \tan^2\theta}$

$= \frac{(\tan^2\theta + 1) - 1 + \tan^2\theta}{(\tan^2\theta + 1) + 1 - \tan^2\theta}$

$= \frac{2\tan^2\theta}{2}$

$= \tan^2\theta$

$= \text{LHS}.$ ✓ *(2 marks)*

ii Using **i**, let $\theta = \frac{\pi}{8}$:

$\tan^2\left(\frac{\pi}{8}\right) = \frac{1-\cos\left(\frac{2\pi}{8}\right)}{1+\cos\left(\frac{2\pi}{8}\right)}$

$= \frac{1-\frac{1}{\sqrt{2}}}{1+\frac{1}{\sqrt{2}}}$

$= \frac{\sqrt{2}-1}{\sqrt{2}} \times \frac{\sqrt{2}}{\sqrt{2}+1}$

$= \frac{\sqrt{2}-1}{\sqrt{2}+1}$

$= \frac{\sqrt{2}-1}{\sqrt{2}+1} \times \frac{\sqrt{2}-1}{\sqrt{2}-1}$

$= \frac{(\sqrt{2}-1)^2}{(\sqrt{2})^2 - 1^2}$

$= (\sqrt{2}-1)^2$

As $0 < \frac{\pi}{8} < \frac{\pi}{2}$, $\tan\frac{\pi}{8} > 0$.

$\therefore \tan\frac{\pi}{8} = \sqrt{2} - 1$ ✓ *(1 mark)*

7 $t = \tan\frac{\theta}{2}$

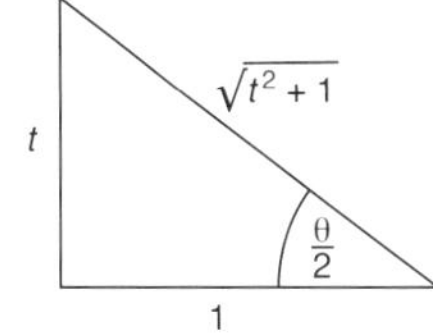

$\therefore \cot\frac{\theta}{2} = \frac{1}{t}$ ✓

$\therefore \sin\frac{\theta}{2} = \frac{t}{\sqrt{t^2+1}} \rightarrow \sin\theta = \frac{2t}{t^2+1}$

$\therefore \cos\frac{\theta}{2} = \frac{1}{\sqrt{t^2+1}} \rightarrow \cos\theta = \frac{1-t^2}{t^2+1}$

$\therefore \tan\theta = \frac{\sin\theta}{\cos\theta} = \frac{2t}{1-t^2}$

Now LHS $= \text{cosec}\,\theta + \cot\theta$

$= \frac{1}{\sin\theta} + \frac{1}{\tan\theta}$

$= \frac{t^2+1}{2t} + \frac{1-t^2}{2t}$

$= \frac{2}{2t}$

$= \frac{1}{t}$

$= \cot\frac{\theta}{2}$

$= \text{RHS}.$ ✓ *(2 marks)*

8 $\frac{\sin 3\theta}{\sin\theta} - \frac{\cos 3\theta}{\cos\theta}$

$= \frac{\sin\theta\cos 2\theta + \cos\theta\sin 2\theta}{\sin\theta} - \frac{\cos\theta\cos 2\theta - \sin\theta\sin 2\theta}{\cos\theta}$ ✓

$= \cos 2\theta + \frac{\cos\theta}{\sin\theta}\sin 2\theta - \cos 2\theta + \frac{\sin\theta}{\cos\theta}\sin 2\theta$

$= \sin 2\theta\left(\frac{\cos\theta}{\sin\theta} + \frac{\sin\theta}{\cos\theta}\right)$ ✓

$= 2\sin\theta\cos\theta\left(\dfrac{\cos^2\theta+\sin^2\theta}{\sin\theta\cos\theta}\right)$

$= 2\sin\theta\cos\theta\left(\dfrac{1}{\sin\theta\cos\theta}\right)$

$= 2.$ ✓

(3 marks)

9 $\dfrac{2\tan A}{1+\tan^2 A} = \dfrac{2\tan A}{\sec^2 A}$

$= \dfrac{2\sin A}{\cos A}\times\cos^2 A$

$= 2\sin A\cos A$

$= \sin 2A.$

(No mark allocation in exam)

10 $\dfrac{\sin A}{\cos A+\sin A}+\dfrac{\sin A}{\cos A-\sin A}$

$= \dfrac{\sin A(\cos A-\sin A)+\sin A(\cos A+\sin A)}{(\cos A+\sin A)(\cos A-\sin A)}$

$= \dfrac{\sin A\cos A-\sin^2 A+\sin A\cos A+\sin^2 A}{\cos^2 A-\sin^2 A}$

$= \dfrac{2\sin A\cos A}{\cos^2 A-\sin^2 A}$

$= \dfrac{\sin 2A}{\cos 2A}$

$= \tan 2A.$

(No mark allocation in exam)

11 $\sin\left(x+\dfrac{\pi}{2}\right)$

$= \sin x\cos\dfrac{\pi}{2}+\cos x\sin\dfrac{\pi}{2}$

$= \sin x\times 0+\cos x\times 1$

$= \cos x$ ✓

(1 mark)

12 $\cos 2x = \cos^2 x-\sin^2 x$

$= \cos^2 x-(1-\cos^2 x)$

$= 2\cos^2 x-1$ ✓

When $\cos x = \dfrac{\sqrt{5}}{3}$,

$\cos 2x = 2\left(\dfrac{\sqrt{5}}{3}\right)^2-1$

$= 2\times\dfrac{5}{9}-1$

$= \dfrac{1}{9}$ ✓

(2 marks)

13 $\sin 2x\cot x-\cos 2x$

$= 2\sin x\cos x\times\dfrac{\cos x}{\sin x}-(\cos^2 x-\sin^2 x)$ ✓

$= 2\cos^2 x-\cos^2 x+\sin^2 x$

$= \cos^2 x+\sin^2 x$

$= 1$ ✓

(2 marks)

14 $\sin 75° = \sin(45°+30°)$

$= \sin 45°\cos 30°+\cos 45°\sin 30°$

$= \dfrac{1}{\sqrt{2}}\times\dfrac{\sqrt{3}}{2}+\dfrac{1}{\sqrt{2}}\times\dfrac{1}{2}$

$= \dfrac{\sqrt{3}+1}{2\sqrt{2}}$ ✓

$= \dfrac{\sqrt{3}+1}{2\sqrt{2}}\times\dfrac{\sqrt{2}}{\sqrt{2}}$

$= \dfrac{\sqrt{6}+\sqrt{2}}{2\times 2}$

$= \dfrac{1}{4}(\sqrt{6}+\sqrt{2})$

$\therefore a = 6$ and $b = 2$. ✓

(2 marks)

15 **i** $\tan\dfrac{\theta}{2} = t$

$\sin\theta = \dfrac{2t}{1+t^2}$ and

$\tan\theta = \dfrac{2t}{1-t^2}$

So $\operatorname{cosec}\theta = \dfrac{1+t^2}{2t}$ and

$\cot\theta = \dfrac{1-t^2}{2t}$

$\operatorname{cosec}\theta-\cot\theta$

$= \dfrac{1+t^2}{2t}-\dfrac{1-t^2}{2t}$ ✓

$= \dfrac{1+t^2-1+t^2}{2t}$

$= \dfrac{2t^2}{2t}$

$= t$

$= \tan\dfrac{\theta}{2}$ ✓

(2 marks)

ii $\tan\dfrac{\pi}{12} = \operatorname{cosec}\dfrac{\pi}{6}-\cot\dfrac{\pi}{6}$

$= \dfrac{1}{\frac{1}{2}}-\dfrac{1}{\frac{1}{\sqrt{3}}}$

$= 2-\sqrt{3}$ ✓

(1 mark)

16 $\tan\left(x+\dfrac{\pi}{6}\right) = \dfrac{\sin\left(x+\frac{\pi}{6}\right)}{\cos\left(x+\frac{\pi}{6}\right)}$

$= \dfrac{\sin x\cos\frac{\pi}{6}+\cos x\sin\frac{\pi}{6}}{\cos x\cos\frac{\pi}{6}-\sin x\sin\frac{\pi}{6}}$ ✓

$= \dfrac{\sin x\times\frac{\sqrt{3}}{2}+\cos x\times\frac{1}{2}}{\cos x\times\frac{\sqrt{3}}{2}-\sin x\times\frac{1}{2}}$

$= \dfrac{\frac{1}{2}(\sqrt{3}\sin x+\cos x)}{\frac{1}{2}(\sqrt{3}\cos x-\sin x)}$

$= \dfrac{\sqrt{3}\sin x+\cos x}{\sqrt{3}\cos x-\sin x}$ ✓

(2 marks)

17 **i** $\tan\dfrac{\theta}{2} = t$

$\sin\theta = \dfrac{2t}{1+t^2}$ and

$\cos\theta = \dfrac{1-t^2}{1+t^2}$

So $\dfrac{\sin x-\cos x+1}{\sin x+\cos x+1}$

$= \dfrac{\frac{2t}{1+t^2}-\frac{1-t^2}{1+t^2}+1}{\frac{2t}{1+t^2}+\frac{1-t^2}{1+t^2}+1}$

$= \dfrac{\frac{2t-1+t^2+1+t^2}{1+t^2}}{\frac{2t+1-t^2+1+t^2}{1+t^2}}$

$= \dfrac{2t+2t^2}{2t+2}$ ✓

$= \dfrac{2t(1+t)}{2(t+1)}$

$= t$

$= \tan\left(\dfrac{x}{2}\right)$ ✓

(2 marks)

ii Let $x = \dfrac{5\pi}{6}$ so $\dfrac{x}{2} = \dfrac{5\pi}{12}$.

Now $\sin\dfrac{5\pi}{6} = \sin\left(\pi-\dfrac{\pi}{6}\right)$

$= \sin\dfrac{\pi}{6} = \dfrac{1}{2}$

and $\cos\dfrac{5\pi}{6} = \cos\left(\pi-\dfrac{\pi}{6}\right)$

$= -\cos\dfrac{\pi}{6} = -\dfrac{\sqrt{3}}{2}$

$\tan\dfrac{5\pi}{12} = \dfrac{\sin\frac{5\pi}{6}-\cos\frac{5\pi}{6}+1}{\sin\frac{5\pi}{6}+\cos\frac{5\pi}{6}+1}$ ✓

$= \dfrac{\frac{1}{2}+\frac{\sqrt{3}}{2}+1}{\frac{1}{2}-\frac{\sqrt{3}}{2}+1}$

$= \dfrac{1+\sqrt{3}+2}{1-\sqrt{3}+2}$

$= \dfrac{3+\sqrt{3}}{3-\sqrt{3}}$ ✓

$= \dfrac{3+\sqrt{3}}{3-\sqrt{3}}\times\dfrac{3+\sqrt{3}}{3+\sqrt{3}}$

$= \dfrac{(3+\sqrt{3})^2}{3^2-(\sqrt{3})^2}$

$= \dfrac{9+6\sqrt{3}+3}{9-3}$

$= \dfrac{12+6\sqrt{3}}{6}$

$= 2+\sqrt{3}$ ✓

(3 marks)

1 The displacement of a particle moving along the x-axis is given by

$$x = \frac{t^3}{3} - 2t^2 + 3t,$$

where x is the displacement from the origin in metres and t is the time in seconds, for $t \geq 0$.

i What is the initial velocity of the particle? *(1 mark)* Easy

ii At which times is the particle stationary? *(2 marks)* Easy

iii Find the position of the particle when the acceleration is zero. *(2 marks)* Easy

(Q12d, **2018 MA HSC**)

2 The displacement of a particle moving along the x-axis is given by

$$x = t - \frac{1}{1+t},$$

where x is the displacement from the origin in metres, t is the time in seconds, and $t \geq 0$.

i Show that the acceleration of the particle is always negative. *(2 marks)* Easy

ii What value does the velocity approach as t increases indefinitely? *(1 mark)* Easy

(Q13c, **2014 MA HSC**)

3 A particle is moving along the x-axis. The displacement of the particle at time t seconds is x metres.

At a certain time, $\dot{x} = -3\ \text{ms}^{-1}$ and $\ddot{x} = 2\ \text{ms}^{-2}$.

Which statement describes the motion of the particle at that time?

A The particle is moving to the right with increasing speed.

B The particle is moving to the left with increasing speed.

C The particle is moving to the right with decreasing speed.

D The particle is moving to the left with decreasing speed. *(1 mark)*

(Q10, **2013 MA HSC**) Medium

4 A particle is moving in a straight line. Its displacement, x metres, from the origin, O, at time t seconds, where $t \geq 0$, is given by

$$x = 1 - \frac{7}{t+4}.$$

i Find the initial displacement of the particle. *(1 mark)* Easy

ii Find the velocity of the particle as it passes through the origin. *(3 marks)* Medium

iii Show that the acceleration of the particle is always negative. *(1 mark)* Easy

iv Sketch the graph of the displacement of the particle as a function of time. *(2 marks)* Easy

(Q8a, **2006 MA HSC**)

5 A tank initially holds 3600 litres of water. The water drains from the bottom of the tank. The tank takes 60 minutes to empty.

A mathematical model predicts that the volume, V litres, of water that will remain in the tank after t minutes is given by

$$V = 3600\left(1 - \frac{t}{60}\right)^2, \text{ where } 0 \leq t \leq 60.$$

i What volume does the model predict will remain after ten minutes? *(1 mark)* Easy

ii At what rate does the model predict that the water will drain from the tank after twenty minutes? *(2 marks)* Easy

iii At what time does the model predict that the water will drain from the tank at its fastest rate? *(2 marks)* Easy

(Q6b, **2005 MA HSC**)

6 A cooler, which is initially full, is drained so that at time t seconds the volume of water V, in litres, is given by

$$V = 25\left(1 - \frac{t}{60}\right)^2 \text{ for } 0 \leq t \leq 60.$$

i How much water was initially in the cooler? *(1 mark)* Easy

ii After how many seconds was the cooler one-quarter full? *(2 marks)* Medium

iii At what rate was the water draining out when the cooler was one-quarter full? *(2 marks)* Medium

(Q7b, **2002 MA HSC**)

7 A particle moves in a straight line so that its displacement, in metres, is given by

$x = \dfrac{t-2}{t+2}$ where t is measured in seconds.

i What is the displacement when $t = 0$? *(1 mark)* Easy

ii Show that $x = 1 - \dfrac{4}{t+2}$.

Hence find expressions for the velocity and the acceleration in terms of t. *(3 marks)* Medium

iii Is the particle ever at rest? Give reasons for your answer. *(1 mark)* Easy

iv What is the limiting velocity of the particle as t increases indefinitely? *(1 mark)* Easy

(Q7c, **2001 MA HSC**)

8 The number N of students logged onto a website at any time over a five-hour period is approximated by the formula

$$N = 175 + 18t^2 - t^4, 0 \leq t \leq 5.$$

i What was the initial number of students logged onto the website?

ii How many students were logged onto the website at the end of the five hours?

iii What was the maximum number of students logged onto the website?

iv When were the students logging onto the website most rapidly?

v Sketch the curve $N = 175 + 18t^2 - t^4$ for $0 \leq t \leq 5$. *(9 marks)*

(Q6b, **2000 MA HSC**) Easy

Year 11 Rates of change with respect to time—Worked Answers

1 **i** $x = \frac{t^3}{3} - 2t^2 + 3t$

$\frac{dx}{dt} = t^2 - 4t + 3$

$\frac{dx}{dt}(0) = 0^2 - 4(0) + 3$

$= 3$

$\therefore$ initial velocity of 3 ms^{-1}. ✓ *(1 mark)*

ii $\frac{dx}{dt} = t^2 - 4t + 3 = 0$ ✓

$(t-1)(t-3) = 0$

$t = 1, 3$

$\therefore$ after 1 second and 3 seconds. ✓ *(2 marks)*

iii $\frac{d^2x}{dt^2} = 2t - 4 = 0$

$2t = 4$

$t = 2$ ✓

Substitute $t = 2$ into x:

$x(2) = \frac{2^3}{3} - 2(2)^2 + 3(2)$

$= \frac{8}{3} - 8 + 6$

$= \frac{2}{3}$

$\therefore$ acceleration zero when particle is $\frac{2}{3}$ m to the right of origin. ✓ *(2 marks)*

2 $x = t - \frac{1}{1+t}$

i $x = t - (1+t)^{-1}$

$\dot{x} = 1 + 1(1+t)^{-2}$

$= 1 + \frac{1}{(1+t)^2}$

$\ddot{x} = -2(1+t)^{-3}$

$= \frac{-2}{(1+t)^3}$ ✓

As $t > 0$, then $(1+t)^3 > 0$,

then $\ddot{x} < 0$ for all t. ✓ *(2 marks)*

ii Use $\dot{x} = 1 + \frac{1}{(1+t)^2}$.

As $t \to \infty, (1+t)^2 \to \infty$,

and $\frac{1}{(1+t)^2} \to 0$.

Hence, $1 + \frac{1}{(1+t)^2} \to 1$.

$\therefore$ velocity approaches 1 m/s. ✓ *(1 mark)*

3 $\dot{x} = -3$ means the particle is moving to the left; as $\ddot{x} = +2$, the particle is slowing down.

$\therefore$ particle is moving to the left with decreasing speed.

Answer D

4 **i** $x = 1 - \frac{7}{t+4}$

When $t = 0$,

$x = 1 - \frac{7}{4}$

$= -\frac{3}{4}$

$\therefore$ The initial displacement of the particle is $\frac{3}{4}$ m to the left of the origin. ✓ *(1 mark)*

ii $v = \frac{dx}{dt}$

$= \frac{d}{dt}\left(1 - \frac{7}{t+4}\right)$

$= \frac{d}{dt}\left(1 - 7(t+4)^{-1}\right)$

$= 7(t+4)^{-2}$

$= \frac{7}{(t+4)^2}$ ✓

Now $x = 1 - \frac{7}{t+4}$

When the particle is at the origin, $x = 0$.

$\therefore 0 = 1 - \frac{7}{t+4}$

$-1 = -\frac{7}{t+4}$

$-1(t+4) = -7$

$-t - 4 = -7$

$\therefore t = 3$ ✓

$\therefore$ The particle is at the origin after 3 seconds.

Now $v = \frac{7}{(t+4)^2}$

When $t = 3$,

$v = \frac{7}{(3+4)^2}$

$= \frac{1}{7}$

$\doteqdot 0.1428\ldots$

$\therefore$ The velocity of the particle as it passes through the origin is $\frac{1}{7}$ ms^{-1}. ✓ *(3 marks)*

iii $a = \frac{dv}{dt}$

$= \frac{d}{dt}\left(\frac{7}{(t+4)^2}\right)$

$= \frac{d}{dt}\left(7(t+4)^{-2}\right)$

$= -14(t+4)^{-3}$

$= -\frac{14}{(t+4)^3}$

Now for $t \geq 0$, $(t+4)^3 > 0$

$\therefore -\frac{14}{(t+4)^3} < 0$

$\therefore$ Acceleration of the particle is always negative. ✓ *(1 mark)*

iv

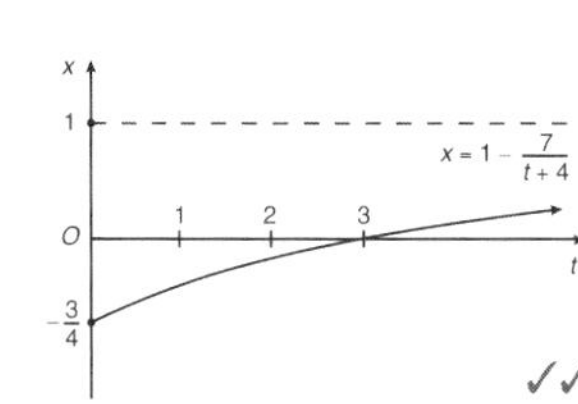

✓✓

N.B.

From **i**, when $t = 0, x = -\frac{3}{4}$

From **ii**, when $t = 3, x = 0$, $v = \frac{1}{7}$

From **ii**, $a < 0$

Also as $t \to \infty, x \to 1$ *(2 marks)*

5 **i** $V = 3600\left(1 - \frac{t}{60}\right)^2$, where $0 \leq t \leq 60$

When $t = 10$,

$V = 3600\left(1 - \frac{10}{60}\right)^2$

$= 2500$ litres ✓ *(1 mark)*

ii $V = 3600\left(1 - \frac{t}{60}\right)^2$

$\frac{dV}{dt} = 2 \times 3600\left(1 - \frac{t}{60}\right) \cdot \frac{d}{dt}\left(1 - \frac{t}{60}\right)$

$= 7200\left(1 - \frac{t}{60}\right) \cdot -\frac{1}{60}$

$= -120\left(1 - \frac{t}{60}\right)$ ✓

When $t = 20$,

$\frac{dV}{dt} = -120\left(1 - \frac{20}{60}\right)$

$= -120 \times \frac{40}{60}$

$= -80$

$\therefore$ After 20 minutes, water is draining at 80 L/min. ✓ *(2 marks)*

iii Now $\frac{dV}{dt} = -120\left(1 - \frac{t}{60}\right)$

$= -120 + 2t$

To find when water will drain fastest, consider $\frac{d^2V}{dt^2}$:

Now $\frac{d^2V}{dt^2} = 2$ ✓

Since, $\frac{d^2V}{dt^2}$ is a constant, check end-points:

When $t = 0$, $\frac{dV}{dt} = -120$

When $t = 60$, $\frac{dV}{dt} = 0$

∴ The water will **drain** at the fastest rate when $t = 0$. ✓ *(2 marks)*

6 **i** $V = 25\left(1 - \frac{t}{60}\right)^2$ for $0 \le t \le 60$

At $t = 0$, $V = 25\left(1 - \frac{0}{60}\right)^2$

$= 25 \times 1$

$= 25$ litres. ✓ *(1 mark)*

ii $V = 25\left(1 - \frac{t}{60}\right)^2$

When $V = \frac{25}{4}$,

$\frac{25}{4} = 25\left(1 - \frac{t}{60}\right)^2$

$\frac{1}{4} = \left(1 - \frac{t}{60}\right)^2$

$\pm\frac{1}{2} = 1 - \frac{t}{60}$

$\frac{t}{60} = 1 \pm \frac{1}{2}$ ✓

$\frac{t}{60} = 1\frac{1}{2}$ or $\frac{t}{60} = \frac{1}{2}$

∴ $t = 90$ or $t = 30$

∴ $t = 30$ seconds

$0 \le t \le 60$. ✓

(2 marks)

iii Now $V = 25\left(1 - \frac{t}{60}\right)^2$

$\frac{dV}{dt} = 50\left(1 - \frac{t}{60}\right) \cdot -\frac{1}{60}$

$= -\frac{5}{6}\left(1 - \frac{t}{60}\right)$ ✓

When $t = 30$,

$\frac{dV}{dt} = -\frac{5}{6}\left(1 - \frac{30}{60}\right)$

$= -\frac{5}{12}$

So, the volume is decreasing by $\frac{5}{12}$ litres per second.

The water was draining out at $\frac{5}{12}$ litres per second when the cooler was one-quarter full. ✓ *(2 marks)*

7 **i** $x = \frac{t-2}{t+2}$

When $t = 0$, $x = \frac{0-2}{0+2}$

$= \frac{-2}{2}$

$= -1$m. ✓

(1 mark)

ii $x = 1 - \frac{4}{t+2}$

RHS $= 1 - \frac{4}{t+2}$

$= \frac{t+2}{t+2} - \frac{4}{t+2}$

$= \frac{t+2-4}{t+2}$

$= \frac{t-2}{t+2}$

$= x$ ✓

$=$ LHS

Now $x = 1 - \frac{4}{t+2}$

$= 1 - 4(t+2)^{-1}$

$v = \frac{dx}{dt} = 0 + 4(t+2)^{-2}$

$= \frac{4}{(t+2)^2}$ ✓

$= 4(t+2)^{-2}$

$a = \frac{dv}{dt} = -8(t+2)^{-3}$

$= \frac{-8}{(t+2)^3}$ ✓

(3 marks)

iii The particle is at rest if

$v = 0$

$\frac{4}{(t+2)^2} = 0$

Since $t \ge 0$, $(t+2)^2 > 0$

∴ $\frac{4}{(t+2)^2} > 0$

∴ Particle is never at rest. ✓

(1 mark)

iv $v = \frac{4}{(t+2)^2}$

As t increases indefinitely, $(t+2)^2$ increases indefinitely

∴ $\frac{4}{(t+2)^2}$ approaches zero.

∴ Limiting velocity is zero. ✓

(1 mark)

8 **i** $N = 175 + 18t^2 - t^4$, $0 \leqslant t \leqslant 5$

When $t = 0$,

$N = 175 + 0 - 0$

$= 175$. ✓

ii When $t = 5$,

$N = 175 + 18(5)^2 - 5^4$

$= 175 + 450 - 625$

$= 0$. ✓

iii $\frac{dN}{dt} = 36t - 4t^3$

$= 4t(9 - t^2)$

$= 4t(3 + t)(3 - t)$

$\frac{d^2N}{dt^2} = 36 - 12t^2$

$= 12(3 - t^2)$

Stationary points occur when

$\frac{dN}{dt} = 0$

∴ $4t(3 + t)(3 - t) = 0$ ✓

∴ $t = 0$ or 3 since $0 \leqslant t \leqslant 5$.

When $t = 0$, $\frac{d^2N}{dt^2} = 36 > 0$.

∴ A minimum turning point at $t = 0$.

When $t = 3$, $\frac{d^2N}{dt^2} = 12(3 - 9)$

$= 12 \times -6$

$= -72 < 0$.

∴ A maximum turning point at $t = 3$. ✓

When $t = 3$,

$N = 175 + 18(3)^2 - 3^4$

$= 175 + 162 - 81$

$= 256$.

∴ Maximum number of students was 256. ✓

iv Students were logging on most rapidly

when $\frac{d^2N}{dt^2} = 0$

∴ $12(3 - t^2) = 0$ ✓

$t^2 = 3$

$t = \pm\sqrt{3}$

∴ $t = \sqrt{3}$

since $0 \leqslant t \leqslant 5$ ✓

v When $t = 0$, $N = 175$.

When $t = 5$, $N = 0$.

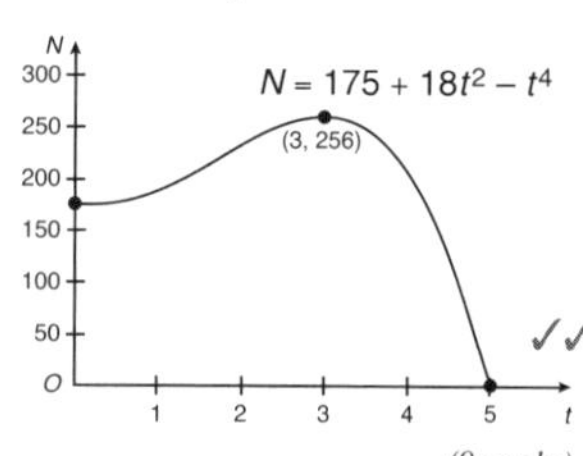

(9 marks)

(Total mark allocation only in exam)

1 A bottle of water, with temperature 5°C, is placed on a table in a room. The temperature of the room remains constant at 25°C. After t minutes, the temperature of the water, in degrees Celsius, is T.

The temperature of the water can be modelled using the differential equation

$$\frac{dT}{dt} = k(T - 25) \qquad \text{(Do NOT prove this.)}$$

where k is the growth constant.

i After 8 minutes, the temperature of the water is 10°C.
By solving the differential equation, find the value of t when the temperature of the water reaches 20°C. Give your answer to the nearest minute. *(3 marks)* **Medium**

ii Sketch the graph of T as a function of t. *(1 mark)* **Hard**

(Q12b, **2021 HSC**)

2 The amount of a drug, A mg, remaining in Emma's bloodstream t hours after taking the drug is given by the differential equation $\frac{dA}{dt} = -0.24A$.
The number of hours needed for the amount A to halve is

A $2\log_e \frac{6}{25}$ **B** $25\log_e \frac{2}{3}$

C $\frac{6}{25}\log_e 2$ **D** $\frac{25}{6}\log_e 2$ *(1 mark)*

Bonus question (see page iv) **Medium**

3 A biologist tested to see how much bacteria was present in a variety of food samples left in the classroom. It is known that after t hours the number of bacteria (N) present in a particular type of food is given by $N(t) = N_0e^{kt}$.

i If initially there were 16 000 bacteria present and after two hours there were 35 000 present, calculate the value of k (correct to 2 decimal places). *(2 marks)* **Easy**

ii How long would it take for the initial number of bacteria to triple in quantity? Give your answer to the nearest minute. *(2 marks)* **Medium**

iii What will be the rate of increase of the bacteria after 8 hours?
Express your answer to the nearest hundred. *(1 mark)* **Easy**

Bonus question

4 A refrigerator has a constant temperature of 3°C. A can of drink with temperature 30°C is placed in the refrigerator.

After being in the refrigerator for 15 minutes, the temperature of the can of drink is 28°C.

The change in the temperature of the can of drink can be modelled by $\frac{dT}{dt} = k(T - 3)$, where T is the temperature of the can of drink, t is the time in minutes after the can is placed in the refrigerator and k is a constant.

i Show that $T = 3 + Ae^{kt}$, where A is a constant, satisfies $\frac{dT}{dt} = k(T - 3)$. *(1 mark)* **Easy**

ii After 60 minutes, at what rate is the temperature of the can of drink changing? *(3 marks)* **Easy**

(Q12d, **2019 HSC**)

5 The diagram shows the number of penguins, $P(t)$, on an island at time t.

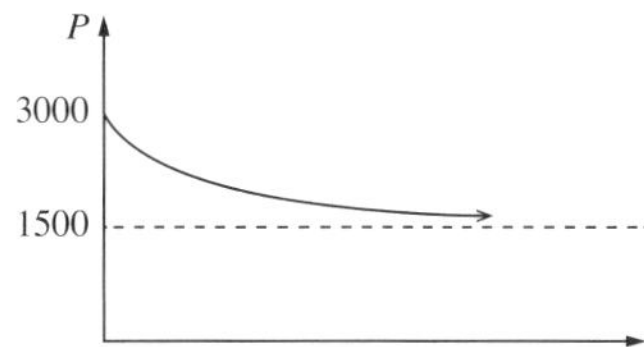

Which equation best represents this graph?

A $P(t) = 1500 + 1500e^{-kt}$

B $P(t) = 3000 - 1500e^{-kt}$

C $P(t) = 3000 + 1500e^{-kt}$

D $P(t) = 4500 - 1500e^{-kt}$ *(1 mark)*

(Q5, **2018 HSC**) **Medium**

6 Carbon-14 is a radioactive substance that decays over time. The amount of carbon-14 present in a kangaroo bone is given by

$$C(t) = Ae^{kt},$$

where A and k are constants, and t is the number of years since the kangaroo died.

i Show that $C(t)$ satisfies $\frac{dC}{dt} = kC$. *(1 mark)* **Easy**

ii After 5730 years, half of the original amount of carbon-14 is present.
Show that the value of k, correct to 2 significant figures, is $-0.000\,12$. *(2 marks)* **Easy**

iii The amount of carbon-14 now present in a kangaroo bone is 90% of the original amount.

Find the number of years since the kangaroo died. Give your answer correct to 2 significant figures. *(2 marks)* Easy

(Q14c, **2017 MA HSC**)

7 In a chemical reaction, a compound X is formed from a compound Y. The mass in grams of X and Y are $x(t)$ and $y(t)$ respectively, where t is the time in seconds after the start of the chemical reaction.

Throughout the reaction the sum of the two masses is 500 g.

At any time t, the rate at which the mass of compound X is increasing is proportional to the mass of compound Y.

At the start of the chemical reaction, $x = 0$ and $\frac{dx}{dt} = 2$.

i Show that $\frac{dx}{dt} = 0.004(500 - x)$. *(3 marks)* Medium

ii Show that $x = 500 - Ae^{-0.004t}$ satisfies the equation in part **i**, and find the value of A. *(2 marks)* Medium

(Q12b, **2016 HSC**)

8 Given that $N = 100 + 80e^{kt}$, which expression is equal to $\frac{dN}{dt}$?

A $k(100 - N)$ **B** $k(180 - N)$
C $k(N - 100)$ **D** $k(N - 180)$ *(1 mark)*

(Q2, **2015 HSC**) Easy

9 The amount of caffeine, C, in the human body decreases according to the equation

$$\frac{dC}{dt} = -0.14C,$$

where C is measured in mg and t is the time in hours.

i Show that $C = Ae^{-0.14t}$ is a solution to $\frac{dC}{dt} = -0.14C$, where A is a constant. *(1 mark)* Easy

When $t = 0$, there are 130 mg of caffeine in Lee's body.

ii Find the value of A. *(1 mark)* Easy

iii What is the amount of caffeine in Lee's body after 7 hours? *(1 mark)* Easy

iv What is the time taken for the amount of caffeine in Lee's body to halve? *(2 marks)* Easy

(Q15a, **2015 MA HSC**)

10 Milk taken out of a refrigerator has a temperature of 2°C. It is placed in a room of constant temperature 23°C. After t minutes the temperature, T°C, of the milk is given by

$$T = A - Be^{-0.03t},$$

where A and B are positive constants.

How long does it take for the milk to reach a temperature of 10°C? *(3 marks)*

(Q12f, **2014 HSC**) Easy

11 A quantity of radioactive material decays according to the equation

$$\frac{dM}{dt} = -km,$$

where M is the mass of the material in kg, t is the time in years and k is a constant.

i Show that $M = Ae^{-kt}$ is a solution to the equation, where A is a constant. *(1 mark)* Easy

ii The time for half of the material to decay is 300 years. If the initial amount of material is 20 kg, find the amount remaining after 1000 years. *(3 marks)* Easy

(Q13b, **2014 MA HSC**)

12 A cup of coffee with an initial temperature of 80°C is placed in a room with a constant temperature of 22°C.

The temperature, T°C, of the coffee after t minutes is given by

$$T = A + Be^{-kt},$$

where A, B and k are positive constants. The temperature of the coffee drops to 60°C after 10 minutes.

How long does it take for the temperature of the coffee to drop to 40°C? Give your answer to the nearest minute. *(3 marks)*

(Q12c, **2013 HSC**) Easy

13 Trout and carp are types of fish. A lake contains a number of trout. At a certain time 10 carp are introduced into the lake and start eating the trout. As a consequence, the number of trout, N, decreases according to

$$N = 375 - e^{0.04t},$$

where t is the time in months after the carp are introduced.

The population of carp, P, increases according to

$$\frac{dP}{dt} = 0.02P.$$

i How many trout were in the lake when the carp were introduced? *(1 mark)* Easy

ii When will the population of trout be zero? *(1 mark)* Easy

iii Sketch the number of trout as a function of time. *(1 mark)* Medium iv When is the rate of increase of carp equal to the rate of decrease of trout? *(3 marks)* Medium

v When is the number of carp equal to the number of trout? *(2 marks)* Hard

(Q16b, **2013 MA HSC**)

14 Professor Smith has a colony of bacteria. Initially there are 1000 bacteria. The number of bacteria, $N(t)$, after t minutes is given by

$$N(t) = 1000e^{kt}.$$

i After 20 minutes there are 2000 bacteria. Show that $k = 0.0347$ correct to four decimal places. *(1 mark)* Easy

ii How many bacteria are there when $t = 120$? *(1 mark)* Easy

iii What is the rate of change of the number of bacteria per minute, when $t = 120$? *(1 mark)* Easy

iv How long does it take for the number of bacteria to increase from 1000 to 100 000? *(2 marks)* Easy

(Q14c, **2012 MA HSC**)

15 To test some forensic science students, an object has been left in a park. At 10 am the temperature of the object is measured to be 30°C. The temperature in the park is a constant 22°C. The object is moved immediately to a room where the temperature is a constant 5°C.

i The temperature of the object in the room can be modelled by the equation

$$T = 5 + 25e^{-kt},$$

where T is the temperature of the object in degrees Celsius, t is the time in hours since the object was placed in the room and k is a constant.

After one hour in the room the temperature of the object is 20°C.

Show that $k = \ln\left(\frac{5}{3}\right)$. *(2 marks)* Easy

ii In a similar manner, the temperature of the object in the park before it was discovered can be modelled by an equation of the form

$$T = A + Be^{-kt},$$

with the same constant $k = \ln\left(\frac{5}{3}\right)$.

Find the time of day when the object had a temperature of 37°C. *(3 marks)* Medium

(Q5b, **2011 HSC**)

16 The mass M of a whale is modelled by

$$M = 36 - 35.5e^{-kt},$$

where M is measured in tonnes, t is the age of the whale in years and k is a positive constant.

i Show that the rate of growth of the mass of the whale is given by the differential equation

$$\frac{dM}{dt} = k(36 - M).$$

(1 mark) Easy

ii When the whale is 10 years old its mass is 20 tonnes.

Find the value of k, correct to three decimal places. *(2 marks)* Medium

iii According to this model, what is the limiting mass of the whale? *(1 mark)* Easy

(Q2b, **2010 HSC**)

17 Assume that the population, P, of cane toads in Australia has been growing at a rate proportional to P. That is, $\frac{dP}{dt} = kP$ where k is a positive constant.

There were 102 cane toads brought to Australia from Hawaii in 1935.

Seventy-five years later, in 2010, it is estimated that there are 200 million cane toads in Australia.

If the population continues to grow at this rate, how many cane toads will there be in Australia in 2035? *(4 marks)*

(Q8a, **2010 MA HSC**) Easy

18 Radium decays at a rate proportional to the amount of radium present. That is, if $Q(t)$ is the amount of radium present at time t, then $Q = Ae^{-kt}$, where k is a positive constant and A is the amount present at $t = 0$. It takes 1600 years for an amount of radium to reduce by half.

i Find the value of k. *(2 marks)* Easy

ii A factory site is contaminated with radium. The amount of radium on the site is currently three times the safe level. How many years will it be before the amount of radium reaches the safe level? *(2 marks)* Easy

(Q6b, **2009 MA HSC**)

19 A turkey is taken from the refrigerator. Its temperature is 5°C when it is placed in an oven preheated to 190°C.

Its temperature, T°C, after t hours in the oven satisfies the equation

$$\frac{dT}{dt} = -k(T - 190).$$

i Show that $T = 190 - 185e^{-kt}$ satisfies both this equation and the initial condition. *(2 marks)* Easy ii The turkey is placed into the oven at 9 am. At 10 am the turkey reaches a temperature of 29°C. The turkey will be cooked when it reaches a temperature of 80°C. At what time (to the nearest minute) will it be cooked? *(3 marks)* Medium

(Q4a, **2008 HSC**)

20 One model for the number of mobile phones in use worldwide is the exponential growth model,

$$N = Ae^{kt},$$

where N is the estimate for the number of mobile phones in use (in millions), and t is the time in years after 1 January 2008.

i It is estimated that at the start of 2009, when $t = 1$, there will be 1600 million mobile phones in use, while at the start of 2010, when $t = 2$, there will be 2600 million. Find A and k. *(3 marks)* Medium

ii According to the model, during which month and year will the number of mobile phones in use first exceed 4000 million? *(2 marks)* Medium

(Q8a, **2007 MA HSC**)

21 Show that $y = 10e^{-0.7t} + 3$ is a solution of $\frac{dy}{dt} = -0.7(y - 3)$. *(2 marks)*

(Q5a, **2006 HSC**) Easy

22 A salad, which is initially at a temperature of 25°C, is placed in a refrigerator that has a constant temperature of 3°C. The cooling rate of the salad is proportional to the difference between the temperature of the refrigerator and the temperature, T, of the salad. That is, T satisfies the equation

$$\frac{dT}{dt} = -k(T - 3),$$

where t is the number of minutes after the salad is placed in the refrigerator.

i Show that $T = 3 + Ae^{-kt}$ satisfies this equation. *(1 mark)* Easy

ii The temperature of the salad is 11°C after 10 minutes. Find the temperature of the salad after 15 minutes. *(3 marks)* Medium

(Q2d, **2005 HSC**)

23 At the beginning of 1991 Australia's population was 17 million. At the beginning of 2004 the population was 20 million.

Assume that the population P is increasing exponentially and satisfies an equation of the form $P = Ae^{kt}$, where A and k are constants, and t is measured in years from the beginning of 1991.

i Show that $P = Ae^{kt}$ satisfies $\frac{dP}{dt} = kP$. *(1 mark)* Easy

ii What is the value of A? *(1 mark)* Easy

iii Find the value of k. *(2 marks)* Easy

iv Predict the year during which Australia's population will reach 30 million. *(2 marks)* Easy

(Q7b, **2004 MA HSC**)

24 Dr Kool wishes to find the temperature of a very hot substance using his thermometer, which only measures up to 100°C. Dr Kool takes a sample of the substance and places it in a room with a surrounding air temperature of 20°C, and allows it to cool.

After 6 minutes the temperature of the substance is 80°C, and after a further 2 minutes it is 50°C. If $T(t)$ is the temperature of the substance after t minutes, then Newton's law of cooling states that T satisfies the equation

$$\frac{dT}{dt} = k(T - A),$$

where k is a constant and A is the surrounding air temperature.

i Verify that $T = A + Be^{kt}$ satisfies the above equation. *(1 mark)* Easy

ii Show that $k = -\frac{\log_e 2}{2}$, and find the value of B. *(3 marks)* Medium

iii Hence find the initial temperature of the substance. *(1 mark)* Easy

(Q5c, **2003 HSC**)

25 A farmer accidentally spread a dangerous chemical on a paddock. The concentration of the chemical in the soil was initially measured at 5 kg/ha. One year later the concentration was found to be 2.8 kg/ha.

It is known that the concentration, C, is given by

$$C = C_0e^{-kt},$$

where C_0 and k are constants, and t is measured in years.

i Evaluate C_0 and k. *(3 marks)* Easy

ii It is safe to use the paddock when the concentration is below 0.2 kg/ha. How long must the farmer wait after the accident before the paddock can be used? Give your answer in years, correct to one decimal place. *(2 marks)* Easy

(Q6c, **2003 MA HSC**)

26 A household iron is cooling in a room of constant temperature 22°C. At time t minutes its temperature T decreases according to the equation

$\frac{dT}{dt} = -k(T - 22)$ where k is a positive constant. The initial temperature of the iron is 80°C and it cools to 60°C after 10 minutes.

i Verify that $T = 22 + Ae^{-kt}$ is a solution of this equation, where A is a constant. *(1 mark)* Easy

ii Find the values of A and k. *(2 marks)* Easy

iii How long will it take for the temperature of the iron to cool to 30°C? Give your answer to the nearest minute. *(2 marks)* Easy

(Q3c, **2002 HSC**)

27 A drug is used to control a medical condition. It is known that the quantity Q of drug remaining in the body after t hours satisfies an equation of the form

$$Q = Q_0 e^{-kt}$$

where Q_0 and k are constants.

The initial dose is 6 milligrams and after 15 hours the amount remaining in the body is half the initial dose.

i Find the values of Q_0 and k. *(3 marks)* Easy

ii When will one-eighth of the initial dose remain? *(2 marks)* Easy

(Q8a, **2002 MA HSC**)

28 In November 1923, 18 koalas were introduced on Kangaroo Island. By November 1993, the number of koalas had increased to 5000.

Assume that the number N of koalas is increasing exponentially and satisfies an equation of the form $N = N_0 e^{kt}$, where N_0 and k are constants and t is measured in years from November 1923.

Find the values of N_0 and k, and predict the number of koalas that will be present on Kangaroo Island in November 2001. *(5 marks)*

(Q8a, **2001 MA HSC**) Medium

29 The population of a certain insect is growing exponentially according to $N = 200e^{kt}$, where t is the time in weeks after the insects are first counted.

At the end of three weeks the insect population has doubled.

i Calculate the value of the constant k. Easy

ii How many insects will there be after 12 weeks? Easy

iii At what rate is the population increasing after three weeks? Easy

(5 marks) (Q5c, **2000 MA HSC**)

30 The mass M kg of a radioactive substance present after t years is given by $M = 10e^{-kt}$, where k is a positive constant. After 100 years the mass has reduced to 5 kg.

i What was the initial mass? Easy

ii Find the value of k. Easy

iii What amount of the radioactive substance would remain after a period of 1000 years? Easy

iv How long would it take for the initial mass to reduce to 8 kg? Easy

(7 marks) (Q6a, **1999 MA HSC**)

31 The population P of a city is growing at a rate that is proportional to the current population. The population at time t years is given by

$$P = Ae^{kt},$$

where A and k are constants.

The population at time $t = 0$ was 1 000 000 and at time $t = 2$ was 1 072 500.

i Find the value of A. Easy

ii Find the value of k. Easy

iii At what time will the population reach 2 000 000? Easy

(4 marks) (Q5b, **1998 MA HSC**)

32 A ball is dropped into a long vertical tube filled with honey. The rate at which the ball decelerates is proportional to its velocity. Thus

$$\frac{dv}{dt} = -kv,$$

where v is the velocity in centimetres per second, t is the time in seconds, and k is a constant.

When the ball first enters the honey, at $t = 0$, $v = 100$. When $t = 0{\cdot}25$, $v = 85$.

i Show that $v = Ce^{-kt}$ satisfies the equation $\frac{dv}{dt} = -kv$. Easy

ii Find the value of the constant C. Easy

iii Find the value of the constant k. Easy

iv Find the velocity when $t = 2$. Easy

(5 marks) (Q7c, **1997 MA HSC**)

33 A cup of hot coffee at temperature T°C loses heat when placed in a cooler environment. It cools according to the law

$$\frac{dT}{dt} = k(T - T_0),$$

where t is time elapsed in minutes, and T_0 is the temperature of the environment in degrees Celsius.

i A cup of coffee at 100°C is placed in an environment at –20°C for 3 minutes, and cools to 70°C. Find k. Medium ii The same cup of coffee, at 70°C, is then placed in an environment at 20°C. Assuming k remains the same, find the temperature of the coffee after a further 15 minutes. Easy

(5 marks) (Q5a, **1996 HSC**)

34 Coal is extracted from a mine at a rate that is proportional to the amount of coal remaining in the mine. Hence the amount R remaining after t years is given by

$$R = R_0e^{-kt},$$

where k is a constant and R_0 is the initial amount of coal.

After 20 years, 50% of the initial amount of coal remains.

i Find the value of k. Easy

ii How many *more* years will elapse before only 30% of the original amount remains? Easy

(4 marks) (Q6c, **1995 MA HSC**)

35 Let T be the temperature inside a room at time t and let A be the constant outside air temperature. Newton's law of cooling states that the rate of change of the temperature T is proportional to $(T - A)$.

i Show that $T = A + Ce^{kt}$ (where C and k are constants) satisfies Newton's law of cooling. Easy

ii The outside air temperature is 5°C and a heating system breakdown causes the inside room temperature to drop from 20°C to 17°C in half an hour. After how many hours is the inside room temperature equal to 10°C? Medium

(Q6a, **1993 HSC**)

Year 11 Exponential growth and decay—Worked Answers

1 **i**
$$\frac{dT}{dt}=k(T-25)$$
$$\frac{1}{T-25}dT=k\,dt$$
$$\int\frac{1}{T-25}dT=\int k\,dt$$
$$\ln|T-25|=kt+C \quad ✓$$

When $t=0$, $T=5$.
$$\ln|5-25|=k(0)+C$$
$$\ln|-20|=C$$

$$\ln|T-25|=kt+\ln|-20|$$
$$kt=\ln|T-25|-\ln|-20|$$
$$kt=\ln\left|\frac{T-25}{-20}\right|$$

When $t=8$, $T=10$
$$8k=\ln\left|\frac{10-25}{-20}\right|$$
$$8k=\ln\left(\frac{3}{4}\right)$$
$$k=\frac{\ln\left(\frac{3}{4}\right)}{8}$$
$$k=-0.03596... \quad ✓$$

$$\frac{1}{8}\ln\left(\frac{3}{4}\right)t=\ln\left|\frac{T-25}{-20}\right|$$
When $T=20$,
$$t=\frac{8\ln\left|\frac{20-25}{-20}\right|}{\ln\left|\frac{3}{4}\right|}$$
$$t=38.55$$
$\therefore$ The bottle of water would reach 20 °C after 39 minutes.

✓ *(3 marks)*

ii

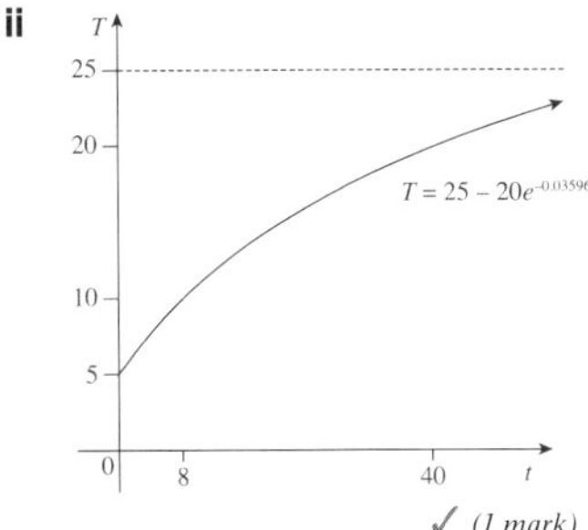

✓ *(1 mark)*

2 As $\frac{dA}{dt}=-0.24A$, then $A=A_0e^{-0.24t}$.

Let $\frac{A}{A_0}=\frac{1}{2}$.
$$\therefore e^{-0.24t}=\frac{1}{2}$$
$$-0.24t=-\log_e 2$$
$$t=\frac{\log_e 2}{0.24}$$
$$=\frac{25}{6}\log_e 2$$

Answer D

3 **i**
$$N=N_0ekt$$
Let $t=0$, $N=16000$:
$$16000=N_0e^{k(0)}$$
$$N_0=16000$$
$$\therefore N=16000e^{kt}$$
Let $t=2$, $N=35000$:
$$35000=16000e^{k(2)} \quad ✓$$
$$e^{2k}=2.1875$$
$$\log_e e^{2k}=\log_e 2.1875$$
$$2k=\log_e 2.1875$$
$$k=\frac{\log_e 2.1875}{2}$$
$$=0.391\,379\,669...$$
$$=0.39\text{ (2 dec. pl.)} \quad ✓$$
(2 marks)

ii
$$N=16000e^{kt}$$
Let $N=48000$:
$$48000=16000e^{kt}$$
$$e^{kt}=3$$
$$\log_e e^{kt}=\log_e 3$$
$$kt=\log_e 3$$
$$t=\frac{\log_e 3}{k} \quad ✓$$
$$=2.807\,024\,416....$$
$$=2\text{ h }48\text{ min} \quad ✓$$
(nearest minute)
(2 marks)

iii Let $t=8$:
$$N=16000e^{k(8)}$$
$$=366\,364\text{ (nearest whole)}$$
$$\frac{dN}{dt}=kN$$
$$=k(366\,364)$$
$$=143\,387.2355...$$
$$=143\,400$$
(nearest hundred)

$\therefore$ increasing at 143 400 per hour. ✓
(1 mark)

4 **i**
$$T=3+Ae^{kt}$$
$$\frac{dT}{dt}=Ae^{kt}\times k$$
But $Ae^{kt}=T-3$
$$\text{So }\frac{dT}{dt}=(T-3)\times k$$
$$=k(T-3) \quad ✓$$
(1 mark)

ii
$$T=3+Ae^{kt}$$
When $t=0$, $T=30$
$$30=3+Ae^{0}$$
$$3+A=30$$
$$A=27$$
So $T=3+27e^{kt}$ ✓

When $t=15$, $T=28$
$$28=3+27e^{k\times 15}$$
$$27e^{15k}=25$$
$$e^{15k}=\frac{25}{27}$$
$$15k=\ln\frac{25}{27}$$
$$k=\frac{1}{15}\ln\frac{25}{27} \quad ✓$$
When $t=60$,
$$T=3+27e^{\left(\frac{1}{15}\ln\frac{25}{27}\right)\times 60}$$
$$=22.845\,8060...$$
$$\frac{dT}{dt}=\frac{1}{15}\ln\frac{25}{27}\times(22.845\,8060...-3)$$
$$=-0.101\,823\,592...$$
So after 60 minutes the temperature of the can is decreasing at about 0.1° C per minute. ✓
(3 marks)

5 $P(0)=3000$

Now $1500e^{0}=1500$

So possible equations are A and D.

As $t\to\infty$, $1500e^{-kt}\to 0$

So the equation that best represents the graph is $P(t)=1500+1500e^{-kt}$.

Answer A

6 **i**
$$C=Ae^{kt}$$
$$\frac{dC}{dt}=k.Ae^{kt}$$
$$=kC$$
$$\therefore C=Ae^{kt}\text{ satisfies }\frac{dC}{dt}=kC \quad ✓$$
(1 mark)

ii Let $t=5730$, $A=1$, $C=0.5$:
$$0.5=1e^{k(5730)}$$
$$e^{5730k}=0.5$$
$$\ln e^{5730k}=\ln 0.5$$
$$5730k\ln e=\ln 0.5$$
$$k=\frac{\ln 0.5}{5730} \quad ✓$$
$$=-0.000\,120\,968...$$
$$=-0.000\,12\text{ (2 sig. figs)} \quad ✓$$
(2 marks)

iii Let $A=1$, $C=0.9$:
$$0.9=1e^{kt}$$
$$e^{kt}=0.9$$
$$\ln e^{kt}=\ln 0.9$$
$$kt\ln e=\ln 0.9$$
$$t=\frac{\ln 0.9}{k} \quad ✓$$
$$=870.977\,7254...$$
$$=870\text{ (2 sig. figs)}$$
$\therefore$ the kangaroo died 870 years ago.

(or 880 years ago if using $k=-0.000\,12$) ✓
(2 marks)

7 **i**
$$x+y=500$$
$$y=500-x$$

Now $\frac{dx}{dt} = ky$ (k is a constant) ✓

$= k(500 - x)$

When $t = 0, x = 0$ and $\frac{dx}{dt} = 2$

So $2 = k(500 - 0)$

$k = \frac{2}{500}$ ✓

$= 0.004$

$\therefore \frac{dx}{dt} = 0.004(500 - x)$ ✓

(3 marks)

ii $x = 500 - Ae^{-0.004t}$

$\frac{dx}{dt} = 0.004Ae^{-0.004t}$

But $x = 500 - Ae^{-0.004t}$

So $Ae^{-0.004t} = 500 - x$

$\therefore \frac{dx}{dt} = 0.004(500 - x)$

$\therefore x = 500 - Ae^{-0.004t}$ satisfies ✓ the equation in **i**

$x = 500 - Ae^{-0.004t}$

When $t = 0, x = 0$

$0 = 500 - A \times 1$

$A = 500$ ✓

(2 marks)

8 $N = 100 + 80e^{kt}$

$\frac{dN}{dt} = k \times 80e^{kt}$

$= k(N - 100)$

Answer C

9 i $C = Ae^{-0.14t}$

$\frac{dC}{dt} = -0.14 \times Ae^{-0.14t}$

$= -0.14C$

$\therefore C = Ae^{-0.14t}$ is a solution. ✓

(1 mark)

ii Substitute $C = 130, t = 0$ in $C = Ae^{-0.14t}$

$130 = Ae^{-0.14(0)}$

$130 = A$

$\therefore A = 130$ ✓

(1 mark)

iii Substitute $t = 7$ in $C = 130e^{-0.14t}$

$C = 130e^{-0.14(7)}$

$= 48.790\,442\,85...$

$= 48.79$ (2 dec. pl.)

$\therefore$ there is 48.79 mg of caffeine in Lee's body. ✓

(1 mark)

iv Substitute $C = 65$ in $C = 130e^{-0.14t}$

$65 = 130e^{-0.14t}$

$e^{-0.14t} = 0.5$

$\log_e e^{-0.14t} = \log_e 0.5$

$-0.14t = \log_e 0.5$

$t = \frac{\log_e 0.5}{-0.14}$ ✓

$= 4.951\,051\,29...$

$= 4.95$ (2 dec. pl.)

$\therefore$ it will take 4.95 hours. ✓

(2 marks)

10 $T = A - Be^{-0.03t}$

Room has constant temperature of 23 °C.

So $A = 23$

$T = 23 - Be^{-0.03t}$

When $t = 0, T = 2$

$2 = 23 - Be^{0}$

$B = 23 - 2$

$= 21$

$\therefore T = 23 - 21e^{-0.03t}$ ✓

When $T = 10$,

$10 = 23 - 21e^{-0.03t}$

$21e^{-0.03t} = 13$

$e^{-0.03t} = \frac{13}{21}$

$-0.03t = \ln\left(\frac{13}{21}\right)$

$t = \frac{\ln\left(\frac{13}{21}\right)}{-0.03}$ ✓

$= 15.985\,769...$

$= 16.0$ (1 d.p.)

It takes 16 minutes for the milk to reach 10 °C. ✓ *(3 marks)*

11 i $M = Ae^{-kt}$

$\frac{dM}{dt} = -k.Ae^{-kt}$

$= -kM$

$\therefore M = Ae^{-kt}$ is a solution. ✓

(1 mark)

ii Let $A = 1$, then $M = 0.5$ and $t = 300$:

$0.5 = 1e^{-k(300)}$

$e^{-300k} = 0.5$

$-300k = \log_e 0.5$

$k = -\frac{\log_e 0.5}{300}$ ✓

Let $A = 20, t = 1000$:

$M = 20e^{-k(1000)}$ ✓

$= 1.984\,251\,315...$

$= 1.98$ (2 dec. pl.)

$\therefore$ 1.98 kg remain. ✓

(3 marks)

12 $T = A + Be^{-kt}$

$A = 22$

When $t = 0, T = 80$

$80 = 22 + Be^{-k \times 0}$

$58 = B$

When $t = 10, T = 60$

$60 = 22 + 58e^{-k \times 10}$ ✓

$38 = 58e^{-10k}$

$e^{-10k} = \frac{38}{58}$

$-10k = \ln\left(\frac{38}{58}\right)$

$k = -\frac{1}{10}\ln\left(\frac{38}{58}\right)$ ✓

$[= 0.042\,2856...]$

When $T = 40$,

$40 = 22 + 58e^{-kt}$

$18 = 58e^{-kt}$

$e^{-kt} = \frac{18}{58}$

$-kt = \ln\left(\frac{18}{58}\right)$

$t = -\frac{1}{k}\ln\left(\frac{18}{58}\right)$

$= 27.670\,623...$

$= 28$ [nearest unit]

To the nearest minute, it will take 28 minutes for the coffee to cool to 40 °C. ✓ *(3 marks)*

13 i $N = 375 - e^{0.04t}$

Let $t = 0$:

$N = 375 - e^{0.04(0)}$

$= 375 - 1$

$= 374$

$\therefore$ 374 trout ✓

(1 mark)

ii Let $N = 0$:

$0 = 375 - e^{0.04t}$

$e^{0.04t} = 375$

$\log_e e^{0.04t} = \log_e 375$

$0.04\,t = \log_e 375$

$t = \frac{\log_e 375}{0.04}$

$= 148.173\,1506\,...$

$= 148$ (nearest whole)

$\therefore$ after about 148 months ✓

(1 mark)

iii

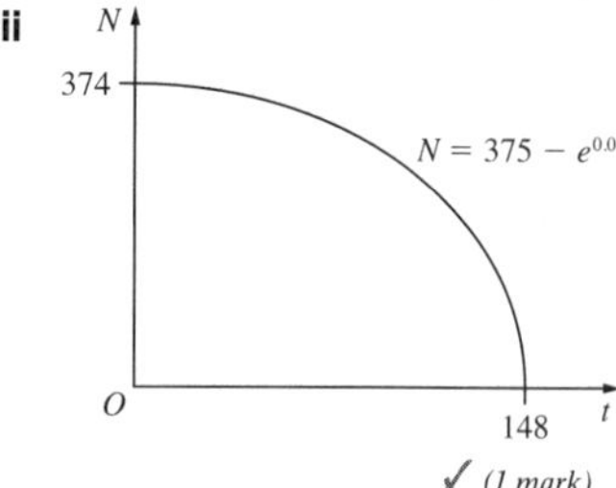

✓ *(1 mark)*

iv $N = 375 - e^{0.04t}$

$\frac{dN}{dt} = -0.04e^{0.04t}$

If $\frac{dP}{dt} = 0.02P$, then

$P = Ae^{0.02t}$

Now, when $t = 0, P = 10$:

$\therefore 10 = Ae^{0.02(0)}$ $\therefore A = 10$

$\therefore P = 10e^{0.02t}$ ✓

Now, $\frac{dP}{dt} = 0.2e^{0.02t}$

If rates equal, then:

$0.04e^{0.04t} = 0.2e^{0.02t}$

$e^{0.04t} = 5e^{0.02t}$

$e^{0.04t} \div e^{0.02t} = 5$

$e^{0.02t} = 5$ ✓

$\log_e e^{0.02t} = \log_e 5$

$0.02t = \log_e 5$

$t = \frac{\log_e 5}{0.02}$

$= 80.471\,895\,62\ldots$

$= 80$ (nearest whole)

$\therefore$ after about 80 months ✓ *(3 marks)*

v $N = P$

$375 - e^{0.04t} = 10e^{0.02t}$

$e^{0.04t} + 10e^{0.02t} - 375 = 0$

Let $m = e^{0.02t}$

$m^2 + 10m - 375 = 0$

$(m + 25)(m - 15) = 0$

$m = -25, 15$

$e^{0.02t} = 15$ (as $e^{0.02t} > 0$) ✓

$\log_e e^{0.02t} = \log_e 15$

$0.02t = \log_e 15$

$t = \frac{\log_e 15}{0.02}$

$= 135.402\,5101\ldots$

$= 135$ (nearest whole)

$\therefore$ after about 135 months ✓ *(2 marks)*

14 i $N = 1000e^{kt}, t = 20, N = 2000$

$2000 = 1000e^{20k}$

$e^{20k} = 2$

$\log_e e^{20k} = \log_e 2$

$20k = \log_e 2$

$k = \frac{\log_e 2}{20}$

$= 0.034\,657\,359\ldots$

$= 0.0347$ (to 4 dec. pl.)

✓ *(1 mark)*

ii $N = 1000e^{kt}, t = 120,$

$= 1000e^{k(120)}$

$= 64\,000$

$\therefore$ 64 000 bacteria ✓

[or, using $k = 0.0347$, $N = 64\,328$] *(1 mark)*

iii $N = 1000e^{kt}$

$\frac{dN}{dt} = k.1000e^{kt}$

$= k.64\,000$

$= 64\,000k$

$= 2218.070\,978\ldots$

$= 2218$ (nearest whole)

$\therefore$ 2218 bacteria per minute ✓

[or, using $k = 0.0347$, 2232 per minute] *(1 mark)*

iv $N = 1000e^{kt}, N = 100\,000$

$100\,000 = 1000e^{kt}$

$e^{kt} = 100$

$\log_e e^{kt} = \log_e 100$

$kt = \log_e 100$

$t = \frac{\log_e 100}{k}$ ✓

$= 132.877\,1238\ldots$

$= 133$ (nearest whole)

$\therefore$ 2h 13 min ✓ *(2 marks)*

15 i $T = 5 + 25e^{-kt}$

When $t = 1, T = 20$

$20 = 5 + 25e^{-k \times 1}$

$15 = 25e^{-k}$

$e^{-k} = \frac{3}{5}$ ✓

$-k = \ln\left(\frac{3}{5}\right)$

$k = -\ln\left(\frac{3}{5}\right)$

$= \ln\left(\frac{3}{5}\right)^{-1}$

$= \ln\left(\frac{5}{3}\right)$ ✓

(2 marks)

ii $T = A + Be^{-kt}$

The temperature in the park is a constant 22 °C.

So $A = 22$

$T = 22 + Be^{-kt}$

When $t = 0, T = 30$

$30 = 22 + Be^{-k \times 0}$

$8 = B$

$\therefore T = 22 + 8e^{-kt}$ ✓

If $T = 37$,

$37 = 22 + 8e^{-kt}$

$15 = 8e^{-kt}$

$\frac{15}{8} = e^{-kt}$

$-kt = \ln\left(\frac{15}{8}\right)$

$t = \ln\left(\frac{15}{8}\right) \div \ln\left(\frac{3}{5}\right)$ ✓

$= -1.230\,573\,86\ldots$ h

$= -1$ h 14 min [nearest minute]

So the object has a temperature of 37 °C one hour and 14 minutes before 10 am i.e. at 8:46 am. ✓ *(3 marks)*

16 i $M = 36 - 35.5e^{-kt}$

$\therefore 35.5e^{-kt} = 36 - M$

$M = 36 - 35.5e^{-kt}$

$\frac{dM}{dt} = k(35.5e^{-kt})$

$= k(36 - M)$ ✓

(1 mark)

ii $M = 36 - 35.5e^{-kt}$

When $t = 10, M = 20$

$\therefore 20 = 36 - 35.5e^{-k(10)}$

$-16 = -35.5e^{-10k}$

$e^{-10k} = \frac{16}{35.5}$ ✓

$-10k = \ln\left(\frac{16}{35.5}\right)$

$k = -\frac{1}{10}\ln\left(\frac{16}{35.5}\right)$

$= 0.079694397\ldots.$

$= 0.080$

[3 decimal places] ✓ *(2 marks)*

iii $M = 36 - 35.5e^{-kt}$

As $t \to \infty, e^{-kt} \to 0$

$\therefore M \to 36$

The limiting mass of the whale is 36 tonnes. ✓ *(1 mark)*

17 $\frac{dP}{dt} = kP$

Let $P = P_0e^{kt}$,

where $P_0 =$ initial population

When $t = 0, P_0 = 102$

$\therefore P = 102e^{kt}$, ✓

From 1935 to 2010 is 75 years.

Let $P = 200\,000\,000, t = 75$:

$P = 102e^{kt}$

$200\,000\,000 = 102e^{75k}$

$e^{75k} = \frac{200\,000\,000}{102}$ ✓

$75k = \log_e\left[\frac{200\,000\,000}{102}\right]$

$k = \frac{\log_e\left[\frac{200\,000\,000}{102}\right]}{75}$ ✓

$= 0.193184734\ldots$

From 2010 to 2035 is another 25 years:

$\therefore$ Let $t = 100$:

$P = 102e^{100k}$

$= 2.503\ldots \times 10^{10}$

$\therefore$ The population will be approximately 2.5×10^{10} cane toads. ✓ *(4 marks)*

18 **i** $Q = Ae^{-kt}$

When $t = 1600$, $Q = \frac{1}{2}A$

$\therefore \frac{1}{2}A = Ae^{-1600k}$

$\frac{1}{2} = e^{-1600k}$

$\ln\frac{1}{2} = -1600k$ ✓

$\therefore k = \dfrac{\ln\frac{1}{2}}{-1600}$

$= 0.0004\ldots$ ✓

(2 marks)

ii $Q = Ae^{-kt}$

Let $A = 3Q$,

$\therefore Q = 3Qe^{-kt}$

$1 = 3e^{-kt}$

$\frac{1}{3} = e^{-kt}$

$\ln\frac{1}{3} = -kt$

$\therefore t = \ln\frac{1}{3} \div -k$ ✓

where $k = -\dfrac{\ln\frac{1}{2}}{-1600} = 0.0004\ldots$

$\therefore t = 2535.94\ldots$

It will take 2536 years for the amount of radium to reach a safe level. ✓

(2 marks)

19 **i** $T = 190 - 185e^{-kt}$

$\frac{dT}{dt} = 185ke^{-kt}$

$= -k(-185e^{-kt})$

$= -k(T - 190)$

which satisfies $T = 190 - 185e^{-kt}$. ✓

When $t = 0$, $T = 190 - 185$

$= 5$

which is the initial condition. ✓

(2 marks)

ii When $t = 1$, $T = 29$

$\therefore 29 = 190 - 185e^{-k}$

$-161 = -185e^{-k}$

$e^{-k} = \frac{161}{185}$

$-k = \ln\left(\frac{161}{185}\right)$

$\therefore k = \ln\left(\frac{185}{161}\right)$ ✓

$= 0.1389\ldots$

When $T = 80$,

$80 = 190 - 185e^{-kt}$

$185e^{-kt} = 110$

$e^{-kt} = \frac{110}{185}$

$-kt = \ln\left(\frac{110}{185}\right)$

$t = \ln\left(\frac{110}{185}\right) \div -\ln\left(\frac{185}{161}\right)$ ✓

(on substitution for k)

$= 3.7414\ldots$

$\doteqdot$ 3 hours 44 minutes

$\therefore$ The turkey will be cooked at 12.44 p.m. ✓

(3 marks)

20 **i** $N = Ae^{kt}$

At $t = 1$, $N = 1600$

$\therefore 1600 = Ae^{k}$ ...(1)

At $t = 2$, $N = 2600$

$\therefore 2600 = Ae^{2k}$ ...(2) ✓

Solving simultaneously:

(2) ÷ (1) $\dfrac{Ae^{2k}}{Ae^{k}} = \dfrac{2600}{1600}$

$e^{k} = \left(\frac{13}{8}\right)$

$\ln(e^{k}) = \ln\left(\frac{13}{8}\right)$

$\therefore k = \ln\left(\frac{13}{8}\right)$ ✓

$\doteqdot 0.4855\ldots$

$= 0.49$

to 2 decimal places.

If $k = \ln\left(\frac{13}{8}\right)$, $Ae^{k} = 1600$

$\therefore Ae^{\ln\left(\frac{13}{8}\right)} = 1600$

$\therefore A \times \left(\frac{13}{8}\right) = 1600$

$\therefore A = \frac{12\,800}{13}$. ✓

(3 marks)

ii $N = \frac{12\,800}{13}e^{kt}$

$\left(\text{where } k = \ln\frac{13}{8}\right)$

If $N = 4000$,

$4000 = \frac{12\,800}{13}e^{kt}$

$\dfrac{4000}{\frac{28\,000}{13}} = e^{kt}$

$\frac{65}{16} = e^{kt}$

$\ln\left(\frac{65}{16}\right) = \ln e^{kt}$

$\ln\left(\frac{65}{16}\right) = kt$

$t = \dfrac{\ln\left(\frac{65}{16}\right)}{k}$ ✓

$= \dfrac{\ln\left(\frac{65}{16}\right)}{\ln\left(\frac{13}{8}\right)}$

$\doteqdot 2.8872\ldots$

$\doteqdot$ 2 years 10.6 months.

$\therefore$ In November 2010, the number of mobile phones will exceed 4000 million. ✓

(2 marks)

21 $y = 10e^{-0.7t} + 3$

$\therefore \frac{dy}{dt} = 10 \times -0.7e^{-0.7t}$

$= -0.7(10e^{-0.7t})$ ✓

$= -0.7(10e^{-0.7t} + 3 - 3)$

$= -0.7(y - 3)$. ✓

(2 marks)

22 **i** $T = 3 + Ae^{-kt}$

$\therefore T - 3 = Ae^{-kt}$

and $\frac{dT}{dt} = -Ake^{-kt}$

$= -k(Ae^{-kt})$

$= -k(T - 3)$

$\therefore T = 3 + Ae^{-kt}$

satisfies this equation.

✓ *(1 mark)*

ii When $t = 0$, $T = 25$

$\therefore 25 = 3 + Ae^{0}$

$25 = 3 + A$

$A = 22$

$\therefore T = 3 + 22e^{-kt}$ ✓

When $t = 10$, $T = 11$

$\therefore 11 = 3 + 22e^{-10k}$

$8 = 22e^{-10k}$

$4 = 11e^{-10k}$

$e^{-10k} = \frac{4}{11}$

$\ln\frac{4}{11} = -10k$

$k = -\frac{1}{10}\ln\frac{4}{11}$ ✓

$= 0.10116\ldots$

When $t = 15$,

$T = 3 + 22e^{-15k}$

$= 3 + 22e^{-15\left(-\frac{1}{10}\ln\frac{4}{11}\right)}$

$= 3 + 22e^{\frac{3}{2}\left(\ln\frac{4}{11}\right)}$

$= 7.8241\ldots$

or $T = 3 + 22e^{-15k}$
$= 3 + 22e^{-15 \times 0.10116\ldots}$
$= 7.8241 \ldots$

$\therefore$ The temperature of the salad after 15 minutes is approximately 8 °C. ✓ *(3 marks)*

23 **i** $P = Ae^{kt}$

$\frac{dP}{dt} = kAe^{kt}$

$= kP$ ✓ *(1 mark)*

ii $P = Ae^{kt}$

At $t = 0$,
$P = 17\,000\,000 = 1.7 \times 10^7$

$\therefore 1.7 \times 10^7 = Ae^0$
$1.7 \times 10^7 = A(1)$
$\therefore A = 1.7 \times 10^7$. ✓ *(1 mark)*

iii $P = 1.7 \times 10^7 e^{kt}$

At $t = 13$,
$P = 20\,000\,000 = 2 \times 10^7$

$\therefore 2 \times 10^7 = 1.7 \times 10^7 e^{13k}$

$\frac{2 \times 10^7}{1.7 \times 10^7} = e^{13k}$ ✓

$\ln\left(\frac{2}{1.7}\right) = \ln e^{13k}$

$= 13k$

$\therefore k = \frac{1}{13}\ln\left(\frac{2}{1.7}\right)$ ✓

$\doteqdot 0.0125\ldots$
$= 0.013$
to 3 decimal places. *(2 marks)*

iv $P = 1.7 \times 10^7 e^{kt}$
(where $k = 0.0125\ldots$)

If $P = 30\,000\,000 = 3 \times 10^7$,

$3 \times 10^7 = 1.7 \times 10^7 e^{kt}$

$\frac{3 \times 10^7}{1.7 \times 10^7} = e^{kt}$

$\ln\left(\frac{3}{1.7}\right) = \ln e^{kt}$

$\ln\left(\frac{3}{1.7}\right) = kt \ln e$

$= kt$

$\therefore t = \frac{\ln\left(\frac{3}{1.7}\right)}{k}$ ✓

$= \frac{\ln\left(\frac{3}{1.7}\right)}{0.0125\ldots}$

(using calc. memory for k)
$= 45.433\ldots$
$\doteqdot 45.4$ years

$\therefore$ Now 45 years from the beginning of 1991 is the beginning of 2036. So in the year 2036 the population will reach 30 million. ✓ *(2 marks)*

24 **i** $T = A + Be^{kt}$ $\therefore Be^{kt} = T - A$ . . . (1)

$\frac{dT}{dt} = k\,Be^{kt}$

$= k(T - A)$ from (1)

$\therefore T = A + Be^{kt}$ satisfies the equation

$\frac{dT}{dt} = k\,(T - A)$. ✓ *(1 mark)*

ii $T = A + Be^{kt}$

Given: $A = 20$
At $t = 6, T = 80$
At $t = 8, T = 50$

$\therefore 80 = 20 + Be^{6k}$
and $50 = 20 + Be^{8k}$

$\therefore 60 = Be^{6k}$ and $30 = Be^{8k}$

$\therefore \frac{Be^{8k}}{Be^{6k}} = \frac{30}{60}$ ✓

$\therefore \frac{e^{8k}}{e^{6k}} = \frac{1}{2}$

$\therefore e^{2k} = \frac{1}{2}$

$\therefore 2k = \log_e \frac{1}{2}$

$\therefore 2k = -\log_e 2$

$\therefore k = -\frac{\log_e 2}{2}$. ✓

Substituting this value of k in $Be^{6k} = 60$:

$Be^{6\left(-\frac{\log_e 2}{2}\right)} = 60$

$\therefore Be^{-3\log_e 2} = 60$

$\therefore Be^{-\log_e 8} = 60$

$\therefore Be^{\log \frac{1}{8}} = 60$

$\therefore \frac{1}{8}B = 60$

(since $e^{\log x} = x$, by definition)

$\therefore B = 480$. ✓ *(3 marks)*

iii $T = A + Be^{kt}$

$A = 20,\ B = 480,\ k = -\frac{\log_e 2}{2}$

$\therefore T = 20 + 480e^{-\frac{\log_e 2}{2}t}$

At $t = 0$,
$T = 20 + 480e^{0}$
$= 20 + 480$
$= 500$

$\therefore$ Initial temperature is 500 °C. ✓ *(1 mark)*

25 **i** $C = C_0 e^{-kt}$

At $t = 0, C = 5$

$\therefore 5 = C_0 e^0$
$\therefore C_0 = 5$ kg/ha
$\therefore C = 5e^{-kt}$ ✓

At $t = 1, C = 2.8$

$\therefore 2.8 = 5e^{-k(1)}$
$\therefore 2.8 = 5e^{-k}$
$\therefore \frac{2.8}{5} = e^{-k}$
$\therefore 0.56 = e^{-k}$ ✓
$\therefore \ln(0.56) = \ln e^{-k}$
$\therefore \ln(0.56) = -k$
$\therefore k = -\ln(0.56)$
$= 0.57981 \ldots$
$\doteqdot 0.5798$ ✓ *(3 marks)*

ii $C = 5e^{-kt}$

If $C = 0.2$,
$0.2 = 5e^{-kt}$
$\frac{0.2}{5} = e^{-kt}$
$0.04 = e^{-kt}$
$\ln(0.04) = \ln(e^{-kt})$
$\ln(0.04) = -kt\ln e$
$\ln(0.04) = -kt$

$t = \frac{\ln(0.04)}{-k}$ ✓

$= \frac{\ln(0.04)}{-(-\ln(0.56))}$

$= \frac{\ln(0.04)}{\ln(0.56)}$

$= 5.551 \ldots$
$= 5.6$ years
to 1 decimal place. ✓ *(2 marks)*

26 $\frac{dT}{dt} = -k(T - 22)$ where $k > 0$

At $t = 0$, $T = 80$°C
At $t = 10$ min, $T = 60$°C

i $T = 22 + Ae^{-kt}$ . . . (1)

$\frac{dT}{dt} = -kAe^{-kt}$

and $Ae^{-kt} = T - 22$ (from (1))

$\therefore \frac{dT}{dt} = -k(T - 22)$

Hence $T = 22 + Ae^{-kt}$ is a solution of the given differential equation. ✓ *(1 mark)*

ii $T = 22 + Ae^{-kt}$

At $t = 0$, $T = 80$

$\therefore 80 = 22 + Ae^0$
$\therefore A = 80 - 22$
$= 58$ ✓
$\therefore T = 22 + 58e^{-kt}$

At $t = 10, T = 60$

$\therefore 60 = 22 + 58e^{-10k}$
$38 = 58e^{-10k}$
$e^{-10k} = \frac{38}{58}$
$-10k = \ln\left(\frac{38}{58}\right)$
$k = \frac{\ln\left(\frac{38}{58}\right)}{-10}$
$= 0.04228 \ldots$
$\doteqdot 0.0423$ ✓
(2 marks)

iii $T = 22 + 58e^{-kt}$
(where $k = 0.4228 \ldots$)

At $T = 30$,
$30 = 22 + 58e^{-kt}$
$8 = 58e^{-kt}$
$\frac{8}{58} = e^{-kt}$
$\ln\left(\frac{8}{58}\right) = -kt$
$t = \frac{\ln\left(\frac{8}{58}\right)}{-k}$ ✓
$= \frac{\ln\left(\frac{8}{58}\right)}{-0.04228 \ldots}$
$= 46.8480 \ldots$
$= 47$ minutes.
(to nearest minute)
✓ *(2 marks)*

27 **i** $Q = Q_0\, e^{-kt}$
When $t = 0$, $Q = 6$
$\therefore 6 = Q_0\, e^{-k(0)}$
$\therefore 6 = Q_0\, e^0$
$\therefore Q_0 = 6$
$\therefore Q = 6\, e^{-kt}$ ✓
When $t = 15$, $Q = 3$
$\therefore 3 = 6\, e^{-k(15)}$
$\therefore \frac{1}{2} = e^{-15k}$
$\ln\left(\frac{1}{2}\right) = \ln e^{-15k}$ ✓
$\ln\left(\frac{1}{2}\right) = -15k \ln e$
$\ln\left(\frac{1}{2}\right) = -15k$
$k = \frac{\ln\left(\frac{1}{2}\right)}{-15}$
$= 0.04620 \ldots$
$\doteqdot 0.0462.$ ✓
(3 marks)

ii Now $Q = 6\, e^{-kt}$
If $Q = \frac{1}{8} \times 6$
$= \frac{3}{4}$
$\therefore \frac{3}{4} = 6e^{-kt}$
$\frac{3}{24} = e^{-kt}$
$\frac{1}{8} = e^{-kt}$ ✓
$\ln\left(\frac{1}{8}\right) = \ln e^{-kt}$
$\ln\left(\frac{1}{8}\right) = -kt \ln e$
$t = \frac{\ln\left(\frac{1}{8}\right)}{-k}$
$= \frac{\ln\left(\frac{1}{8}\right)}{-0.04620\ldots}$
$= 45$ hours ✓
(2 marks)

28 $N = N_0 e^{kt}$
When $t = 0$, $N = 18$
$\therefore 18 = N_0 e^{k(0)}$
$\therefore 18 = N_0 e^0$
$\therefore N_0 = 18.$ ✓

Now when $t = 70$, $N = 5000$
$\therefore 5000 = 18e^{k(70)}$
$\frac{5000}{18} = e^{70k}$ ✓
$\ln\left(\frac{5000}{18}\right) = \ln e^{70k}$
$\ln\left(\frac{5000}{18}\right) = 70k \ln e$
$\ln\left(\frac{5000}{18}\right) = 70k$ ✓
$k = \frac{\ln\left(\frac{5000}{18}\right)}{70}$
$= 0.080\,38\ldots$
$\doteqdot 0.0804$ ✓

In November 2001, $t = 78$
$\therefore N = 18e^{0.080\,38\ldots \times 78}$
$= 9511.5154\ldots$
$= 9512$ koalas. ✓
(5 marks)

29 **i** $N = 200e^{kt}$

When $t = 0$, $N = 200$
$\therefore$ When $t = 3$, $N = 400$

$\therefore 400 = 200e^{3k}$
$2 = e^{3k}$ ✓
$\ln 2 = \ln e^{3k}$
$\ln 2 = 3k \ln e$
$\ln 2 = 3k$
$k = \frac{\ln 2}{3}$
$= 0.2310 \ldots$
$\doteqdot 0.231.$ ✓

ii After 12 weeks,
$N = 200e^{0.2310\ldots \times 12}$
$= 3200.$ ✓

iii $N = 200e^{kt}$
$\frac{dN}{dt} = 200ke^{kt}$ ✓
When $t = 3$,
$\frac{dN}{dt} = 200(0.2310\ldots)e^{0.2310\ldots \times 3}$
$= 92.419\ldots$
$= 92$ insects per week. ✓
(5 marks)
(Total mark allocation only in exam)

30 **i** $M = 10e^{-kt}$

When $t = 0$, $M = 10e^{-k(0)}$
$= 10e^0$
$= 10$

$\therefore$ Initial mass was 10 kg. ✓

ii When $t = 100$,
$M = 5$
$\therefore 5 = 10e^{-k(100)}$
$\frac{5}{10} = e^{-100k}$ ✓
$\ln\left(\frac{5}{10}\right) = \ln e^{-100k}$
$\ln 0.5 = -100k \ln e$
$\ln 0.5 = -100k$
$k = \frac{\ln 0.5}{-100}$
$= 0.006\,931\ldots$
$\doteqdot 0.00693.$ ✓

iii When $t = 1000$,
$M = 10e^{-0.006\,931\ldots \times 1000}$
$= 0.009765 \ldots$
$\doteqdot 0.00977$ kg
$\doteqdot 9.77$ g ✓

iv When $M = 8$,
$8 = 10e^{-kt}$
$\frac{8}{10} = e^{-kt}$ ✓
$\ln 0.8 = \ln e^{-kt}$
$\ln 0.8 = -kt$
$t = \frac{\ln 0.8}{-k}$ ✓
$= \frac{\ln 0.8}{-0.006931 \ldots}$
(using calc. memory for k)
$\doteqdot 32.19$ years ✓
(7 marks)
(Total mark allocation only in exam)

31 **i** $P = Ae^{kt}$

At $t = 0$, $P = 1\,000\,000$,
$\therefore 1\,000\,000 = Ae^0$
$1\,000\,000 = A(1)$
$\therefore A = 1\,000\,000$ ✓

ii $P = 1\,000\,000e^{kt}$

At $t = 2$, $P = 1\,072\,500$

$\therefore 1\,072\,500 = 1\,000\,000e^{2k}$
$\frac{1\,072\,500}{1\,000\,000} = e^{2k}$
$\ln \frac{1\,072\,500}{1\,000\,000} = \ln e^{2k}$
$\ln 1.072\,500 = 2k \ln e$
$\ln 1.072\,500 = 2k$
$k = \frac{\ln 1.072\,500}{2}$
$= 0.03499\ldots$
$= 0.035$ ✓
to 3 decimal places

iii $2\,000\,000 = 1\,000\,000e^{kt}$
(where $k = 0.03499\ldots$)
$\frac{2\,000\,000}{1\,000\,000} = e^{kt}$
$2 = e^{kt}$
$\ln 2 = \ln e^{kt}$
$\ln 2 = kt \ln e$
$= kt$
$\therefore t = \frac{\ln 2}{k}$ ✓
$= \frac{\ln 2}{(0.03499\ldots)}$
(using calc. memory for k)
$= 19.806\ldots$
$\doteqdot 19.8$ years ✓

(4 marks)
(Total mark allocation only in exam)

32 **i** $v = Ce^{-kt}$
$\frac{dv}{dt} = -kCe^{-kt}$
$= -k(v)$
$= -kv$ ✓

ii $t = 0, v = 100$
$\therefore 100 = Ce^0$
$\therefore 100 = C(1)$
$\therefore C = 100$ ✓

iii $t = 0.25, v = 85$
$\therefore 85 = 100e^{-0.25k}$
$\frac{85}{100} = e^{-0.25k}$
$\ln \frac{85}{100} = \ln e^{-0.25k}$
$\ln 0.85 = -0.25k \ln e$
$\ln 0.85 = -0.25k$ ✓
$k = \frac{\ln 0.85}{-0.25}$
$\doteqdot 0.650\,075\ldots$
$= 0.6501$ ✓
to 4 decimal places.

iv When $t = 2$,
$v = 100e^{-0.650075\ldots \times 2}$
$= 27.2490\ldots$ cm/s
$= 27.2$ cm/s ✓
to one decimal place

(5 marks)
(Total mark allocation only in exam)

33 **i** $\frac{dT}{dt} = k(T - T_0)$
$\therefore T = T_0 + Ae^{kt}$

Now $T_0 = -20$ and
when $t = 0, T = 100$

$\therefore 100 = -20 + Ae^{k(0)}$
$\therefore 120 = A$

$\therefore T = -20 + 120e^{kt}$ ✓

When $t = 3, T = 70$,

$\therefore 70 = -20 + 120e^{3k}$
$90 = 120e^{3k}$
$0.75 = e^{3k}$ ✓
$3k = \ln 0.75$
$k = \frac{\ln 0.75}{3}$
$= -0.09589\ldots$
$\doteqdot -0.0959$. ✓

ii Now $T_0 = 20$ and
when $t = 0, T = 70$

$\therefore 70 = 20 + Ae^{k(0)}$
$50 = A$

$\therefore T = 20 + 50e^{-0.09589\ldots \times t}$ ✓

When $t = 15$,
$T = 20 + 50e^{-0.09589\ldots \times 15}$
$= 31.8652\ldots$
$\doteqdot 32°$ C ✓

(5 marks)
(Total mark allocation only in exam)

34 **i** $R = R_0e^{-kt}$
When $t = 20$, $R = 0.5R_0$
$\therefore 0.5R_0 = R_0e^{-20k}$
$0.5 = e^{-20k}$ ✓
$\ln 0.5 = -20k$
$k = \frac{\ln 0.5}{-20}$
$= 0.03465\ldots$
$\doteqdot 0.0347$. ✓

ii If $R = 0.3R_0$
$0.3R_0 = R_0e^{-kt}$
$0.3 = e^{-kt}$
$\ln 0.3 = -kt$
$t = \frac{\ln 0.3}{-k}$ ✓
$= \frac{\ln 0.3}{-\left(\frac{\ln 0.5}{-20}\right)}$
$= \frac{\ln 0.3 \times 20}{\ln 0.5}$
$= 34.7393\ldots$ years
$\doteqdot 34.7$ years

$\therefore$ Additional elapsed time
$= 34.7 - 20 = 14.7$ years. ✓

(4 marks)
(Total mark allocation only in exam)

35 **i** Newton's law of cooling:
$\frac{dT}{dt} = k(T - A)$
where k is a constant.
If $T = A + Ce^{kt}$ . . . (1).
$\frac{dT}{dt} = kCe^{kt}$
But $Ce^{kt} = T - A$ from (1)
$\therefore \frac{dT}{dt} = k(T - A)$
$\therefore T = A + Ce^{kt}$ satisfies
Newton's law of cooling.

ii $T = A + Ce^{kt}$
$A = 5°$ C; At $t = 0$, $T = 20°$ C

$\therefore 20 = 5 + Ce^0$
$20 = 5 + C$
$\therefore C = 15$
$\therefore T = 5 + 15e^{kt}$

At $t = 0.5$ hours, $T = 17°$C

$\therefore 17 = 5 + 15e^{0.5k}$
$15e^{0.5k} = 12$
$e^{0.5k} = \frac{12}{15}$
$e^{0.5k} = \frac{4}{5}$
$0.5k = \ln 0.8$
$k = \frac{\ln 0.8}{0.5}$
$= -0.44628\ldots$
$\doteqdot 0.4463$

For $T = 10°$C,

$10 = 5 + 15e^{kt}$
$15e^{kt} = 5$
$e^{kt} = \frac{1}{3}$
$kt = \ln \frac{1}{3}$
$t = \frac{\ln \frac{1}{3}}{k}$
(using calc. memory for k)
$= 2.4616\ldots$
$= 2.46$ hours
to 2 decimal places.

(No mark allocation in exam)

1 A spherical bubble is moving up through a liquid. As it rises, the bubble gets bigger and its radius increases at the rate of 0.2 mm /s. At what rate is the volume of the bubble increasing when its radius reaches 0.6 mm? Express your answer in mm^3/s rounded to one decimal place. *(2 marks)*

(Q11e, **2021 HSC**) Easy

2 The quantities P, Q and R are connected by the related rates,

$$\frac{dR}{dt} = -k^2$$

$$\frac{dP}{dt} = -l^2 \times \frac{dR}{dt}$$

$$\frac{dP}{dt} = m^2 \times \frac{dQ}{dt}$$

where k, l and m are non-zero constants.

Which of the following statements is true?

A P is increasing and Q is increasing
B P is increasing and Q is decreasing
C P is decreasing and Q is increasing
D P is decreasing and Q is decreasing *(1 mark)*

(Q10, **2020 HSC**) Easy

3 Distance A is inversely proportional to distance B, such that $A = \frac{9}{B}$, where A and B are measured in metres. The two distances vary with respect to time.

Distance B is increasing at a rate of 0.2 ms^{-1}.

What is the value of $\frac{dA}{dt}$ when $A = 12$? *(3 marks)*

(Q12a, **2019 HSC**) Easy

4 A stone drops into a pond, creating a circular ripple. The radius of the ripple increases from 0 cm, at a constant rate of 5 cm s^{-1}.
At what rate is the area enclosed within the ripple increasing when the radius is 15 cm?

A 25π $\text{cm}^2\ \text{s}^{-1}$ **B** 30π $\text{cm}^2\ \text{s}^{-1}$
C 150π $\text{cm}^2\ \text{s}^{-1}$ **D** 225π $\text{cm}^2\ \text{s}^{-1}$

(Q8, **2017 HSC**) Easy

5 The diagram shows a conical soap dispenser of radius 5 cm and height 20 cm.

At any time t seconds, the top surface of the soap in the container is a circle of radius r cm and its height is h cm.

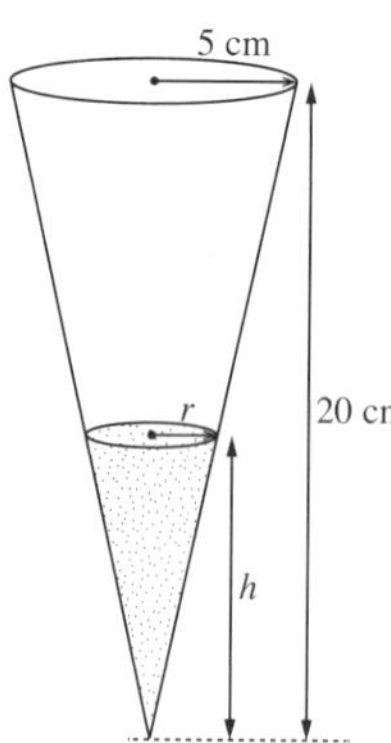

The volume of the soap is given by $v = \frac{1}{3}\pi r^2 h$.

i Explain why $r = \frac{h}{4}$. *(1 mark)* Easy

ii Show that $\frac{dv}{dh} = \frac{\pi}{16}h^2$. *(1 mark)* Easy

The dispenser has a leak which causes soap to drip from the container. The area of the circle formed by the top surface of the soap is decreasing at a constant rate of 0.04 $\text{cm}^2\ \text{s}^{-1}$.

iii Show that $\frac{dh}{dt} = \frac{-0.32}{\pi h}$. *(2 marks)* Medium

iv What is the rate of change of the volume of the soap, with respect to time, when $h = 10$? *(2 marks)* Easy

(Q12a, **2016 HSC**)

6 One end of a rope is attached to a truck and the other end to a weight. The rope passes over a small wheel located at a vertical distance of 40 m above the point where the rope is attached to the truck.

The distance from the truck to the small wheel is L m, and the horizontal distance between them is x m. The rope makes an angle θ with the horizontal at the point where it is attached to the truck.
The truck moves to the right at a constant speed of 3 ms^{-1}, as shown in the diagram.

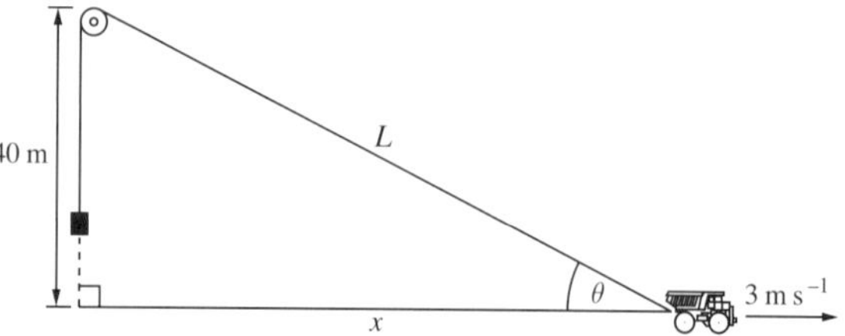

i Using Pythagoras' Theorem, or otherwise, show that $\frac{dL}{dx} = \cos\theta$. *(2 marks)* Medium

ii Show that $\frac{dL}{dt} = 3\cos\theta$. *(1 mark)* Easy

(Q13b, **2014 HSC**)

7 A spherical raindrop of radius r metres loses water through evaporation at a rate that depends on its surface area. The rate of change of the volume V of the raindrop is given by

$$\frac{dV}{dt} = -10^{-4} A,$$

where t is time in seconds and A is the surface area of the raindrop. The surface area and the volume of the raindrop are given by $A = 4\pi r^2$ and $V = \frac{4}{3}\pi r^3$ respectively.

i Show that $\frac{dr}{dt}$ is constant. *(1 mark)* Easy

ii How long does it take for a raindrop of volume 10^{-6} m^3 to completely evaporate? *(2 marks)* Hard

(Q13a, **2013 HSC**)

8 A plane P takes off from a point B. It flies due north at a constant angle α to the horizontal. An observer is located at A, 1 km from B, at a bearing 060° from B. Let u km be the distance from B to the plane and let r km be the distance from the observer to the plane. The point G is on the ground directly below the plane.

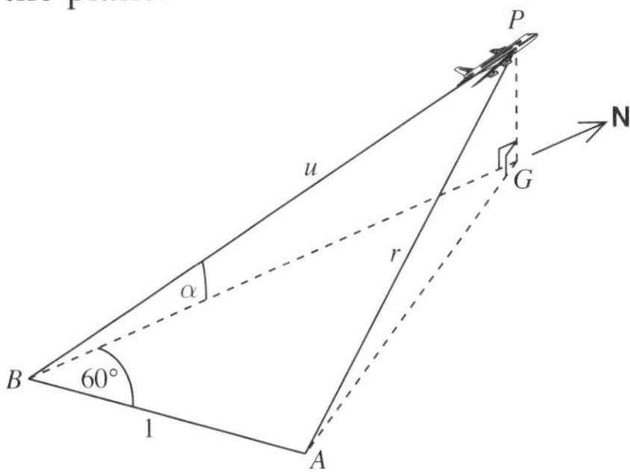

i Show that $r = \sqrt{1 + u^2 - u\cos\alpha}$. *(3 marks)* Hard

ii The plane is travelling at a constant speed of 360 km/h.
At what rate, in terms of α, is the distance of the plane from the observer changing 5 minutes after take-off? *(2 marks)* Hard

(Q14c, **2012 HSC**)

9 The diagram shows two identical circular cones with a common vertical axis.
Each cone has height h cm and semi-vertical angle 45°.

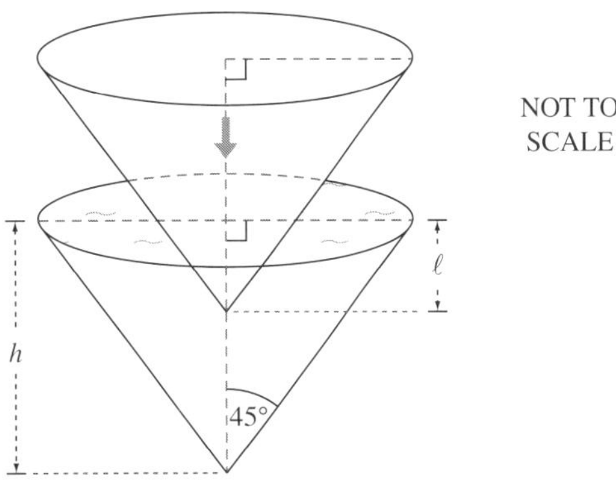

The lower cone is completely filled with water. The upper cone is lowered vertically into the water as shown in the diagram. The rate at which it is lowered is given by

$$\frac{d\ell}{dt} = 10,$$

where ℓ cm is the distance the upper cone has descended into the water after t seconds.

As the upper cone is lowered, water spills from the lower cone. The volume of water remaining in the lower cone at time t is V cm^3.

i Show that $V = \frac{\pi}{3}\left(h^3 - \ell^3\right)$. *(1 mark)* Medium

ii Find the rate at which V is changing with respect to time when $\ell = 2$. *(2 marks)* Medium

iii Find the rate at which V is changing with respect to time when the lower cone has lost $\frac{1}{8}$ of its water. Give your answer in terms of h. *(2 marks)* Hard

(Q7a, **2011 HSC**)

10 A radio transmitter M is situated 6 km from a straight road. The closest point on the road to the transmitter is S.

A car is travelling away from S along the road at a speed of 100 km h^{-1}. The distance from the car to S is x km and from the car to M is r km.

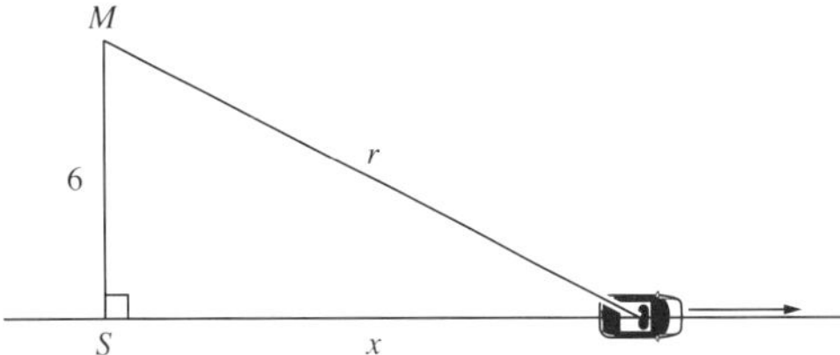

Find an expression in terms of x for $\frac{dr}{dt}$, where t is time in hours. *(3 marks)*

(Q2d, **2010 HSC**) Medium

11 The cross-section of a 10 metre long tank is an isosceles triangle, as shown in the diagram. The top of the tank is horizontal.

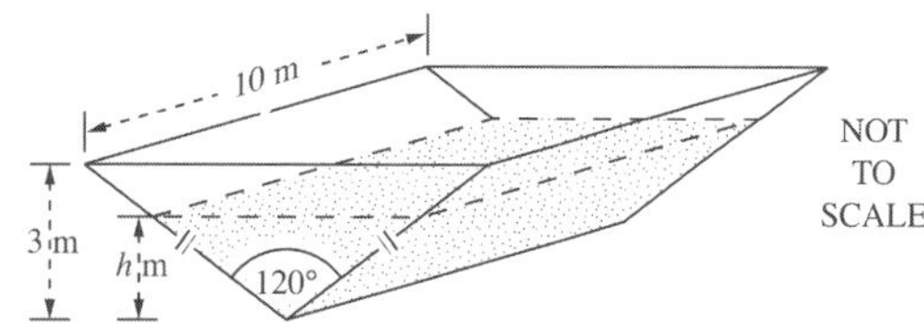

When the tank is full, the depth of water is 3 m. The depth of water at time t days is h metres.

i Find the volume, V, of water in the tank when the depth of water is h metres. *(1 mark)* Easy

ii Show that the area, A, of the top surface of the water is given by $A = 20\sqrt{3}h$. *(1 mark)* Easy

iii The rate of evaporation of the water is given by

$$\frac{dV}{dt} = -kA,$$

where k is a positive constant.
Find the rate at which the depth of water is changing at time t. *(2 marks)* Medium

iv It takes 100 days for the depth to fall from 3 m to 2 m. Find the time taken for the depth to fall from 2 m to 1 m. *(1 mark)* Easy

(Q5b, **2009 HSC**)

12 A hemispherical bowl of radius r cm is initially empty. Water is poured into it at a constant rate of k cm^3 per minute.

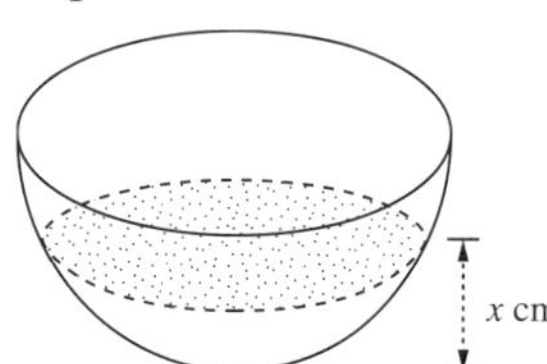

When the depth of water in the bowl is x cm, the volume, V cm^3, of water in the bowl is given by

$V = \frac{\pi}{3}x^2(3r - x)$. (Do NOT prove this.)

i Show that $\frac{dx}{dt} = \frac{k}{\pi x(2r - x)}$. *(2 marks)* Medium

ii Hence, or otherwise, show that it takes 3.5 times as long to fill the bowl to the point where $x = \frac{2}{3}r$ as it does to fill the bowl to the point where $x = \frac{1}{3}r$. *(2 marks)* Hard

(Q5c, **2006 HSC**)

13 An oil tanker at T is leaking oil which forms a circular oil slick. An observer is measuring the oil slick from a position P, 450 metres above sea level and 2 kilometres horizontally from the centre of the oil slick.

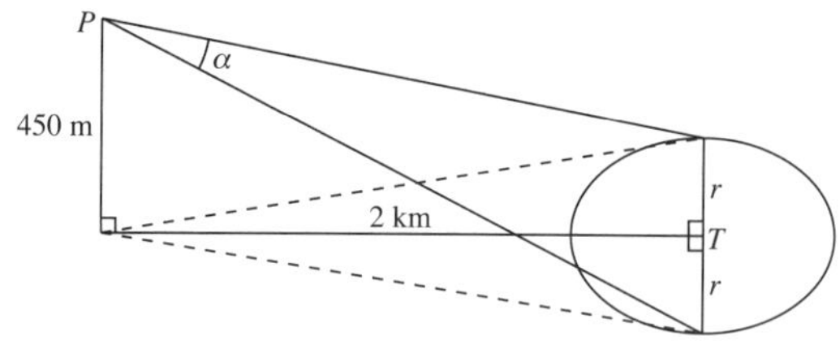

i At a certain time the observer measures the angle, α, subtended by the diameter of the oil slick, to be 0.1 radians. What is the radius, r, at this time? *(2 marks)* Medium

ii At this time, $\frac{d\alpha}{dt} = 0.02$ radians per hour.
Find the rate at which the radius of the oil slick is growing. *(2 marks)* Medium

(Q7a, **2005 HSC**)

14 A ferry wharf consists of a floating pontoon linked to a jetty by a 4 metre long walkway. Let h metres be the difference in height between the top of the pontoon and the top of the jetty and let x metres be the horizontal distance between the pontoon and the jetty.

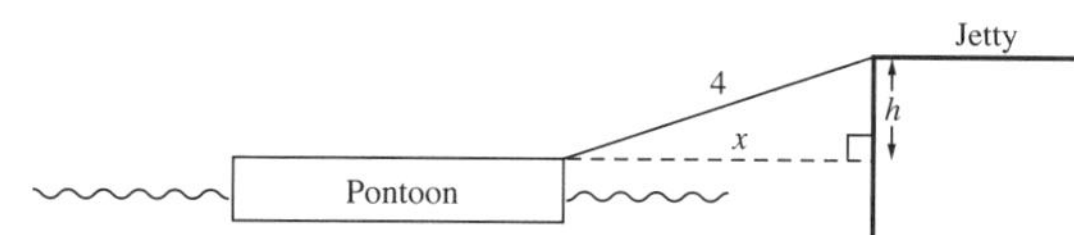

i Find an expression for x in terms of h. *(1 mark)* Easy

ii When the top of the pontoon is 1 metre lower than the top of the jetty, the tide is rising at a rate of 0.3 metres per hour.
At what rate is the pontoon moving away from the jetty? *(3 marks)* Medium

(Q3c, **2004 HSC**)

15 The diagram shows a conical drinking cup of height 12 cm and radius 4 cm. The cup is being filled with water at the rate of 3 cm^3 per second. The height of water at time t seconds is h cm and the radius of the water's surface is r cm.

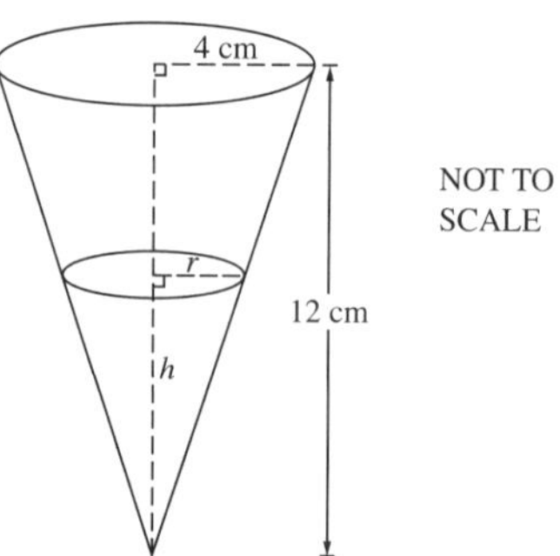

i Show that $r = \frac{1}{3}h$. *(1 mark)* Easy

ii Find the rate at which the height is increasing when the height of water is 9 cm. (Volume of cone $= \frac{1}{3}\pi r^2 h$.)

(3 marks) **Medium**

(Q5b, **2002 HSC**)

16 A searchlight on the ground at S detects and tracks a plane P that is due east of the searchlight. The plane is flying due west at a constant velocity of 240 kilometres per hour and maintains a constant height of 900 metres above ground level.

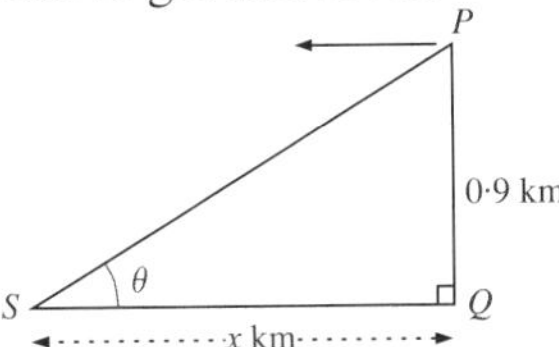

Let $\theta(t)$ radians be the angle of elevation of the plane at time t seconds and let $x(t)$ kilometres be the distance from S to the point Q on the ground directly below P.

i Show that $\dfrac{dx}{d\theta} = -\dfrac{0\cdot 9}{\sin^2\theta}$. **Medium**

ii Show that the rate of change of the angle of elevation of the plane when $\theta = \dfrac{\pi}{4}$ is equal to $\dfrac{1}{27}$ radians per second. **Medium**

(4 marks) (Q4c, **1997 HSC**)

17 Grain is poured at a constant rate of 0·5 cubic metres per second. It forms a conical pile, with the angle at the apex of the cone equal to 60°. The height of the pile is h metres, and the radius of the base is r metres.

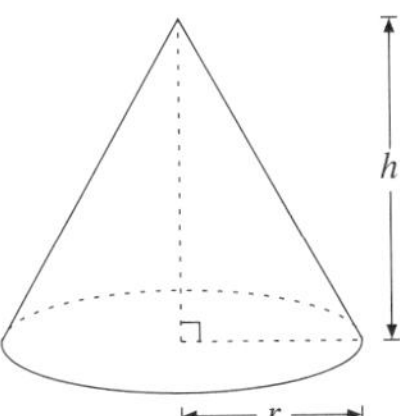

i Show that $r = \dfrac{h}{\sqrt{3}}$. **Easy**

ii Show that V, the volume of the pile, is given by $V = \dfrac{\pi h^3}{9}$. **Easy**

iii Hence find the rate at which the height of the pile is increasing when the height of the pile is 3 metres. **Easy**

(4 marks) (Q4b, **1996 HSC**)

18

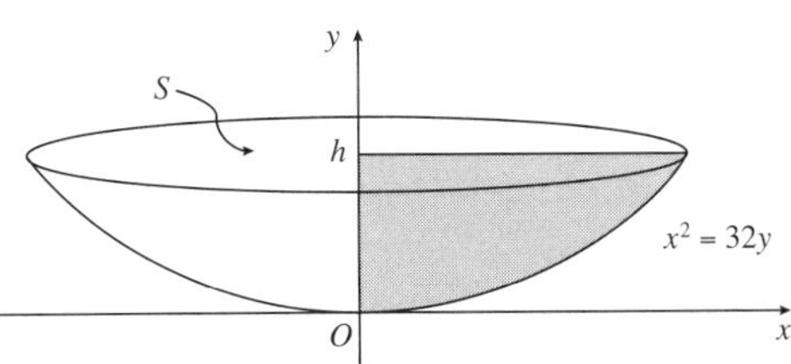

The diagram represents the water in a dam on a farm. The depth of the water is h metres, the volume of water in the dam is V m^3, and the area of the surface of the water is S m^2. The water in the dam evaporates according to the rule

$$\frac{dV}{dt} = -kS,$$

where k is a positive constant, and t is the time in hours.

i Describe in words what the rule says about the rate of evaporation.  **Easy**

ii Show that $\dfrac{dh}{dt} = -k$. **Medium**

[You may use the result that the volume is given by $V = 16\pi h^2$.]

iii Initially the dam contains 64π m^3 of water. Calculate how long it will take for the dam to empty by evaporation when $k = 0\cdot 001$. **Hard**

(Q5c, **1994 HSC**)

Year 11 Related rates of change—Worked Answers

1 The bubble's radius increases at a rate of 0.2 mm/s.

So, $\frac{dr}{dt} = 0.2$

$V = \frac{4}{3}\pi r^3$

$\frac{dV}{dr} = 4\pi r^2$

Now, $\frac{dV}{dt} = \frac{dV}{dr} \times \frac{dr}{dt}$ (Chain Rule)

$\frac{dV}{dt} = 4\pi r^2 \times 0.2$

$\frac{dV}{dt} = 0.8\pi r^2$ ✓

For the rate when radius is 0.6 mm, substitute $r = 0.6$.

$\therefore \frac{dV}{dt} = 0.8\pi(0.6)^2$

$= 0.9$ mm^3/s, correct to one ✓ decimal place

(2 marks)

2 $\frac{dR}{dt} = -k^2$

So $\frac{dR}{dt} < 0$

$\frac{dP}{dt} = -l^2 \times \frac{dR}{dt}$

So $\frac{dP}{dt} > 0$

$\therefore P$ is increasing.

$\frac{dP}{dt} = m^2 \times \frac{dQ}{dt}$

So $\frac{dQ}{dt} > 0$

$\therefore Q$ is increasing.

Answer A

3 $A = \frac{9}{B}$

$= 9B^{-1}$

$\frac{dA}{dB} = -9B^{-2}$

$= -\frac{9}{B^2}$

Now $\frac{dA}{dt} = \frac{dA}{dB} \times \frac{dB}{dt}$

$= -\frac{9}{B^2} \times 0.2$

$= -\frac{1.8}{B^2}$ ✓

When $A = 12$,

$12 = \frac{9}{B}$

$12B = 9$

$B = 0.75$ ✓

So $\frac{dA}{dt} = -\frac{1.8}{(0.75)^2}$

$= -3.2$ ms^{-1} ✓

(3 marks)

4 $\frac{dr}{dt} = 5$

$A = \pi r^2$

$\frac{dA}{dr} = 2\pi r$

$\frac{dA}{dt} = \frac{dA}{dr} \cdot \frac{dr}{dt}$

$= 2 \times \pi \times 15 \times 5$

$= 150\pi$

The area is increasing at 150π cm^2 s^{-1} when the radius is 15 cm.

Answer C

5 **i** The right-angled triangles formed by the radii and height also have a common angle, so are equiangular and similar.

So $\frac{r}{5} = \frac{h}{20}$ (sides are in proportion)

$r = \frac{h}{4}$

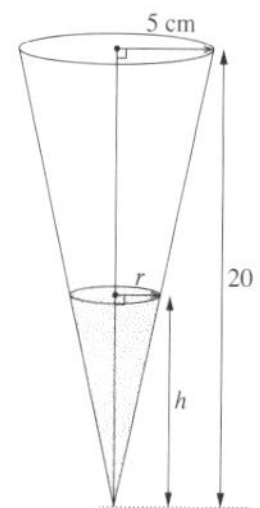

✓ *(1 mark)*

ii $v = \frac{1}{3}\pi r^2 h$

$= \frac{1}{3}\pi\left(\frac{h}{4}\right)^2 h$

$= \frac{\pi h^3}{48}$

So $\frac{dv}{dh} = \frac{3\pi h^2}{48}$

$= \frac{\pi}{16}h^2$ ✓ *(1 mark)*

iii $\frac{dA}{dt} = -0.04$

$A = \pi r^2$

$= \pi\left(\frac{h}{4}\right)^2$

$= \frac{\pi h^2}{16}$

$\frac{dA}{dh} = \frac{2\pi h}{16}$

$= \frac{\pi h}{8}$ ✓

$\frac{dA}{dt} = \frac{dA}{dh} \cdot \frac{dh}{dt}$

$-0.04 = \frac{\pi h}{8} \cdot \frac{dh}{dt}$

So $\frac{dh}{dt} = -0.04 \times \frac{8}{\pi h}$

$= \frac{-0.32}{\pi h}$ ✓

(2 marks)

iv $\frac{dv}{dt} = \frac{dv}{dh} \cdot \frac{dh}{dt}$

$= \frac{\pi}{16}h^2 \cdot \frac{-0.32}{\pi h}$

$= -0.02h$ ✓

When $h = 10$,

$\frac{dv}{dt} = -0.02 \times 10 = -0.2$

The rate of change of the volume of the soap is -0.2 cm^{3}s^{-1}. ✓

(2 marks)

6 **i** By Pythagoras' theorem:

$L^2 = x^2 + 40^2$

$= x^2 + 1600$

$L = \sqrt{x^2 + 1600} \quad (L > 0)$

$= (x^2 + 1600)^{\frac{1}{2}}$ ✓

$\frac{dL}{dx} = \frac{1}{2}(x^2 + 1600)^{-\frac{1}{2}} \times 2x$

$= \frac{x}{\sqrt{x^2 + 1600}}$

$= \frac{x}{L}$

$= \cos\theta$ ✓

(2 marks)

ii

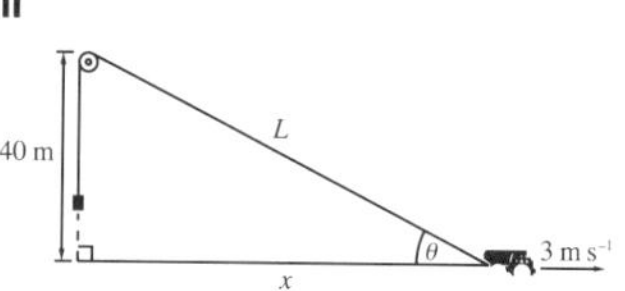

The truck moves right at 3 m s^{-1}

So $\frac{dx}{dt} = 3$

$\frac{dL}{dt} = \frac{dL}{dx} \cdot \frac{dx}{dt}$

$= (\cos\theta) \times 3$

$= 3\cos\theta$ ✓

(1 mark)

7 **i** $\frac{dV}{dt} = -10^{-4}A$

$V = \frac{4}{3}\pi r^3$

$\frac{dV}{dr} = 4\pi r^2$

$= A$

$\frac{dr}{dt} = \frac{dr}{dV} \cdot \frac{dV}{dt}$

$= \frac{1}{A}(-10^{-4}A)$

$= -10^{-4}$

$\therefore \frac{dr}{dt}$ is a constant. ✓

(1 mark)

ii $\frac{dr}{dt} = -10^{-4}$

$r = -10^{-4}t + C$

When $t = 0$, $V = 10^{-6}$

$\frac{4}{3}\pi r^3 = 10^{-6}$

$r^3 = \frac{3 \times 10^{-6}}{4\pi}$

$r = \sqrt[3]{\frac{3 \times 10^{-6}}{4\pi}}$

$\therefore C = \sqrt[3]{\frac{3 \times 10^{-6}}{4\pi}}$

$\therefore r = -10^{-4}t + \sqrt[3]{\frac{3 \times 10^{-6}}{4\pi}}$ ✓

When the raindrop has completely evaporated $r = 0$.

i.e $-10^{-4}t + \sqrt[3]{\frac{3 \times 10^{-6}}{4\pi}} = 0$

$10^{-4}t = \sqrt[3]{\frac{3 \times 10^{-6}}{4\pi}}$

$t = 10^4 \times \sqrt[3]{\frac{3 \times 10^{-6}}{4\pi}}$

$= 62.035\,049\ldots$

$\therefore$ it will take 62 seconds, to the nearest second, for the raindrop to completely evaporate. ✓

(2 marks)

8

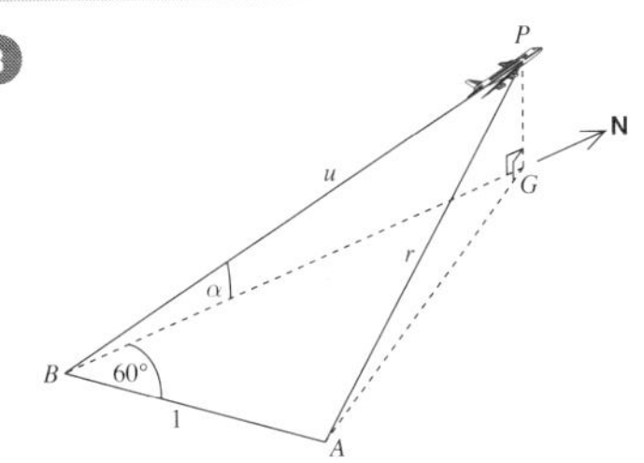

i In ΔPBG,

$\sin\alpha = \frac{PG}{u}$

$\therefore PG = u\sin\alpha$

$\cos\alpha = \frac{BG}{u}$

$\therefore BG = u\cos\alpha$ ✓

In ΔABG, by the cosine rule,

$AG^2 = (u\cos\alpha)^2 + 1^2 - 2 \times u\cos\alpha \times 1 \times \cos 60°$

$= u^2\cos^2\alpha + 1 - u\cos\alpha$ ✓

In ΔPGA, by Pythagoras' theorem,

$PA^2 = PG^2 + AG^2$

$r^2 = u^2\sin^2\alpha + u^2\cos^2\alpha + 1 - u\cos\alpha$

$= u^2(\sin^2\alpha + \cos^2\alpha) + 1 - u\cos\alpha$

$= u^2 + 1 - u\cos\alpha$

$= 1 + u^2 - u\cos\alpha$

$\therefore r = \sqrt{1 + u^2 - u\cos\alpha}$ ✓ $(r > 0)$

(3 marks)

ii $r = \sqrt{1 + u^2 - u\cos\alpha}$

$= (1 + u^2 - u\cos\alpha)^{\frac{1}{2}}$

$\frac{dr}{du} = \frac{1}{2}(1 + u^2 - u\cos\alpha)^{-\frac{1}{2}}(2u - \cos\alpha)$

$= \frac{2u - \cos\alpha}{2\sqrt{1 + u^2 - u\cos\alpha}}$

Now 360 km/h = 6 km/min

So $\frac{du}{dt} = 6$

$\frac{dr}{dt} = \frac{dr}{du} \cdot \frac{du}{dt}$

$= \frac{2u - \cos\alpha}{2\sqrt{1 + u^2 - u\cos\alpha}} \times 6$

$= \frac{3(2u - \cos\alpha)}{\sqrt{1 + u^2 - u\cos\alpha}}$ ✓

5 minutes after take off

$u = 5 \times 6$

$= 30$

$\therefore \frac{dr}{dt} = \frac{3(2 \times 30 - \cos\alpha)}{\sqrt{1 + 30^2 - 30\cos\alpha}}$

$= \frac{3(60 - \cos\alpha)}{\sqrt{901 - 30\cos\alpha}}$ km/min

[or $\frac{60 - \cos\alpha}{20\sqrt{901 - 30\cos\alpha}}$ km/s] ✓

(2 marks)

9 i Each cone has semi-vertical angle 45°.

$\therefore$ the radius is equal to the height of the cone.

Volume of each cone $= \frac{1}{3}\pi r^2 h$

$= \frac{\pi}{3}h^3$

When the upper cone is lowered vertically to a depth of l cm, the volume of displaced water is $\frac{\pi}{3}l^3$.

$\therefore V = \frac{\pi}{3}(h^3 - l^3)$ ✓

(1 mark)

ii $V = \frac{\pi}{3}(h^3 - l^3)$

$\frac{dV}{dl} = -\pi l^2$

$\frac{dV}{dt} = \frac{dV}{dl} \times \frac{dl}{dt}$

$= -\pi l^2 \times 10$

$= -10\pi l^2$ ✓

When $l = 2$,

$\frac{dV}{dt} = -10\pi \times 2^2$

$= -40\pi$

$\therefore V$ is decreasing at the rate of 40π cm^3 per second. ✓

(2 marks)

iii The volume of displaced water is $\frac{1}{8}$ of the volume of the cone.

$\therefore \frac{\pi}{3}l^3 = \frac{1}{8}\left(\frac{\pi}{3}h^3\right)$

$8l^3 = h^3$

$l^3 = \frac{h^3}{8}$

$l = \frac{h}{2}$ ✓

$\frac{dV}{dt} = -10\pi l^2$

When $l = \frac{h}{2}$,

$\frac{dV}{dt} = -10\pi\left(\frac{h}{2}\right)^2$

$= -10\pi\left(\frac{h^2}{4}\right)$

$= -\frac{5\pi h^2}{2}$

$\therefore V$ is decreasing at the rate of $\frac{5\pi h^2}{2}$ cm^3 per second. ✓

(2 marks)

10

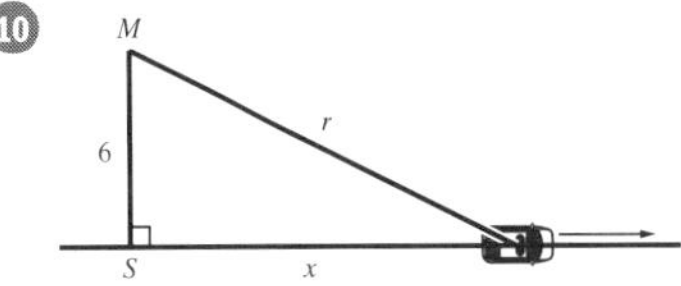

The car is travelling away from S at a speed of 100 km/h.

So $\frac{dx}{dt} = 100$ ✓

By Pythagoras' theorem,

$r^2 = x^2 + 6^2$

$r^2 = x^2 + 36$

$r = \sqrt{x^2 + 36}$ $(r > 0)$

$= (x^2 + 36)^{\frac{1}{2}}$

$\frac{dr}{dx} = \frac{1}{2}(x^2 + 36)^{-\frac{1}{2}} \times 2x$

$= \frac{x}{\sqrt{x^2 + 36}}$ ✓

Now $\frac{dr}{dt} = \frac{dr}{dx} \times \frac{dx}{dt}$

$= \frac{x}{\sqrt{x^2 + 36}} \times 100$

$= \frac{100x}{\sqrt{x^2 + 36}}$ ✓

(3 marks)

11

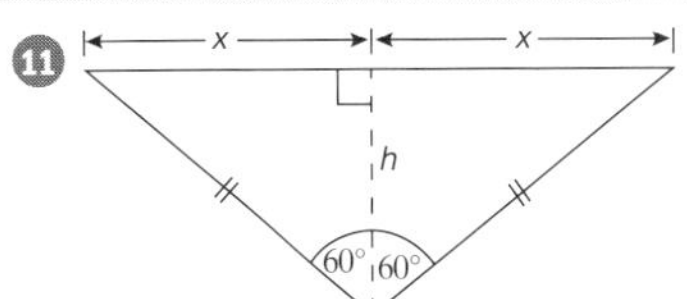

i Consider cross-section above:

$\tan 60° = \frac{x}{h}$

$\therefore x = h \tan 60°$
$= \sqrt{3}\,h$

V = Area of triangle × length of tank

$= \frac{1}{2} \times b \times h \times 10$

$= \frac{1}{2} \times 2x \times h \times 10$

$= \frac{1}{2} \times 2\sqrt{3}\,h \times h \times 10$

$= \sqrt{3}\,h \times h \times 10$

$= 10\sqrt{3}\,h^2$. ✓

(1 mark)

ii $A = 2x \times 10$
$= 2(\sqrt{3}\,h) \times 10$
$= 20\sqrt{3}\,h$. ✓

(1 mark)

iii Now $\frac{dh}{dt} = \frac{dh}{dV} \times \frac{dV}{dt}$

and $\frac{dV}{dt} = -kA = -k(20\sqrt{3}\,h)$

and $\frac{dV}{dh} = \frac{d}{dh} 10\sqrt{3}\,h^2 = 20\sqrt{3}\,h$ ✓

$\therefore \frac{dh}{dV} = \frac{1}{20\sqrt{3}h}$

$\therefore \frac{dh}{dt} = \frac{1}{20\sqrt{3}h} \times -k(20\sqrt{3}\,h)$
$= -k$. ✓

(2 marks)

iv Since $\frac{dh}{dt} = -k$ is a constant, it will take another 100 days for the depth to fall another metre.

(1 mark)

12 i $V = \frac{\pi}{3}x^2(3r - x)$

$= \frac{\pi}{3}x^2.3r - \frac{\pi}{3}x^2.x$

$= \pi r x^2 - \frac{\pi}{3}x^3$

$\frac{dV}{dx} = 2\pi r x - \pi x^2$ ✓

Now $\frac{dx}{dt} = \frac{dV}{dt} \times \frac{dx}{dV}$

and $\frac{dV}{dt} = k$

$\therefore \frac{dx}{dt} = k \times \frac{1}{2\pi r x - \pi x^2}$

$= \frac{k}{\pi x(2r - x)}$. ✓

(2 marks)

ii Since $\frac{dV}{dt} = k$

$\therefore V = kt + C$

When $t = 0, V = 0 \;\therefore C = 0$

$\therefore V = kt$

$t = \frac{V}{k}$

$= \frac{\frac{\pi}{3}x^2(3r - x)}{k}$

$= \frac{\pi x^2(3r - x)}{3k}$

When $x = \frac{1}{3}r = \frac{r}{3}$

$t = \frac{\pi\left(\frac{r}{3}\right)^2\left(3r - \frac{r}{3}\right)}{3k}$

$= \frac{\pi \times \frac{r^2}{9} \times \frac{8r}{3}}{3k}$

$= \frac{8\pi r^3}{81k}$. ✓

When $x = \frac{2}{3}r = \frac{2r}{3}$

$t = \frac{\pi\left(\frac{2r}{3}\right)^2\left(3r - \frac{2r}{3}\right)}{3k}$

$= \frac{\pi \times \frac{4r^2}{9} \times \frac{7r}{3}}{3k}$

$= \frac{28\pi r^3}{81k}$

Now $\frac{28\pi r^3}{81k} \div \frac{8\pi r^3}{81k}$

$= \frac{28\pi r^3}{81k} \times \frac{81k}{8\pi r^3}$

$= \frac{28}{8}$

$= 3.5$

$\therefore$ It takes 3.5 times as long to fill the bowl. ✓

(2 marks)

13

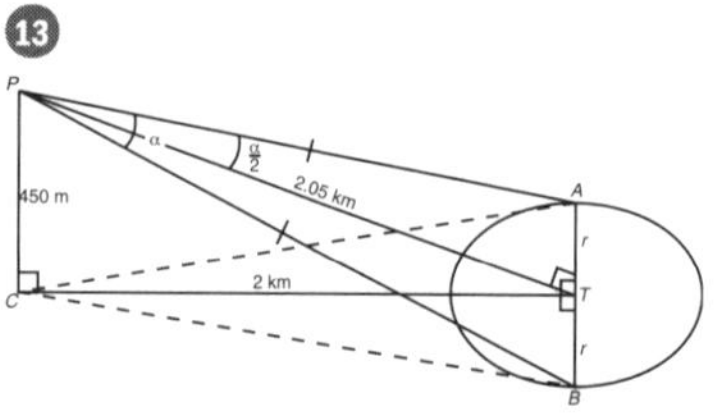

i *Construct PT*

Note PC = 450 m = 0.45 km.

In ΔPCT,

$PT^2 = 2^2 + 0.45^2$ (Pythagoras' theorem)

$\therefore PT = \sqrt{2^2 + 0.45^2}$

$\therefore PT = 2.05$ km

Now ΔPAB is isosceles with $PT = 2.05$

and $\angle APT = \frac{\alpha}{2}$

and $\angle PTA = 90°$

In ΔPAT,

$\tan\left(\frac{\alpha}{2}\right) = \frac{r}{2.05}$ ✓

$\therefore r = 2.05 \tan\left(\frac{\alpha}{2}\right)$

$= 2.05 \tan\left(\frac{0.1}{2}\right)$

$= 0.10258\ldots$ km
$= 102.5855\ldots$ m

$\therefore$ Radius is approximately 103 m. ✓

(2 marks)

ii $\frac{d\alpha}{dt} = 0.02$ radians per hour

$\frac{dr}{dt} = ?$ when $\alpha = 0.1$

$r = 2.05 \tan\left(\frac{\alpha}{2}\right)$ (from **i**)

$\therefore \frac{dr}{d\alpha} = \frac{2.05}{2}\sec^2\left(\frac{\alpha}{2}\right)$ ✓

Now

$\frac{dr}{dt} = \frac{d\alpha}{dt} \times \frac{dr}{d\alpha}$

$= 0.02 \times \frac{2.05}{2}\sec^2\left(\frac{0.1}{2}\right)$

(where $\alpha = 0.1$)

$= 0.02055\ldots$

$\therefore$ Radius is growing at approximately 0.0206 km/h or 20.6 m/h. ✓

(2 marks)

14 i $x^2 + h^2 = 4^2$

$\therefore x = \sqrt{16 - h^2}$. ✓

(1 mark)

ii When $h = 1$, $\frac{dh}{dt} = -0.3$ m/h

Now $\frac{dx}{dt} = \frac{dx}{dh} \times \frac{dh}{dt}$

and $x = (16 - h^2)^{\frac{1}{2}}$ ✓

$\therefore \frac{dx}{dh} = \frac{1}{2}(16 - h^2)^{-\frac{1}{2}}.-2h$

$= -h(16 - h^2)^{-\frac{1}{2}}$ ✓

$\therefore$ When $h = 1$,

$\frac{dx}{dt} = -1(16 - 1)^{-\frac{1}{2}} \times -0.3$

$= \frac{0.3}{\sqrt{15}}$

$\doteqdot 0.077$ metres per hour. ✓

(3 marks)

15

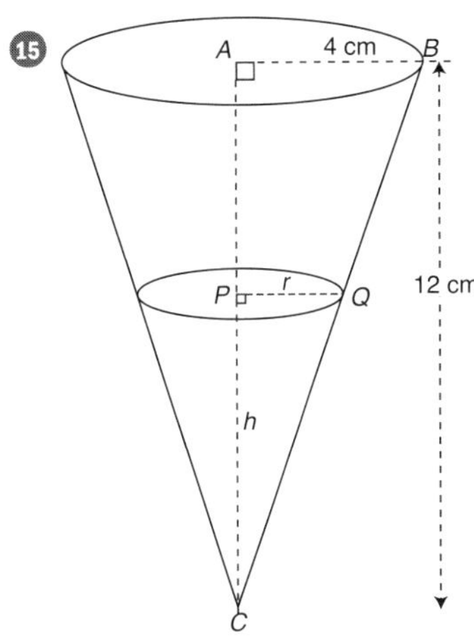

i In triangles ABC and PQC:

$\angle BAC = \angle QPC$ (given)
$\angle ACB$ is common
$\therefore \Delta ABC \parallel\!| \Delta PQC$
(2 equal angles)

$\therefore \frac{PQ}{PC} = \frac{AB}{AC}$
(corr. sides in similar

$\therefore \frac{r}{h} = \frac{4}{12}$

$\therefore \frac{r}{h} = \frac{1}{3}$

$\therefore r = \frac{1}{3}h.$ ✓

(1 mark)

ii $V = \frac{1}{3}\pi r^2 h$

$= \frac{1}{3}\pi\left(\frac{1}{3}h\right)^2 h$

$= \frac{1}{27}\pi h^3$

$\frac{dV}{dh} = \frac{3\pi h^2}{27}$

$= \frac{\pi h^2}{9}$ ✓

Now $\frac{dV}{dt} = \frac{dV}{dh}\cdot\frac{dh}{dt}$

and $\frac{dV}{dt} = 3\text{ cm}^3$ per second

$\therefore 3 = \frac{\pi h^2}{9}\cdot\frac{dh}{dt}$

$\therefore 27 = \pi h^2\cdot\frac{dh}{dt}$

$\therefore \frac{dh}{dt} = \frac{27}{\pi h^2}$ ✓

At $h = 9$ cm,

$\frac{dh}{dt} = \frac{27}{81\pi}$

$= \frac{1}{3\pi}$ cm/s
(≈ 0.106 cm/s). ✓

(3 marks)

16 i $\tan\theta = \frac{0.9}{x}$

$\therefore x = \frac{0.9}{\tan\theta}$

$= 0.9(\tan\theta)^{-1}$

$\frac{dx}{d\theta} = -0.9(\tan\theta)^{-2}.\sec^2\theta$ ✓

$= -0.9 \times \frac{1}{\tan^2\theta} \times \sec^2\theta$

$= -0.9 \times \frac{\cos^2\theta}{\sin^2\theta} \times \frac{1}{\cos^2\theta}$

$= -\frac{0.9}{\sin^2\theta}$ ✓

ii Now $\frac{dx}{dt} = -240$ km/h

and $\frac{dx}{d\theta} = -\frac{0.9}{\sin^2\theta}$ km/radian

and $\frac{d\theta}{dt} = \frac{d\theta}{dx}\cdot\frac{dx}{dt}$

$= -\frac{\sin^2\theta}{0.9} \times (-240)$

$= \frac{240\sin^2\theta}{0.9}$ ✓

At $\theta = \frac{\pi}{4}$,

$\frac{d\theta}{dt} = \frac{240 \times \frac{1}{2}}{0.9}$ radians per hour

$= \frac{1200}{9}$ radians per hour

$= \frac{400}{3}$ radians per hour

$= \frac{400}{3 \times 60 \times 60}$ radians per second

$= \frac{4}{108}$ radians per second

$= \frac{1}{27}$ radians per second. ✓

(4 marks)
(Total mark allocation only in exam)

17 i

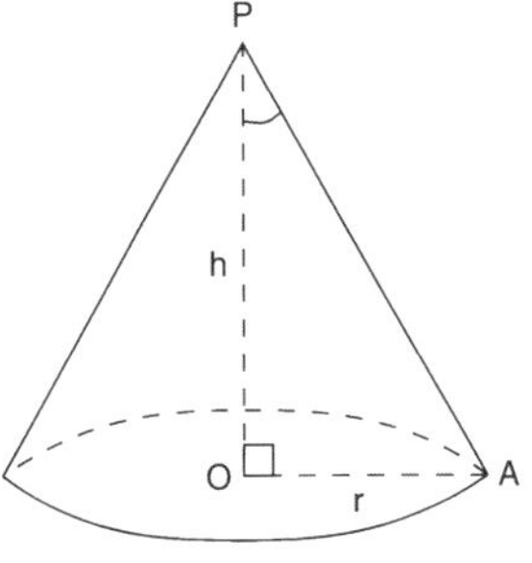

In ΔOPA,

$\angle OPA = \frac{1}{2} \times 60°$
$= 30°$

$\therefore \tan 30° = \frac{r}{h}$

$\therefore r = h\tan 30°$

$= \frac{h}{\sqrt{3}}.$ ✓

ii Volume of cone:

$V = \frac{1}{3}\pi r^2 h$

$= \frac{1}{3}\pi\left(\frac{h}{\sqrt{3}}\right)^2 h$

$= \frac{1}{3}\pi\left(\frac{h^2}{3}\right)h$

$= \frac{\pi h^3}{9}.$ ✓

iii $\frac{dV}{dt} = \frac{d}{dh}\left(\frac{\pi h^3}{9}\right)\frac{dh}{dt}$

$= \frac{3\pi h^2}{9}\cdot\frac{dh}{dt}$

$= \frac{\pi h^2}{3}\cdot\frac{dh}{dt}$ ✓

Now $\frac{dV}{dt} = 0.5$ (given)

$\therefore 0.5 = \frac{\pi h^2}{3}\cdot\frac{dh}{dt}$

$\therefore \frac{dh}{dt} = 0.5 \times \frac{3}{\pi h^2}$

$= \frac{1.5}{\pi h^2}$

$= \frac{1.5}{9\pi}$ when $h = 3$

$= \frac{1}{6\pi}$

$\therefore$ The height of the pile is increasing at a rate of $\frac{1}{6\pi}$ m/s when $h = 3$ m. ✓

(4 marks)
(Total mark allocation only in exam)

18 i The rule says that the rate at which the water evaporates is directly proportional to the surface area of the water

ii $S = \pi r^2$
$= \pi x^2$
$= \pi.32h$
$= 32\pi h$

Now $V = 16\pi h^2$

$\therefore \frac{dV}{dt} = 32\pi h\frac{dh}{dt}$

$= S\frac{dh}{dt}$
(since $S = 32\pi h$)

But $\frac{dV}{dt} = -kS$

$\therefore S\frac{dh}{dt} = -kS$

$\therefore \frac{dh}{dt} = \frac{-kS}{S}$

$\therefore \frac{dh}{dt} = -k.$

iii $V = 16\pi h^2$

At $t = 0$, $V = 64\pi$,

$\therefore 16\pi h^2 = 64\pi$

$h^2 = 4$

$h = 2$

$\therefore$ At $t = 0$, $h = 2$ m

Now $\frac{dh}{dt} = -k$ (from **ii**)

$= -0.001$

$\therefore h = \int -0.001\, dt$

$= -0.001t + C$

From above, at $t = 0$, $h = 2$:

$\therefore 2 = 0 + C$

$\therefore C = 2$

$\therefore h = -0.001t + 2$

When the dam is empty, $h = 0$:

$\therefore 0 = -0.001t + 2$

$0.001t = 2$

$t = \frac{2}{0.001}$

$= 2000$

$\therefore$ It will take 2000 hours for the dam to empty.

(No mark allocation in exam)

1 The diagram shows triangle ABC with points chosen on each of the sides. On side AB, 3 points are chosen. On side AC, 4 points are chosen. On side BC, 5 points are chosen.

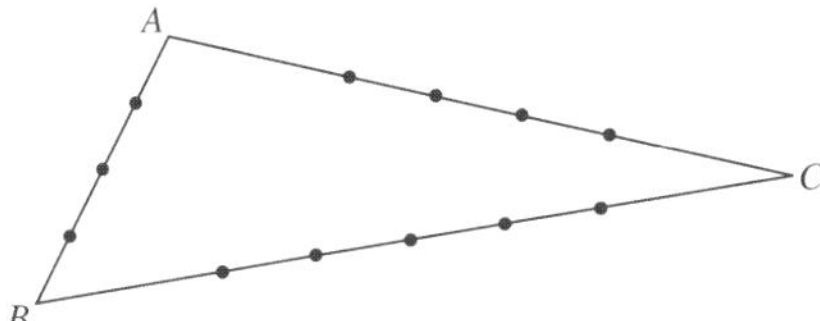

How many triangles can be formed using the chosen points as vertices?

A 60 **B** 145
C 205 **D** 220 *(1 mark)*

(Q7, **2022 HSC**) Medium

2 A sports association manages 13 junior teams. It decides to check the age of all players. Any team that has more than 3 players above the age limit will be penalised.

A total of 41 players are found to be above the age limit.

Will any team be penalised? Justify your answer. *(2 marks)*

(Q12b, **2022 HSC**) 

3 The members of a club voted for a new president. There were 15 candidates for the position of president and 3543 members voted. Each member voted for one candidate only.

One candidate received more votes than anyone else and so became the new president. What is the smallest number of votes the new president could have received?

A 236 **B** 237
C 238 **D** 239 *(1 mark)*

(Q10, **2021 HSC**) Hard

4 A committee containing 5 men and 3 women is to be formed from a group of 10 men and 8 women.

In how many different ways can the committee be formed? *(1 mark)*

(Q11d, **2021 HSC**) Easy

5 Out of 10 contestants, six are to be selected for the final round of a competition. Four of those six will be placed 1st, 2nd, 3rd and 4th. In how many ways can this process be carried out?

A $\dfrac{10!}{6!4!}$ **B** $\dfrac{10!}{6!}$
C $\dfrac{10!}{4!2!}$ **D** $\dfrac{10!}{4!4!}$ *(1 mark)*

(Q8, **2020 HSC**) Medium

6 To complete a course, a student must choose and pass exactly three topics.

There are eight topics from which to choose.

Last year 400 students completed the course.

Explain, using the pigeonhole principle, why at least eight students passed exactly the same three topics. *(2 marks)* Medium

(Q12c, **2020 HSC**)

7 A drawer contains many socks.

There are 6 blue socks, 10 red socks, 5 black socks, 4 yellow socks and 7 white socks.

What is the minimum number of randomly chosen socks to be taken out of the drawer until it is guaranteed that 4 socks are of the same colour?

A 11 **B** 16
C 20 **D** 21 *(1 mark)*

Bonus question (see page iv) Easy

8 What is the smallest group of people that can be assembled where it is guaranteed that at least two people have the same first and last initials in their name?

A 27 **B** 53
C 675 **D** 677 *(1 mark)*

Bonus question Easy

9 In how many ways can all the letters of the word PARALLEL be placed in a line with the three Ls together?

A $\dfrac{6!}{2!}$ **B** $\dfrac{6!}{2!3!}$
C $\dfrac{8!}{2!}$ **D** $\dfrac{8!}{2!3!}$ *(1 mark)*

(Q8, **2019 HSC**) Medium

10 Six men and six women are to be seated at a round table.

In how many different ways can they be seated if men and women alternate?

A 5! 5! **B** 5! 6!
C 2! 5! 5! **D** 2! 5! 6! *(1 mark)*

(Q8, **2018 HSC**) 

11 Three squares are chosen at random from the 3×3 grid below, and a cross is placed in each chosen square.

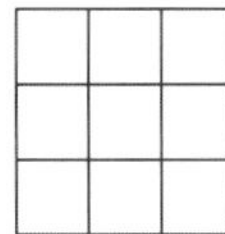

What is the probability that all three crosses lie in the same row, column or diagonal?

A $\frac{1}{28}$ **B** $\frac{2}{21}$

C $\frac{1}{3}$ **D** $\frac{8}{9}$ *(1 mark)*

(Q10, **2017 HSC**) 

12 A team of 11 students is to be formed from a group of 18 students. Among the 18 students are 3 students who are left-handed.

What is the number of possible teams containing at least 1 student who is left-handed?

A 19448 **B** 30459
C 31824 **D** 58344 *(1 mark)*

(Q8, **2016 HSC**) 

13 A rowing team consists of 8 rowers and a coxswain.

The rowers are selected from 12 students in Year 10.

The coxswain is selected from 4 students in Year 9.

In how many ways could the team be selected?

A ${}^{12}C_8 + {}^4C_1$ **B** ${}^{12}P_8 + {}^4P_1$
C ${}^{12}C_8 \times {}^4C_1$ **D** ${}^{12}P_8 \times {}^4P_1$ *(1 mark)*

(Q4, **2015 HSC**) 

14 Two players A and B play a series of games against each other to get a prize. In any game, either of the players is equally likely to win.

To begin with, the first player who wins a total of 5 games gets the prize.

i Explain why the probability of player A getting the prize in exactly 7 games is $\binom{6}{4}\left(\frac{1}{2}\right)^7$. *(1 mark)* 

ii Write an expression for the probability of player A getting the prize in at most 7 games. *(1 mark)* **Medium**

iii Suppose now that the prize is given to the first player to win a total of $(n + 1)$ games, where n is a positive integer.
By considering the probability that A gets the prize, prove that

$$\binom{n}{n}2^n + \binom{n+1}{n}2^{n-1} + \binom{n+2}{n}2^{n-2} + \ldots + \binom{2n}{n} = 2^{2n}.$$

(2 marks) **Hard**

(Q14c, **2015 HSC**)

15 In how many ways can 6 people from a group of 15 people be chosen and then arranged in a circle?

A $\frac{14!}{8!}$ **B** $\frac{14!}{8!6}$

C $\frac{15!}{9!}$ **D** $\frac{15!}{9!6}$ *(1 mark)*

(Q8, **2014 HSC**) **Medium**

16 Two players A and B play a game that consists of taking turns until a winner is determined. Each turn consists of spinning the arrow on a spinner once. The spinner has three sectors P, Q and R. The probabilities that the arrow stops in sectors P, Q and R are p, q and r respectively.

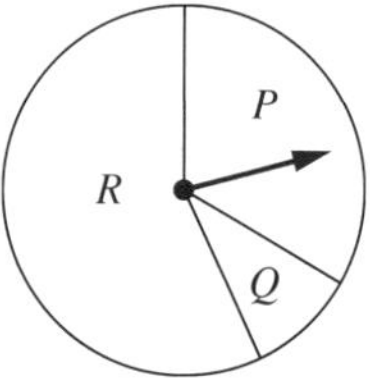

The rules of the game are as follows:
- If the arrow stops in sector P, then the player having the turn wins.
- If the arrow stops in sector Q, then the player having the turn loses and the other player wins.
- If the arrow stops in sector R, then the other player takes a turn.

Player A takes the first turn.

i Show that the probability of player A winning on the first or the second turn of the game is $(1 - r)(p + r)$. *(2 marks)* **Hard**

ii Show that the probability that player A eventually wins the game is $\frac{p+r}{1+r}$.

(3 marks) **Hard**

(Q14b, **2014 HSC**)

17 A family of eight is seated randomly around a circular table.

What is the probability that the two youngest members of the family sit together?

A $\frac{6!2!}{7!}$ **B** $\frac{6!}{7!2!}$

C $\frac{6!2!}{8!}$ **D** $\frac{6!}{8!2!}$ *(1 mark)*

(Q7, **2013 HSC**) Easy

18 How many arrangements of the letters of the word OLYMPIC are possible if the C and the L are to be together in any order?

A 5! **B** 6!

C $2 \times 5!$ **D** $2 \times 6!$ *(1 mark)*

(Q5, **2012 HSC**) Easy

19 In how many ways can a committee of 3 men and 4 women be selected from a group of 8 men and 10 women? *(1 mark)*

(Q11e, **2012 HSC**) Easy

20 Alex's playlist consists of 40 different songs that can be arranged in any order.

i How many arrangements are there for the 40 songs? *(1 mark)* Easy

ii Alex decides that she wants to play her three favourite songs first, in any order. How many arrangements of the 40 songs are now possible? *(1 mark)* Easy

(Q2e, **2011 HSC**)

21 At the front of a building there are five garage doors. Two of the doors are to be painted red, one is to be painted green, one blue and one orange.

i How many possible arrangements are there for the colours on the doors? *(1 mark)* Medium

ii How many possible arrangements are there for the colours on the doors if the two red doors are next to each other? *(1 mark)* Medium

(Q3a, **2010 HSC**)

22 **i** A box contains n identical red balls and n identical blue balls. A selection of r balls is made from the box, where $0 \leq r \leq n$. Explain why the number of possible colour combinations is $r + 1$. *(1 mark)* Hard

ii Another box contains n white balls labelled consecutively from 1 to n. A selection of $n - r$ balls is made from the box, where $0 \leq r \leq n$.

Explain why the number of different selections is $\binom{n}{r}$. *(1 mark)* Hard

iii The n red balls, the n blue balls and the n white labelled balls are all placed into one box, and a selection of n balls is made.

Using part (b), or otherwise, show that the number of different selections is $(n + 2)2^{n-1}$.

[Note: Part b is not in this topic. Use these results instead.

$$2^n = \binom{n}{0} + \binom{n}{1} + \binom{n}{2} + \binom{n}{3} + \ldots + \binom{n}{n}$$

and

$$n(2)^{n-1} = \binom{n}{1} + 2\binom{n}{2} + 3\binom{n}{3} + 4\binom{n}{4} + \ldots + n\binom{n}{n}.]$$

(3 marks) Hard

(Q7c, **2010 HSC**)

23 Barbara and John and six other people go through a doorway one at a time.

i In how many ways can the eight people go through the doorway if John goes through the doorway after Barbara with no-one in between? *(1 mark)* Easy

ii Find the number of ways in which the eight people can go through the doorway if John goes through the doorway after Barbara. *(1 mark)* Easy

(Q4b, **2008 HSC**)

24 Mr and Mrs Roberts and their four children go to the theatre. They are randomly allocated six adjacent seats in a single row.

What is the probability that the four children are allocated seats next to each other? *(2 marks)* Medium

(Q5b, **2007 HSC**)

25 Sophie has five coloured blocks: one red, one blue, one green, one yellow and one white. She stacks two, three, four or five blocks on top of one another to form a vertical tower.

i How many different towers are there that she could form that are three blocks high? *(1 mark)* Easy

ii How many different towers can she form in total? *(2 marks)* Medium

(Q3c, **2006 HSC**)

26 A four-person team is to be chosen at random from nine women and seven men.

i In how many ways can this team be chosen? *(1 mark)* Easy

ii What is the probability that the team will consist of four women? *(1 mark)* Easy

(Q2e, **2004 HSC**)

27 How many nine-letter arrangements can be made using the letters of the word ISOSCELES? *(2 marks)*

(Q3a, **2003 HSC**) Easy

28 A committee of 6 is to be chosen from 14 candidates. In how many different ways can this be done? *(1 mark)*

(Q4a, **2003 HSC**) Easy

29 Seven people are to be seated at a round table.

i How many seating arrangements are possible? *(1 mark)* Easy

ii Two people, Kevin and Jill, refuse to sit next to each other. How many seating arrangements are then possible? *(2 marks)* Easy

(Q3a, **2002 HSC**)

30 The letters $A, E, I, O,$ and U are vowels.

i How many arrangements of the letters in the word ALGEBRAIC are possible? *(1 mark)* Easy

ii How many arrangements of the letters in the word ALGEBRAIC are possible if the vowels must occupy the 2nd, 3rd, 5th and 8th positions? *(2 marks)* Easy

(Q2c, **2001 HSC**)

31 How many arrangements of the letters of the word HOCKEYROO are possible? *(2 marks)*

(Q2a, **2000 HSC**) Easy

32 A standard pack of 52 cards consists of 13 cards of each of the four suits: spades, hearts, clubs and diamonds.

i In how many ways can six cards be selected without replacement so that exactly two are spades and four are clubs? (Assume that the order of selection of the six cards is not important.) Easy

ii In how many ways can six cards be selected without replacement if at least five must be of the same suit? (Assume that the order of selection of the six cards is not important.) Medium

(4 marks) (Q6b, **2000 HSC**)

33 The staff in an office consists of 4 males and 7 females. How many committees of 5 staff can be chosen which contain exactly 3 females? *(2 marks)* Easy

(Q2a, **1999 HSC**)

34 A committee of 3 men and 4 women is to be formed from a group of 8 men and 6 women. Write an expression for the number of ways this can be done. *(1 mark)*

(Q1d, **1996 HSC**) Easy

35 Mice are placed in the centre of a maze which has five exits. Each mouse is equally likely to leave the maze through any one of the five exits. Thus, the probability of any given mouse leaving by a particular exit is $\frac{1}{5}$.

Four mice, $A, B, C,$ and D, are put into the maze and behave independently.

i What is the probability that $A, B, C,$ and D all come out the same exit? Easy

ii What is the probability that $A, B,$ and C come out the same exit, and D comes out a different exit? Easy

iii What is the probability that *any* three of the four mice come out the same exit, and the other comes out a different exit? Easy

iv What is the probability that no more than two mice come out the same exit? Medium

(5 marks) (Q5c, **1996 HSC**)

36 A security lock has 8 buttons labelled as shown. Each person using the lock is given a 3-letter code.

A	B	C	D	E	F	G	H

i How many different codes are possible if letters can be repeated and their order is important? Easy

ii How many different codes are possible if letters cannot be repeated and their order is important? Easy

iii Now suppose that the lock operates by holding 3 buttons down together, so that order is NOT important. How many different codes are possible? Easy

(3 marks) (Q3a, **1995 HSC**)

37

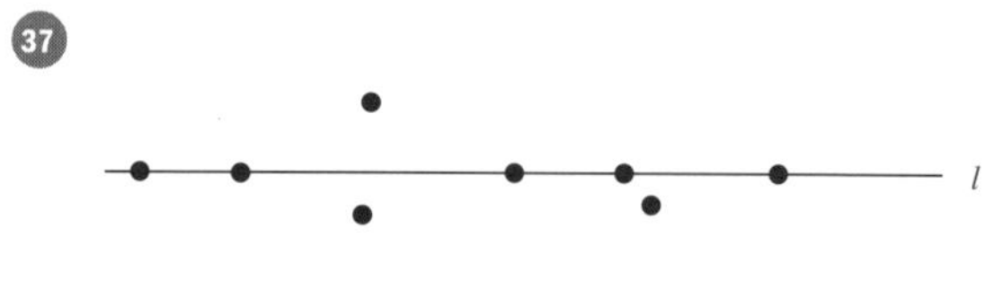

The figure shows 9 points lying in the plane, 5 of which lie on the line l. No other set of 3 of these points is collinear.

- **i** How many sets of 3 points can be chosen from the 5 points lying on l? **Easy**
- **ii** How many different triangles can be formed using the 9 points as vertices? **Medium**

(Q7b, **1994 HSC**)

---

38 A class consists of 10 girls and 12 boys. How many ways are there of selecting a committee of 3 girls and 2 boys from this class?

(Q1e, **1993 HSC**) 

---

39 Five travellers arrive in a town where there are five hotels.

- **i** How many different accommodation arrangements are there if there are no restrictions on where the travellers stay? **Easy**
- **ii** How many different accommodation arrangements are there if each traveller stays at a different hotel? **Easy**
- **iii** Suppose two of the travellers are husband and wife and must go to the same hotel. How many different accommodation arrangements are there if the other three can go to any of the *other* hotels? **Medium**

(Q4b, **1993 HSC**)

---

40 A total of five players is selected at random from four sporting teams. Each of the teams consists of ten players numbered from **1** to **10**.

- **i** What is the probability that of the five selected players, three are numbered '**6**' and two are numbered '**8**'? **Medium**
- **ii** What is the probability that the five selected players contain at least four players from the same team? **Medium**

(Q6b, **1992 HSC**)

Year 11 Permutations and combinations—Worked Answers

1 Triangles are formed by choosing three vertices that are not all on the one line segment.
N = Total number of combinations without restrictions—three points on same line segment.
$N = {}^{12}C_3 - ({}^5C_3 + {}^4C_3 + {}^3C_3)$
$N = 205$
Answer C

2 Let n = number of over-age players and k = number of teams
$\therefore \frac{n}{k} = \frac{41}{13}$
$= 3\frac{2}{13}$ ✓
$\therefore$ *Each* team can have 3 over-age players with 2 over-age players yet to be placed.
$\therefore$ At least one team will be penalised. ✓ *(2 marks)*

3 n votes are to be given to m candidates.
If each candidate receives the same number of votes:
$3543 \div 15 = 236\frac{3}{15}$
$\therefore$ Three candidates could receive 237 votes and the others 236.
$\therefore$ The smallest number of votes for a majority is 238 votes.
Answer C

4 Number of ways $= {}^{10}C_5 \times {}^8C_3$
$= 14\,112$ ✓ *(1 mark)*

5 The six contestants can be selected in ${}^{10}C_6$ ways.
The four positions can be organised in ${}^6P_4$ ways.
Number of ways $= {}^{10}C_6 \times {}^6P_4$
$= \frac{10!}{6!(10-6)!} \times \frac{6!}{(6-4)!}$
$= \frac{10!}{6!4!} \times \frac{6!}{2!}$
$= \frac{10!}{4!2!}$
Answer C

6 Number of choices $= {}^8C_3$
$= 56$
Now $400 \div 56 = 7.1428\ldots$ ✓
So if exactly seven students passed every possible choice, that doesn't account for all 400 students. By the pigeonhole principle, at least eight students must have passed exactly the same three topics. ✓ *(2 marks)*

7 As there are 4 socks to be chosen, then $n = 4$.
The number of colours $(k) = 5$
Smallest group $= (n-1)k + 1$
$= 3 \times 5 + 1$
$= 16$
$\therefore$ at least 16 socks need to be selected.
Answer B

8 As there are 2 initials, then $n = 2$.
The number of initial pairings $(k) = 26^2$
$= 676$
Smallest group $= (n-1)k + 1$
$= 1 \times 676 + 1$
$= 677$
$\therefore$ the group must have at least 677 people.
Answer D

9 Together the three Ls make one letter unit.
So there are 6 letter units but two As.
So number of arrangements $= \frac{6!}{2!}$
Answer A

10 The men can sit in $(6-1)!$ or $5!$ ways.
For each of those the women can sit in $6!$ ways.
Number of ways $= 5!\ 6!$
Answer B

11 Combinations of 3 squares from 9
$= {}^9C_3$
$= 84$
Total rows, columns or diagonals = 8
$P(\text{form a line}) = \frac{8}{84}$
$= \frac{2}{21}$
Answer B

12 Total number of possible teams
$= {}^{18}C_{11}$
$= 31\,824$
Number with no left-handed students
$= {}^{15}C_{11}$
$= 1365$
Number of teams with at least 1 left-handed student
$= 31\,824 - 1365$
$= 30\,459$
Answer B

13 Unordered selection of 8 rowers from 12 students.
Ways rowers can be selected $= {}^{12}C_8$
For each way they can be selected there are ${}^4C_1$ ways of choosing the coxswain.
So the number of ways is ${}^{12}C_8 \times {}^4C_1$
Answer C

14 **i** $P(A \text{ wins a game})$
$= P(B \text{ wins a game}) = \frac{1}{2}$
If player A gets the prize in exactly 7 games then A's 5th win must be the 7th game. The other 4 wins can come from any of the previous 6 games. The number of ways this can occur is $\binom{6}{4}$.
$P(A \text{ wins in exactly 7 games})$
$= \binom{6}{4}\left(\frac{1}{2}\right)^4\left(\frac{1}{2}\right)^2 \times \frac{1}{2}$
$= \binom{6}{4}\left(\frac{1}{2}\right)^7$ ✓ *(1 mark)*

ii $P(A \text{ wins in at most 7 games})$
$= P(A \text{ wins in 5 games}) + P(A \text{ wins in 6 games}) + P(A \text{ wins in 7 games})$
$= \binom{4}{4}\left(\frac{1}{2}\right)^5 + \binom{5}{4}\left(\frac{1}{2}\right)^6 + \binom{6}{4}\left(\frac{1}{2}\right)^7$ ✓ *(1 mark)*

iii The maximum number of games for someone to win $= n + 1 + n = 2n + 1$
$P(A \text{ gets the prize})$
$= P(A \text{ wins in } (n+1) \text{ games}) + P(A \text{ wins in } (n+2) \text{ games} + \ldots + P(A \text{ wins in } (2n+1))$ games
$= \binom{n}{n}\left(\frac{1}{2}\right)^{n+1} + \binom{n+1}{n}\left(\frac{1}{2}\right)^{n+2} + \binom{n+2}{n}\left(\frac{1}{2}\right)^{n+3} + \ldots + \binom{2n}{n}\left(\frac{1}{2}\right)^{2n+1}$
But A and B have an equal chance of getting the prize so $P(A \text{ gets the prize}) = \frac{1}{2}$

So $\binom{n}{n}\left(\frac{1}{2}\right)^{n+1}+\binom{n+1}{n}\left(\frac{1}{2}\right)^{n+2}+\binom{n+2}{n}\left(\frac{1}{2}\right)^{n+3}$

$+\ldots+\binom{2n}{n}\left(\frac{1}{2}\right)^{2n+1}=\frac{1}{2}$ ✓

$2^{2n+1}\times\left(\binom{n}{n}\left(\frac{1}{2}\right)^{n+1}+\binom{n+1}{n}\left(\frac{1}{2}\right)^{n+2}+\binom{n+2}{n}\left(\frac{1}{2}\right)^{n+3}\right.$

$\left.+\ldots+\binom{2n}{n}\left(\frac{1}{2}\right)^{2n+1}\right)=2^{2n+1}\times\frac{1}{2}$

$\binom{n}{n}\left(\frac{2^{2n+1}}{2^{n+1}}\right)+\binom{n+1}{n}\left(\frac{2^{2n+1}}{2^{n+2}}\right)+\binom{n+2}{n}\left(\frac{2^{2n+1}}{2^{n+3}}\right)$

$+\ldots+\binom{2n}{n}\left(\frac{2^{2n+1}}{2^{2n+1}}\right)=\frac{2^{2n+1}}{2}$

$\binom{n}{n}2^{n}+\binom{n+1}{n}2^{n-1}+\binom{n+2}{n}2^{n-2}+\ldots+\binom{2n}{n}=2^{2n}$ ✓ *(2 marks)*

15 Combinations of 6 people from 15

$=\frac{15!}{9!6!}$

6 people can be arranged in a circle in 5! ways.

Total arrangements $=\frac{15!}{9!6!}\times 5!$

$=\frac{15!}{9!6}$

Answer D

16 **i**

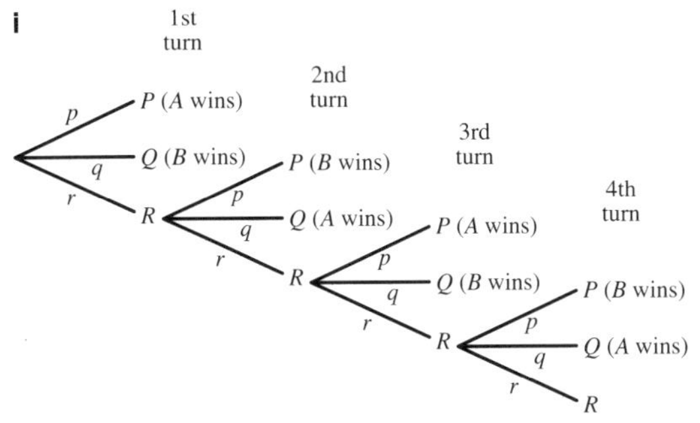

1st turn: p → P (A wins); q → Q (B wins); r → R

2nd turn: from R: p → P (B wins); q → Q (A wins); r → R

3rd turn: from R: p → P (A wins); q → Q (B wins); r → R

4th turn: from R: p → P (B wins); q → Q (A wins); r → R

$P(A \text{ wins on 1st or 2nd turn}) = p + rq$ ✓

Now $p + q + r = 1$

$q = 1 - p - r$

So $P(A$ wins on 1st or 2nd turn)

$= p + r(1 - p - r)$

$= p + r - rp - r^2$

$= p - pr + r - r^2$

$= p(1 - r) + r(1 - r)$

$= (1 - r)(p + r)$ ✓

(2 marks)

ii $P(A$ eventually wins)

$= (1 - r)(p + r) + r^2(1 - r)(p + r) + r^4(1 - r)(p + r) + \ldots$

$= (1 - r)(p + r)(1 + r^2 + r^4 + \ldots)$ ✓

Now $1 + r^2 + r^4 + \ldots$ is an infinite geometric series with first term 1 and common ratio r^2.

$S = \frac{a}{1 - r}$

$= \frac{1}{1 - r^2}$

$= \frac{1}{(1 + r)(1 - r)}$ ✓

So $P(A$ eventually wins)

$= (1 - r)(p + r) \times \frac{1}{(1 + r)(1 - r)}$

$= \frac{p + r}{1 + r}$ ✓

(3 marks)

17 The youngest family member can sit anywhere. The next youngest can sit in any of 7 places, 2 of which are beside the youngest.

$\therefore P(2 \text{ youngest sit together}) = \frac{2}{7}$

Of the options, only $\frac{6!2!}{7!} = \frac{2}{7}$

[The first person can sit anywhere. There are 7! possible arrangements of the remaining family members. Two particular members can sit together in 2! ways, while the remaining 6 can sit in 6! ways, so the number of arrangements of the family members with the 2 youngest sitting together is 6!2!]

Answer A

18 If the C and L are kept together they count as a single element. But the C and L can be arranged in 2 ways.

Total arrangements = 2 × 6!

Answer D

19 Number of ways $= {}^{8}C_3 \times {}^{10}C_4$

$= 11\,760$ ✓ *(1 mark)*

20 **i** Number of arrangements of 40 songs

$= 40! \quad [\approx 8.2 \times 10^{47}]$ ✓ *(1 mark)*

ii New number of arrangements

$= 3! \times 37! \quad [\approx 8.3 \times 10^{43}]$ ✓ *(1 mark)*

21 **i** Number of arrangements $= \frac{5!}{2!}$

$= 60$ ✓ *(1 mark)*

ii Two red doors can be placed next to each other in 4 ways.

The remaining 3 doors can be arranged in 3! ways.

Number of arrangements with 2 adjacent red doors = 4 × 3!

= 24 ✓ *(1 mark)*

22 **i** If one ball is chosen there are two possibilities; red or blue. So the number of red balls could be 0 or 1.

If two balls are chosen there are three possibilities; both red, both blue or one red and one blue. So the number of red balls could be 0, 1 or 2.

If three balls are chosen there are four possibilities; all red, all blue, two red and one blue or two blue and one red. So the number of red balls could be 0, 1, 2 or 3.

If r balls are chosen the number of red balls could be $0, 1, 2, 3, \ldots, r$.

So the number of possible colour combinations is $r + 1$. ✓

(1 mark)

ii Choosing $n - r$ balls from n balls:

$$\text{Number of selections} = \binom{n}{n-r} = \frac{n!}{(n-r)!(n-(n-r))!} = \frac{n!}{(n-r)!r!} = \binom{n}{r}$$ ✓

(1 mark)

iii In a particular selection, r red or blue balls can be chosen in $(r + 1)$ ways. (From part **i**.)
The remaining $(n - r)$ balls will be white and can be chosen in $\binom{n}{r}$ ways. (From part **ii**.)
So each selection of r red or blue balls and $(n - r)$ white balls can be made in $(r + 1)\binom{n}{r}$ ways.
The number of red or blue balls can range from 0 to n, so the number of possible selections

$$= 1\binom{n}{0} + 2\binom{n}{1} + 3\binom{n}{2} + 4\binom{n}{3} + \ldots + (n+1)\binom{n}{n}$$ ✓

$$= 1\binom{n}{1} + 2\binom{n}{2} + 3\binom{n}{3} + \ldots + n\binom{n}{n} + \binom{n}{0} + \binom{n}{1} + \binom{n}{2} + \binom{n}{3} + \ldots + \binom{n}{n}$$

$$= n2^{n-1} + 2^n$$

$$= (n + 2)2^{n-1}$$ ✓

(3 marks)

23 i Barbara and John will have to remain together with John always behind Barbara.
∴ There are only 7 places to fill:

[BJ] __ __ __ __ __ __

∴ Number of ways = 7!
= 5040. ✓

(1 mark)

ii The total number of ways eight people can go through the door is 8!
Barbara can walk through ahead of John half of these number of times.

∴ Number of ways = $\frac{8!}{2}$

∴ Number of ways = 20 160. ✓

OR

The possibilities are:

None between them: 7! (from **i**)

One between them: 6 × 6!

[BXJ] __ __ __ __ __

Two between them: $5! \times {}^6P_4$

[BXYJ] __ __ __ __

Three between them: $4! \times {}^6P_3$

[BXYZJ] __ __ __

Four between them: $3! \times {}^6P_4$
Five between them: $2! \times {}^6P_5$
Six between them: ${}^6P_6$

The total of all these possibilities is 20 160.

(1 mark)

24 The number of ways they can be seated with no restrictions is 6! = 720.
The 4 children can be seated together in 4! ways.
This group of 4 children (X) can be combined with 2 parents in 3! ways. ✓
(E.g. P_1 P_2 X)
∴ The total no. of ways is 3! × 4! = 144.

∴ Required probability = $\frac{144}{720}$
= 0.2. ✓

(2 marks)

25 i ${}^5P_3 = 60$. ✓

(1 mark)

ii Number of different towers in total
$= {}^5P_2 + {}^5P_3 + {}^5P_4 + {}^5P_5$ ✓
= 20 + 60 + 120 + 120
= 320. ✓

(2 marks)

26 i Number of ways
$= {}^{16}C_4$
= 1820 ✓

(1 mark)

ii P(team consists of 4 women)

$$= \frac{{}^9C_4}{1820} = \frac{126}{1820} = \frac{9}{130}.$$ ✓

(1 mark)

27 The word ISOSCELES has 9 letters with "S" appearing 3 times and "E" appearing twice. ✓
Therefore the number of nine-letter arrangements of these letters is
$\frac{9!}{3!2!} = 30\,240$. ✓

(2 marks)

28 This represents the number of combinations of 14 objects taken 6 at a time,
i.e. ${}^{14}C_6 = 3003$

(1 mark)

29 i Number of circular permutations of 7 objects
= (7 – 1)!
= 6!
= 720. ✓

(1 mark)

ii Number of arrangements when Kevin and Jill sit together, Kevin on Jill's left, is 5!
Number of arrangements when Kevin and Jill sit together, Kevin on Jill's right, is 5!

Hence, number of arrangements when they do sit next to each other is $2 \times 5!$ ✓

$\therefore$ Number of seating arrangements required

$= 6! - (2 \times 5!)$
$= 6 \times 5! - 2 \times 5!$
$= (6 - 2) \times 5!$
$= 4 \times 5!$
$= 480.$ ✓

(2 marks)

30 **i** There are 9 letters in ALGEBRAIC of which 2 are alike, i.e. A. The number of arrangements of these letters is:

$\frac{9!}{2!} = 181\,440.$ ✓

(1 mark)

ii The 4 vowels (2 alike) can occupy these positions in $\frac{4!}{2!} = 12$ different ways.

For each of these arrangements, the 5 consonants can be placed in $5! = 120$ different ways. ✓

$\therefore$ The total number of arrangements is $12 \times 120 = 1440.$ ✓

(2 marks)

31 Number of permutations of 9 letters of which 3 are alike:

$\frac{9!}{3!} = 4 \times 5 \times 6 \times 7 \times 8 \times 9$ ✓

$= 60\,480.$ ✓

(2 marks)

32 **i** The number of combintions of 6 cards of which 2 are spades and 4 are clubs is

${}^{13}C_2 \times {}^{13}C_4 = 78 \times 715$
$= 55\,770$ ✓

ii If all 6 cards are of the same suit, the number of combinations is

$4 \times {}^{13}C_6 = 4 \times 1716 = 6864.$ ✓

If exactly 5 cards are of the same suit, the number of combinations is

$4 \times {}^{13}C_5 \times (3 \times 13)$
$= 4 \times 1287 \times 39$
$= 200\,772$ ✓

$\therefore$ The total number of ways is $6864 + 200\,772 = 207\,636.$ ✓

(4 marks)
(Total mark allocation only in exam)

33 Each committee will consist of 2 males and 3 females. The number of committees of 5 staff with 3 females will be

${}^4C_2 \times {}^7C_3 = 6 \times 35$ ✓
$= 210.$ ✓

(2 marks)

34 3 men can be chosen from 8 men in ${}^8C_3$ different ways.

4 women can be chosen from 6 women in ${}^6C_4$ different ways.

The total number of ways the committee can be formed is ${}^8C_3 \times {}^6C_4$ (i.e. 840 ways) ✓

(1 mark)

35 **i** Let A come out by any exit. The probability of B coming out by this exit is $\frac{1}{5}$; and the same applies to C and D.

$\therefore$ Required probability

$= \frac{1}{5} \times \frac{1}{5} \times \frac{1}{5}$

$= \frac{1}{125}.$ ✓

ii Let A come out by any exit. The probability of B coming out by this exit is $\frac{1}{5}$ and that of C is also $\frac{1}{5}$. The probability of D not getting out by this exit is $\frac{4}{5}$.

$\therefore$ Required probability

$= \frac{1}{5} \times \frac{1}{5} \times \frac{4}{5} = \frac{4}{125}$ ✓

iii This event will occur if A, B, C come out the same exit and D comes out a different exit (as in **ii**) or if A, B, D use the same exit and C a different exit or if A, C, D use the same exit and B a different one or finally if B, C, D come out the same exit and A a different one. Each one of these cases has the same probability

$\therefore$ Required probability

$= 4 \times \frac{4}{125} = \frac{16}{125}$ ✓

iv *More* than 2 mice come out the same exit if all 4 come out the same exit or if 3 come out the same exit and the other a different exit.

The probability of this occuring

$= \frac{1}{125} + \frac{16}{125} = \frac{17}{125}$ ✓

(from **i** and **ii**)

$\therefore$ Probability that *no more* than 2 mice come out the same exit

$= 1 - \frac{17}{125}$

$= \frac{108}{125}$ ✓

(5 marks)
(Total mark allocation only in exam)

36 **i** Number of ordered triples with repetitions $= 8^3 = 512.$ ✓

ii Number of permutations

$= {}^8P_3$
$= 8 \times 7 \times 6$
$= 336.$ ✓

iii Number of combinations

$= {}^8C_3$
$= \frac{8 \times 7 \times 6}{1 \times 2 \times 3}$
$= 56.$ ✓

(3 marks)
(Total mark allocation only in exam)

37 **i** ${}^5C_3 = \frac{5 \times 4}{1 \times 2}$
$= 10$

ii The number of sets of 3 points that can be chosen from the 9 points is:

${}^9C_3 = \frac{9 \times 8 \times 7}{1 \times 2 \times 3}$
$= 84$

Of these, 10 sets of 3 points are on a straight line and hence cannot form triangles.

$\therefore$ The number of different triangles formed using the 9 points as vertices is $84 - 10 = 74.$

(No mark allocation in exam)

38 There are ${}^{10}C_3$ ways of selecting the 3 girls and ${}^{12}C_2$ ways of selecting the 2 boys.

$\therefore$ Number of ways of selecting committee

$= {}^{10}C_3 \times {}^{12}C_2$
$= \frac{10 \times 9 \times 8}{1 \times 2 \times 3} \times \frac{12 \times 11}{1 \times 2}$
$= 120 \times 66$
$= 7920.$

(No mark allocation in exam)

39 **i** Each traveller can be accommodated in 5 different ways.

$\therefore$ The total number of ways the 5 travellers can be accommodated is

$5 \times 5 \times 5 \times 5 \times 5 = 5^5$
$= 3125$

ii The first traveller can be accommodated in 5 different ways; the second in 4 different ways, and so on.

$\therefore$ The total number of ways the 5 travellers can be accommodated is

$5 \times 4 \times 3 \times 2 \times 1 = 5!$
$= 120$

iii The husband and wife can choose their hotel in 5 ways. The other 3 travellers can be accommodated in
$4 \times 4 \times 4 = 4^3$ ways.
$\therefore$ The total number of different arrangements is
$5 \times 4^3 = 320$

(No mark allocation in exam)

---

40 **i** There are 4 players numbered '6' and 4 players numbered '8' from a total of forty players.
Three '6's can be selected in ${}^4C_3$ ways.
Two '8's can be selected in ${}^4C_2$ ways.
Five players can be selected in ${}^{40}C_5$ ways.

$\therefore$ Required probability

$$= \frac{{}^4C_3 \times {}^4C_2}{{}^{40}C_5}$$

$$= \frac{4 \times 6}{658\,008}$$

$$= \frac{1}{27\,417}.$$

ii 'At least 4 players' means 4 or 5 players:
5 players from one team can be selected in ${}^{10}C_5$ ways. But there are 4 teams, hence 5 players from the same team can be selected in ${}^4C_1 \times {}^{10}C_5$ ways.
4 players from one team and one player from the remaining teams (30 players) can be selected in
${}^4C_1 \times {}^{10}C_4 \times {}^{30}C_1$ ways.

$\therefore$ Required probability

$$= \frac{{}^4C_1 \times {}^{10}C_5}{{}^{40}C_5} + \frac{{}^4C_1 \times {}^{10}C_4 \times {}^{30}C_1}{{}^{40}C_5}$$

$$= \frac{{}^4C_1({}^{10}C_5 + {}^{10}C_4 \times {}^{30}C_1)}{{}^{40}C_5}$$

$$= \frac{4(252 + 210 \times 30)}{658\,008}$$

$$= \frac{252 + 6300}{164\,502}$$

$$= \frac{6552}{164\,502}$$

$$= \frac{28}{703}.$$

(No mark allocation in exam)

1 Find the coefficients of x^2 and x^3 in the expansion of $\left(1 - \frac{x}{2}\right)^8$. *(2 marks)*

(Q11c, **2022 HSC**) Easy

2 Expand and simplify $(2a - b)^4$. *(2 marks)*

(Q11b, **2021 HSC**) Easy

3 **i** Use the identity $(1 + x)^{2n} = (1 + x)^n(1 + x)^n$ to show that

$$\binom{2n}{n} = \binom{n}{0}^2 + \binom{n}{1}^2 + \cdots + \binom{n}{n}^2,$$

where n is a positive integer. *(2 marks)* Hard

ii A club has $2n$ members, with n women and n men.
A group consisting of an even number $(0, 2, 4, \ldots, 2n)$ of members is chosen, with the number of men equal to the number of women.
Show, giving reasons, that the number of ways to do this is $\binom{2n}{n}$. *(2 marks)* Hard

iii From the group chosen in part **ii**, one of the men and one of the women are selected as leaders.
Show, giving reasons, that the number of ways to choose the even number of people and then the leaders is

$$1^2\binom{n}{1}^2 + 2^2\binom{n}{2}^2 + \cdots + n^2\binom{n}{n}^2.$$

(2 marks) Hard

iv The process is now reversed so that the leaders, one man and one woman, are chosen first. The rest of the group is then selected, still made up of an equal number of women and men.
By considering this reversed process and using part **ii**, find a simple expression for the sum in part **iii**. *(2 marks)* Hard

(Q14a, **2020 HSC**)

4 In the expansion of $(5x + 2)^{20}$, the coefficients of x^k and x^{k+1} are equal.

What is the value of k? *(3 marks)*

(Q13b, **2019 HSC**) Medium

5 When expanded, which expression has a non-zero constant term?

A $\left(x + \frac{1}{x^2}\right)^7$ **B** $\left(x^2 + \frac{1}{x^3}\right)^7$

C $\left(x^3 + \frac{1}{x^4}\right)^7$ **D** $\left(x^4 + \frac{1}{x^5}\right)^7$ *(1 mark)*

(Q9, **2017 HSC**) Easy

6 Consider the binomial expansion

$$\left(2x + \frac{1}{3x}\right)^{18} = a_0x^{18} + a_1x^{16} + a_2x^{14} + \ldots$$

where $a_0, a_1, a_2, \ldots$ are constants.

i Find an expression for a_2. *(2 marks)* Easy

ii Find an expression for the term independent of x. *(2 marks)* Easy

(Q13b, **2015 HSC**)

7 What is the constant term in the binomial expansion of $\left(2x - \frac{5}{x^3}\right)^{12}$?

A $\binom{12}{3}2^9 5^3$ **B** $\binom{12}{9}2^3 5^9$

C $-\binom{12}{3}2^9 5^3$ **D** $-\binom{12}{9}2^3 5^9$ *(1 mark)*

(Q3, **2014 HSC**) Medium

8 **i** Use the binomial theorem to find an expression for the constant term in the expansion of $\left(2x^3 - \frac{1}{x}\right)^{12}$. *(2 marks)* Easy

ii For what values of n does $\left(2x^3 - \frac{1}{x}\right)^n$ have a non-zero constant term? *(1 mark)* Hard

(Q11f, **2012 HSC**)

9 Find an expression for the coefficient of x^2 in the expansion of $\left(3x - \frac{4}{x}\right)^8$. *(2 marks)*

(Q2c, **2011 HSC**) Easy

10 Find an expression for the coefficient of x^8y^4 in the expansion of $(2x + 3y)^{12}$. *(2 marks)*

(Q1d, **2008 HSC**) Easy

11 Use the binomial theorem to find the term independent of x in the expansion of $\left(2x - \frac{1}{x^2}\right)^{12}$. *(3 marks)*

(Q2b, **2005 HSC**) Easy

12 Find the coefficient of x^4 in the expansion of $(2 + x^2)^5$. *(2 marks)*

(Q2d, **2003 HSC**) Easy

13 Find the term independent of x in the binomial expansion of $\left(x^2 - \frac{1}{x}\right)^9$. *(3 marks)*

(Q2d, **2001 HSC**) Easy

14 By using the binomial expansion, show that

$$(q+p)^n - (q-p)^n = 2\binom{n}{1}q^{n-1}p + 2\binom{n}{3}q^{n-3}p^3 + \cdots$$

What is the last term in the expansion when n is odd? What is the last term in the expansion when n is even? *(3 marks)*

(Q5b, **2001 HSC**) Hard

15 Find the coefficient of x^6 in the expansion of $(5 + 2x^2)^7$. *(3 marks)*

(Q2b, **2000 HSC**) Easy

16 Using the fact that $(1+x)^4(1+x)^9 = (1+x)^{13}$, show that

$$^4C_0\,^9C_4 + {}^4C_1\,^9C_3 + {}^4C_2\,^9C_2 + {}^4C_3\,^9C_1 + {}^4C_4\,^9C_0 = {}^{13}C_4.$$

(2 marks)

(Q7a, **1996 HSC**) Medium

17 Find the value of the term that does not depend on x in the expansion of $\left(x^2 + \frac{3}{x}\right)^6$. *(3 marks)*

(Q3b, **1995 HSC**) Medium

Year 11 Binomial expansions and Pascal's triangle—Worked Answers

1 $\left(1-\frac{x}{2}\right)^8 = \binom{8}{0}+\binom{8}{1}\left(-\frac{x}{2}\right)^1+\binom{8}{2}\left(-\frac{x}{2}\right)^2+\binom{8}{3}\left(-\frac{x}{2}\right)^3+\dots$

$\therefore$ coefficient of x^2 is given by $\binom{8}{2}\left(\frac{1}{4}\right)=7$, ✓

and coefficient of x^3 is given by $\binom{8}{3}\left(-\frac{1}{8}\right)=-7$ ✓ *(2 marks)*

2 $(2a-b)^4 = {}^4C_0(2a)^4+{}^4C_1(2a)^3(-b)+{}^4C_2(2a)^2(-b)^2+{}^4C_3(2a)^1(-b)^3+{}^4C_4(-b)^4$ ✓

$= 16a^4-32a^3b+24a^2b^2-8ab^3+b^4$ ✓ *(2 marks)*

3 **i** $(1+x)^{2n}=(1+x)^n(1+x)^n$

Now $(1+x)^n=\binom{n}{0}+\binom{n}{1}x+\binom{n}{2}x^2+\dots+\binom{n}{n}x^n$

$(1+x)^{2n}=\binom{2n}{0}+\binom{2n}{1}x+\binom{2n}{2}x^2+\dots+\binom{2n}{n}x^n+\dots+\binom{2n}{2n}x^{2n}$

The coefficient of x^n is $\binom{2n}{n}$.

$(1+x)^n(1+x)^n=\left(\binom{n}{0}+\binom{n}{1}x+\binom{n}{2}x^2+\dots+\binom{n}{n}x^n\right)\left(\binom{n}{0}+\binom{n}{1}x+\binom{n}{2}x^2+\dots+\binom{n}{n}x^n\right)$ ✓

Term involving x^n is $\binom{n}{0}\binom{n}{n}x^n+\binom{n}{1}x\binom{n}{n-1}x^{n-1}+\binom{n}{2}x^2\binom{n}{n-2}x^{n-2}+\dots$

So the coefficient of x^n is $\binom{n}{0}\binom{n}{n}+\binom{n}{1}\binom{n}{n-1}+\binom{n}{2}\binom{n}{n-2}+\dots+\binom{n}{n}\binom{n}{n-n}$.

But $\binom{n}{n-k}=\binom{n}{k}$.

So the coefficient of x^n is $\binom{n}{0}^2+\binom{n}{1}^2+\dots+\binom{n}{n}^2$.

Equating coefficients $\binom{2n}{n}=\binom{n}{0}^2+\binom{n}{1}^2+\dots+\binom{n}{n}^2$. ✓ *(2 marks)*

ii n women and n men

If k women are chosen, the number of ways the women can be chosen is $\binom{n}{k}$.

As there are equal numbers of men and women to be chosen, the number of ways that the men can be chosen will be equal to the number of ways that the women can be chosen.

So for each way that the women can be chosen there are $\binom{n}{k}$ ways that the men can be chosen.

So the number of ways the people can be chosen is $\binom{n}{k}^2$. ✓

The total number of ways is $\binom{n}{0}^2+\binom{n}{1}^2+\dots+\binom{n}{n}^2$.

So, from part **i**, the number of ways is $\binom{2n}{n}$. ✓ *(2 marks)*

iii If k women are chosen, the number of ways that one leader can be chosen is $\binom{k}{1}$ or k.

For each way that a female leader can be chosen, there are k ways that a male leader can be chosen.

So the number of ways to choose k women and k men and then the leaders is $k^2\binom{n}{k}^2$. ✓

The total number of ways to choose the people and then the leaders $=0^2\binom{n}{0}^2+1^2\binom{n}{1}^2+2^2\binom{n}{2}^2+\dots+n^2\binom{n}{n}^2$

$=1^2\binom{n}{1}^2+2^2\binom{n}{2}^2+\dots+n^2\binom{n}{n}^2$ ✓ *(2 marks)*

iv If the leader is chosen first, then there is one leader chosen from n women so there are n ways that woman can be chosen.
For each of those there are n ways that the male leader can be chosen.
So the number of ways the two leaders can be chosen is n^2.
Once the leader has been chosen there are $n-1$ women, and men, left to choose from.

From part **ii** the number of ways to choose those is $\binom{2(n-1)}{n-1}$.

Total ways to choose the leaders first and then the rest of the group $=n^2\binom{2n-2}{n-1}$. ✓

The total number of ways must be the same whether the leaders are chosen first or last.

So $1^2\binom{n}{1}^2 + 2^2\binom{n}{2}^2 + \ldots + n^2\binom{n}{n}^2 = n^2\binom{2n-2}{n-1}$. ✓ *(2 marks)*

4 $(5x+2)^{20} = \binom{20}{0}(5x)^0 2^{20-0} + \binom{20}{1}(5x)^1 2^{20-1} + \binom{20}{2}(5x)^2 2^{20-2} + \ldots$

The coefficients of x^k and x^{k+1} are equal.

$$\text{So} \binom{20}{k}(5)^k 2^{20-k} = \binom{20}{k+1}(5)^{k+1} 2^{20-(k+1)} \quad ✓$$

$$\frac{20!}{k!(20-k)!} \times 5^k \times 2^{20-k} = \frac{20!}{(k+1)!(20-(k+1))!} \times 5^{k+1} \times 2^{19-k}$$

$$\frac{2}{k!(20-k)!} = \frac{5}{(k+1)!(19-k)!} \quad ✓$$

$$2(k+1) = 5(20-k)$$
$$2k + 2 = 100 - 5k$$
$$7k = 98$$
$$k = 14 \quad ✓$$

(3 marks)

5 Consider $\left(x^3 + \frac{1}{x^4}\right)^7$.

General term $= \binom{7}{k}(x^3)^{7-k}(x^{-4})^k$

$= \binom{7}{k}x^{21-3k}x^{-4k}$

$= \binom{7}{k}x^{21-7k}$

So when $k = 3$ there is a non-zero constant term.

[This is not the case with any of the other options.]

Answer C

6 **i** $\left(2x + \frac{1}{3x}\right)^{18} = a_0x^{18} + a_1x^{16} + a_2x^{14} + \ldots$

But $\left(2x + \frac{1}{3x}\right)^{18} = \binom{18}{0}(2x)^{18} + \binom{18}{1}(2x)^{17}\left(\frac{1}{3x}\right) + \binom{18}{2}(2x)^{16}\left(\frac{1}{3x}\right)^2 + \ldots$

So $a_2x^{14} = \binom{18}{2}(2x)^{16}\left(\frac{1}{3x}\right)^2$ ✓

$= \binom{18}{2}2^{16}\left(\frac{1}{3}\right)^2 x^{14}$

$a_2 = \binom{18}{2}\frac{2^{16}}{3^2}$ $\quad [= 17 \times 2^{16} = 1114112]$ ✓ *(2 marks)*

ii Term independent of x

$= \binom{18}{9}(2x)^9\left(\frac{1}{3x}\right)^9$ ✓

$= \binom{18}{9}2^9\left(\frac{1}{3}\right)^9$

$= \binom{18}{9}\left(\frac{2}{3}\right)^9$ ✓ *(2 marks)*

7 $\left(2x - \frac{5}{x^3}\right)^{12}$

$= \binom{12}{0}(2x)^{12} + \binom{12}{1}(2x)^{11}\left(\frac{-5}{x^3}\right) + \binom{12}{2}(2x)^{10}\left(\frac{-5}{x^3}\right)^2 + \ldots$

General term $= \binom{12}{k}(2x)^{12-k}\left(\frac{-5}{x^3}\right)^k$

$= \binom{12}{k}2^{12-k}x^{12-k}(-5)^k x^{-3k}$

$= \binom{12}{k}2^{12-k}(-5)^k x^{12-4k}$

Constant term when $12 - 4k = 0$

$k = 3$

Constant term $= \binom{12}{3}2^9(-5)^3$

$= -\binom{12}{3}2^9 5^3$

Answer C

8 **i** $\left(2x^3 - \frac{1}{x}\right)^{12}$

General term $= {}^{12}C_k(2x^3)^k\left(-\frac{1}{x}\right)^{12-k}$

$= {}^{12}C_k\, 2^k x^{3k}(-1)^{12-k} x^{k-12}$

$= {}^{12}C_k\,(-1)^k 2^k x^{4k-12}$ ✓

Constant term occurs when $4k - 12 = 0$

$4k = 12$

$k = 3$

Constant term $= {}^{12}C_3(-2)^3$

$[= -1760]$ ✓

(2 marks)

ii $\left(2x^3 - \frac{1}{x}\right)^n$

General term $= {}^nC_k(2x^3)^k\left(-\frac{1}{x}\right)^{n-k}$

$= {}^nC_k\,(-1)^{n-k} 2^k x^{4k-n}$

Constant term occurs when $4k - n = 0$

$n = 4k$

$\therefore \left(2x^3 - \frac{1}{x}\right)^n$ will have a non-zero constant term when n is a multiple of 4. ✓ *(1 mark)*

9 $\left(3x - \frac{4}{x}\right)^8$

$= {}^8C_0(3x)^8 - {}^8C_1(3x)^7\left(\frac{4}{x}\right) + {}^8C_2(3x)^6\left(\frac{4}{x}\right)^2 - \ldots$

General term $= {}^8C_k(-1)^k(3x)^{8-k}\left(\frac{4}{x}\right)^k$

$= {}^8C_k(-1)^k 3^{8-k}x^{8-k}4^k(x^{-1})^k$

$= {}^8C_k(-1)^k 3^{8-k}x^{8-k}4^k x^{-k}$

$= {}^8C_k(-1)^k 3^{8-k}4^k x^{8-2k}$ ✓

For the term involving x^2,

$8 - 2k = 2$

$2k = 6$

$k = 3$

Coefficient of $x^2 = {}^8C_3(-1)^3 3^5 4^3$
$= -{}^8C_3 3^5 4^3$ [= −870 912] ✓

(2 marks)

10 $u_{r+1} = {}^nC_r a^{n-r} b^r$

For $(2x + 3y)^{12}$,
$u_{r+1} = {}^{12}C_r(2x)^{12-r}(3y)^r$
$\therefore u_5 = {}^{12}C_4(2x)^8(3y)^4$
$= {}^{12}C_4 . 2^8 . x^8 . 3^4 . y^4$
$= {}^{12}C_4 . 2^8 . 3^4 . x^8 . y^4$ ✓

$\therefore$ The coefficient of x^8y^4
$= {}^{12}C_4 . 2^8 . 3^4$
$= 10\,264\,320$ ✓

(2 marks)

11 ${}^{12}C_r(2x)^{12-r}\left(\frac{-1}{x^2}\right)^r$

$= {}^{12}C_r(2x)^{12-r} . \frac{(-1)^r}{x^{2r}}$

$= {}^{12}C_r 2^{12-r}(-1)^r . \frac{x^{12-r}}{x^{2r}}$

$= {}^{12}C_r 2^{12-r}(-1)^r x^{12-r-2r}$
$= {}^{12}C_r 2^{12-r}(-1)^r x^{12-3r}$ ✓

This term is independent of x when
$12 - 3r = 0$
$12 = 3r$
$\therefore r = 4$ ✓

The term independent of x
$= {}^{12}C_4 2^{12-4}(-1)^4$
$= {}^{12}C_4 2^8 (-1)^4$
$= 126\,720.$ ✓

(3 marks)

12 $(2 + x^2)^5$

General term $= {}^5C_r 2^{5-r}(x^2)^r$
$= {}^5C_r 2^{5-r} x^{2r}$ ✓

The term in x^4 in this expansion
occurs when $2r = 4 \therefore r = 2$

Its coefficient is ${}^5C_2 2^{5-2} = {}^5C_2 2^3$
$= 10 \times 8$
$= 80.$ ✓

(2 marks)

13 $\left(x^2 - \frac{1}{x}\right)^9$

General term $= {}^9C_i (x^2)^{9-i}\left(-\frac{1}{x}\right)^i$
$= {}^9C_i x^{18-2i}(-1)^i x^{-i}$
$= {}^9C_i (-1)^i x^{18-3i}$ ✓

The term independent of x will
occur when $18 - 3i = 0$
$\therefore i = 6$ ✓

$\therefore$ The term independent of x
$= {}^9C_6(-1)^6x^0$
$= {}^9C_6$
$= 84.$ ✓

(3 marks)

14 $(q + p)^n - (q - p)^n$

$$= \left[\binom{n}{0}q^n + \binom{n}{1}q^{n-1}p + \binom{n}{2}q^{n-2}p^2 + \binom{n}{3}q^{n-3}p^3 + \ldots\right] - \left[\binom{n}{0}q^n - \binom{n}{1}q^{n-1}p + \binom{n}{2}q^{n-2}p^2 - \binom{n}{3}q^{n-3}p^3 + \ldots\right]$$

$$= \binom{n}{0}q^n + \binom{n}{1}q^{n-1}p + \binom{n}{2}q^{n-2}p^2 + \binom{n}{3}q^{n-3}p^3 + \ldots - \binom{n}{0}q^n + \binom{n}{1}q^{n-1}p - \binom{n}{2}q^{n-2}p^2 + \binom{n}{3}q^{n-3}p^3 - \ldots$$ ✓

$$= 2\binom{n}{1}q^{n-1}p + 2\binom{n}{3}q^{n-3}p^3 + \ldots$$

If n is an odd number, then the last term of the expansion will be $2\binom{n}{n}q^0p^n = 2p^n$ ✓

If n is even, then the last term will be $2\binom{n}{n-1}q^1p^{n-1} = 2nqp^{n-1}$. ✓

(3 marks)

15 $(5 + 2x^2)$

General term $= {}^7C_i\, 5^{7-i}\,(2x^2)^i$

$= {}^7C_i\, 5^{7-i}\, 2^i x^{2i}$ ✓

The term in x^6 is the term with $i = 3$: i.e. ${}^7C_3\, 5^{7-3}\, 2^3 x^6$. ✓

Its coefficient is ${}^7C_3\, 5^4\, 2^3 = 5000\, {}^7C_3 = 175\,000$. ✓

(3 marks)

---

16 $(1 + x)^4(1 + x)^9 = ({}^4C_0 + {}^4C_1x + {}^4C_2x^2 + \ldots + {}^4C_4x^4)({}^9C_0 + {}^9C_1x + {}^9C_2x^2 + \ldots + {}^9C_9x^9)$

The term in x^4 of this expansion has the coefficient:

${}^4C_0{}^9C_4 + {}^4C_1{}^9C_3 + {}^4C_2{}^9C_2 + {}^4C_3{}^9C_1 + {}^4C_4{}^9C_0$ ✓

Now $(1 + x)^4\,(1 + x)^9 = (1 + x)^{13}$ and the coefficient of the term in x^4 in the expansion $(1+ x)^{13}$ is ${}^{13}C_4$

$\therefore\ {}^4C_0{}^9C_4 + {}^4C_1{}^9C_3 + {}^4C_2{}^9C_2 + {}^4C_3{}^9C_1 + {}^4C_4{}^9C_0 = {}^{13}C_4$ ✓

(2 marks)

---

17 $\left(x^2 + \dfrac{3}{x}\right)^6$

The general term

$= \binom{6}{k} x^{12-2k}.\dfrac{3^k}{x^k}$

$= \binom{6}{k} 3^k.\dfrac{x^{12-2k}}{x^k}$

$= \binom{6}{k} 3^k.x^{12-3k}$ ✓

For the term independent of x:

$12 - 3k = 0$

$12 = 3k$

$k = 4$ ✓

$\therefore$ The term independent of x

$= \binom{6}{4} 3^4.x^0$

$= \dfrac{6 \times 5}{1 \times 2} \times 3^4$

$= 15 \times 3^4$

$= 1215.$ ✓

(3 marks)

1 Use mathematical induction to prove that $15^n + 6^{2n+1}$ is divisible by 7 for all integers $n \geq 0$. *(3 marks)*

(Q12f, **2022 HSC**) Medium

2 Use mathematical induction to prove that
$$\frac{1}{1 \times 2 \times 3} + \frac{1}{2 \times 3 \times 4} + \cdots + \frac{1}{n(n+1)(n+2)} = \frac{1}{4} - \frac{1}{2(n+1)(n+2)}$$
for all integers $n \geq 1$. *(3 marks)*

(Q12c, **2021 HSC**) Easy

3 Use the principle of mathematical induction to show that for all integers $n \geq 1$,
$1 \times 2 + 2 \times 5 + 3 \times 8 + \ldots + n(3n - 1) = n^2(n + 1)$. *(3 marks)*

(Q12a, **2020 HSC**) Easy

4 Prove by mathematical induction that, for all integers $n \geq 1$,
$1(1!) + 2(2!) + 3(3!) + \ldots + n(n!) = (n + 1)! - 1$. *(3 marks)*

(Q14a, **2019 HSC**) Easy

5 Prove by mathematical induction that, for $n \geq 1$,
$$2 - 6 + 18 - 54 + \ldots + 2(-3)^{n-1} = \frac{1-(-3)^n}{2}.$$
(3 marks)

(Q13a, **2018 HSC**) Easy

6 Prove by mathematical induction that $8^{2n+1} + 6^{2n-1}$ is divisible by 7, for any integer $n \geq 1$. *(3 marks)*

(Q14a, **2017 HSC**) Medium

7 **i** Show that $4n^3 + 18n^2 + 23n + 9$ can be written as $(n + 1)(4n^2 + 14n + 9)$. *(1 mark)* Easy

ii Using the result in part **i**, or otherwise, prove by mathematical induction that, for $n \geq 1$,
$1 \times 3 + 3 \times 5 + 5 \times 7 + \ldots + (2n - 1)(2n + 1) = \frac{1}{3}n(4n^2 + 6n - 1)$. *(3 marks)* Easy

(Q14a, **2016 HSC**)

8 Prove by mathematical induction that for all integers $n \geq 1$,
$$\frac{1}{2!} + \frac{2}{3!} + \frac{3}{4!} + \ldots + \frac{n}{(n+1)!} = 1 - \frac{1}{(n+1)!}.$$
(3 marks)

(Q13c, **2015 HSC**) Medium

9 Use mathematical induction to prove that $2n + (-1)^{n+1}$ is divisible by 3 for all integers $n \geq 1$. *(3 marks)*

(Q13a, **2014 HSC**) Medium

10 Use mathematical induction to prove that $2^{3n} - 3^n$ is divisible by 5 for $n \geq 1$. *(3 marks)*

(Q12a, **2012 HSC**) Medium

11 Use mathematical induction to prove that, for $n \geq 1$,
$$1 \times 5 + 2 \times 6 + 3 \times 7 + \cdots + n(n+4) = \frac{1}{6}n(n+1)(2n+13).$$
(3 marks)

(Q6a, **2011 HSC**) Easy

12 Prove by induction that
$47^n + 53 \times 147^{n-1}$
is divisible by 100 for all integers $n \geq 1$. *(3 marks)*

(Q7a, **2010 HSC**) Medium

13 Use mathematical induction to prove that, for integers $n \geq 1$,
$$1 \times 3 + 2 \times 4 + 3 \times 5 + \ldots + n(n+2) = \frac{n}{6}(n+1)(2n+7).$$
(3 marks)

(Q3b, **2008 HSC**) Easy

14 Use mathematical induction to prove that $7^{2n-1} + 5$ is divisible by 12, for all integers $n \geq 1$. *(3 marks)*

(Q4b, **2007 HSC**) Medium

15 **i** Use the fact that $\tan(\alpha - \beta) = \frac{\tan\alpha - \tan\beta}{1 + \tan\alpha\tan\beta}$ to show that
$1 + \tan n\theta \tan(n+1)\theta = \cot\theta(\tan(n+1)\theta - \tan n\theta)$. *(1 mark)* Medium

ii Use mathematical induction to prove that, for all integers $n \geq 1$,

$\tan\theta \tan 2\theta + \tan 2\theta \tan 3\theta + \dots$
$+ \tan n\theta \tan(n+1)\theta$
$= -(n+1) + \cot\theta \tan(n+1)\theta.$

(3 marks) Medium

(Q5d, **2006 HSC**)

16 Use mathematical induction to prove that

$$\frac{1}{1\times 3} + \frac{1}{3\times 5} + \frac{1}{5\times 7} + \dots + \frac{1}{(2n-1)(2n+1)} = \frac{n}{2n+1}$$

for all positive integers n. *(3 marks)*

(Q3d, **2003 HSC**) Medium

17 Use the principle of mathematical induction to show that

$2 \times 1! + 5 \times 2! + 10 \times 3! + \dots + (n^2 + 1)n!$
$= n(n+1)!$

for all positive integers n. *(3 marks)*

(Q5a, **2002 HSC**) Medium

18 Prove by induction that

$n^3 + (n+1)^3 + (n+2)^3$

is divisible by 9 for $n = 1, 2, 3, \dots$ *(3 marks)*

(Q6a, **2001 HSC**) Medium

19 Use mathematical induction to prove that

$1 + 3 + 6 + \dots + \frac{1}{2}n(n+1) = \frac{1}{6}n(n+1)(n+2)$

for all integers $n = 1, 2, 3, \dots$. *(3 marks)*

(Q4a, **2000 HSC**) Easy

20 Prove by induction that, for all integers $n \geq 1$,

$(n+1)(n+2) \dots (2n-1)2n$
$= 2^n[1 \times 3 \times \dots \times (2n-1)]$ *(3 marks)*

(Q5a, **1999 HSC**) Easy

21 Use the method of mathematical induction to prove that $4^n + 14$ is a multiple of 6 for $n \geq 1$.

(3 marks)

(Q3a, **1998 HSC**) Medium

22 Prove by mathematical induction that $n^3 + 2n$ is divisible by 3, for all positive integers n.

(Q3c, **1994 HSC**) Medium

23 For $n = 1, 2, 3, \dots$, let

$S_n = 1^2 + 2^2 + \dots + n^2.$

i Use mathematical induction to prove that, for $n = 1, 2, 3, \dots$,

$S_n = \frac{1}{6}n(n+1)(2n+1).$ Easy

ii By using the result of **i** estimate the least n such that $S_n \geq 10^9$. Medium

(Q5a, **1993 HSC**)

24 Let $S_n = 1 \times 2 + 2 \times 3 + \dots + (n-1) \times n.$

Use mathematical induction to prove that, for all integers n with $n \geq 2$,

$S_n = \frac{1}{3}(n-1)\,n(n+1).$

(Q4b, **1992 HSC**) Medium

Year 12 Proof by mathematical induction—Worked Answers

1 Prove true for $n = 0$

$15^0 + 6^{2(0)+1} = 1 + 6$
$= 7$

Since 7 is divisible by 7, then true for $n = 0$. ✓

Assume true for $n = k$.

$15^k + 6^{2k+1} = 7p$ where p is an integer

$\therefore 6^{2k+1} = 7p - 15^k$

Prove true for $n = k + 1$.

$15^{k+1} + 6^{2(k+1)+1} = 7q$ where q is an integer

LHS $= 15^{k+1} + 6^{2k+2+1}$
$= 15^{k+1} + 6^{2k+1+2}$
$= 15 \times 15^k + 6^2 \times 6^{2k+1}$
$= 15 \times 15^k + 6^2(7p - 15^k)$ from Assumption step
$= 15 \times 15^k + 252p - 36 \times 15^k$
$= 252p - 21 \times 15^k$
$= 7(36p - 3 \times 15^k)$
$= 7q$ for some integer q ✓
(since p and k are integers)

$\therefore 15^{k+1} + 62^{(k+1)+1}$ is divisible by 7.

The statement is true for $n = 0$ and true for $n = k + 1$.
So by mathematical induction it is true for all integers $n \geq 0$. ✓

(3 marks)

2 We want to prove that:

$$\frac{1}{1\times2\times3}+\frac{1}{2\times3\times4}+\ldots+\frac{1}{n(n+1)(n+2)}=\frac{1}{4}-\frac{1}{2(n+1)(n+2)}$$

Prove true for $n = 1$.

LHS $= \frac{1}{1\times2\times3}$
$= \frac{1}{6}$

RHS $= \frac{1}{4}-\frac{1}{2(2)(3)}$
$= \frac{1}{4}-\frac{1}{12}$
$= \frac{1}{6}$

$\therefore$ the statement is true for $n = 1$. ✓

Assume true for $n = k$,
i.e. assume

$$\frac{1}{1\times2\times3}+\frac{1}{2\times3\times4}+\ldots+\frac{1}{k(k+1)(k+2)}=\frac{1}{4}-\frac{1}{2(k+1)(k+2)}$$

We want to prove true for $n = k + 1$.
If $n = k + 1$, then required to prove:

$$\frac{1}{1\times2\times3}+\frac{1}{2\times3\times4}+\ldots+\frac{1}{k(k+1)(k+2)}+\frac{1}{(k+1)(k+2)(k+3)}=\frac{1}{4}-\frac{1}{2(k+2)(k+3)}$$

$$\text{LHS}=\frac{1}{4}-\frac{1}{2(k+1)(k+2)}+\frac{1}{(k+1)(k+2)(k+3)}$$
$$=\frac{1}{4}-\left(\frac{k+3}{2(k+1)(k+2)(k+3)}-\frac{2}{2(k+1)(k+2)(k+3)}\right)$$
$$=\frac{1}{4}-\left(\frac{k+1}{2(k+1)(k+2)(k+3)}\right)$$
$$=\frac{1}{4}-\frac{1}{2(k+2)(k+3)} \quad ✓$$
$= \text{RHS}$

The statement is true for $n = k + 1$.
Since the statement is true for $n =1$ and $n = k +1$, then it is true for all integers $n \geq 1$. ✓ *(3 marks)*

3 $1 \times 2 + 2 \times 5 + 3 \times 8 + \ldots + n(3n - 1) = n^2(n + 1)$

If $n = 1$,
LHS $= 1 \times 2 = 2$
RHS $= 1^2(1 + 1) = 2$

The statement is true for $n = 1$. ✓

Assume it is true for $n = k$,
i.e. $1 \times 2 + 2 \times 5 + 3 \times 8 + \ldots + k(3k - 1) = k^2(k + 1)$

We need to show that it is also true for $n = k + 1$.
We wish to prove that:

$1 \times 2 + 2 \times 5 + 3 \times 8 + \ldots + k(3k - 1) + (k + 1)(3(k + 1) - 1) = (k + 1)^2(k + 1 + 1)$
i.e. $1 \times 2 + 2 \times 5 + 3 \times 8 + \ldots + k(3k - 1) + (k + 1)(3k + 2) = (k + 1)^2(k + 2)$

LHS $= k^2(k + 1) + (k + 1)(3k + 2)$
$= (k + 1)(k^2 + 3k + 2)$
$= (k + 1)(k + 2)(k + 1)$
$= (k + 1)^2(k + 2)$ ✓
$= \text{RHS}$

So, if the statement is true for $n = k$, it is also true for $n = k + 1$.

It is true for $n = 1$.
So it is true for $n = 1 + 1 = 2$.
So it is true for $n = 2 + 1 = 3$.
And so on.
By the process of induction it is true for all integers n, $n \geq 1$. ✓

(3 marks)

4 $1(1!) + 2(2!) + 3(3!) + \ldots + n(n!) = (n + 1)! - 1$

If $n = 1$,

$$\begin{aligned} \text{LHS} &= 1(1!) \\ &= 1 \\ \text{RHS} &= (1 + 1)! - 1 \\ &= 2! - 1 \\ &= 1 \end{aligned}$$

The statement is true for $n = 1$. ✓

Assume it is true for $n = k$.
i.e. assume $1(1!) + 2(2!) + 3(3!) + \ldots + k(k!)$
$= (k + 1)! - 1$

We need to show that it is also true for $n = k + 1$.
We wish to prove that:
$1(1!) + 2(2!) + 3(3!) + \ldots + k(k!) + (k + 1)(k + 1)!$
$= (k + 2)! - 1$

Now $1(1!) + 2(2!) + 3(3!) + \ldots + k(k!) + (k + 1)(k + 1)!$
$= (k + 1)! - 1 + (k + 1)(k + 1)!$
$= (k + 1)! - 1 + k(k + 1)! + (k + 1)!$
$= k(k + 1)! + 2(k + 1)! - 1$
$= (k + 1)!(k + 2) - 1$
$= (k + 2)! - 1$ ✓

So if the statement is true for $n = k$ it is also true for $n = k + 1$.
It is true for $n = 1$.
So it is true for $n = 1 + 1 = 2$.
So it is true for $n = 2 + 1 = 3$.
And so on.
By the process of induction it is true for all integers n, $n \geq 1$. ✓ *(3 marks)*

5 To prove $2 - 6 + 18 - 54 + \ldots + 2(-3)^{n-1} = \dfrac{1-(-3)^n}{2}$.

When $n = 1$,

$$\begin{aligned} \text{LHS} &= 2(-3)^{1-1} \\ &= 2 \\ \text{RHS} &= \frac{1-(-3)^1}{2} \\ &= 2 \end{aligned}$$

$\therefore$ it is true for $n = 1$. ✓

Assume true for $n = k$

i.e. assume $2 - 6 + 18 - 54 + \ldots + 2(-3)^{k-1} = \dfrac{1-(-3)^k}{2}$

We want to prove that it is true for $n = k + 1$.

i.e. $2 - 6 + 18 - 54 + \ldots + 2(-3)^k = \dfrac{1-(-3)^{k+1}}{2}$

$$\begin{aligned} \text{LHS} &= 2 - 6 + 18 - 54 + \ldots + 2(-3)^{k-1} + 2(-3)^k \\ &= \frac{1-(-3)^k}{2} + 2(-3)^k \\ &= \frac{1-(-3)^k + 4(-3)^k}{2} \\ &= \frac{1+3(-3)^k}{2} \\ &= \frac{1-(-3)^{k+1}}{2} \\ &= \text{RHS} \end{aligned}$$ ✓

So, if true for $n = k$, it is also true for $n = k + 1$.
It is true for $n = 1$, so it is true for $n = 1 + 1 = 2$.
It is true for $n = 2$, so it is true for $n = 2 + 1 = 3$.
And so on.
By the process of induction it is true for all integers $n \geq 1$.

✓ *(3 marks)*

6 To prove $8^{2n+1} + 6^{2n-1}$ is divisible by 7 for all $n \geq 1$

When $n = 1$,

$$\begin{aligned} 8^{2 \times 1+1} + 6^{2 \times 1-1} &= 8^3 + 6 \\ &= 518 \\ &= 7 \times 74. \end{aligned}$$

It is true for $n = 1$. ✓

Assume it is true for $n = k$.
i.e. assume $8^{2k+1} + 6^{2k-1} = 7m$ for some positive integer m.

When $n = k + 1$,

$$\begin{aligned} &8^{2(k+1)+1} + 6^{2(k+1)-1} \\ &= 8^{2k+3} + 6^{2k+1} \\ &= 8^2 \times 8^{2k+1} + 6^2 \times 6^{2k-1} \\ &= 64 \times 8^{2k+1} + 36 \times 6^{2k-1} \\ &= 28 \times 8^{2k+1} + 36 \times 8^{2k+1} + 36 \times 6^{2k-1} \\ &= 7 \times 4 \times 8^{2k+1} + 36(7m) \\ &= 7[4 \times 8^{2k+1} + 36m] \end{aligned}$$ ✓

So if it is true for $n = k$, it is also true for $n = k + 1$.
It is true for $n = 1$, so it is true for $n = 1 + 1 = 2$.
It is true for $n = 2$, so it is true for $n = 2 + 1 = 3$.
And so on.
By the process of induction it is true for all integers $n \geq 1$.
$8^{2n+1} + 6^{2n-1}$ is divisible by 7 for all $n \geq 1$. ✓

(3 marks)

7 **i**

$$\begin{aligned} &(n + 1)(4n^2 + 14n + 9) \\ &= n(4n^2 + 14n + 9) + 1(4n^2 + 14n + 9) \\ &= 4n^3 + 14n^2 + 9n + 4n^2 + 14n + 9 \\ &= 4n^3 + 18n^2 + 23n + 9 \end{aligned}$$

So $4n^3 + 18n^2 + 23n + 9$ can be written as $(n + 1)(4n^2 + 14n + 9)$. ✓

(1 mark)

ii We want to prove, for $n \geq 1$, that
$1 \times 3 + 3 \times 5 + 5 \times 7 + \ldots + (2n - 1)(2n + 1)$
$= \frac{1}{3}n(4n^2 + 6n - 1)$

If $n = 1$,
$\text{LHS} = 1 \times 3 = 3$

$\text{RHS} = \frac{1}{3} \times 1 \times (4 \times 1^2 + 6 \times 1 - 1) = 3$

$\therefore$ it is true for $n = 1$. ✓

Assume true for $n = k$,
i.e. assume $1 \times 3 + 3 \times 5 + 5 \times 7 + \ldots + (2k - 1)(2k + 1) = \frac{1}{3}k(4k^2 + 6k - 1)$

When $n = k + 1$,

$$\begin{aligned} \text{LHS} &= 1 \times 3 + 3 \times 5 + \ldots + (2k - 1)(2k + 1) + (2(k + 1) - 1)(2(k + 1) + 1) \\ &= \frac{1}{3}k(4k^2 + 6k - 1) + (2k + 1)(2k + 3) \\ &= \frac{1}{3}(4k^3 + 6k^2 - k) + 4k^2 + 8k + 3 \\ &= \frac{1}{3}(4k^3 + 6k^2 - k + 12k^2 + 24k + 9) \\ &= \frac{1}{3}(4k^3 + 18k^2 + 23k + 9) \end{aligned}$$

$= \frac{1}{3}(k+1)(4k^2+14k+9)$ (from part **i**) ✓

$\text{RHS} = \frac{1}{3}(k+1)(4(k+1)^2+6(k+1)-1)$

$= \frac{1}{3}(k+1)(4k^2+8k+4+6k+6-1)$

$= \frac{1}{3}(k+1)(4k^2+14k+9)$

$= \text{LHS}$

So, if true for $n = k$ it is also true for $n = k + 1$.

It is true for $n = 1$, so it is true for $n = 2$ and so for $n = 3$, and so on.

By the process of induction it is true for all $n \geq 1$. ✓ *(3 marks)*

8 To prove for all integers $n \geq 1$:

$$\frac{1}{2!}+\frac{2}{3!}+\frac{3}{4!}+\ldots+\frac{n}{(n+1)!}=1-\frac{1}{(n+1)!}$$

When $n = 1$,

$\text{LHS} = \frac{1}{2!} = \frac{1}{2}$

$\text{RHS} = 1 - \frac{1}{2!} = 1 - \frac{1}{2} = \frac{1}{2}$

So the statement is true for $n = 1$. ✓

Assume the statement is true for $n = k$

i.e. assume $\frac{1}{2!}+\frac{2}{3!}+\frac{3}{4!}+\ldots+\frac{k}{(k+1)!}=1-\frac{1}{(k+1)!}$

We need to show that it is true for $n = k + 1$

i.e. that $\frac{1}{2!}+\frac{2}{3!}+\frac{3}{4!}+\ldots+\frac{k+1}{(k+2)!}=1-\frac{1}{(k+2)!}$

When $n = k + 1$

$\text{LHS} = \frac{1}{2!}+\frac{2}{3!}+\frac{3}{4!}+\ldots+\frac{k}{(k+1)!}+\frac{k+1}{(k+2)!}$

$= 1 - \frac{1}{(k+1)!}+\frac{k+1}{(k+2)!}$

$= 1 - \frac{k+2}{(k+2)(k+1)!}+\frac{k+1}{(k+2)!}$

$= 1 - \left(\frac{k+2}{(k+2)!}-\frac{k+1}{(k+2)!}\right)$

$= 1 - \frac{1}{(k+2)!}$ ✓

$= \text{RHS}$

$\therefore$ if it is true for $n = k$, it is also true for $n = k + 1$.

As it is true for $n = 1$, it is true for $n = 2$.

As it is true for $n = 2$, it is true for $n = 3$.

By the process of induction it is true for all integers $n \geq 1$. ✓

(3 marks)

9 To prove $2^n + (-1)^{n+1}$ is divisible by 3 for $n \geq 1$.

When $n = 1$,

$2^1 + (-1)^2 = 3$

$\therefore$ it is true when $n = 1$. ✓

Assume $2^k + (-1)^{k+1} = 3m$ where m is an integer.

If $n = k + 1$,

$2^{k+1} + (-1)^{k+1+1}$

$= 2 \times 2^k + (-1) \times (-1)^{k+1}$

$= 3 \times 2^k - 1 \times 2^k - (-1)^{k+1}$

$= 3 \times 2^k - (2^k + (-1)^{k+1})$

$= 3 \times 2^k - 3m$

$= 3(2^k - m)$ ✓

So if true for $n = k$, it is also true for $n = k + 1$.

It is true for $n = 1$, so it is true for $n = 2$ and hence it is true for $n = 3$, and so on.

By the process of mathematical induction it is true for all integers $n \geq 1$. ✓

(3 marks)

10 Aim to prove that $2^{3n} - 3^n$ is divisible by 5 for $n \geq 1$.

When $n = 1$,

$2^3 - 3^1 = 8 - 3 = 5$

So it is true for $n = 1$. ✓

Assume true for $n = k$.

i.e. assume $2^{3k} - 3^k = 5m$

If $n = k + 1$,

$2^{3(k+1)} - 3^{k+1} = 2^{3k+3} - 3^{k+1}$

$= 2^{3k}.2^3 - 3^k.3^1$

$= 8(2^{3k}) - 3(3^k)$

$= (5 + 3)(2^{3k}) - 3(3^k)$

$= 5(2^{3k}) + 3(2^{3k}) - 3(3^k)$

$= 5(2^{3k}) + 3(2^{3k} - 3^k)$

$= 5(2^{3k}) + 3(5m)$

$= 5(2^{3k} + 3m)$ ✓

So, if true for $n = k$ it is also true for $n = k + 1$.

It is true for $n = 1$, so it is true for $n = 1 + 1 = 2$, so it is true for $n = 2 + 1 = 3$ and so on.

By the process of induction, $2^{3n} - 3^n$ is divisible by 5 for all integers n greater than or equal to 1. ✓

(3 marks)

11 We want to prove that:

$$1 \times 5 + 2 \times 6 + 3 \times 7 + \ldots + n(n+4) = \frac{1}{6}n(n+1)(2n+13)$$

If $n = 1$,

$\text{LHS} = 1 \times 5$

$= 5$

$\text{RHS} = \frac{1}{6} \times 1 \times (1+1) \times (2 \times 1 + 13)$

$= 5$

$\therefore$ the statement is true for $n = 1$. ✓

Assume it is true for $n = k$

i.e. assume

$$1 \times 5 + 2 \times 6 + 3 \times 7 + \ldots + k(k+4) = \frac{1}{6}k(k+1)(2k+13)$$

We want to prove true for $n = k + 1$

If $n = k + 1$,

$$\text{RHS} = \frac{1}{6}(k+1)((k+1)+1)(2(k+1)+13)$$
$$= \frac{1}{6}(k+1)(k+2)(2k+15)$$

$$\text{LHS} = 1 \times 5 + 2 \times 6 + \ldots + k(k+4) + (k+1)(k+5)$$
$$= \frac{1}{6}k(k+1)(2k+13) + (k+1)(k+5)$$
$$= \frac{1}{6}(k+1)[k(2k+13) + 6(k+5)]$$
$$= \frac{1}{6}(k+1)(2k^2 + 13k + 6k + 30)$$
$$= \frac{1}{6}(k+1)(2k^2 + 19k + 30)$$
$$= \frac{1}{6}(k+1)(k+2)(2k+15) \quad ✓$$
$$= \text{RHS}$$

So, if true for $n = k$ it is also true for $n = k + 1$.

It is true for $n = 1$, so it also true for $n = 2$, and because it is true for $n = 2$ it is true for $n = 3$ and so on.

By the process of induction it is true for all integers $n \geq 1$. ✓

(3 marks)

12 $47^n + 53 \times 147^{n-1}$

When $n = 1$,

$$47^1 + 53 \times 147^{1-1} = 47 + 53 \times 1$$
$$= 100$$

$\therefore$ $47^n + 53 \times 147^{n-1}$ is divisible by 100 when $n = 1$. ✓

Assume $47^n + 53 \times 147^{n-1}$ is divisible by 100 when $n = k$

i.e. assume $47^k + 53 \times 147^{k-1} = 100M$ where M is an integer.

Let $n = k + 1$.

$$47^{k+1} + 53 \times 147^k$$
$$= 47(47^k) + 147(53 \times 147^{k-1})$$
$$= 47(47^k) + 47(53 \times 147^{k-1}) + 100(53 \times 147^{k-1})$$
$$= 47(47^k + 53 \times 147^{k-1}) + 100(53 \times 147^{k-1})$$
$$= 47(100M) + 100(53 \times 147^{k-1})$$
$$= 100(47M + 53 \times 147^{k-1}) \quad ✓$$

$\therefore$ If $47^n + 53 \times 147^{n-1}$ is divisible by 100 when $n = k$ it is also divisible by 100 when $n = k + 1$.

$47^n + 53 \times 147^{n-1}$
is divisible by 100 when $n = 1$,
so it is also divisible by 100 when $n = 2$,
so it is also divisible by 100 when $n = 3$ and so on.

By the principle of mathematical induction $47^n + 53 \times 147^{n-1}$ is divisible by 100 for all positive integers $n \geq 1$. ✓

(3 marks)

13 To prove:

$$1 \times 3 + 2 \times 4 + 3 \times 5 + \ldots + n(n+2) = \frac{n}{6}(n+1)(2n+7) \quad \text{for integers } n \geqslant 1,$$

where $T_n = n(n+2)$

and $S_n = \frac{n}{6}(n+1)(2n+7)$

Prove the statement is true for $n = 1$:

$T_1 = 1(1+2) = 3$

$$S_1 = \frac{1}{6}(1+1)(2 \times 1 + 7)$$
$$= \frac{1}{6}(2)(9)$$
$$= 3$$

$\therefore$ The statement is true for $n = 1$. ✓

Assume the statement is true for $n = k$:

i.e. $1 \times 3 + 2 \times 4 + \ldots + k(k+2) = \frac{k}{6}(k+1)(2k+7)$

Prove that the statement is true for $n = k + 1$:

$$\text{i.e. } S_{k+1} = \frac{k+1}{6}(k+1+1)(2(k+1)+7)$$
$$= \frac{k+1}{6}(k+2)(2k+9)$$

$$\text{Now } S_{k+1} = T_{k+1} + S_k$$
$$= (k+1)(k+3) + \frac{k}{6}(k+1)(2k+7)$$
$$= (k+1)\left[(k+3) + \frac{k}{6}(2k+7)\right]$$
$$= \frac{1}{6}(k+1)\left[6(k+3) + k(2k+7)\right]$$
$$= \frac{1}{6}(k+1)\left[6k + 18 + 2k^2 + 7k\right]$$
$$= \frac{1}{6}(k+1)(2k^2 + 13k + 18)$$
$$= \frac{1}{6}(k+1)(k+2)(2k+9) \quad ✓$$

$\therefore$ The statement is true for $n = k + 1$

Since the statement is true for $n = 1$ and $n = k + 1$, then it is true for $n = 2, 3, 4 \ldots$ for all integers $n \geqslant 1$. ✓ *(3 marks)*

14 To prove:

$7^{2n-1} + 5$ is divisible by 12 for all integers $n \geqslant 1$.

Prove that the statement is true for $n = 1$:

i.e. $7^1 + 5 = 12$ which is divisible by 12.

$\therefore$ The statement is true when $n = 1$. ✓

Assume the statement is true for $n = k$.

i.e. $\frac{7^{2k-1} - 5}{12} = I$ where I is any integer.

$\therefore$ $7^{2k-1} + 5 = 12I$
or $7^{2k-1} = 12I - 5 \quad \ldots(1)$

Prove that the statement is true for $n = k + 1$:

Now $7^{2(k+1)-1} + 5$
$= 7^{2k+1} + 5$
$= 7^2 . 7^{2k-1} + 5$
$= 49(12I - 5) + 5$ (using (1))
$= 49(12I) - 245 + 5$
$= 12(49I) - 240$
$= 12(49I - 20)$ which is divisible by 12. ✓

$\therefore$ The statement is true for $n = k + 1$.

$\therefore$ By mathematical induction, the statement is true for all integers, $n \geqslant 1$. ✓

(3 marks)

15 **i** $\tan(\alpha - \beta) = \dfrac{\tan\alpha - \tan\beta}{1 + \tan\alpha\tan\beta}$

$\therefore 1 + \tan\alpha\tan\beta = \dfrac{\tan\alpha - \tan\beta}{\tan(\alpha - \beta)}$

Compare LHS of above with:
$1 + \tan n\theta \tan(n+1)\theta$
Let $\alpha = n\theta$ and $\beta = (n+1)\theta$.

$\therefore 1 + \tan n\theta . \tan(n+1)\theta$

$= \dfrac{\tan n\theta - \tan(n+1)\theta}{\tan[n\theta - (n+1)\theta]}$

$= \dfrac{\tan n\theta - \tan(n+1)\theta}{\tan(-\theta)}$

$= \dfrac{\tan n\theta - \tan(n+1)\theta}{-\tan\theta}$

$= \dfrac{-[\tan(n+1)\theta - \tan n\theta]}{-\tan\theta}$

$= \dfrac{\tan(n+1)\theta - \tan n\theta}{\tan\theta}$

$= \cot\theta\left(\tan(n+1)\theta - \tan n\theta\right)$. ✓

(1 mark)

ii Let $S_n = \tan\theta\tan 2\theta + \tan 2\theta\tan 3\theta + \ldots + \tan n\theta\tan(n+1)\theta$

To prove: $S_n = -(n+1) + \cot\theta\tan(n+1)\theta$

For $n = 1$:

$$\begin{aligned}\text{LHS} &= \tan\theta\tan 2\theta \\ &= \cot\theta\left(\tan 2\theta - \tan\theta\right) - 1 \quad \text{(from i)} \\ &= \cot\theta\tan 2\theta - 1 - 1 \\ &= \cot\theta\tan 2\theta - 2\end{aligned}$$

$$\begin{aligned}\text{RHS} &= -2 + \cot\theta\tan 2\theta \\ &= \cot\theta\tan 2\theta - 2 \\ &= \text{LHS}\end{aligned}$$

$\therefore$ The statement is true for $n = 1$. ✓

Assume the statement is true for $n = k$:
i.e. $S_k = -(k+1) + \cot\theta\tan(k+1)\theta$

Prove the statement is true for $n = k + 1$:
i.e. $S_{k+1} = -(k+2) + \cot\theta\tan(k+2)\theta$

Now $S_{k+1} = S_k + T_{k+1}$

$= -(k+1) + \cot\theta\tan(k+1)\theta + \tan(k+1)\theta\tan(k+2)\theta$

$= -(k+1) + \cot\theta\tan(k+1)\theta + \cot\theta\left[\tan(k+2)\theta - \tan(k+1)\theta\right] - 1$ (from i)

$= -(k+1) + \cot\theta\tan(k+1)\theta + \cot\theta\tan(k+2)\theta - \cot\theta\tan(k+1)\theta - 1$ ✓

$= -(k+2) + \cot\theta\tan(k+2)\theta$

$\therefore$ The statement is true for $n = k + 1$.
Since the statement is true for $n = 1$ and for $n = k + 1$, then it is true for $n = 2, 3, \ldots$ and for all integers $n \geqslant 1$. ✓

(3 marks)

16 To prove:

$$\frac{1}{1\times 3} + \frac{1}{3\times 5} + \frac{1}{5\times 7} + \ldots + \frac{1}{(2n-1)(2n+1)} = \frac{n}{2n+1}$$ for all positive integers n.

For $n = 1$,

$\text{LHS} = \dfrac{1}{1\times 3} = \dfrac{1}{3}$

$\text{RHS} = \dfrac{1}{2(1)+1} = \dfrac{1}{3}$

$\therefore$ The statement is true for $n = 1$. ✓

Assume that the statement is true for $n = k$.:

i.e. $\dfrac{1}{1\times 3} + \dfrac{1}{3\times 5} + \ldots + \dfrac{1}{(2k-1)(2k+1)} = \dfrac{k}{2k+1}$

Show that the statement is true for $n = k + 1$:

i.e. $\dfrac{1}{1\times 3} + \dfrac{1}{3\times 5} + \ldots + \dfrac{1}{(2k-1)(2k+1)} + \dfrac{1}{[2(k+1)-1][2(k+1)+1]} = \dfrac{k+1}{2(k+1)+1}$

i.e. $\dfrac{1}{1\times 3} + \dfrac{1}{3\times 5} + \ldots + \dfrac{1}{(2k-1)(2k+1)} + \dfrac{1}{(2k+1)(2k+3)} = \dfrac{k+1}{2k+3}$

Now $\dfrac{1}{1\times 3} + \dfrac{1}{3\times 5} + \ldots + \dfrac{1}{(2k-1)(2k+1)} + \dfrac{1}{(2k+1)(2k+3)}$

$= \dfrac{k}{2k+1} + \dfrac{1}{(2k+1)(2k+3)}$ (using the assumption above)

$= \dfrac{k(2k+3)+1}{(2k+1)(2k+3)}$

$= \dfrac{2k^2+3k+1}{(2k+1)(2k+3)}$

$= \dfrac{(2k+1)(k+1)}{(2k+1)(2k+3)}$

$= \dfrac{k+1}{2k+3}$ ✓

Therefore, if the statement is true for $n = k$, it is also true for $n = k + 1$.
Therefore the statement is true for all positive integers n, by mathematical induction. ✓

(3 marks)

17 To prove:
$2 \times 1! + 5 \times 2! + 10 \times 3! + \ldots + (n^2 + 1)n! = n(n + 1)!$
for all positive integers n.

For $n = 1$, LHS $= 2 \times 1! = 2$
RHS $= 1(1 + 1)!$
$= 2$
$\therefore$ The statement is true for $n = 1$. ✓

Assume that the statement is true for $n = k$:
i.e. $2 \times 1! + 5 \times 2! + \ldots + (k^2 + 1)k! = k(k + 1)!$

Show that the statement is true for $n = k + 1$:
i.e. $2 \times 1! + 5 \times 2! + \ldots + (k^2 + 1)k! + [(k + 1)^2 + 1](k + 1)! = (k + 1)(k + 1 + 1)!$
$= (k + 1)(k + 2)!$

Now $2 \times 1! + 5 \times 2! + \ldots + (k^2 + 1)k! + [(k + 1)^2 + 1](k + 1)!$
$= k(k + 1)! + [(k + 1)^2 + 1](k + 1)!$
$= (k + 1)!\,[k + (k + 1)^2 + 1]$ (on factorising)
$= (k + 1)!\,[k + k^2 + 2k + 1 + 1]$
$= (k + 1)!\,(k^2 + 3k + 2)$
$= (k + 1)!\,(k + 2)(k + 1)$
$= (k + 2)!\,(k + 1)$ (since $(k + 1)!(k + 2) = (k + 2)!$)
$= (k + 1)(k + 2)!$ ✓

Therefore, if the statement is true for $n = k$, it is also true for $n = k + 1$.
And if the statement is true for $n = 1$, it is also true for $n = 2, 3 \ldots$ etc.
Therefore the statement is true for all positive integers n, by mathematical induction. ✓

(3 marks)

18 To prove:

$n^3 + (n + 1)^3 + (n + 2)^3 = 9M$
for some integer M for $n = 1, 2, 3 \ldots$

For $n = 1$, $1^3 + 2^3 + 3^3 = 36$
$= 9 \times 4$
$\therefore$ The statement is true for $n = 1$. ✓

Assume the statement is true for $n = k$
i.e. $k^3 + (k + 1)^3 + (k + 2)^3 = 9P$
(for some integer P)
$\therefore (k + 1)^3 + (k + 2)^3 = 9P - k^3 \quad \ldots(1)$

Show that the statement is true for $n = k + 1$
i.e. Show that $(k + 1)^3 + (k + 2)^3 + (k + 3)^3 = 9Q$
(for some integer Q)

Now $(k + 1)^3 + (k + 2)^3 + (k + 3)^3$
$= 9P - k^3 + (k + 3)^3$ from (1)
$= 9P - k^3 + k^3 + 9k^2 + 27k + 27$
$= 9P + 9k^2 + 27k + 27$
$= 9(P + k^2 + 3k + 3)$
$= 9Q$ ✓

where Q is the integer $(P + k^2 + 3k + 3)$

$\therefore n^3 + (n + 1)^3 + (n + 2)^3$ is divisible by 9
for $n = 1, 2, 3, \ldots$ by mathematical induction. ✓

(3 marks)

19 For $n = 1$, $1 = \dfrac{1}{6}1(2)(3)$
$= \dfrac{1}{6} \times 6$
$= 1$
$\therefore$ The statement is true for $n = 1$. ✓

Assume the statement is true for $n = k$:
i.e. $1 + 3 + 6 + \ldots + \frac{1}{2}k(k + 1) = \frac{1}{6}k(k + 1)(k + 2)$

Show that the statement is true for $n = k + 1$;
i.e. Show that:

i.e. $1 + 3 + 6 + \ldots + \frac{1}{2}k(k + 1) + \frac{1}{2}(k + 1)(k + 2)$
$= \frac{1}{6}(k + 1)(k + 2)(k + 3)$

Now $1 + 3 + 6 + \ldots + \frac{1}{2}k(k+1) + \frac{1}{2}(k+1)(k+2)$

$= \frac{1}{6}k(k+1)(k+2) + \frac{1}{2}(k+1)(k+2)$

$= \frac{1}{2}(k+1)(k+2)[\frac{1}{3}k + 1]$

$= \frac{1}{2}(k+1)(k+2)[\frac{1}{3}(k+3)]$

$= \frac{1}{6}(k+1)(k+2)(k+3)$. ✓

∴ If the statement is true for $n = k$, it is true for $n = k + 1$. Hence the statement is true for all integers $n = 1, 2, 3 \ldots$ by mathematical induction. ✓ *(3 marks)*

20 To prove:
$(n+1)(n+2)\ldots(2n-1)2n$
$= 2^n[1 \times 3 \times \ldots \times (2n-1)]$ for all integers $n \geqslant 1$

For $n = 1$, $1 \times 2 = 2^1$
∴ The statement is true for $n = 1$. ✓

Assume the statement is true for $n = k$:
i.e. $(k+1)(k+2)\ldots(2k-1)2k = 2^k[1 \times 3 \times \ldots \times (2k-1)]$

Show that the statement is true for $n = k + 1$; i.e. Show that:
$(k+2)(k+3)\ldots[2(k+1)-1]2(k+1) = 2^{k+1}[1 \times 3 \times \ldots \times (2(k+1)-1)]$

$(k+2)(k+3)\ldots(2k+1)2(k+1) = 2^{k+1}[1 \times 3 \times \ldots \times (2k+1)]$

Now $(k+2)(k+3)\ldots 2k(2k+1)2(k+1)$

$= \dfrac{(k+1)(k+2)(k+3)\ldots 2k(2k+1)2(k+1)}{(k+1)}$

(on multiplying numerator and denominator by $(k+1)$)

$= (k+1)(k+2)\ldots(2k-1)2k \times \dfrac{(2k+1)2(k+1)}{(k+1)}$

$= 2^k[1 \times 3 \times \ldots \times (2k-1)] \times \dfrac{(2k+1)2(k+1)}{(k+1)}$

$= 2^k[1 \times 3 \times \ldots \times (2k-1)] \times (2k+1) \times 2$

$= 2 \times 2^k[1 \times 3 \times \ldots (2k-1) \times (2k+1)]$

$= 2^{k+1}[1 \times 3 \times \ldots (2k-1)(2k+1)]$

$= 2^{k+1}[1 \times 3 \times \ldots \times (2k+1)]$ ✓

∴ If the statement is true for $n = k$, it is true for $n = k + 1$. Hence the statement is true for all integers $n \geq 1$ by mathematical induction. ✓ *(3 marks)*

21 To prove:
$4^n + 14 = 6m$ for some integer m,
for all $n \geqslant 1$.

For $n = 1$,
$4 + 14 = 18$
$= 6 \times 3$
∴ The statement is true for $n = 1$. ✓

Assume the statement is true for $n = k$:
i.e. $4^k + 14 = 6P$ for some integer P
i.e. $4^k = 6P - 14$

Show that the statement is true for $n = k + 1$
i.e. Show that $4^{k+1} + 14 = 6Q$ for some integer Q

Now $4^{k+1} + 14 = 4 \times 4^k + 14$
$= 4(6P - 14) + 14$
$= 24P - 56 + 14$
$= 24P - 42$
$= 6(4P - 7)$
$= 6Q$ ✓
where Q is the integer $(4P - 7)$

∴ $4n + 14$ is divisible by 6 for all $n \geqslant 1$ by mathematical induction. ✓ *(3 marks)*

22 To prove: $n^3 + 2n = 3M$
where M is an integer.

For $n = 1$, $n^3 + 2n = 1 + 2 = 3$
∴ The statement is true for $n = 1$.

Assume the statement is true for $n = k$:
i.e. $k^3 + 2k = 3P$
where P is an integer.

Show that the statement is true for $n = k + 1$:
i.e. Show that
$(k+1)^3 + 2(k+1) = 3Q$
where Q is an integer.

Now $(k+1)^3 + 2(k+1)$
$= k^3 + 3k^2 + 3k + 1 + 2k + 2$
$= k^3 + 3k^2 + 5k + 3$
$= (3P - 2k) + 3k^2 + 5k + 3$
(since $k^3 = 3P - 2k$)
$= 3P + 3k^2 + 3k + 3$
$= 3(P + k^2 + k + 1)$
$= 3Q$
where Q is an integer.

∴ If the statement is true for $n = k$, it is also true for $n = k + 1$.
∴ $n^3 + 2n$ is divisible by 3 for all positive integers n by Mathematical Induction.

(No mark allocation in exam)

23 **i** To prove: $S_n = \frac{1}{6}n(n+1)(2n+1)$

For $n = 1$,
LHS $= 1^2 = 1$
RHS $= \frac{1}{6}(1)(1+1)(2+1)$
$= \frac{6}{6}$
$= 1$
∴ The statement is true for $n = 1$.

Assume the statement is true for $n = k$:
$S_k = \frac{1}{6}k(k+1)(2k+1)$

Show that the statement is true for $n = k + 1$:
i.e. Show that
$S_{k+1} = \frac{1}{6}(k+1)(k+1+1)[2(k+1)+1]$
$= \frac{1}{6}(k+1)(k+2)(2k+3)$

Now,
$S_{k+1} = S_k + (k+1)^2$
$= \frac{1}{6}k(k+1)(2k+1) + (k+1)^2$
$= (k+1)[\frac{1}{6}k(2k+1) + (k+1)]$
$= (k+1)\left(\dfrac{k(2k+1) + 6(k+1)}{6}\right)$
$= \frac{1}{6}(k+1)(2k^2 + k + 6k + 6)$
$= \frac{1}{6}(k+1)(2k^2 + 7k + 6)$
$= \frac{1}{6}(k+1)(k+2)(2k+3)$

∴ If the statement is true for $n = k$, it is also true for $n = k + 1$.
∴ The statement is true for all natural numbers, n, by mathematical induction.

ii $\frac{1}{6}n(n + 1)(2n + 1) \geqslant 10^9$

$\frac{1}{6}(n^2 + n)(2n + 1) \geqslant 10^9$

$2n^3 + 3n^2 + n \geqslant 6 \times 10^9$

$n^3 + \frac{3}{2}n^2 + \frac{1}{2}n \geqslant 3 \times 10^9$

For large numbers, n; n^2 and n are very small compared to n^3.

$\therefore n^3 \doteqdot 3 \times 10^9$

$n \doteqdot \sqrt[3]{3} \times 10^3$

$\doteqdot 1442$

Now $1442^3 + \frac{3}{2}(1442)^2 + \frac{1}{2}(1442) > 3 \times 10^9$

and $1441^3 + \frac{3}{2}(1441)^2 + \frac{1}{2}(1441) < 3 \times 10^9$

$\therefore$ The least n such that $S_n \geqslant 10^9$ is $n \approx 1442$.

(No mark allocation in exam)

24 To prove: $S_n = \frac{1}{3}(n - 1)n(n + 1)$

For $n = 2$,

LHS $= 1 \times 2 = 2$

RHS $= \frac{1}{3}(2 - 1)2(2 + 1)$

$= \frac{1}{3}(1)2(3)$

$= \frac{6}{3}$

$= 2$

Hence for $n = 2$, the statement is valid.

Assume the statement is valid for $n = k$:

$S_k = \frac{1}{3}(k - 1)k(k + 1)$

Show that the statement is true for $n = k + 1$:

i.e. Show that:

$S_{k+1} = \frac{1}{3}(k + 1 - 1)(k + 1)(k + 1 + 1)$

$= \frac{1}{3}k(k + 1)(k + 2)$

Now $S_{k+1} = 1 \times 2 + 2 \times 3 + \ldots + (k - 1)k + (k + 1 - 1)(k + 1)$

$= 1 \times 2 + 2 \times 3 + \ldots + (k - 1)k + k(k + 1)$

$= S_k + k(k + 1)$

$= \frac{1}{3}(k - 1)k(k + 1) + k(k + 1)$

$= k(k + 1)\left[\frac{1}{3}(k - 1) + 1\right]$

$= k(k + 1)\left[\frac{k - 1}{3} + \frac{3}{3}\right]$

$= k(k + 1)\left[\frac{k + 2}{3}\right]$

$= \frac{1}{3}k(k + 1)(k + 2)$

$\therefore$ The statement is true for $n = k + 1$.

$\therefore$ The statement is true for all $n \geqslant 2$ by mathematical induction.

(No mark allocation in exam)

1 For the vectors $\underset{\sim}{u} = \underset{\sim}{i} - \underset{\sim}{j}$ and $\underset{\sim}{v} = 2\underset{\sim}{i} + \underset{\sim}{j}$, evaluate each of the following.

i $\underset{\sim}{u} + 3\underset{\sim}{v}$ *(1 mark)* Easy

ii $\underset{\sim}{u} \cdot \underset{\sim}{v}$ *(1 mark)* Easy

(Q11a, **2022 HSC**)

2 Given that $\overrightarrow{OP} = \begin{pmatrix} -3 \\ 1 \end{pmatrix}$ and $\overrightarrow{OQ} = \begin{pmatrix} 2 \\ 5 \end{pmatrix}$, what is $\overrightarrow{PQ}$?

A $\begin{pmatrix} 1 \\ -6 \end{pmatrix}$ **B** $\begin{pmatrix} -1 \\ 6 \end{pmatrix}$

C $\begin{pmatrix} 5 \\ 4 \end{pmatrix}$ **D** $\begin{pmatrix} -5 \\ -4 \end{pmatrix}$ *(1 mark)*

(Q1, **2021 HSC**) Easy

3 Find $(\underset{\sim}{i} + 6\underset{\sim}{j}) + (2\underset{\sim}{i} - 7\underset{\sim}{j})$. *(1 mark)*

(Q11a, **2021 HSC**) Easy

4 The vectors $\underset{\sim}{a}$ and $\underset{\sim}{b}$ are shown.

Which diagram below shows the vector $\underset{\sim}{v} = \underset{\sim}{a} - \underset{\sim}{b}$?

A

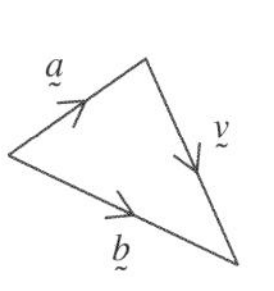

B

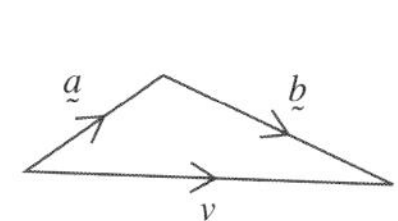

C

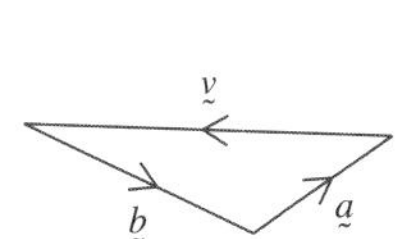

D

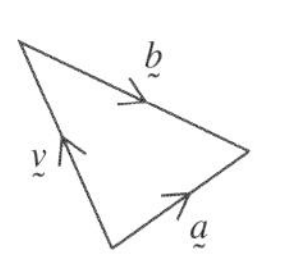

(1 mark)

(Q6, **2020 HSC**) Easy

5 The projection of the vector $\begin{pmatrix} 6 \\ 7 \end{pmatrix}$ onto the line $y = 2x$ is $\begin{pmatrix} 4 \\ 8 \end{pmatrix}$.

The point (6, 7) is reflected in the line $y = 2x$ to a point A.

What is the position vector of the point A?

A $\begin{pmatrix} 6 \\ 12 \end{pmatrix}$ **B** $\begin{pmatrix} 2 \\ 9 \end{pmatrix}$

C $\begin{pmatrix} -6 \\ 7 \end{pmatrix}$ **D** $\begin{pmatrix} -2 \\ 1 \end{pmatrix}$ *(1 mark)*

(Q9, **2020 HSC**) Medium

6 Given $\underset{\sim}{a} = 3\underset{\sim}{i} + 2\underset{\sim}{j}$ and $\underset{\sim}{b} = 4\underset{\sim}{i} - \underset{\sim}{j}$, find:

i $\underset{\sim}{a} - \underset{\sim}{b}$ *(1 mark)* Easy

ii $2\underset{\sim}{a}$ *(1 mark)* Easy

iii $4\underset{\sim}{a} - 3\underset{\sim}{b}$ *(1 mark)* Easy

Bonus question (see page iv)

7 Given $\underset{\sim}{a} = \begin{bmatrix} 1 \\ -3 \end{bmatrix}$ and $\underset{\sim}{b} = \begin{bmatrix} -2 \\ 2 \end{bmatrix}$, find:

i $\underset{\sim}{a} - \underset{\sim}{b}$ *(1 mark)* Easy

ii $2\underset{\sim}{b} - 3\underset{\sim}{a}$ *(1 mark)* Easy

iii k if $k(\underset{\sim}{a} + \underset{\sim}{b}) = \begin{bmatrix} -3 \\ -3 \end{bmatrix}$. *(2 marks)* Easy

Bonus question

8 Write the component vector for $\overrightarrow{AB}$, given:

a A (1,6) and B (3,2).

b

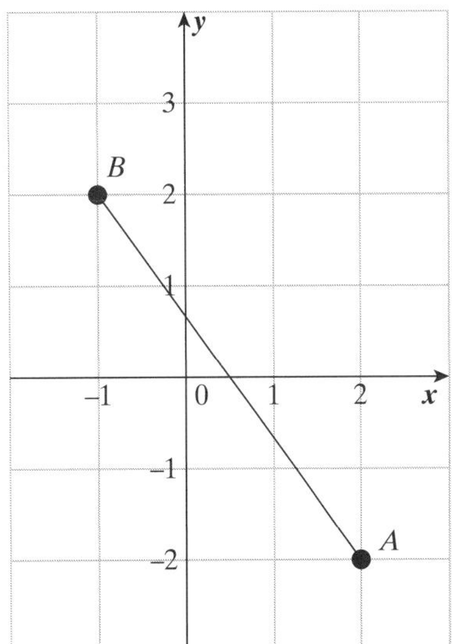

(1 mark)

Bonus question Easy

9 Given $k\underset{\sim}{a} + m\underset{\sim}{b} = -\underset{\sim}{i} + 5\underset{\sim}{j}$, $\underset{\sim}{a} = \underset{\sim}{i} + 2\underset{\sim}{j}$ and $\underset{\sim}{b} = 3\underset{\sim}{i} - \underset{\sim}{j}$ find the values of k and m. *(3 marks)*

Bonus question Easy

10 $ABCD$ is a quadrilateral, with $2 \times AE = DE$. Given $\overrightarrow{AB} = \underset{\sim}{a}$, $\overrightarrow{BC} = \underset{\sim}{b}$, and $\overrightarrow{CD} = \underset{\sim}{c}$:

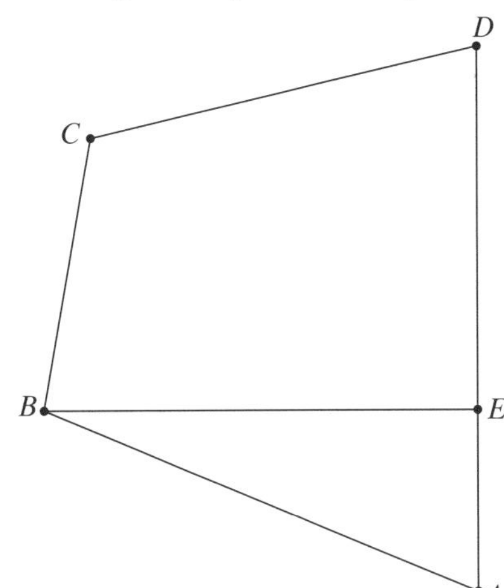

i Show that $\overrightarrow{AE} = \frac{1}{3}(\underset{\sim}{a} + \underset{\sim}{b} + \underset{\sim}{c})$. *(2 marks)* Easy

ii Hence find $\overrightarrow{CE}$. *(2 marks)* Easy

Bonus question

---

11 Given $BE:EC = 1:3$, $\overrightarrow{AB} = 2\underset{\sim}{a}$ and $\overrightarrow{AC} = 4\underset{\sim}{c}$.

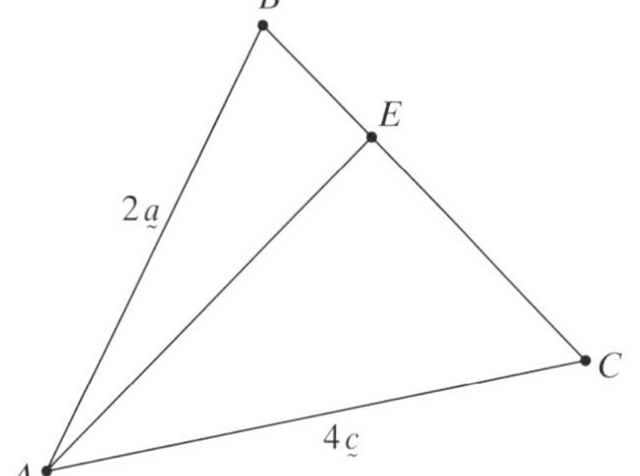

Find $\overrightarrow{AE}$. *(3 marks)*

Bonus question Easy

---

12 $PQRS$ is a parallelogram whose diagonals are $2\underset{\sim}{a}$ and $2\underset{\sim}{b}$. A is the point of intersection of the two diagonals.

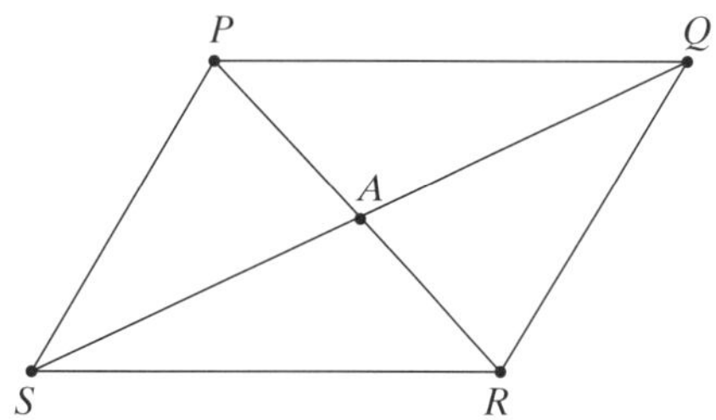

Show that the sides of the parallelogram can be represented as $\underset{\sim}{a} + \underset{\sim}{b}$ and $\underset{\sim}{a} - \underset{\sim}{b}$. *(2 marks)*

Bonus question Medium

---

13 E is the midpoint of DC. $AF:FE = 2:1$. Let $\overrightarrow{DA} = \underset{\sim}{a}$ and $\overrightarrow{DE} = \underset{\sim}{b}$.

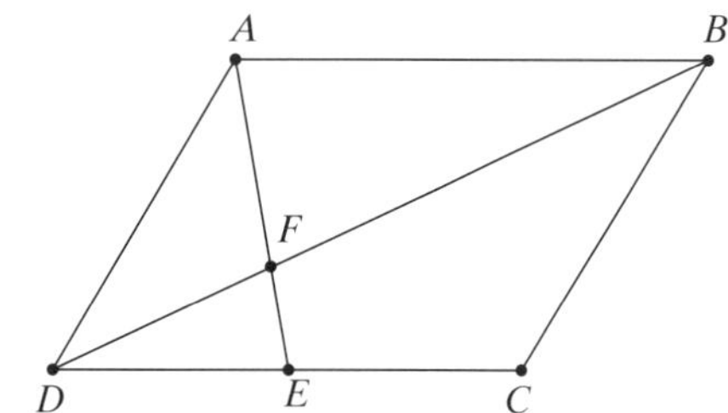

Find:

i $\overrightarrow{AE}$ *(1 mark)* Easy

ii $\overrightarrow{DB}$ *(1 mark)* Easy

iii Hence, prove $\overrightarrow{DB} = 3 \times \overrightarrow{DF}$. *(2 marks)* Medium

Bonus question

---

14 $ABCDEF$ is a regular hexagon. Let $\overrightarrow{AB} = \underset{\sim}{u}$, $\overrightarrow{BC} = \underset{\sim}{v}$ and $\overrightarrow{CD} = \underset{\sim}{w}$. G is the midpoint of CD. M is the midpoint of AD.

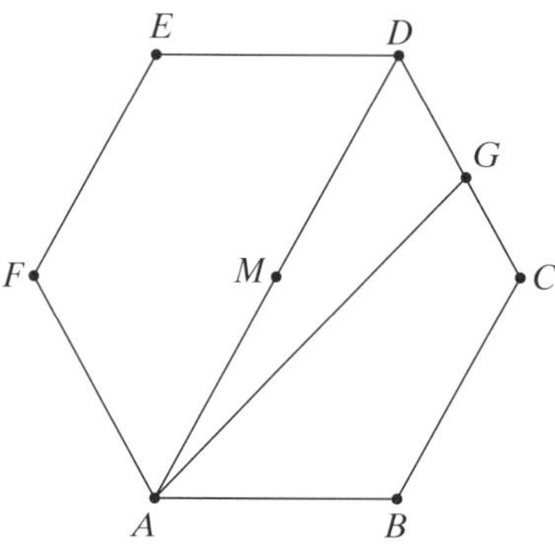

Find:

i $\overrightarrow{AG}$ *(1 mark)* Medium

ii Hence, find $\overrightarrow{MG}$. *(3 marks)* Hard

Bonus question

---

15 Given $\overrightarrow{AD} = k \times \underset{\sim}{q}$, $\overrightarrow{DF} = m \times \underset{\sim}{q}$, $\overrightarrow{BC} = k \times \underset{\sim}{r}$, $\overrightarrow{CE} = m \times \underset{\sim}{r}$, $\overrightarrow{AB} = \underset{\sim}{p}$ and $\overrightarrow{DC} = 2\underset{\sim}{p}$

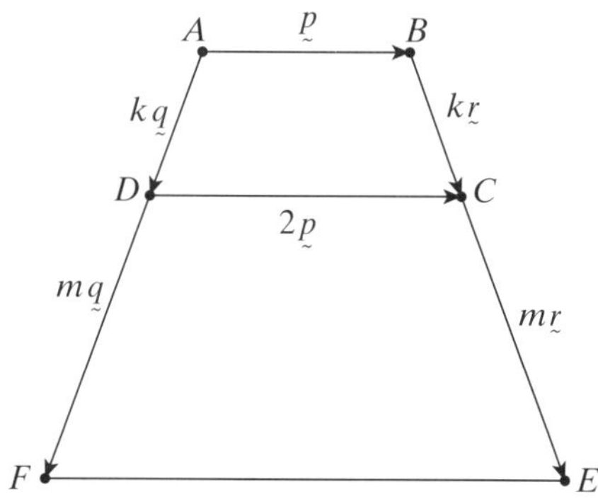

i Show that $\underset{\sim}{p} = k(\underset{\sim}{r} - \underset{\sim}{q})$. *(2 marks)* Medium

ii Find an expression for $\overrightarrow{FE}$ in terms of $\underset{\sim}{p}$, m and k. *(2 marks)* Medium

iii If $\overrightarrow{FE} = 3 \times \overrightarrow{DC}$ find $m:k$. *(3 marks)* Hard

Bonus question

---

16 M is the midpoint of BE. $\overrightarrow{BA} \parallel \overrightarrow{MC}$ and $\overrightarrow{BA} = 2 \times \overrightarrow{MC}$. Let $\overrightarrow{CM} = \underset{\sim}{a}$ and $\overrightarrow{BC} = \underset{\sim}{b}$.

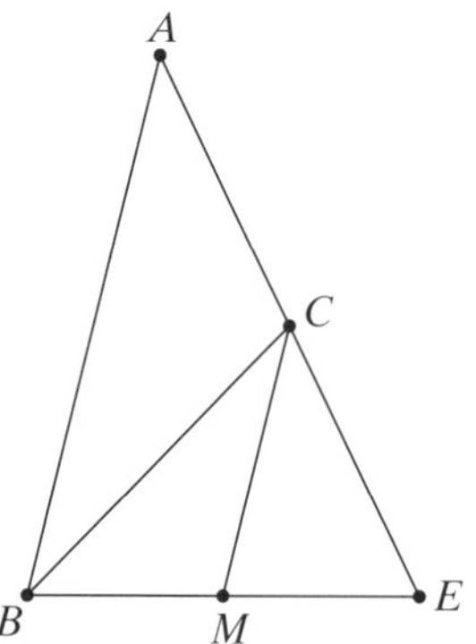

Find:

i $\overrightarrow{AB}$ *(1 mark)* Easy

ii $\overrightarrow{BE}$ *(1 mark)* Medium

iii $\overrightarrow{CE}$ *(2 marks)* 

iv Hence, explain why C bisects AE. *(3 marks)* Hard

Bonus question

Year 12 Introduction to vectors—Worked Answers

1 **i** $\underset{\sim}{u}+\underset{\sim}{v}=(\underset{\sim}{i}-\underset{\sim}{j})+3(2\underset{\sim}{i}+\underset{\sim}{j})$
$=\underset{\sim}{i}-\underset{\sim}{j}+6\underset{\sim}{i}+3\underset{\sim}{j}$
$=7\underset{\sim}{i}+2\underset{\sim}{j}$ ✓ *(1 mark)*

ii $\underset{\sim}{u}\bullet\underset{\sim}{v}=1\times2+(-1)\times1$
$=1$ ✓ *(1 mark)*

2 $\overrightarrow{OP}+\overrightarrow{PQ}=\overrightarrow{OQ}$
$\overrightarrow{PQ}=\overrightarrow{OQ}-\overrightarrow{OP}$
$\overrightarrow{PQ}=\begin{pmatrix}2\\5\end{pmatrix}-\begin{pmatrix}-3\\1\end{pmatrix}$
$\overrightarrow{PQ}=\begin{pmatrix}5\\4\end{pmatrix}$

Answer C

3 $(\underset{\sim}{i}+6\underset{\sim}{j})+(2\underset{\sim}{i}-7\underset{\sim}{j})$
$=(1+2)\underset{\sim}{i}+(6-7)\underset{\sim}{j}$
$=3\underset{\sim}{i}-\underset{\sim}{j}$ ✓ *(1 mark)*

4 $\underset{\sim}{v}=\underset{\sim}{a}-\underset{\sim}{b}$

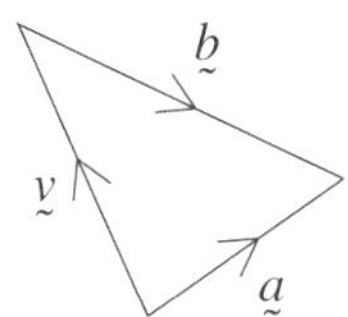

Answer D

5 A is at (2, 9).

Answer B

6 **i** $(3-4)\underset{\sim}{i}+(2-(-1))\underset{\sim}{j}$
$=-\underset{\sim}{i}+3\underset{\sim}{j}$ ✓ *(1 mark)*

ii $2(3\underset{\sim}{i}+2\underset{\sim}{j})$
$=6\underset{\sim}{i}+4\underset{\sim}{j}$ ✓ *(1 mark)*

iii $4(3\underset{\sim}{i}+2\underset{\sim}{j})-3(4\underset{\sim}{i}-\underset{\sim}{j})$
$=(12-12)\underset{\sim}{i}+(8+3)\underset{\sim}{j}$
$=11\underset{\sim}{j}$ ✓ *(1 mark)*

7 **i** $\begin{bmatrix}1-(-2)\\-3-2\end{bmatrix}$
$=\begin{bmatrix}3\\-5\end{bmatrix}$ ✓ *(1 mark)*

ii $\begin{bmatrix}2(-2)-3\times1\\2\times2-3(-3)\end{bmatrix}$
$=\begin{bmatrix}-7\\13\end{bmatrix}$ ✓ *(1 mark)*

iii $k(\underset{\sim}{a}+\underset{\sim}{b})=\begin{bmatrix}k(1+(-2))\\k((-3)+2)\end{bmatrix}$
$=\begin{bmatrix}-k\\-k\end{bmatrix}$ ✓

But $k(\underset{\sim}{a}+\underset{\sim}{b})=\begin{bmatrix}-3\\-3\end{bmatrix}$

$\therefore -k=-3$, so $k=3$. ✓ *(2 marks)*

8 **a** $\overrightarrow{AB}=2\underset{\sim}{i}-4\underset{\sim}{j}$ ✓ *(1 mark)*

b $\overrightarrow{AB}=-3\underset{\sim}{i}+4\underset{\sim}{j}$ ✓ *(1 mark)*

9 $k\underset{\sim}{a}+m\underset{\sim}{b}=k(\underset{\sim}{i}+2\underset{\sim}{j})+m(3\underset{\sim}{i}-\underset{\sim}{j})$
$=(k+3m)\underset{\sim}{i}+(2k-m)\underset{\sim}{j}$ ✓

Hence,
$k+3m=-1$ ← (1)
$2k-m=5$ ← (2)
(1) × 2
$2k+6m=-2$ ← (3)
(3) − (2)
$7m=-7$
$\therefore m=-1$, hence $k=2$. ✓
✓ *(3 marks)*

10 **i** $\overrightarrow{AD}=\overrightarrow{AE}+\overrightarrow{ED}$
$=\overrightarrow{AE}+2\times\overrightarrow{AE}$
$=3\times\overrightarrow{AE}$ or $\overrightarrow{AE}=\frac{1}{3}\times\overrightarrow{AD}$ ✓

Now $\overrightarrow{AD}=\overrightarrow{AB}+\overrightarrow{BC}+\overrightarrow{CD}$
$=\underset{\sim}{a}+\underset{\sim}{b}+\underset{\sim}{c}$
$\therefore\overrightarrow{AE}=1/3\,(\underset{\sim}{a}+\underset{\sim}{b}+\underset{\sim}{c})$ ✓ *(2 marks)*

ii $\overrightarrow{CE}=\overrightarrow{CB}+\overrightarrow{BA}+\overrightarrow{AE}$ ✓
$=-\underset{\sim}{b}-\underset{\sim}{a}+\frac{1}{3}\underset{\sim}{a}+\frac{1}{3}\underset{\sim}{b}+\frac{1}{3}\underset{\sim}{c}$
$=\frac{1}{3}(\underset{\sim}{c}-2\underset{\sim}{b}-2\underset{\sim}{a})$ ✓ *(2 marks)*

11 $\overrightarrow{BC}=\overrightarrow{BA}+\overrightarrow{AC}$
$=4\underset{\sim}{c}-2\underset{\sim}{a}$ ✓
$\overrightarrow{BE}=\frac{1}{4}\times\overrightarrow{BC}$ ✓
$\overrightarrow{AE}=\overrightarrow{AB}+\overrightarrow{BE}$
$=2\underset{\sim}{a}+\frac{1}{4}(4\underset{\sim}{c}-2\underset{\sim}{a})$
$=\frac{1}{2}(3\underset{\sim}{a}+2\underset{\sim}{c})$ ✓ *(3 marks)*

12 Let $SQ=2\underset{\sim}{a}$ and $RP=2\underset{\sim}{b}$.
Diagonals of a parallelogram bisect each other.
$\therefore\overrightarrow{SA}=\overrightarrow{AQ}=\underset{\sim}{a}$ and
$\overrightarrow{RA}=\overrightarrow{AP}=\underset{\sim}{b}$ ✓
$\overrightarrow{SP}=\overrightarrow{SA}+\overrightarrow{AP}$
$=\underset{\sim}{a}+\underset{\sim}{b}$
$\overrightarrow{PQ}=\overrightarrow{AQ}+\overrightarrow{PA}$
$=\underset{\sim}{a}-\underset{\sim}{b}$ ✓ *(2 marks)*

13 **i** $\overrightarrow{AE}=\overrightarrow{AD}+\overrightarrow{DE}$
$=\underset{\sim}{b}-\underset{\sim}{a}$ ✓ *(1 mark)*

ii $\overrightarrow{DB}=\overrightarrow{DA}+\overrightarrow{AB}$
$=\underset{\sim}{a}+2\underset{\sim}{b}$ ✓ *(1 mark)*

iii $\overrightarrow{DF}=\overrightarrow{DA}+\overrightarrow{AF}$
$\overrightarrow{AF}=\frac{2}{3}\times\overrightarrow{AE}$
$\therefore\overrightarrow{DF}=\underset{\sim}{a}+\frac{2}{3}\times(\underset{\sim}{b}-\underset{\sim}{a})$ ✓
$=\frac{1}{3}(2\underset{\sim}{b}+\underset{\sim}{a})$
$=\frac{1}{3}\times\overrightarrow{DB}$ ✓
Or $\overrightarrow{DB}=3\times\overrightarrow{DF}$ *(2 marks)*

14 **i** $\overrightarrow{AG}=\overrightarrow{AB}+\overrightarrow{BC}+\frac{1}{2}\overrightarrow{CD}$
$=\underset{\sim}{u}+\underset{\sim}{v}+\frac{1}{2}\underset{\sim}{w}$ ✓ *(1 mark)*

ii $\overrightarrow{MG}=\overrightarrow{MA}+\overrightarrow{AG}$
$\overrightarrow{MA}=\frac{1}{2}(\overrightarrow{DC}+\overrightarrow{CB}+\overrightarrow{BA})$
$=-\frac{1}{2}\underset{\sim}{u}-\frac{1}{2}\underset{\sim}{v}-\frac{1}{2}\underset{\sim}{w}$ ✓
$\therefore\overrightarrow{MG}=(-\frac{1}{2}\underset{\sim}{u}-\frac{1}{2}\underset{\sim}{v}-\frac{1}{2}\underset{\sim}{w})$
$+(\underset{\sim}{u}+\underset{\sim}{v}+\frac{1}{2}\underset{\sim}{w})$ ✓
$=\frac{1}{2}(\underset{\sim}{u}+\underset{\sim}{v})$ ✓ *(3 marks)*

15 **i** $\overrightarrow{DC}=\overrightarrow{DA}+\overrightarrow{AB}+\overrightarrow{BC}$
$2\underset{\sim}{p}=-k\underset{\sim}{q}+\underset{\sim}{p}+k\underset{\sim}{r}$ ✓
$\underset{\sim}{p}=k\underset{\sim}{r}-k\underset{\sim}{q}$
$\underset{\sim}{p}=k(\underset{\sim}{r}-\underset{\sim}{q})$ ✓ *(2 marks)*

ii $\overrightarrow{FE}=\overrightarrow{FD}+\overrightarrow{DC}+\overrightarrow{CE}$
$=-m\underset{\sim}{q}+2\underset{\sim}{p}+m\underset{\sim}{r}$ ✓
$=2\underset{\sim}{p}+m(\underset{\sim}{r}-\underset{\sim}{q})$
$=2\underset{\sim}{p}+m\left(\frac{\underset{\sim}{p}}{k}\right)$
$=\underset{\sim}{p}\left(2+\frac{m}{k}\right)$ ✓ *(2 marks)*

iii Since $\overrightarrow{FE}=3\times\overrightarrow{DC}$ ✓
$\therefore 6\underset{\sim}{p}=\underset{\sim}{p}\left(2+\frac{m}{k}\right)$ ✓
$\frac{m}{k}=4$
Hence $m:k=4:1$. ✓ *(3 marks)*

16 **i** $\overrightarrow{AB}=2\times\overrightarrow{CM}$
$=2\underset{\sim}{a}$ ✓ *(1 mark)*

ii $\overrightarrow{BE}=2\times\overrightarrow{BM}$
$=2\times(\overrightarrow{BC}+\overrightarrow{CM})$
$=2\underset{\sim}{a}+2\underset{\sim}{b}$ ✓ *(1 mark)*

iii $\overrightarrow{CE}=\overrightarrow{CB}+\overrightarrow{BE}$
$=-\underset{\sim}{b}+(2\underset{\sim}{a}+2\underset{\sim}{b})$ ✓
$=2\underset{\sim}{a}+\underset{\sim}{b}$ ✓ *(2 marks)*

iv $\overrightarrow{AE}=\overrightarrow{AB}+\overrightarrow{BE}$
$=2\underset{\sim}{a}+(2\underset{\sim}{a}+2\underset{\sim}{b})$ ✓
$=4\underset{\sim}{a}+2\underset{\sim}{b}$
$=2(2\underset{\sim}{a}+\underset{\sim}{b})$
$=2\times\overrightarrow{CE}$ ✓
Therefore, C bisects AE. ✓ *(3 marks)*

1 The following diagram shows the vector $\underset{\sim}{u}$ and the vectors $\underset{\sim}{i} + \underset{\sim}{j}, -\underset{\sim}{i} + \underset{\sim}{j}, -\underset{\sim}{i} - \underset{\sim}{j}$ and $\underset{\sim}{i} - \underset{\sim}{j}$.

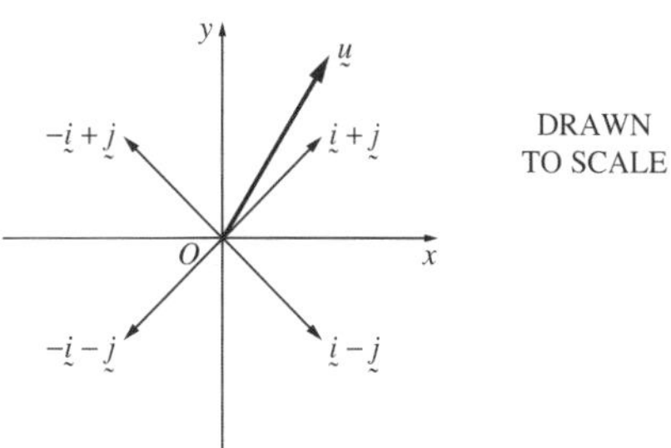

Which statement regarding this diagram could be true?

A The projection of $\underset{\sim}{u}$ onto $\underset{\sim}{i} + \underset{\sim}{j}$ is the vector $1.1\underset{\sim}{i} + 1.8\underset{\sim}{j}$.

B The projection of $\underset{\sim}{u}$ onto $-\underset{\sim}{i} + \underset{\sim}{j}$ is the vector $-0.4\underset{\sim}{i} + 0.4\underset{\sim}{j}$.

C The projection of $\underset{\sim}{u}$ onto $-\underset{\sim}{i} - \underset{\sim}{j}$ is the vector $3.2\underset{\sim}{i} + 3.2\underset{\sim}{j}$.

D The projection of $\underset{\sim}{u}$ onto $\underset{\sim}{i} - \underset{\sim}{j}$ is the vector $0.5\underset{\sim}{i} - 0.5\underset{\sim}{j}$. *(1 mark)*

(Q6, **2022 HSC**) **Medium**

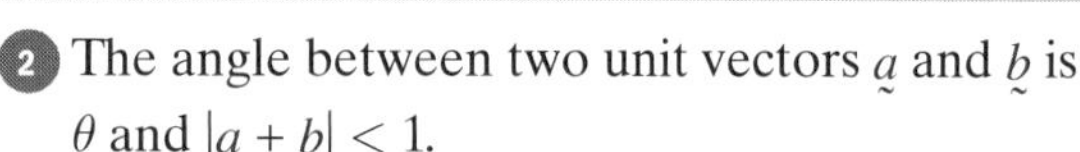

2 The angle between two unit vectors $\underset{\sim}{a}$ and $\underset{\sim}{b}$ is θ and $|\underset{\sim}{a} + \underset{\sim}{b}| < 1$.

Which of the following best describes the possible range of values of θ?

A $0 \le \theta < \frac{\pi}{3}$ **B** $0 \le \theta < \frac{2\pi}{3}$

C $\frac{\pi}{3} < \theta \le \pi$ **D** $\frac{2\pi}{3} < \theta \le \pi$ *(1 mark)*

(Q8, **2022 HSC**) **Medium**

3 The vectors $\underset{\sim}{u} = \begin{pmatrix} a \\ 2 \end{pmatrix}$ and $\underset{\sim}{v} = \begin{pmatrix} a-7 \\ 4a-1 \end{pmatrix}$ are perpendicular.

What are the possible values of a? *(2 marks)*

(Q11d, **2022 HSC**) **Easy**

4 Three different points A, B and C are chosen on a circle centred at O.

Let $\underset{\sim}{a} = \overrightarrow{OA}$, $\underset{\sim}{b} = \overrightarrow{OB}$ and $\underset{\sim}{c} = \overrightarrow{OC}$.

Let $\underset{\sim}{h} = \underset{\sim}{a} + \underset{\sim}{b} + \underset{\sim}{c}$ and let H be the point such that $\overrightarrow{OH} = \underset{\sim}{h}$, as shown in the diagram.

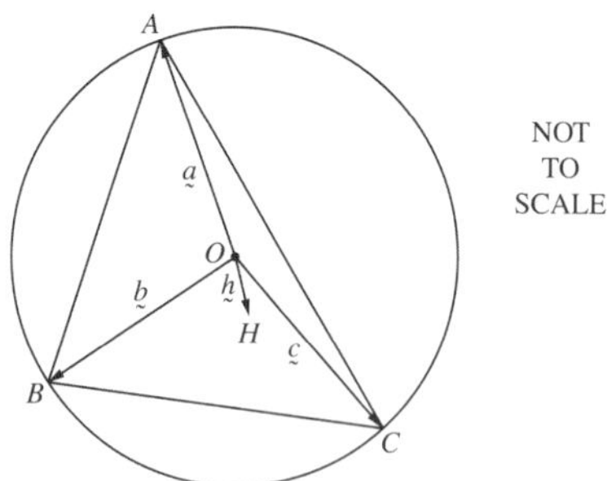

Show that $\overrightarrow{BH}$ and $\overrightarrow{CA}$ are perpendicular. *(3 marks)*

(Q13a, **2022 HSC**) **Medium**

5 The vectors $\vec{u}$ and $\vec{v}$ are not parallel. The vector $\vec{p}$ is the projection of $\vec{u}$ onto the vector $\vec{v}$.

The vector $\vec{p}$ is parallel to $\vec{v}$ so it can be written $\lambda_0\vec{v}$ for some real number λ_0. (Do NOT prove this.)

Prove that $|\vec{u} - \lambda\vec{v}|$ is smallest when $\lambda = \lambda_0$ by showing that, for all real numbers λ, $|\vec{u} - \lambda_0\vec{v}| \le |\vec{u} - \lambda\vec{v}|$. *(3 marks)*

(Q14b, **2022 HSC**) **Hard**

6 For the two vectors $\overrightarrow{OA}$ and $\overrightarrow{OB}$ it is known that $\overrightarrow{OA} \cdot \overrightarrow{OB} < 0$.

Which of the following statements MUST be true?

A Either, $\overrightarrow{OA}$ is negative and $\overrightarrow{OB}$ is positive, or, $\overrightarrow{OA}$ is positive and $\overrightarrow{OB}$ is negative.

B The angle between $\overrightarrow{OA}$ and $\overrightarrow{OB}$ is obtuse.

C The product $|\overrightarrow{OA}||\overrightarrow{OB}|$ is negative.

D The points O, A and B are collinear. *(1 mark)*

(Q5, **2021 HSC**) **Medium**

7 A plane needs to travel to a destination that is on a bearing of 063°. The engine is set to fly at a constant 175 km/h. However, there is a wind from the south with a constant speed of 42 km/h.

On what constant bearing, to the nearest degree, should the direction of the plane be set in order to reach the destination? *(3 marks)*

(Q14a, **2021 HSC**) **Hard**

8 **i** For vector $\underset{\sim}{v}$, show that $\underset{\sim}{v} \cdot \underset{\sim}{v} = |\underset{\sim}{v}|^2$. *(1 mark)* Medium

ii In the trapezium $ABCD$, BC is parallel to AD and $|\overrightarrow{AC}| = |\overrightarrow{BD}|$.

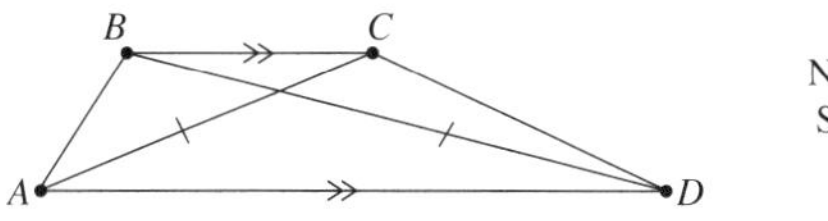

Let $\underset{\sim}{a} = \overrightarrow{AB}$, $\underset{\sim}{b} = \overrightarrow{BC}$ and $\overrightarrow{AD} = k\overrightarrow{BC}$, where $k > 0$.
Using part **i**, or otherwise, show

$2\underset{\sim}{a} \cdot \underset{\sim}{b} + (1 - k)|\underset{\sim}{b}|^2 = 0$. *(3 marks)* Hard

(Q14c, **2021 HSC**)

9 Maria starts at the origin and walks along all of the vector $2\underset{\sim}{i} + 3\underset{\sim}{j}$, then walks along all of the vector $3\underset{\sim}{i} - 2\underset{\sim}{j}$ and finally along all of the vector $4\underset{\sim}{i} - 3\underset{\sim}{j}$.

How far from the origin is she?

A $\sqrt{77}$ **B** $\sqrt{85}$

C $2\sqrt{13} + \sqrt{5}$ **D** $\sqrt{5} + \sqrt{7} + \sqrt{13}$

(Q4, **2020 HSC**) Easy

10 For what value(s) of a are the vectors $\begin{pmatrix} a \\ -1 \end{pmatrix}$ and $\begin{pmatrix} 2a-3 \\ 2 \end{pmatrix}$ perpendicular? *(3 marks)*

(Q11b, **2020 HSC**) Easy

11 An equilateral triangle of side 4 units with vectors $\underset{\sim}{a}$ and $\underset{\sim}{b}$ are shown.

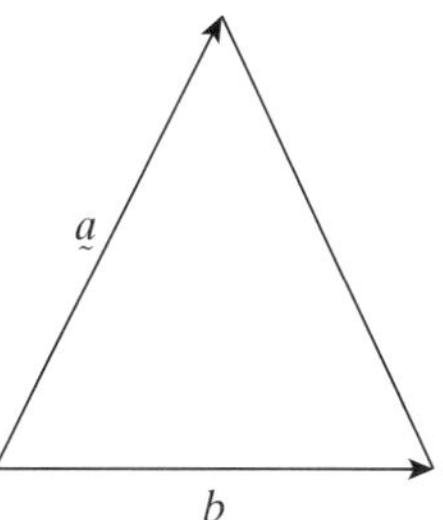

What is the value of $\underset{\sim}{a} \cdot \underset{\sim}{b}$? *(1 mark)*

Bonus question (see page iv) Easy

12 Points A and B have position vectors $\overrightarrow{OA} = \underset{\sim}{i} + 4\underset{\sim}{j}$ and $\overrightarrow{OB} = 5\underset{\sim}{i} - \underset{\sim}{j}$.

a Find the unit vector along $\overrightarrow{AB}$. *(2 marks)* Easy

b Suppose N is a point on AB such that $\overrightarrow{ON} \perp \overrightarrow{AB}$. Find $\overrightarrow{ON}$ *(3 marks)* Medium

Bonus question

13 Find the size of the smallest angle between vectors $\underset{\sim}{a}$ and $\underset{\sim}{b}$, where $\underset{\sim}{a} = 4\underset{\sim}{i} + 2\underset{\sim}{j}$ and $\underset{\sim}{b} = n\underset{\sim}{i} + 3n\underset{\sim}{j}$. *(3 marks)*

Bonus question Medium

14 $\underset{\sim}{a} = 2\underset{\sim}{i} + 5\underset{\sim}{j}$, $\underset{\sim}{b} = -5\underset{\sim}{i} + 2\underset{\sim}{j}$, the point R is $(-3, 1)$ and the point S is $(5, -8)$.

Given $\overrightarrow{RS} = u\underset{\sim}{a} + v\underset{\sim}{b}$, where u and v are constants, find the value of u and v. *(3 marks)*

Bonus question Easy

15 Given that $\underset{\sim}{u} = (a + 5)\underset{\sim}{i} + \underset{\sim}{j}$ and $\underset{\sim}{v} = 4\underset{\sim}{i} + (3a + 2)\underset{\sim}{j}$, find the values of a if $|\underset{\sim}{u}| = |\underset{\sim}{v}|$. *(3 marks)*

Bonus question Easy

16 Given $\underset{\sim}{u} = -3\underset{\sim}{i} + 4\underset{\sim}{j}$, $\underset{\sim}{v} = a\underset{\sim}{i} - 10\underset{\sim}{j}$ and $\underset{\sim}{w} = 6\underset{\sim}{i} + c\underset{\sim}{j}$:

a find the value of a and k for which $2k\underset{\sim}{u} + \underset{\sim}{v} = -3\underset{\sim}{i} + 2\underset{\sim}{j}$ *(2 marks)* Easy

b find the value of c such that $\underset{\sim}{u}$ is parallel to $\underset{\sim}{w}$. *(2 marks)* Medium

Bonus question

17 $ABCD$ is a trapezium, such that $AD \| BC$.
$\overrightarrow{BC} = \frac{2}{3} \times \overrightarrow{AD}$, $\overrightarrow{AE} = \frac{1}{3} \times \overrightarrow{AC}$, $\overrightarrow{AC} = 15\underset{\sim}{a}$ and $\overrightarrow{AB} = 12\underset{\sim}{b}$.

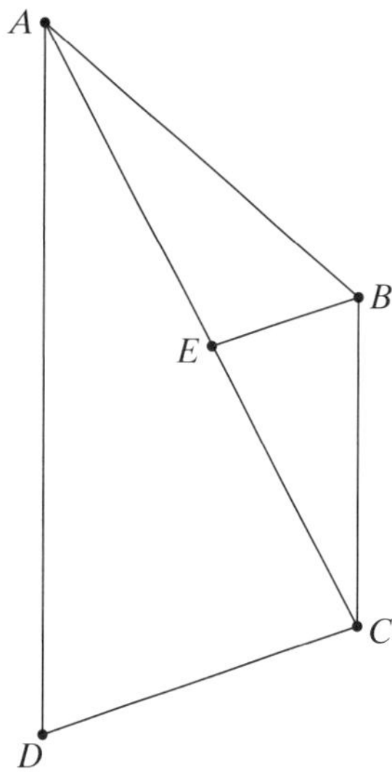

a Express each of the following in terms of $\underset{\sim}{a}$ and $\underset{\sim}{b}$.

i $\overrightarrow{BC}$ *(1 mark)* Easy

ii $\overrightarrow{BE}$ *(1 mark)* Easy

iii $\overrightarrow{CD}$ *(1 mark)* Easy

b Hence prove $BE \| CD$. *(2 marks)* Medium

Bonus question

18 O is the origin, A, B and C are 3 points such that $\overrightarrow{OA} = 2\underset{\sim}{a}$, $\overrightarrow{OB} = 4\underset{\sim}{a} + \underset{\sim}{b}$ and $\overrightarrow{OC} = m\underset{\sim}{a} + 3\underset{\sim}{b}$, where m is a constant.

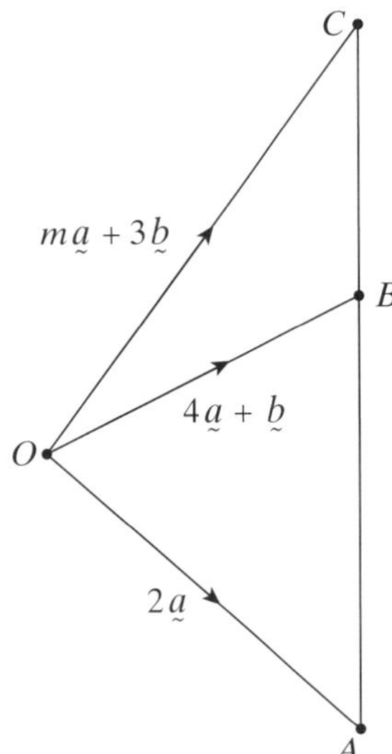

Find the value of m if A, B and C are collinear. *(3 marks)*

Bonus question Medium

19 Given that $\underset{\sim}{a} = 3\underset{\sim}{i} + 2\underset{\sim}{j}$, $\underset{\sim}{b} = k\underset{\sim}{i} - 6\underset{\sim}{j}$, $\underset{\sim}{c} = \underset{\sim}{a} + \underset{\sim}{b}$ and $\hat{\underset{\sim}{c}} = \frac{1}{5}\underset{\sim}{c}$, find all possible values of k. *(3 marks)*

Bonus question Medium

20 Barney runs 1 km due east, then 1 km due north-east, then 1.6 km due north, at a constant speed of 4 m/s as shown in the diagram.

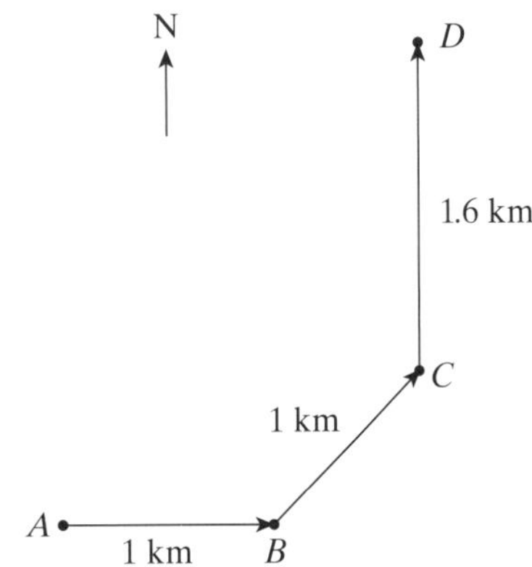

a How long did it take Barney to complete his run? *(1 mark)* Easy

b Find Barney's displacement $\overrightarrow{AD}$ in the form $a\underset{\sim}{i} + b\underset{\sim}{j}$. *(2 marks)* Medium

c Hence, calculate Barney's average velocity in metres per second correct to one decimal place. *(2 marks)* Medium

Bonus question

21 A plane is flying due north from S to F which are 600 km apart. In still air, the plane flies at a speed of 400 km/h, but the wind is blowing at 40 km/h on a bearing of 330°T.

To account for the wind, the pilot flies at a bearing that is east of north such that the resultant vector is due north.

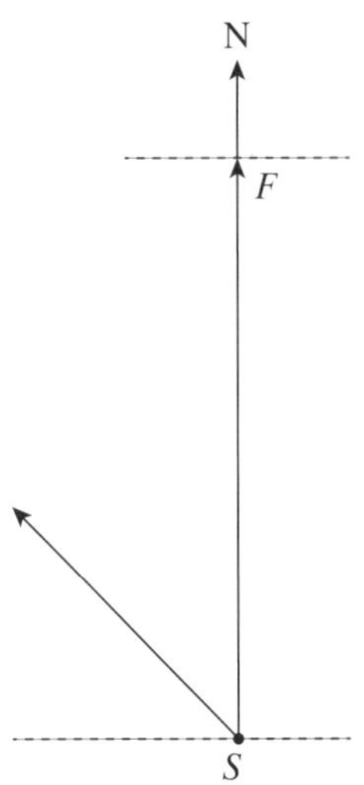

a Calculate the true bearing the pilot flies, to the nearest degree. *(4 marks)* Hard

b Hence, calculate how long the journey will take to the nearest minute. *(3 marks)* Hard

Bonus question

22 $ABCD$ is a quadrilateral.

W, X, Y and Z are the midpoints of sides AB, BC, CD and DA respectively.

$\overrightarrow{DC} = \underset{\sim}{a}$, $\overrightarrow{CB} = \underset{\sim}{b}$, $\overrightarrow{DA} = \underset{\sim}{c}$, $\overrightarrow{AB} = \underset{\sim}{d}$.

Prove, using vectors, that $WXYZ$ is a parallelogram. *(3 marks)*

Bonus question Medium

23 Find the magnitude of the projection of $\underset{\sim}{a}$ onto $\underset{\sim}{b}$ if $\underset{\sim}{a} = 2\underset{\sim}{i} + 6\underset{\sim}{j}$ and $\underset{\sim}{b} = -7\underset{\sim}{i} - \underset{\sim}{j}$. *(2 marks)* Medium

Bonus question

24 $ABCD$ is a parallelogram.

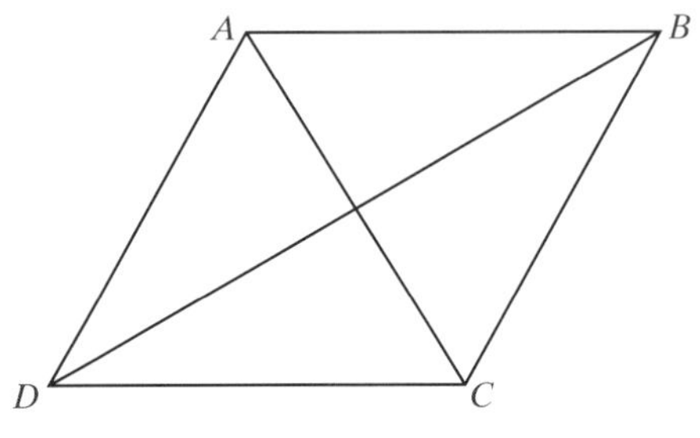

$\overrightarrow{AB} = \underset{\sim}{a}$ and $\overrightarrow{BC} = \underset{\sim}{b}$

If $AC \perp DB$, prove, using vectors, that $ABCD$ is a rhombus. *(3 marks)*

Bonus question **Medium**

25 Two boats are out on the water. Boat A's position is given by $x_A(t) = 5 - t$, $y_A(t) = 5 + 2t$ and Boat B's position is given by $x_B(t) = 2 + 4t$, $y_B(t) = 3 + t$.

The distance units are kilometres, and the time units are hours.

a Find the initial position of each boat. *(2 marks)* **Easy**

b Find the velocity vector of each boat. *(2 marks)* **Medium**

c Find the angle between the paths of the boats. *(3 marks)* **Hard**

d At what time are the boats closest to each other? *(4 marks)* **Hard**

Bonus question

26 The change of velocity of a body (over time taken) is given by $\underset{\sim}{a}$, $\underset{\sim}{b}$ and $\underset{\sim}{c}$ where:

$\underset{\sim}{a} = \dfrac{\underset{\sim}{c} - \underset{\sim}{b}}{(time\ taken)}$

$\underset{\sim}{b}$, is 3.2 m/s due south and $\underset{\sim}{c}$ is 8 m/s due west, and the time taken is 8 seconds.

Find the magnitude and direction of $\underset{\sim}{a}$. *(4 marks)*

Bonus question **Medium**

27 In the figure $A\hat{B}C = 90°$, $AP = 2 \times BP$, $AM = 2 \times CM$.

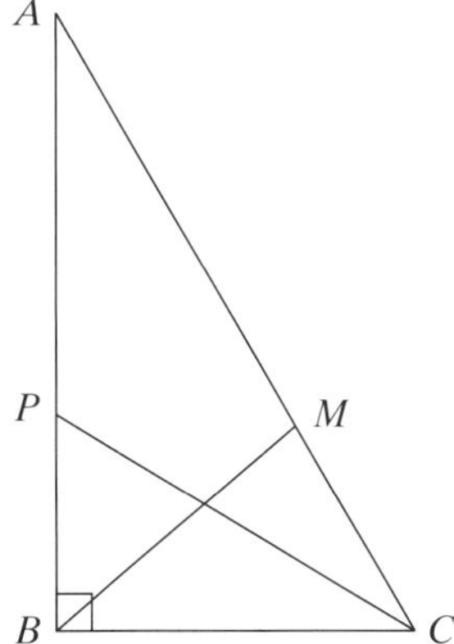

Let $\overrightarrow{BA} = \underset{\sim}{v}$ and $\overrightarrow{BC} = \underset{\sim}{u}$.

a Express $\overrightarrow{BM}$ and $\overrightarrow{CP}$ in terms of $\underset{\sim}{u}$ and $\underset{\sim}{v}$. *(3 marks)* **Medium**

b If $BM \perp CP$, find $\dfrac{BA}{BC}$. *(4 marks)* **Hard**

Bonus question

28 It is known that

$$\frac{1+y-x}{2\sqrt{(1-x)(1+y)}} = \frac{1+y+x}{2\sqrt{(1+x)(1+y)}} \text{ when}$$

$x^2 + y^2 = 1$ $(x, y \neq \pm 1)$ DO NOT PROVE THIS.

$A(-1, 0)$, $B(0, -1)$, $C(-x, y)$ and $D(x, y)$ are four points that lie on the unit circle. Using vector methods, prove $A\hat{C}B = A\hat{D}B$. *(5 marks)*

Bonus question **Hard**

29 Let $\underset{\sim}{r}(t) = (2 + \sqrt{a-1}\sin t)\underset{\sim}{i} + (-1 + \frac{1}{b}\cos t)\underset{\sim}{j}$ for $t \geq 0$ be the path of a particle moving in the cartesian plane.

The path of the particle will always be a circle if

A $b^2\sqrt{a-1} = 1$ **B** $b^2(a-1) = 1$

C $\dfrac{\sqrt{a-1}}{b} = 1$ **D** $\dfrac{\sqrt{a-1}}{2} - \dfrac{1}{b} = 1$ *(1 mark)*

Bonus question **Easy**

30 ABC is a triangle. E and D are the midpoints of AC and AB respectively.

Vectors $\overrightarrow{AC}$, $\overrightarrow{AD}$ and $\overrightarrow{CB}$ are shown.

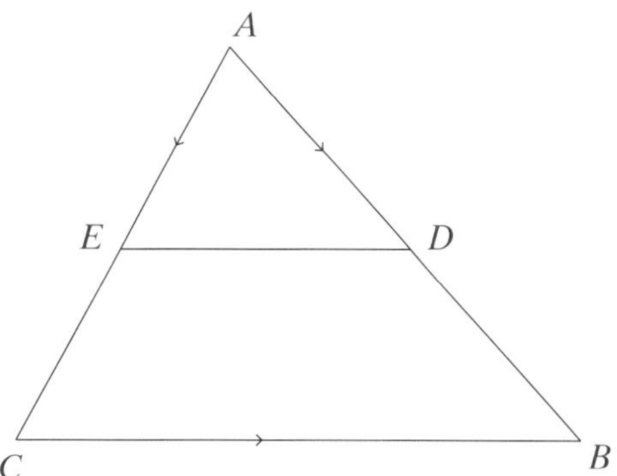

Prove, using vectors, that ED is parallel to CB and half its length. *(3 marks)*

Bonus question **Medium**

31 Consider the vectors given by $\underset{\sim}{a} = k\underset{\sim}{i} + \underset{\sim}{j}$ and $\underset{\sim}{b} = \underset{\sim}{i} + k\underset{\sim}{j}$ where k is a real number.

If the acute angle between the two vectors is 60°, find the two possible values for k. *(2 marks)*

Bonus question **Medium**

32 Express the vector $\underset{\sim}{u}$ in the form $x\underset{\sim}{i} + y\underset{\sim}{j}$, if the vector has a magnitude 6 and makes an angle of 120° with the horizontal. *(2 marks)*

Bonus question **Easy**

33 If $\underset{\sim}{u} = -4\underset{\sim}{i} + 2\underset{\sim}{j}$ and $\underset{\sim}{v} = -3\underset{\sim}{i} + \underset{\sim}{j}$, find the vector projection of $\underset{\sim}{v}$ in the direction of $\underset{\sim}{u}$. *(2 marks)*

Bonus question Easy

34 Consider the quadrilateral $ABCD$.

The midpoints of AB, BC, CD and DA are M, N, P and Q respectively.

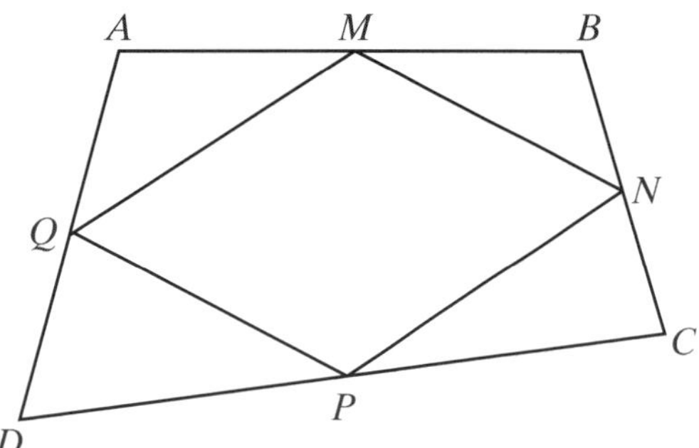

Use vector methods to prove that $MNPQ$ is a parallelogram. *(3 marks)*

Bonus question Medium

35 Find the value of m if the vectors $\underset{\sim}{u} = -2\underset{\sim}{i} + 4\underset{\sim}{j}$ and $\underset{\sim}{v} = (1 - 5m)\underset{\sim}{i} + \underset{\sim}{j}$ are perpendicular.

A $-\frac{7}{5}$ **B** $-\frac{1}{5}$

C $\frac{1}{5}$ **D** $\frac{7}{5}$ *(1 mark)*

Bonus question Easy

36 ABC is a triangle. P, Q and R are the midpoints of AB, BC and AC respectively.

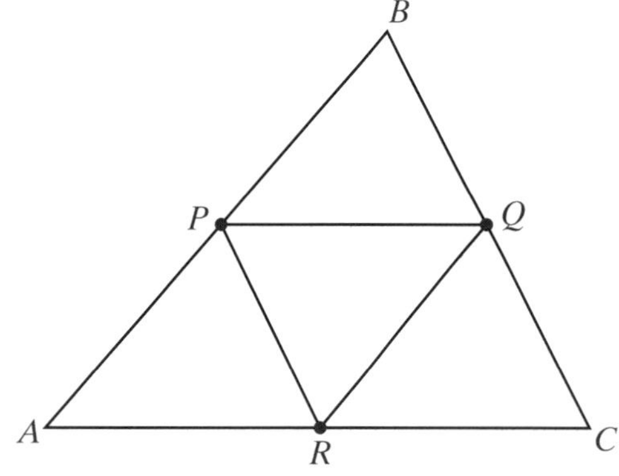

Let $\overrightarrow{AB} = \underset{\sim}{a}$, $\overrightarrow{BC} = \underset{\sim}{b}$ and $\overrightarrow{AC} = \underset{\sim}{c}$.

Which of these is equal to $\overrightarrow{QR}$?

A $\frac{1}{2}(\underset{\sim}{c} - \underset{\sim}{b})$ **B** $\underset{\sim}{a} - \underset{\sim}{b} + \frac{1}{2}\underset{\sim}{c}$

C $\underset{\sim}{a} + \underset{\sim}{b} - \frac{1}{2}\underset{\sim}{c}$ **D** $\frac{1}{2}(\underset{\sim}{b} - \underset{\sim}{c})$ *(1 mark)*

Bonus question Easy

37 In the parallelogram, $|\underset{\sim}{a}| = 2|\underset{\sim}{b}|$.

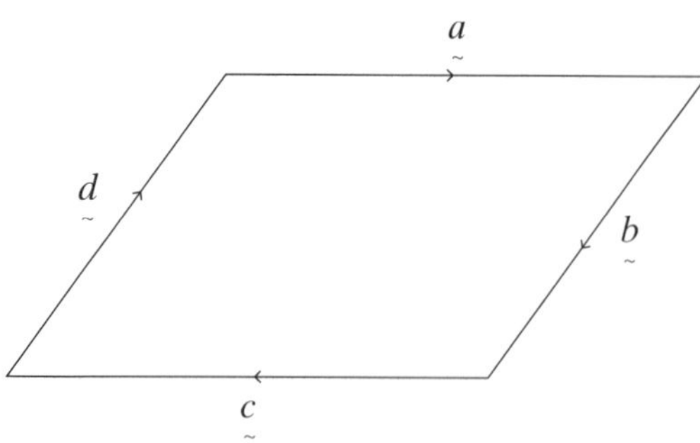

Which of the following statements is true?

A $\underset{\sim}{a} = 2\underset{\sim}{b}$ **B** $\underset{\sim}{a} + \underset{\sim}{b} = \underset{\sim}{c} + \underset{\sim}{d}$

C $\underset{\sim}{a} - \underset{\sim}{c} = 0$ **D** $\underset{\sim}{b} + \underset{\sim}{d} = 0$ *(1 mark)*

Bonus question Easy

38 Consider the two vectors $\underset{\sim}{a} = \begin{pmatrix} 3 \\ 2 \end{pmatrix}$ and $\underset{\sim}{b} = \begin{pmatrix} -2 \\ 4 \end{pmatrix}$.

Determine the smallest angle between these two vectors, to the nearest degree. *(2 marks)*

Bonus question Easy

39 A particle moves so that its position is given by $\underset{\sim}{r}(t) = (t - 2)\underset{\sim}{i} + (2t + 1)\underset{\sim}{j}$.

How far is the particle from the origin after 3 seconds?

A $4\sqrt{2}$ units **B** 7 units

C $5\sqrt{2}$ units **D** 8 units *(1 mark)*

Bonus question Easy

40 $PQRS$ is a parallelogram with $\overrightarrow{PQ} = \underset{\sim}{u}$ and $\overrightarrow{QR} = \underset{\sim}{v}$.

If M is the midpoint of RS, which of these is vector $\overrightarrow{MP}$ in terms of $\underset{\sim}{u}$ and $\underset{\sim}{v}$?

A $\frac{1}{2}\underset{\sim}{u} + \underset{\sim}{v}$ **B** $\underset{\sim}{u} + \frac{1}{2}\underset{\sim}{v}$

C $\underset{\sim}{u} - \frac{1}{2}\underset{\sim}{v}$ **D** $-\frac{1}{2}\underset{\sim}{u} - \underset{\sim}{v}$ *(1 mark)*

Bonus question Easy

41 Two particles P and Q move so that their vector equations are given by

$\underset{\sim}{r}_P(t) = (3t - 2)\underset{\sim}{i} + (t + 1)\underset{\sim}{j}$ and

$\underset{\sim}{r}_Q(t) = t^2\underset{\sim}{i} + (5 - 2t)\underset{\sim}{j}$.

Find the distance between the particles when $t = 3$. *(2 marks)*

Bonus question Medium

42 If $\underset{\sim}{a} = 2\underset{\sim}{i} - 4\underset{\sim}{j}$ and $\underset{\sim}{b} = -3\underset{\sim}{i} + \underset{\sim}{j}$, show that $\underset{\sim}{a} \cdot \underset{\sim}{b} = -10$ and hence find the vector projection of $\underset{\sim}{a}$ onto $\underset{\sim}{b}$. *(3 marks)*

Bonus question Medium

43 Let $O = (0, 0)$, $A = (4, -3)$, $B = (-2, 7)$ and $C = (5, 6)$.

Which of these is $\overrightarrow{BC}$?

A $7\underset{\sim}{i} - \underset{\sim}{j}$ **B** $5\underset{\sim}{i} - \underset{\sim}{j}$

C $7\underset{\sim}{i} - 13\underset{\sim}{j}$ **D** $-7\underset{\sim}{i} + 13\underset{\sim}{j}$ *(1 mark)*

Bonus question Easy

44 Determine whether the vectors $\underset{\sim}{a} = 4\underset{\sim}{i} - 10\underset{\sim}{j}$ and $\underset{\sim}{b} = 5\underset{\sim}{i} + 2\underset{\sim}{j}$ are parallel, perpendicular or neither. *(2 marks)*

Bonus question Easy

45 If $\underset{\sim}{a} = 3\underset{\sim}{i} - 2\underset{\sim}{j}$ and $\underset{\sim}{b} = -\underset{\sim}{i} + 5\underset{\sim}{j}$, find $2\underset{\sim}{a} - 3\underset{\sim}{b}$.

A $6\underset{\sim}{i} - 19\underset{\sim}{j}$ **B** $9\underset{\sim}{i} - 19\underset{\sim}{j}$

C $9\underset{\sim}{i} - 11\underset{\sim}{j}$ **D** $6\underset{\sim}{i} - 11\underset{\sim}{j}$ *(1 mark)*

Bonus question Easy

46 In the diagram $\overrightarrow{OA} = \underset{\sim}{a}$ and $\overrightarrow{OB} = \underset{\sim}{b}$.

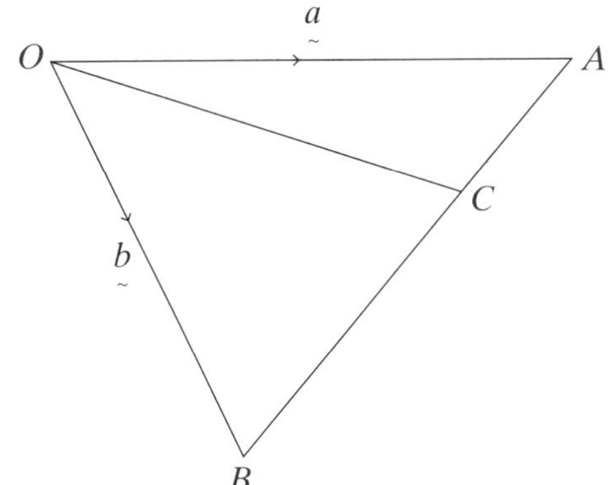

C lies on AB where $AC:CB = p:q$.

Express OC in terms of $\underset{\sim}{a}$, $\underset{\sim}{b}$, p and q. *(2 marks)*

Bonus question Medium

47 P and Q are defined by the position vectors $\underset{\sim}{u}$ and $\underset{\sim}{v}$ respectively, where $\underset{\sim}{u} = 2\underset{\sim}{i} - 3\underset{\sim}{j}$ and $\underset{\sim}{v} = 4\underset{\sim}{i} + \underset{\sim}{j}$. If O is the origin, find the size of $\angle POQ$, to the nearest degree. *(2 marks)*

Bonus question Easy

48 In the diagram AB is a diameter of a circle, centre O.

Radius OC and chords AC and CB are drawn.

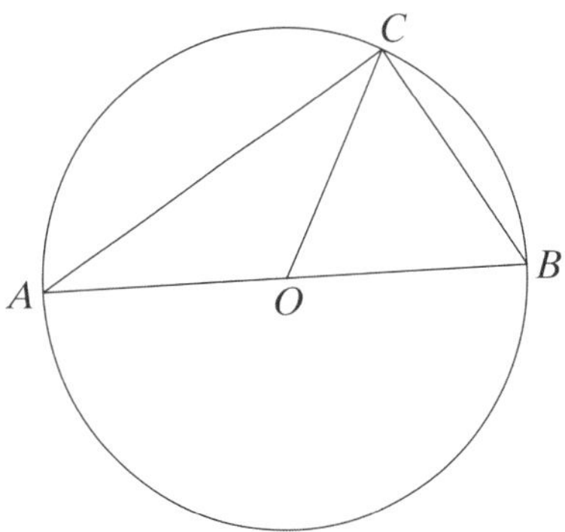

Use vectors to prove $\angle ACB$ is a right angle. *(4 marks)*

Bonus question Medium

49 Find the scalars a and b where $a\begin{pmatrix} 3 \\ -4 \end{pmatrix} + b\begin{pmatrix} 1 \\ -6 \end{pmatrix} = \begin{pmatrix} 3 \\ 10 \end{pmatrix}$. *(2 marks)*

Bonus question Easy

50 In the diagram triangle ABC is right-angled where $\angle CAB = 90°$.

D is the midpoint of side BC.

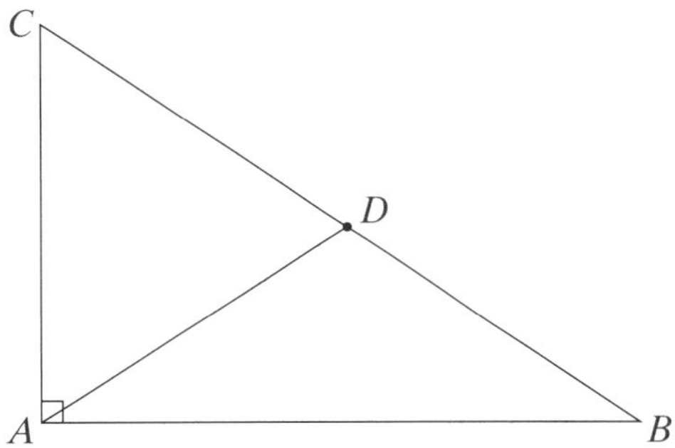

Use vectors to prove that D is equidistant from each of the vertices of the triangle. *(3 marks)*

Bonus question Medium

51 Two vectors are given as $\underset{\sim}{u} = -3\underset{\sim}{i} + m\underset{\sim}{j}$ and $\underset{\sim}{v} = 8\underset{\sim}{i} + n\underset{\sim}{j}$ where $m, n > 0$.

i If $|\underset{\sim}{u}| = 5$, and $\underset{\sim}{u}$ and $\underset{\sim}{v}$ are perpendicular, find the values of m and n. *(2 marks)* Easy

ii Hence find the unit vector $\hat{\underset{\sim}{v}}$. *(1 mark)* Easy

Bonus question

52 Find the values of m and n if $\underset{\sim}{a} = \begin{pmatrix} 3 \\ n-2 \end{pmatrix}$, $\underset{\sim}{b} = \begin{pmatrix} -1 \\ 3 \end{pmatrix}$ and $2\underset{\sim}{a} - 3\underset{\sim}{b} = \begin{pmatrix} m+5 \\ -1 \end{pmatrix}$. *(2 marks)*

Bonus question Medium

Year 12 Further operations with vectors—Worked Answers

1 Let $\underset{\sim}{v} = -\underset{\sim}{i} + \underset{\sim}{j}$ such that the projection of $\underset{\sim}{u}$ onto $\underset{\sim}{v}$ is given by $\frac{(\underset{\sim}{u} \bullet \underset{\sim}{v})}{|\underset{\sim}{v}|}\underset{\sim}{v}$.

The length of v is given by $|\underset{\sim}{v}| = \sqrt{2}$ and $(\underset{\sim}{u} \bullet \underset{\sim}{v}) > 0$ since the angle θ between the vectors is acute.

Therefore the projection of $\underset{\sim}{u}$ onto $\underset{\sim}{v}$ would have the same direction as $\underset{\sim}{v}$ but a scalar of $\underset{\sim}{v}$.

$\therefore -0.4\underset{\sim}{i} + 0.4\underset{\sim}{j}$.

Answer B

2
$$|\underset{\sim}{a}+\underset{\sim}{b}| < 1$$
$$|\underset{\sim}{a}+\underset{\sim}{b}|^2 < 1$$
$$|\underset{\sim}{a}|^2 + |\underset{\sim}{b}|^2 + 2\underset{\sim}{a}\bullet\underset{\sim}{b} < 1$$
$$|\underset{\sim}{a}|^2 + |\underset{\sim}{b}|^2 + 2|\underset{\sim}{a}||\underset{\sim}{b}|\cos\theta < 1$$

Since $\underset{\sim}{a}$ and $\underset{\sim}{b}$ are unit vectors, then $|\underset{\sim}{a}| = |\underset{\sim}{b}| = 1$.

$2 + 2\cos\theta < 1$

$2\cos\theta < -1$

$\cos\theta < -\frac{1}{2}$

$\therefore \theta > \frac{2\pi}{3}$

Alternatively, the tail of the vector $\underset{\sim}{a} + \underset{\sim}{b}$ must remain inside the unit circle. Therefore the angle θ between $\underset{\sim}{a}$ and $\underset{\sim}{b}$ must not be less than $\frac{2\pi}{3}$.

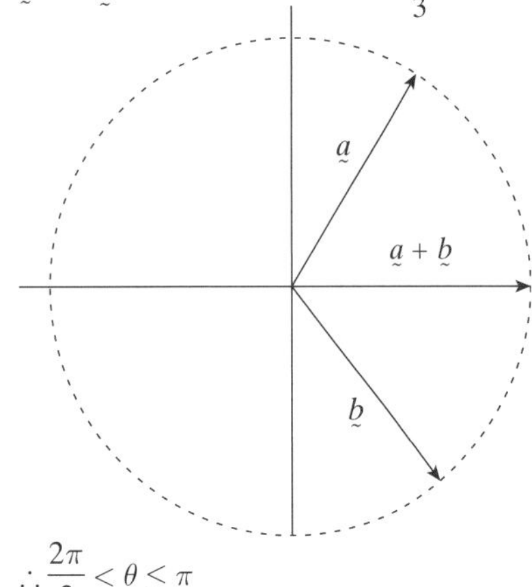

$\therefore \frac{2\pi}{3} < \theta \le \pi$

Answer D

3 Vectors perpendicular if $\underset{\sim}{u}\bullet\underset{\sim}{v} = 0$

$\underset{\sim}{u}\bullet\underset{\sim}{v} = a(a-7)+2(4a-1)$ ✓

$0 = a^2 - 7a + 8a - 2$

$= a^2 + a - 2$

$= (a+2)(a-1)$

$\therefore a = -2$ or $a = 1$ ✓ *(2 marks)*

4 $\underset{\sim}{h} = \underset{\sim}{a}+\underset{\sim}{b}+\underset{\sim}{c}$

$\therefore -\underset{\sim}{b}+\underset{\sim}{h} = \underset{\sim}{a}+\underset{\sim}{c}$

$\overrightarrow{BH} = -\underset{\sim}{b}+\underset{\sim}{h}$

$= \underset{\sim}{a}+\underset{\sim}{c}$

$\overrightarrow{CA} = -\underset{\sim}{c}+\underset{\sim}{a}$ ✓

Vectors are perpendicular if $\overrightarrow{BH}\bullet\overrightarrow{CA} = 0$

$\overrightarrow{BH}\bullet\overrightarrow{CA} = (\underset{\sim}{a}+\underset{\sim}{c})(-\underset{\sim}{c}+\underset{\sim}{a})$

$= -\underset{\sim}{a}\bullet c + \underset{\sim}{a}\bullet\underset{\sim}{a} - \underset{\sim}{c}\bullet\underset{\sim}{c} + \underset{\sim}{a}\bullet\underset{\sim}{c}$ ✓

$= \underset{\sim}{a}\bullet\underset{\sim}{a} - \underset{\sim}{c}\bullet\underset{\sim}{c}$

$= |\underset{\sim}{a}|^2 - |\underset{\sim}{c}|^2$

$= 0$ (since $|\underset{\sim}{a}|^2 = |\underset{\sim}{c}|^2$ as radii)

$\therefore \overrightarrow{BH}$ and $\overrightarrow{CA}$ are perpendicular. ✓ *(3 marks)*

5 The projection of $\vec{u}$ onto $\vec{v}$ is given by $\frac{\vec{u}\bullet\vec{v}}{|\underset{\sim}{v}|^2}\vec{v}$

$\therefore \vec{p} = \frac{\vec{u}\bullet\vec{v}}{|\underset{\sim}{v}|^2}\vec{v}$

$\lambda_0\vec{v} = \frac{\vec{u}\bullet\vec{v}}{|\underset{\sim}{v}|^2}\vec{v}$

$\therefore \lambda_0 = \frac{\vec{u}\bullet\vec{v}}{|\underset{\sim}{v}|^2}$ ✓

$|\vec{u}-\lambda\vec{v}|^2 = (\vec{u}-\lambda\vec{v})\bullet(\vec{u}-\lambda\vec{v})$

$= |\vec{u}|^2 - 2\lambda\vec{u}\bullet\vec{v} + \lambda^2|\vec{u}|^2$ ✓

$\therefore$ The magnitude is given by a quadratic in terms of λ with minimum value at $\lambda = -\frac{b}{2a}$.

$\lambda = -\frac{b}{2a}$

$= -\frac{-2\vec{u}\bullet\vec{v}}{2|\vec{v}|^2}$

$= \frac{\vec{u}\bullet\vec{v}}{|\vec{v}|^2}$

$= \lambda_0$

$\therefore |\vec{u}-\lambda_0\vec{v}|^2 \le |\vec{u}-\lambda\vec{v}|^2$

$\therefore |\vec{u}-\lambda_0\vec{v}| \le |\vec{u}-\lambda\vec{v}|$ ✓

(3 marks)

6 If $\overrightarrow{OA}\cdot\overrightarrow{OB} < 0$, then

$|\overrightarrow{OA}|\cdot|\overrightarrow{OB}|\cos\theta < 0$

Since $|\overrightarrow{OA}|\cdot|\overrightarrow{OB}| > 0$, then $\cos\theta < 0$

$\therefore \theta$ is obtuse.

Answer B

7 Let t = time, in hours.

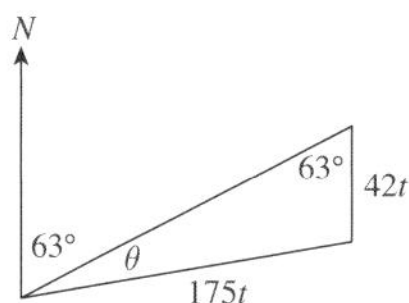

Using the sine rule:

$\frac{\sin\theta}{42t} = \frac{\sin 63^\circ}{175t}$ ✓

$= \frac{42\sin 63^\circ}{175}$

$= 0.21834...$

$\therefore \theta = 12$ (nearest degree) ✓

Bearing is $63 + 12 = 075°$. ✓ *(3 marks)*

8 **i** Let $\underset{\sim}{v} = x\underset{\sim}{i} + y\underset{\sim}{j}$, then

$\underset{\sim}{v}\cdot\underset{\sim}{v} = x^2 + y^2$

$= \left(\sqrt{x^2+y^2}\right)^2$

$= |\underset{\sim}{v}|^2$

Alternatively, $\underset{\sim}{v}\cdot\underset{\sim}{v} = |\underset{\sim}{v}||\underset{\sim}{v}|\cos\theta$

Since $\theta = 0$, $\cos\theta = 1$.

$\underset{\sim}{v}\cdot\underset{\sim}{v} = |\underset{\sim}{v}|^2$ ✓ *(1 mark)*

ii $\overrightarrow{AC} = \overrightarrow{AB} + \overrightarrow{BC}$

$= \underset{\sim}{a} + \underset{\sim}{b}$

$\overrightarrow{BD} = \overrightarrow{AD} - \overrightarrow{AB}$

$= k\underset{\sim}{b} - \underset{\sim}{a}$ ✓

We know that $|\overrightarrow{AC}| = |\overrightarrow{BD}|$, so therefore:

$$|\overrightarrow{AC}|^2 = |\overrightarrow{BD}|^2$$
$$(\underset{\sim}{a}+\underset{\sim}{b})\cdot(\underset{\sim}{a}+\underset{\sim}{b}) = (k\underset{\sim}{b}-\underset{\sim}{a})\cdot(k\underset{\sim}{b}-\underset{\sim}{a})$$
$$\underset{\sim}{a}\cdot\underset{\sim}{a}+2\underset{\sim}{a}\cdot\underset{\sim}{b}+\underset{\sim}{b}\cdot\underset{\sim}{b} = k^2\underset{\sim}{b}\cdot\underset{\sim}{b}-2k\underset{\sim}{a}\cdot\underset{\sim}{b}+\underset{\sim}{a}\cdot\underset{\sim}{a} \quad ✓$$
$$2\underset{\sim}{a}\cdot\underset{\sim}{b}+|\underset{\sim}{b}|^2 = k^2|\underset{\sim}{b}|^2-2k\underset{\sim}{a}\cdot\underset{\sim}{b}$$
$$2\underset{\sim}{a}\cdot\underset{\sim}{b}+2k\underset{\sim}{a}\cdot\underset{\sim}{b} = k^2|\underset{\sim}{b}|^2-|\underset{\sim}{b}|^2$$
$$2(1+k)\underset{\sim}{a}\cdot\underset{\sim}{b} = (k^2-1)|\underset{\sim}{b}|^2$$
$$2(1+k)\underset{\sim}{a}\cdot\underset{\sim}{b} = -(1+k)(1-k)|\underset{\sim}{b}|^2$$
$$2\underset{\sim}{a}\cdot\underset{\sim}{b} = -(1-k)|\underset{\sim}{b}|^2$$
$$2\underset{\sim}{a}\cdot\underset{\sim}{b}+(1-k)|\underset{\sim}{b}|^2 = 0 \quad ✓$$

(3 marks)

9 Maria's position is $(2+3+4)\underset{\sim}{i}+(3-2-3)\underset{\sim}{j}$ or $9\underset{\sim}{i}-2\underset{\sim}{j}$.

Distance from origin $= \sqrt{9^2+(-2)^2}$
$= \sqrt{85}$

Answer B

10 $\begin{pmatrix} a \\ -1 \end{pmatrix}, \begin{pmatrix} 2a-3 \\ 2 \end{pmatrix}$

If perpendicular,

$a(2a-3)+(-1)\times 2 = 0$ ✓
$2a^2-3a-2 = 0$
$(2a+1)(a-2) = 0$
$2a+1 = 0$ or $a-2 = 0$
$a = -\frac{1}{2}$ ✓ or $a = 2$ ✓

(3 marks)

11 $\underset{\sim}{a}\cdot\underset{\sim}{b} = 4\times 4\times\cos 60°$
$= 8$ ✓

(1 mark)

12 **a** $\overrightarrow{AB} = \overrightarrow{AO}+\overrightarrow{OB}$
$= (-1+5)\underset{\sim}{i}+(-4+(-1))\underset{\sim}{j}$
$= 4\underset{\sim}{i}-5\underset{\sim}{j}$ ✓

$|\overrightarrow{AB}| = \sqrt{4^2+(-5)^2}$
$= \sqrt{41}$

$\therefore$ unit vector along $\overrightarrow{AB} = \frac{1}{\sqrt{41}}(4\underset{\sim}{i}-5\underset{\sim}{j})$ ✓

(2 marks)

b Let N have co-ordinates (x, y)

$\overrightarrow{ON}\cdot\overrightarrow{AB} = 0$
$\therefore 4\times x-5\times y = 0$
$y = \frac{4x}{5} \leftarrow (1)$ ✓

$\overrightarrow{NB} = k\times\overrightarrow{AB}$

Hence slope $\overrightarrow{NB}$ = slope $\overrightarrow{AB}$

$\frac{y-(-1)}{x-5} = \frac{4-(-1)}{1-5} = \frac{5}{-4}$
$-4y-4 = 5x-25$
$5x+4y = 21 \leftarrow (2)$ ✓

Subs. (1) into (2)

$5x+4\times\frac{4x}{5} = 21$
$\frac{41x}{5} = 21$
$\therefore x = \frac{105}{41}$ and $y = \frac{84}{41}$

$\overrightarrow{ON} = \frac{1}{41}(105\underset{\sim}{i}+84\underset{\sim}{j})$ ✓

(3 marks)

13 $\underset{\sim}{a}\cdot\underset{\sim}{b} = 4\times n+2\times 3n$
$= 10n$
$|\underset{\sim}{a}| = \sqrt{4^2+2^2}$
$= \sqrt{20}$
$|\underset{\sim}{b}| = \sqrt{n^2+(3n)^2}$
$= \sqrt{10n^2}$
$= n\sqrt{10}$ ✓

$\cos\theta = \frac{\underset{\sim}{a}\cdot\underset{\sim}{b}}{|\underset{\sim}{a}|\cdot|\underset{\sim}{b}|}$
$= \frac{10n}{\sqrt{20}\times n\sqrt{10}}$
$= \frac{1}{\sqrt{2}}$
$\theta = \cos^{-1}\left(\frac{1}{\sqrt{2}}\right)$ ✓
$\theta = 45°$ ✓

(3 marks)

14 $\overrightarrow{RS} = u(2\underset{\sim}{i}+5\underset{\sim}{j})+v(-5\underset{\sim}{i}+2\underset{\sim}{j})$
$= (2u-5v)\underset{\sim}{i}+(5u+2v)\underset{\sim}{j}$

But $\overrightarrow{RS} = 8\underset{\sim}{i}-9\underset{\sim}{j}$ ✓

Therefore,

$2u-5v = 8 \quad \leftarrow(1)$
$5u+2v = -9 \quad \leftarrow(2)$ ✓
$(1)\times 5 \quad 10u-25v = 40 \quad \leftarrow(3)$
$(2)\times 2 \quad 10u+4v = -18 \quad \leftarrow(4)$
$(4)-(3)$
$29v = -58$
$\therefore v = -2$ and $u = -1$ ✓

(3 marks)

15 If $|\underset{\sim}{u}| = |\underset{\sim}{v}|$

$\sqrt{(a+5)^2+1^2} = \sqrt{4^2+(3a+2)^2}$ ✓
$a^2+10a+25+1 = 16+9a^2+12a+4$
$8a^2+2a-6 = 0$ ✓
$2(a+1)(4a-3) = 0$
$\therefore a = -1$ or $\frac{3}{4}$ ✓

(3 marks)

16 **a** $2k\underset{\sim}{u}+\underset{\sim}{v} = 2k(-3\underset{\sim}{i}+4\underset{\sim}{j})+a\underset{\sim}{i}-10\underset{\sim}{j}$
$= (-6k+a)\underset{\sim}{i}+(8k-10)\underset{\sim}{j}$

Therefore, $8k-10 = 2$

$\therefore k = \frac{3}{2}$ ✓

So $-6\times\frac{3}{2}+a = -3$

$\therefore a = 6$ ✓

(2 marks)

b Let $\underset{\sim}{w} = m\times\underset{\sim}{u}$

$\therefore 6\underset{\sim}{i}+c\underset{\sim}{j} = m(-3\underset{\sim}{i}+4\underset{\sim}{j})$ ✓

So $m = -2$, hence $c = -8$ ✓

$\underset{\sim}{u}$ is parallel to $\underset{\sim}{w}$ when $c = -8$.

(2 marks)

17 **a** **i** $\overrightarrow{BC} = \overrightarrow{BA}+\overrightarrow{AC}$
$= -12\underset{\sim}{b}+15\underset{\sim}{a}$ ✓

(1 mark)

ii $\overrightarrow{BE} = \overrightarrow{BA}+\frac{1}{3}\times\overrightarrow{AC}$
$= -12\underset{\sim}{b}+\frac{1}{3}\times 15\underset{\sim}{a}$
$= 5\underset{\sim}{a}-12\underset{\sim}{b}$ ✓

(1 mark)

iii $\overrightarrow{CD} = \overrightarrow{CA} + \frac{3}{2} \times \overrightarrow{BC}$

$= -15\underset{\sim}{a} + \frac{3}{2} \times (15\underset{\sim}{a} - 12\underset{\sim}{b})$

$= \frac{15}{2}\underset{\sim}{a} - 18\underset{\sim}{b}$ ✓

(1 mark)

b $\overrightarrow{CD} = \frac{15}{2}\underset{\sim}{a} - 18\underset{\sim}{b}$

$= \frac{3}{2} \times (5\underset{\sim}{a} - 12\underset{\sim}{b})$ ✓

$= \frac{3}{2} \times \overrightarrow{BE}$

$\therefore BE \| CD.$ ✓

(2 marks)

18 $\overrightarrow{AB} = \overrightarrow{AO} + \overrightarrow{OB}$

$= -2\underset{\sim}{a} + (4\underset{\sim}{a} + \underset{\sim}{b})$

$= 2\underset{\sim}{a} + \underset{\sim}{b}$

$\overrightarrow{AC} = \overrightarrow{AO} + \overrightarrow{OC}$

$= -2\underset{\sim}{a} + (m\underset{\sim}{a} + 3\underset{\sim}{b})$

$= (m-2)\underset{\sim}{a} + 3\underset{\sim}{b}$ ✓

Since A, B and C are collinear, $\overrightarrow{AC} = \lambda \times \overrightarrow{AB}$.

$(m-2)\underset{\sim}{a} + 3\underset{\sim}{b} = \lambda \times (2\underset{\sim}{a} + \underset{\sim}{b})$ ✓

$\therefore 3\underset{\sim}{b} = \lambda\underset{\sim}{b}$

$\lambda = 3$

So $(m-2) = 2 \times 3$

$m = 8$ ✓

(3 marks)

19 $\underset{\sim}{c} = (3\underset{\sim}{i} + 2\underset{\sim}{j}) + (k\underset{\sim}{i} - 6\underset{\sim}{j})$

$= (k+3)\underset{\sim}{i} - 4\underset{\sim}{j}$

$|\underset{\sim}{c}| = \sqrt{(k+3)^2 + (-4)^2}$ ✓

Since $\hat{\underset{\sim}{c}} = \frac{\underset{\sim}{c}}{|\underset{\sim}{c}|} = \frac{1}{5}\underset{\sim}{c}$

$\therefore \sqrt{(k+3)^2 + (-4)^2} = 5$ ✓

$k^2 + 6k + 25 = 25$

$k^2 + 6k = 0$

$k(k+6) = 0$

$\therefore k = 0$ or -6 ✓

(3 marks)

20 **a** $3600 \div 4 = 900$ seconds

$\therefore$ 15 minutes ✓

(1 mark)

b

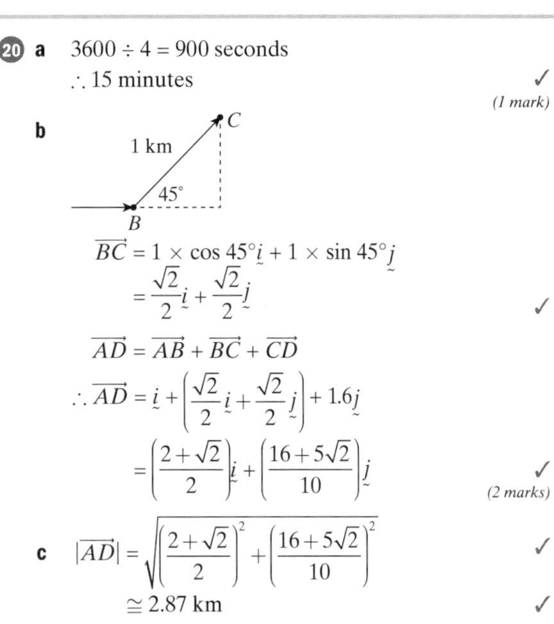

$\overrightarrow{BC} = 1 \times \cos 45°\underset{\sim}{i} + 1 \times \sin 45°\underset{\sim}{j}$

$= \frac{\sqrt{2}}{2}\underset{\sim}{i} + \frac{\sqrt{2}}{2}\underset{\sim}{j}$ ✓

$\overrightarrow{AD} = \overrightarrow{AB} + \overrightarrow{BC} + \overrightarrow{CD}$

$\therefore \overrightarrow{AD} = \underset{\sim}{i} + \left(\frac{\sqrt{2}}{2}\underset{\sim}{i} + \frac{\sqrt{2}}{2}\underset{\sim}{j}\right) + 1.6\underset{\sim}{j}$

$= \left(\frac{2+\sqrt{2}}{2}\right)\underset{\sim}{i} + \left(\frac{16+5\sqrt{2}}{10}\right)\underset{\sim}{j}$ ✓

(2 marks)

c $|\overrightarrow{AD}| = \sqrt{\left(\frac{2+\sqrt{2}}{2}\right)^2 + \left(\frac{16+5\sqrt{2}}{10}\right)^2}$ ✓

$\cong 2.87$ km ✓

$\therefore$ Average velocity $= \frac{|\overrightarrow{AD}|}{900}$

$= 3.1888995\ldots$

$= 3.2$ m/s

(2 marks)

21 **a**

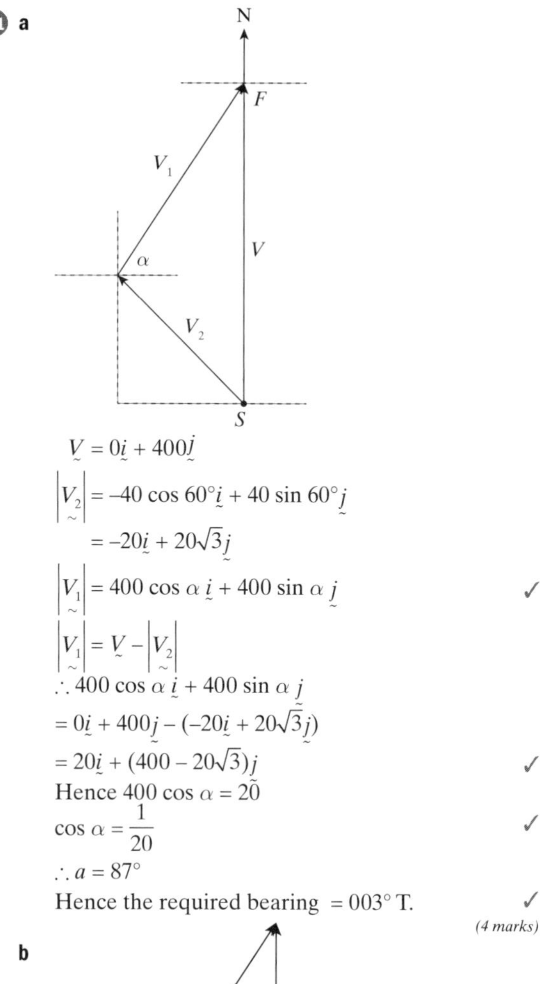

$\underset{\sim}{V} = 0\underset{\sim}{i} + 400\underset{\sim}{j}$

$|\underset{\sim}{V_2}| = -40 \cos 60°\underset{\sim}{i} + 40 \sin 60°\underset{\sim}{j}$

$= -20\underset{\sim}{i} + 20\sqrt{3}\underset{\sim}{j}$

$|\underset{\sim}{V_1}| = 400 \cos \alpha\, \underset{\sim}{i} + 400 \sin \alpha\, \underset{\sim}{j}$ ✓

$|\underset{\sim}{V_1}| = \underset{\sim}{V} - |\underset{\sim}{V_2}|$

$\therefore 400 \cos \alpha\, \underset{\sim}{i} + 400 \sin \alpha\, \underset{\sim}{j}$

$= 0\underset{\sim}{i} + 400\underset{\sim}{j} - (-20\underset{\sim}{i} + 20\sqrt{3}\underset{\sim}{j})$

$= 20\underset{\sim}{i} + (400 - 20\sqrt{3})\underset{\sim}{j}$ ✓

Hence $400 \cos \alpha = 20$

$\cos \alpha = \frac{1}{20}$ ✓

$\therefore a = 87°$

Hence the required bearing $= 003°$ T. ✓

(4 marks)

b

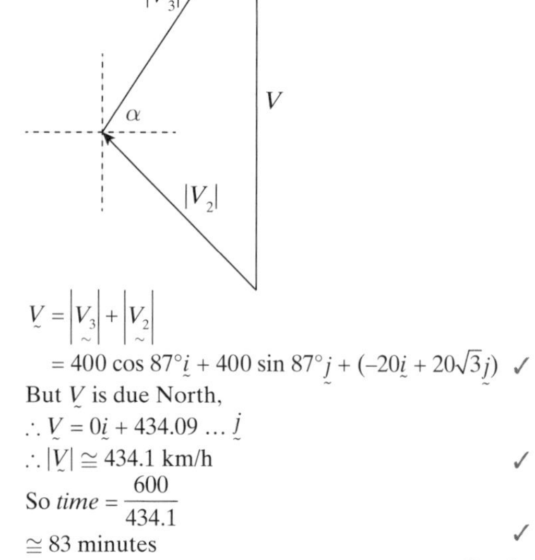

$\underset{\sim}{V} = |\underset{\sim}{V_3}| + |\underset{\sim}{V_2}|$

$= 400 \cos 87°\underset{\sim}{i} + 400 \sin 87°\underset{\sim}{j} + (-20\underset{\sim}{i} + 20\sqrt{3}\underset{\sim}{j})$ ✓

But $\underset{\sim}{V}$ is due North,

$\therefore \underset{\sim}{V} = 0\underset{\sim}{i} + 434.09 \ldots \underset{\sim}{j}$

$\therefore |\underset{\sim}{V}| \cong 434.1$ km/h ✓

So *time* $= \frac{600}{434.1}$

$\cong 83$ minutes ✓

(3 marks)

22

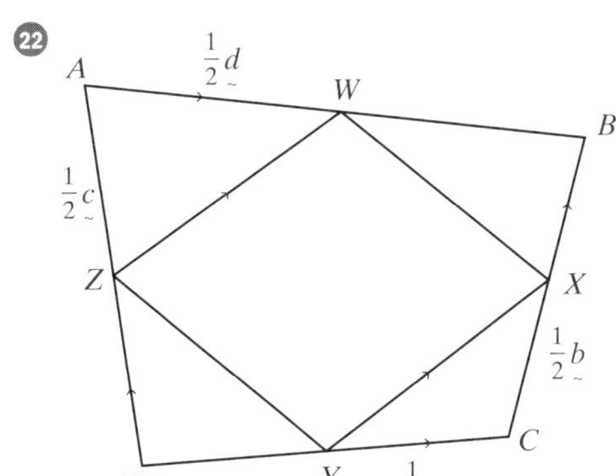

$$\vec{YX} = \frac{1}{2}\underset{\sim}{a} + \frac{1}{2}\underset{\sim}{b}$$
$$= \frac{1}{2}(\underset{\sim}{a} + \underset{\sim}{b})$$
$$\vec{ZW} = \frac{1}{2}\underset{\sim}{c} + \frac{1}{2}\underset{\sim}{d}$$
$$= \frac{1}{2}(\underset{\sim}{c} + \underset{\sim}{d})$$ ✓

$\underset{\sim}{a} + \underset{\sim}{b} - \underset{\sim}{c} - \underset{\sim}{d} = 0$ ($ABCD$ is a quadrilateral) ✓

$\therefore \underset{\sim}{a} + \underset{\sim}{b} = \underset{\sim}{c} + \underset{\sim}{d}$

$\therefore \vec{YX} = \vec{ZW}$

$\therefore WXYZ$ is a parallelogram (opposite sides equal and parallel) ✓

(3 marks)

23 $proj_{\underset{\sim}{b}}\underset{\sim}{a} = \dfrac{2\times(-7)+6\times(-1)}{(-7)\times(-7)+(-1)\times(-1)} \times (-7\underset{\sim}{i} - \underset{\sim}{j})$

$$= \frac{-2}{5}(-7\underset{\sim}{i} - \underset{\sim}{j})$$ ✓

$$\therefore |proj_{\underset{\sim}{b}}\underset{\sim}{a}| = \frac{2}{5} \times \sqrt{(-7)^2 + (-1)^2}$$
$$= \frac{2}{5} \times 5\sqrt{2}$$
$$= 2\sqrt{2}$$ ✓

(2 marks)

24

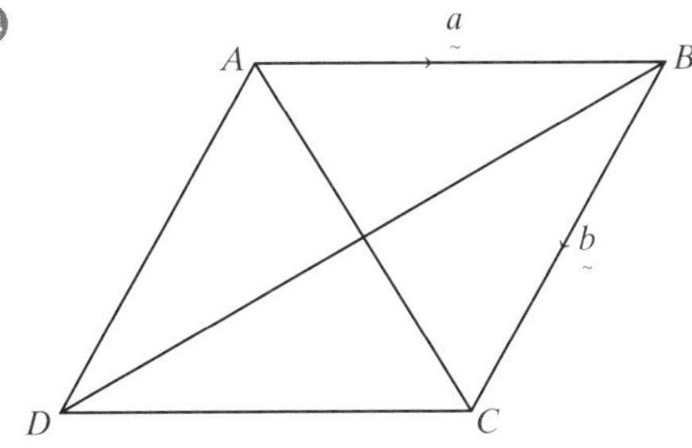

As diagonals are AC and DB, where $AC \perp DB$, then $\vec{AC} \cdot \vec{DB} = 0$

$(\underset{\sim}{a} + \underset{\sim}{b})(\underset{\sim}{a} - \underset{\sim}{b}) = 0$ ✓

$\underset{\sim}{a} \cdot \underset{\sim}{a} - \underset{\sim}{b} \cdot \underset{\sim}{b} = 0$

$\underset{\sim}{a} \cdot \underset{\sim}{a} = \underset{\sim}{b} \cdot \underset{\sim}{b}$

$\therefore |\underset{\sim}{a}| = |\underset{\sim}{b}|$ ✓

$\therefore$ shape is a rhombus (adjacent sides equal) ✓

(3 marks)

25 **a** At $t = 0$,

$$x_A(0) = 5 - 0$$
$$= 5$$
$$y_A(0) = 5 + 2 \times 0$$
$$= 5$$
$$\therefore A(5,5)$$ ✓
$$x_B(0) = 2 + 4 \times 0$$
$$= 2$$
$$y_B(0) = 3 + 0$$
$$= 3$$
$$\therefore B(2,3)$$ ✓

(2 marks)

b $\underset{\sim}{v} = \dfrac{d}{dt}[x]$ where x is the displacement vector.

$x_A(t) = 5 - t$ $y_A(t) = 5 + 2t$

$\therefore \dot{x}_A = -1$ $\therefore \dot{y}_A = 2$

$\therefore \underset{\sim}{v_A} = -\underset{\sim}{i} + 2\underset{\sim}{j}$ ✓

$x_B(t) = 2 + 4t$ $y_B(t) = 3 + t$

$\therefore \dot{x}_B = 4$ $\therefore \dot{y}_B = 1$

$\therefore \underset{\sim}{v_B} = 4\underset{\sim}{i} + \underset{\sim}{j}$ ✓

(2 marks)

c $\underset{\sim}{v_A} \cdot \underset{\sim}{v_B} = |\underset{\sim}{v_A}||\underset{\sim}{v_B}| \cos\theta$

$(-1)(4) + (2)(1)$
$= \sqrt{(-1)^2 + (2)^2} \times \sqrt{(4)^2 + (1)^2} \times \cos\theta$ ✓

$\cos\theta = \dfrac{-2}{\sqrt{85}}$ ✓

$\theta = 102.529....°$

$\therefore$ the angle is 103° to the nearest degree. ✓

(3 marks)

d Boats are closest to each other when the distance between them is a minimum.

$$d^2 = (x_A - x_B)^2 + (y_A - y_B)^2$$
$$= [(5 - t) - (2 + 4t)]^2 + [(5 + 2t) - (3 + t)]^2$$
$$= [3 - 5t]^2 + [2 + t]^2$$
$$= 9 - 30t + 25t^2 + 4 + 4t + t^2$$
$$= 26t^2 - 26t + 13$$ ✓

Stationary points at $\dfrac{d}{dt}[d^2]$,

$\dfrac{d}{dt}[d^2] = 52t - 26$ ✓

$52t - 26 = 0$

$t = \dfrac{1}{2}$ ✓

$\dfrac{d^2}{dt^2}[d^2] = 52 > 0$

$\therefore$ minimum turning point when $t = \dfrac{1}{2}$ ✓

$\therefore$ The boats are closest at $t = 0.5$ hours.

(4 marks)

26 $\underset{\sim}{b} = 0\underset{\sim}{i} - 3.2\underset{\sim}{j}$

$\underset{\sim}{c} = -8\underset{\sim}{i} + 0\underset{\sim}{j}$ ✓

$$\underset{\sim}{a} = \frac{(-8\underset{\sim}{i} + 0\underset{\sim}{j}) - (0\underset{\sim}{i} - 3.2\underset{\sim}{j})}{8}$$
$$= -\underset{\sim}{i} + 0.4\underset{\sim}{j}$$ ✓
$$|\underset{\sim}{a}| = \sqrt{(-1)^2 + (0.4)^2}$$
$$\cong 1.08 \text{ (3 sig. fig.)}$$ ✓

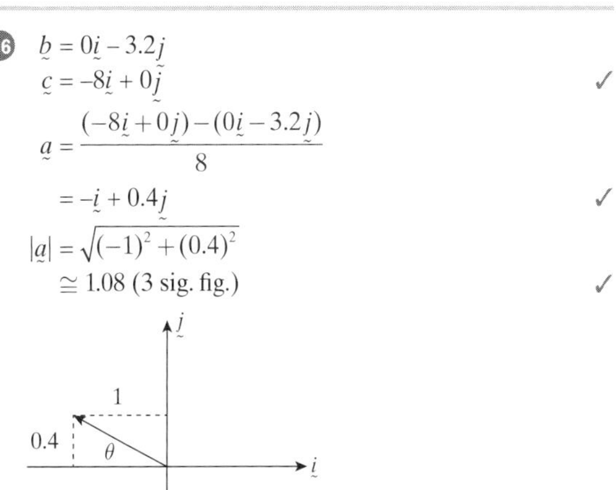

$\tan\theta = \dfrac{0.4}{1}$

$\theta = 21.801\ldots$

$\cong 22°$ ✓

$\therefore |\underset{\sim}{a}| \cong 1.08$ m/s, bearing 292°T

(4 marks)

27 **a** $\overrightarrow{AC} = \overrightarrow{AB} + \overrightarrow{BC}$

$= -\underset{\sim}{v} + \underset{\sim}{u}$

$\therefore \overrightarrow{AM} = \dfrac{2}{3}(\underset{\sim}{u} - \underset{\sim}{v})$ ✓

$\overrightarrow{BM} = \overrightarrow{BA} + \overrightarrow{AM}$

$= \underset{\sim}{v} + \dfrac{2}{3}(\underset{\sim}{u} - \underset{\sim}{v})$

$= \dfrac{2}{3}\underset{\sim}{u} + \dfrac{1}{3}\underset{\sim}{v}$ ✓

$\overrightarrow{CP} = \overrightarrow{CB} + \overrightarrow{BP}$

$= -\underset{\sim}{u} + \dfrac{1}{3}\underset{\sim}{v}$ ✓

(3 marks)

b $BM \perp CP \therefore \overrightarrow{BM}\cdot\overrightarrow{CP} = 0$

We know $\underset{\sim}{u}\cdot\underset{\sim}{v} = 0$ as $A\hat{B}C = 90°$.

$\left(\dfrac{2}{3}\underset{\sim}{u} + \dfrac{1}{3}\underset{\sim}{v}\right)\cdot\left(\dfrac{1}{3}\underset{\sim}{v} - \underset{\sim}{u}\right) = 0$ ✓

$\dfrac{2}{9}\underset{\sim}{u}\cdot\underset{\sim}{v} - \dfrac{2}{3}\underset{\sim}{u}\cdot\underset{\sim}{u} + \dfrac{1}{9}\underset{\sim}{v}\cdot\underset{\sim}{v} - \dfrac{1}{3}\underset{\sim}{u}\cdot\underset{\sim}{v} = 0$

$0 - \dfrac{2|\underset{\sim}{u}|^2}{3} + \dfrac{1|\underset{\sim}{v}|^2}{9} - 0 = 0$ ✓

$\dfrac{|\underset{\sim}{v}|^2}{9} = \dfrac{2|\underset{\sim}{u}|^2}{3}$

$|\underset{\sim}{v}|^2 = 6|\underset{\sim}{u}|^2$ ✓

$\therefore \dfrac{|\underset{\sim}{v}|^2}{|\underset{\sim}{u}|^2} = \dfrac{6}{1}$

So $\dfrac{|\underset{\sim}{v}|}{|\underset{\sim}{u}|} = \dfrac{BA}{BC} = \sqrt{6}$ ✓

(4 marks)

28

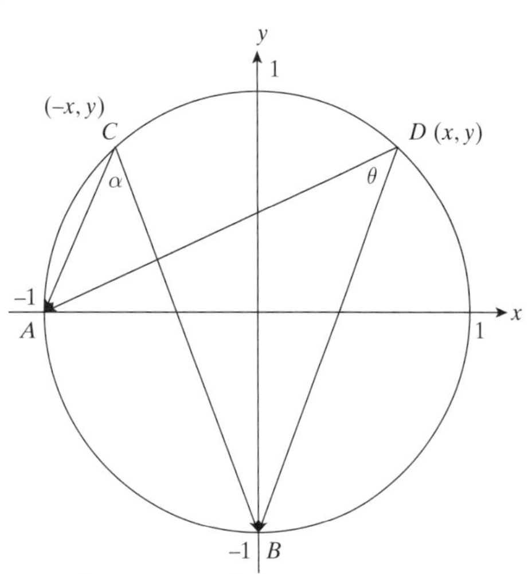

Let $A\hat{C}B = \alpha$ and $A\hat{D}B = \theta$.

Using $C(-x,y)$

$\overrightarrow{CA} = (-1-(-x))\underset{\sim}{i} + (0-y)\underset{\sim}{j}$

$= (x-1)\underset{\sim}{i} - y\underset{\sim}{j}$

$\overrightarrow{CB} = (0-(-x))\underset{\sim}{i} + (-1-y)\underset{\sim}{j}$

$= x\underset{\sim}{i} - (1+y)\underset{\sim}{j}$ ✓

$\cos\alpha = \dfrac{\overrightarrow{CA}\cdot\overrightarrow{CB}}{|CA||CB|}$

$= \dfrac{x(x-1)+(-y)\times[-(1+y)]}{\sqrt{(x-1)^2+(-y)^2}\times\sqrt{x^2+[-(1+y)]^2}}$

$= \dfrac{x^2-x+y+y^2}{\sqrt{x^2-2x+1+y^2}\times\sqrt{x^2+1+2y+y^2}}$ ✓

But $x^2 + y^2 = 1$ as (x,y) lies on the circle.

$\therefore \cos\alpha = \dfrac{1-x+y}{\sqrt{2-2x}\times\sqrt{2+2y}}$

$= \dfrac{1-x+y}{\sqrt{2}\sqrt{1-x}\times\sqrt{2}\sqrt{1+y}}$

$= \dfrac{1-x+y}{2\sqrt{(1-x)(1+y)}}$ ✓

Using $D(x,y)$

$\overrightarrow{DA} = (-1-x)\underset{\sim}{i} + (0-y)\underset{\sim}{j}$

$= -(x+1)\underset{\sim}{i} - y\underset{\sim}{j}$

$\overrightarrow{DB} = (0-x)\underset{\sim}{i} + (-1-y)\underset{\sim}{j}$

$= -x\underset{\sim}{i} - (1+y)\underset{\sim}{j}$

$\cos\theta = \dfrac{\overrightarrow{DA}\cdot\overrightarrow{DB}}{|DA||DB|}$

$= \dfrac{[-(x+1)](-x)+(-y)\times[-(1+y)]}{\sqrt{[-(x+1)]^2+(-y)^2}\times\sqrt{(-x)^2+[-(1+y)]^2}}$

$= \dfrac{x^2+x+y+y^2}{\sqrt{x^2+2x+1+y^2}\times\sqrt{x^2+1+2y+y^2}}$ ✓

But $x^2 + y^2 = 1$ as (x,y) lies on the circle.

$\therefore \cos\alpha = \dfrac{1+x+y}{\sqrt{2+2x}\times\sqrt{2+2y}}$

$= \dfrac{1+x+y}{\sqrt{2}\sqrt{(1+x)}\times\sqrt{2}(1+y)}$

$= \dfrac{1-x+y}{2\sqrt{(1+x)(1+y)}}$

Since we know $\dfrac{1+y-x}{2\sqrt{(1-x)(1+y)}} = \dfrac{1+y+x}{2\sqrt{(1+x)(1+y)}}$

$\therefore \cos\alpha = \cos\theta$

Hence $A\hat{C}B = A\hat{D}B$ since α and $\theta < \pi$ ✓

(5 marks)

29 Path is a circle when $\sqrt{a-1} = \dfrac{1}{b}$

$a - 1 = \dfrac{1}{b^2}$

$b^2(a-1) = 1$

Answer B

30

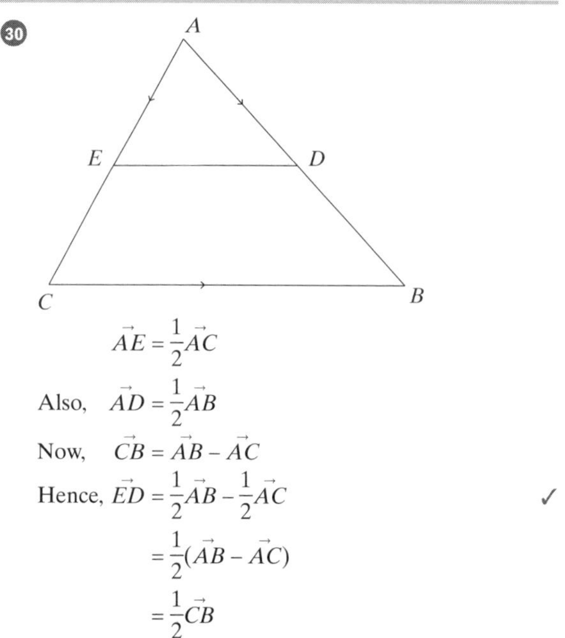

$\vec{AE} = \dfrac{1}{2}\vec{AC}$

Also, $\vec{AD} = \dfrac{1}{2}\vec{AB}$

Now, $\vec{CB} = \vec{AB} - \vec{AC}$

Hence, $\vec{ED} = \dfrac{1}{2}\vec{AB} - \dfrac{1}{2}\vec{AC}$ ✓

$= \dfrac{1}{2}(\vec{AB} - \vec{AC})$

$= \dfrac{1}{2}\vec{CB}$

$\therefore \vec{ED} = \frac{1}{2}\vec{CB}$ ✓

Hence, $CB \parallel ED$.

Also, $|\vec{ED}| = |\frac{1}{2}\vec{CB}| = \frac{1}{2}|\vec{CB}|$ and so CB is half the length of ED.

$\therefore ED$ is parallel to CB and half its length. ✓

(3 marks)

31 $\underset{\sim}{a} \cdot \underset{\sim}{b} = k \times 1 + 1 \times k$

$= 2k$

$|\underset{\sim}{a}||\underset{\sim}{b}|\cos\theta = \sqrt{k^2+1^2} \cdot \sqrt{1^2+k^2} \cdot \cos 60°$

$= \frac{1}{2}(k^2+1)$ ✓

Using $\underset{\sim}{a} \cdot \underset{\sim}{b} = |\underset{\sim}{a}||\underset{\sim}{b}|\cos\theta$,

$2k = \frac{1}{2}(k^2+1)$

$k^2 + 1 = 4k$

$k^2 - 4k + 1 = 0$

$k = \frac{4 \pm \sqrt{16 - 4(1)(1)}}{2(1)}$

$= \frac{4 \pm \sqrt{12}}{2}$

$= 2 \pm \sqrt{3}$ ✓ *(2 marks)*

32 $x = |\underset{\sim}{u}|\cos\theta$

$= 6\cos 120°$

$= 6 \times -\frac{1}{2}$

$= -3$

$y = |\underset{\sim}{u}|\sin\theta$ ✓

$= 6\sin 120°$

$= 6 \times \frac{\sqrt{3}}{2}$

$= 3\sqrt{3}$

$\therefore \underset{\sim}{u} = x\underset{\sim}{i} + y\underset{\sim}{j}$

$= -3\underset{\sim}{i} + 3\sqrt{3}\underset{\sim}{j}$ ✓ *(2 marks)*

33 $\underset{\sim}{v} \cdot \underset{\sim}{u} = (-3) \times (-4) + 1 \times 2$

$= 14$

$|\underset{\sim}{u}|^2 = (-4)^2 + 2^2$

$= 20$ ✓

$\therefore \underset{\sim}{v}_{\underset{\sim}{u}} = \frac{14}{20}(-4\underset{\sim}{i} + 2\underset{\sim}{j})$

$= \frac{7}{5}(-2\underset{\sim}{i} + \underset{\sim}{j})$ ✓ *(2 marks)*

34 Let $\vec{AB} = \underset{\sim}{a}, \vec{BC} = \underset{\sim}{b}, \vec{CD} = \underset{\sim}{c}, \vec{DA} = \underset{\sim}{d}$.

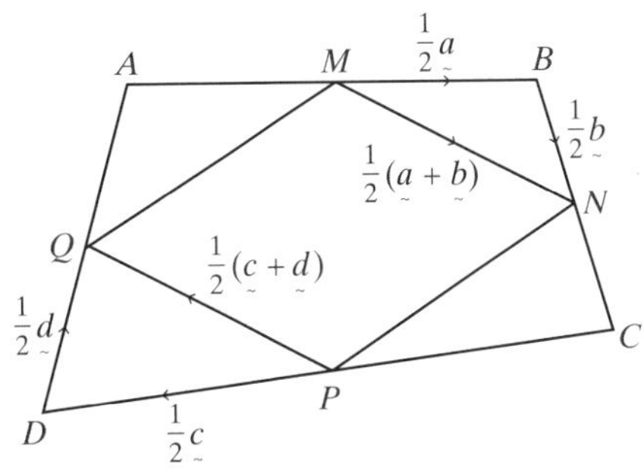

As M, N, P and Q are midpoints, then

$\vec{MB} = \frac{1}{2}\underset{\sim}{a}, \vec{BN} = \frac{1}{2}\underset{\sim}{b}, \vec{PD} = \frac{1}{2}\underset{\sim}{c}, \vec{DQ} = \frac{1}{2}\underset{\sim}{d}$. ✓

Hence, $\vec{MN} = \frac{1}{2}(\underset{\sim}{a} + \underset{\sim}{b})$ and $\vec{QP} = \frac{1}{2}(-\underset{\sim}{c} - \underset{\sim}{d})$.

But $\underset{\sim}{a} + \underset{\sim}{b} + \underset{\sim}{c} + \underset{\sim}{d} = 0$. ✓

Hence, $\underset{\sim}{a} + \underset{\sim}{b} = -(\underset{\sim}{c} + \underset{\sim}{d})$.

$\therefore MN = QP$ and $MN \parallel PQ$.

$\therefore MNPQ$ is a parallelogram. ✓

(3 marks)

35 $-2(1 - 5m) + 4 \times 1 = 0$

$-2 + 10m + 4 = 0$

$10m + 2 = 0$

$m = -\frac{1}{5}$

Answer B

36 As $\vec{BC} = \underset{\sim}{b}$, then $\vec{QC} = \frac{1}{2}\underset{\sim}{b}$.

As $\vec{AC} = \underset{\sim}{c}$, then $\vec{CA} = -\underset{\sim}{c}$, and

hence, $\vec{CR} = -\frac{1}{2}\underset{\sim}{c}$.

Using $\vec{QR} = \vec{QC} + \vec{CR}$

$= \frac{1}{2}\underset{\sim}{b} + \left(-\frac{1}{2}\underset{\sim}{c}\right)$

$= \frac{1}{2}(\underset{\sim}{b} - \underset{\sim}{c})$

Answer D

37 In a parallelogram, opposite sides equal.

As $\underset{\sim}{b} = -\underset{\sim}{d}$ hence $\underset{\sim}{b} + \underset{\sim}{d} = 0$.

Answer D

38 $\underset{\sim}{a} \bullet \underset{\sim}{b} = 3 \times (-2) + 2 \times 4$

$= 2$

$|\underset{\sim}{a}||\underset{\sim}{b}|\cos\theta = \sqrt{3^2+2^2} . \sqrt{(-2)^2+4^2} . \cos\theta$

$= \sqrt{260}\cos\theta$ ✓

Using $\underset{\sim}{a} \bullet \underset{\sim}{b} = |\underset{\sim}{a}||\underset{\sim}{b}|\cos\theta$,

$2 = \sqrt{260}\cos\theta$

$\cos\theta = \frac{2}{\sqrt{260}}$

$= \frac{1}{\sqrt{65}}$

$\theta = 82.874\,983\,65\ldots$

$= 83$ (nearest whole)

$\therefore$ the smallest angle is 83°. ✓

(2 marks)

39 $\underset{\sim}{r}(3) = (3-2)\underset{\sim}{i} + (2(3)+1)\underset{\sim}{j}$

$= \underset{\sim}{i} + 7\underset{\sim}{j}$

$|\underset{\sim}{r}(3)| = \sqrt{1^2+7^2}$

$= \sqrt{50}$

$= 5\sqrt{2}$

$\therefore$ the particle is $5\sqrt{2}$ units from origin.

Answer C

40 As $\overrightarrow{PQ} = \underset{\sim}{u}$ then $\overrightarrow{RS} = -\underset{\sim}{u}$.

Hence, $\overrightarrow{MS} = -\frac{1}{2}\underset{\sim}{u}$.

Also, as $\overrightarrow{QR} = \underset{\sim}{v}$, then $\overrightarrow{SP} = -\underset{\sim}{v}$.

Now, $\overrightarrow{MP} = \overrightarrow{MS} + \overrightarrow{SP}$,

$$= -\frac{1}{2}\underset{\sim}{u} + -\underset{\sim}{v}$$

$$= -\frac{1}{2}\underset{\sim}{u} - \underset{\sim}{v}$$

Answer D

41 $\underset{\sim}{r}_P(3) = (3(3) - 2)\underset{\sim}{i} + (3 + 1)\underset{\sim}{j}$

$= 7\underset{\sim}{i} + 4\underset{\sim}{j}$

$\underset{\sim}{r}_Q(3) = 3^2\underset{\sim}{i} + (5 - 2(3))\underset{\sim}{j}$.

$= 9\underset{\sim}{i} - \underset{\sim}{j}$

$\underset{\sim}{r}_P(3) - \underset{\sim}{r}_Q(3) = (7\underset{\sim}{i} + 4\underset{\sim}{j}) - (9\underset{\sim}{i} - \underset{\sim}{j})$

$= -2\underset{\sim}{i} + 5\underset{\sim}{j}$ ✓

Distance $= \sqrt{(-2)^2 + 5^2}$

$= \sqrt{29}$

$\therefore$ the particles are $\sqrt{29}$ units apart. ✓

(2 marks)

42 $\underset{\sim}{a} \bullet \underset{\sim}{b} = 2(-3) + (-4)1$

$= -10$ ✓

$\frac{\underset{\sim}{a} \bullet \underset{\sim}{b}}{|\underset{\sim}{b}|^2}.\underset{\sim}{b} = \frac{2(-3) + (-4)1}{(-3)^2 + 1^2} \times (-3\underset{\sim}{i} + \underset{\sim}{j})$ ✓

$= \frac{-10}{10} \times (-3\underset{\sim}{i} + \underset{\sim}{j})$

$= -(-3\underset{\sim}{i} + \underset{\sim}{j})$

$= 3\underset{\sim}{i} - \underset{\sim}{j}$ ✓

(3 marks)

43 $\overrightarrow{BC} = \overrightarrow{BO} + \overrightarrow{OC}$

$= -(-2\underset{\sim}{i} + 7\underset{\sim}{j}) + (5\underset{\sim}{i} + 6\underset{\sim}{j})$

$= 2\underset{\sim}{i} - 7\underset{\sim}{j} + 5\underset{\sim}{i} + 6\underset{\sim}{j}$

$= 7\underset{\sim}{i} - \underset{\sim}{j}$

Answer A

44 $\underset{\sim}{a} \bullet \underset{\sim}{b} = 4 \times 5 + (-10) \times 2$

$= 20 - 20$

$= 0$ ✓

As $\underset{\sim}{a} \bullet \underset{\sim}{b} = 0$ then the vectors are perpendicular. ✓ *(2 marks)*

45 $2\underset{\sim}{a} - 3\underset{\sim}{b} = 2(3\underset{\sim}{i} - 2\underset{\sim}{j}) - 3(-\underset{\sim}{i} + 5\underset{\sim}{j})$

$= 6\underset{\sim}{i} - 4\underset{\sim}{j} + 3\underset{\sim}{i} - 15\underset{\sim}{j}$

$= 9\underset{\sim}{i} - 19\underset{\sim}{j}$

Answer B

46 $\overrightarrow{OA} + \overrightarrow{AB} = \overrightarrow{OB}$

$\underset{\sim}{a} + \overrightarrow{AB} = \underset{\sim}{b}$

$\overrightarrow{AB} = \underset{\sim}{b} - \underset{\sim}{a}$

Now, $AC:CB = p:q$ ✓

$\therefore AC:AB = p:(p + q)$

$\therefore \overrightarrow{AC} = \frac{p}{p+q}(\underset{\sim}{b} - \underset{\sim}{a})$ ✓

Now, as $\overrightarrow{OC} = \underset{\sim}{a} + \overrightarrow{AC}$

$\therefore \overrightarrow{OC} = \underset{\sim}{a} + \frac{p}{p+q}(\underset{\sim}{b} - \underset{\sim}{a})$ ✓

$= \frac{1}{p+q}(p\underset{\sim}{a} + q\underset{\sim}{a} + p\underset{\sim}{b} - p\underset{\sim}{a})$

$= \frac{1}{p+q}(q\underset{\sim}{a} + p\underset{\sim}{b})$

(2 marks)

47 $\cos\theta = \frac{\underset{\sim}{u} \cdot \underset{\sim}{v}}{|\underset{\sim}{u}||\underset{\sim}{v}|}$

$= \frac{2 \times 4 + (-3) \times 1}{\sqrt{2^2 + (-3)^2} \bullet \sqrt{4^2 + 1^2}}$ ✓

$= \frac{5}{\sqrt{13} \bullet \sqrt{17}}$

$= 0.336\,336\,397\ldots$

$\theta = 70.346\,175\,94\ldots$

$= 70$ (nearest whole)

$\therefore \angle POQ = 70^\circ$. ✓ *(2 marks)*

48 $|\overrightarrow{OA}| = |\overrightarrow{OB}| = |\overrightarrow{OC}|$ (equal radii)

Let $\overrightarrow{OA} = \underset{\sim}{a}$ and $\overrightarrow{OC} = \underset{\sim}{c}$.

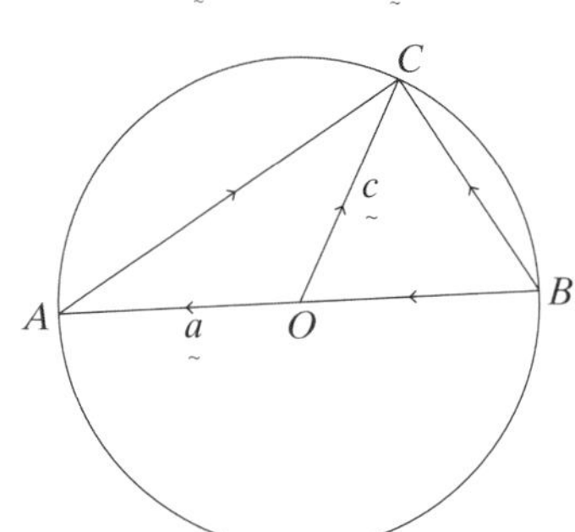

Now, $\overrightarrow{OB} = -\underset{\sim}{a}$, and hence $\overrightarrow{BO} = \underset{\sim}{a}$

As $\overrightarrow{AC} = \overrightarrow{AO} + \overrightarrow{OC}$

$= -\underset{\sim}{a} + \underset{\sim}{c}$

Also, $\overrightarrow{BC} = \overrightarrow{BO} + \overrightarrow{OC}$

$= \underset{\sim}{a} + \underset{\sim}{c}$ ✓

Now, $\overrightarrow{AC} \cdot \overrightarrow{BC} = (-\underset{\sim}{a} + \underset{\sim}{c}) \cdot (\underset{\sim}{a} + \underset{\sim}{c})$

$= -\underset{\sim}{a} \cdot \underset{\sim}{a} + \underset{\sim}{c} \cdot \underset{\sim}{c}$

$= -|\underset{\sim}{a}|^2 + |\underset{\sim}{c}|^2$ ✓

But $|\underset{\sim}{a}|^2 = |\underset{\sim}{c}|^2$, then $\overrightarrow{AC} \cdot \overrightarrow{BC} = 0$. ✓

Hence, $AC \perp BC$.

$\therefore \angle ACB$ is a right angle ✓ *(4 marks)*

49 $3a + b = 3 \ldots.①$

$-4a - 6b = 10 \ldots.②$

$6 \times ①: 18a + 6b = 18 \ldots.③$

$② + ③: 14a = 28$

$a = 2$ ✓

Subs in ①: $3(2) + b = 3$

$b = 3 - 6$

$= -3$

$\therefore a = 2, b = -3$. ✓

(2 marks)

50 Let $\overrightarrow{AB} = \underset{\sim}{b}$ and $\overrightarrow{AC} = \underset{\sim}{c}$

$AB \perp AC$ so $\underset{\sim}{b} \cdot \underset{\sim}{c} = 0$

$\overrightarrow{CB} = \underset{\sim}{b} - \underset{\sim}{c}$

$\therefore \overrightarrow{CD} = \overrightarrow{DB} = \frac{1}{2}(\underset{\sim}{b} - \underset{\sim}{c})$

$\overrightarrow{AD} = \overrightarrow{AB} + \overrightarrow{BD}$

$= \underset{\sim}{b} + \frac{1}{2}(\underset{\sim}{c} - \underset{\sim}{b})$

$= \frac{1}{2}(\underset{\sim}{b} + \underset{\sim}{c})$ ✓

$|\overrightarrow{AD}|^2 = \frac{1}{4}(\underset{\sim}{b}^2 + 2\underset{\sim}{b} \cdot \underset{\sim}{c} + \underset{\sim}{c}^2)$

$= \frac{1}{4}(\underset{\sim}{b}^2 + \underset{\sim}{c}^2)$ ✓

$|\overrightarrow{CD}|^2 = \frac{1}{4}(\underset{\sim}{b}^2 - 2\underset{\sim}{b} \cdot \underset{\sim}{c} + \underset{\sim}{c}^2)$

$= \frac{1}{4}(\underset{\sim}{b}^2 + \underset{\sim}{c}^2)$

$\therefore AD = CD = BD$

$\therefore D$ is equidistant from each of the vertices of the triangle. ✓

(3 marks)

---

51 **i** As $|\underset{\sim}{u}| = 5$, then $(-3)^2 + m^2 = 5^2$

$m^2 = 25 - 9$

$\therefore m = 4 \quad (m > 0)$ ✓

$\therefore \underset{\sim}{u} = -3\underset{\sim}{i} + 4\underset{\sim}{j}$

As $\underset{\sim}{u}$ and $\underset{\sim}{v}$ are perpendicular, then $\underset{\sim}{u} \cdot \underset{\sim}{v} = 0$.

$-3 \times 8 + 4 \times n = 0$

$4n = 24$

$n = 6$ ✓

(2 marks)

ii As $\underset{\sim}{v} = 8\underset{\sim}{i} + 6\underset{\sim}{j}$:

$\hat{\underset{\sim}{v}} = \frac{\underset{\sim}{v}}{|\underset{\sim}{v}|}$

$\hat{\underset{\sim}{v}} = \frac{8\underset{\sim}{i} + 6\underset{\sim}{j}}{\sqrt{8^2 + 6^2}}$

$= \frac{8\underset{\sim}{i} + 6\underset{\sim}{j}}{\sqrt{100}}$

$= \frac{8\underset{\sim}{i} + 6\underset{\sim}{j}}{10}$

$= \frac{1}{5}\left(4\underset{\sim}{i} + 3\underset{\sim}{j}\right)$ ✓

(1 mark)

---

52 $2(3) - 3(-1) = m + 5$

$6 + 3 = m + 5$

$m = 9 - 5$

$m = 4$ ✓

Also, $2(n - 2) - 3(3) = -1$

$2n - 4 - 9 = -1$

$2n - 13 = -1$

$2n = -1 + 13$

$2n = 12$

$n = 6$

$\therefore m = 4, n = 6$ ✓

(2 marks)

Note: Vectors were not a specific part of this course before 2020. So there may be a different approach to older projectile motion questions.

1 A video game designer wants to include an obstacle in the game they are developing. The player will reach one side of a pit and must shoot a projectile to hit a target on the other side of the pit in order to be able to cross. However, the instant the player shoots, the target begins to move away from the player at a constant speed that is half the initial speed of the projectile shot by the player, as shown in the diagram below.

The initial distance between the player and the target is d, the initial speed of the projectile is $2u$ and it is launched at an angle of θ to the horizontal. The acceleration due to gravity is g. The launch angle is the ONLY parameter that the player can change.

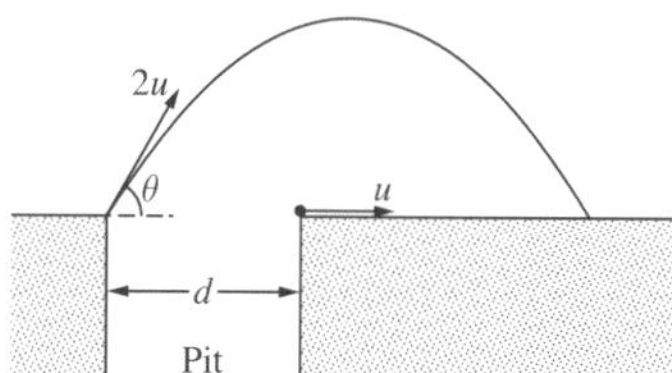

Taking the position of the player when the projectile is launched as the origin, the positions of the projectile and target at time t after the projectile is launched are as follows.

$$\vec{r}_P = \begin{pmatrix} 2ut\cos\theta \\ 2ut\sin\theta - \frac{g}{2}t^2 \end{pmatrix} \quad \text{Projectile}$$

(Do NOT prove these.)

$$\vec{r}_T = \begin{pmatrix} d + ut \\ 0 \end{pmatrix} \quad \text{Target}$$

Show that, for the player to have a chance of hitting the target, d must be less than 37% of the maximum possible range of the projectile (to 2 significant figures). *(4 marks)*

(Q14c, **2022 HSC**) Medium

2 When an object is projected from a point h metres above the origin with initial speed V m /s at an angle of $\theta°$ to the horizontal, its displacement vector, t seconds after projection, is

$$\underset{\sim}{r}(t) = \left(Vt\cos\theta\right)\underset{\sim}{i} + \left(-5t^2 + Vt\sin\theta + h\right)\underset{\sim}{j}.$$

(Do NOT prove this).

A person, standing in an empty room which is 3 m high, throws a ball at the far wall of the room. The ball leaves their hand 1 m above the floor and 10 m from the far wall.

The initial velocity of the ball is 12 m /s at an angle of 30° to the horizontal.

Show that the ball will NOT hit the ceiling of the room but that it will hit the far wall without hitting the floor. *(4 marks)*

(Q13b, **2021 HSC**) Medium

3 A particle is projected from a point so that its position at time t is given by the position vector:

$$r(t) = 30\sqrt{3}t\underset{\sim}{i} + (30t - 5t^2)\underset{\sim}{j}$$

where t is in seconds and $r(t)$ is in metres.
Find:

a an expression for the velocity vector *(1 mark)* Easy

b the initial velocity and its direction. *(2 marks)* Easy

Bonus question (see page iv)

4 A projectile is fired from the ground with an angle of projection given by $\alpha = \tan^{-1}\left(\frac{12}{5}\right)$ and an initial velocity V.

It just clears a wall that is 19 metres high 10 metres away.

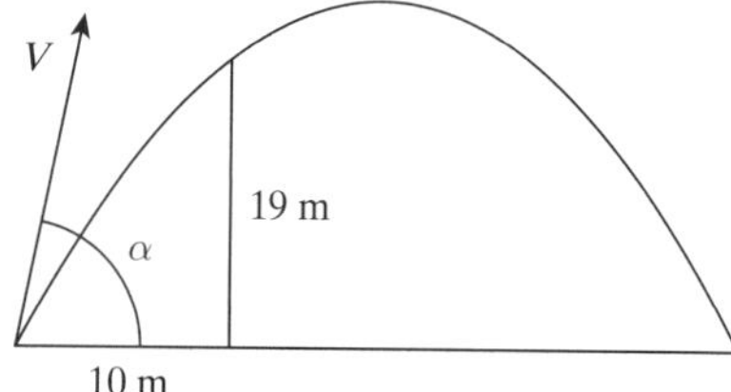

Let acceleration due to gravity be $g = 10$ m/s².

a Show the displacement vector can be written as $\underset{\sim}{r}(t) = \frac{1}{13}[5Vt\underset{\sim}{i} + (12Vt - 65t^2)\underset{\sim}{j}]$. *(3 marks)* Easy

b Find the initial velocity of the projectile. *(2 marks)* Medium

c At what speed is the projectile travelling at the instant it clears the wall. *(2 marks)* Medium

Bonus question

5 A particle is fired from a point O on the floor of a horizontal tunnel of height 10 metres at an initial velocity of 24 m/s, and at an angle of θ above the horizontal, where $0° \leq \theta \leq 90°$.

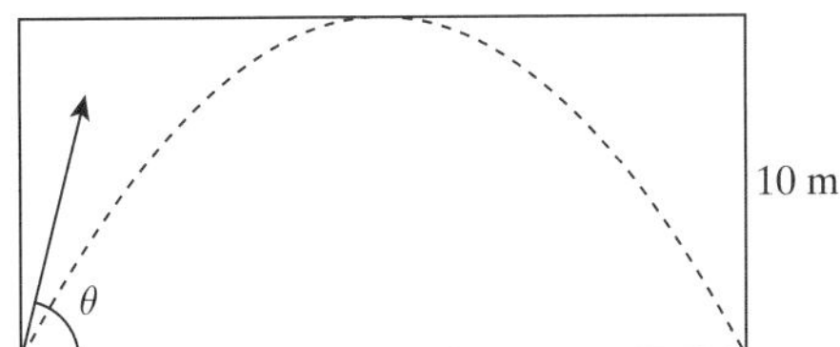

Assume the position vector of the particle from O at time t seconds after firing is given by:

$$r(t) = 24t\cos\theta \underset{\sim}{i} + (24t\sin\theta - 5t^2)\underset{\sim}{j}$$

Find the maximum horizontal range of the particle along the tunnel. *(4 marks)* **Hard**

Bonus question

6 From a window 8 metres above the ground, an object is projected at an angle of 30° to the horizontal at an initial velocity of 36 m/s.

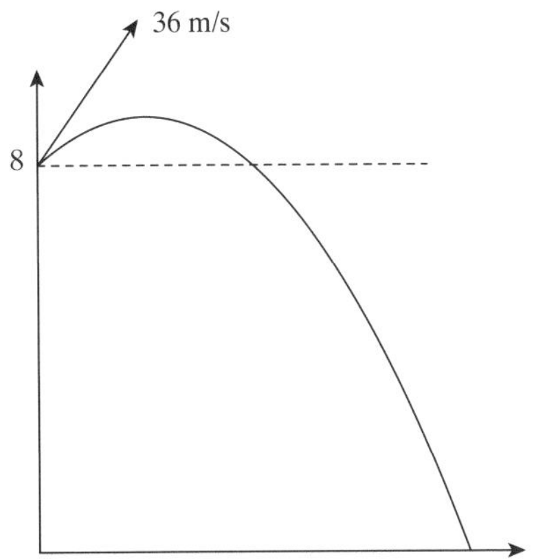

Taking the window as the point of origin, the position vector of the object is given by:

$$r(t) = 18\sqrt{3}t\underset{\sim}{i} + (18t - 5t^2)\underset{\sim}{j}.$$

Find:

a the maximum height of the ball *(2 marks)* **Easy**

b the distance of the point at which the object hits the ground from the window, to the nearest metre. *(3 marks)* **Medium**

Bonus question

7 Two objects are projected from the same point. The first has a position vector of $r_1 = 12t\underset{\sim}{i} + (12t - 5t^2)\underset{\sim}{j}$.

The second is projected 1 second after the first with an initial velocity vector of $24\underset{\sim}{i} + V\underset{\sim}{j}$.

a Show that the position vector of the second object can be written as

$r_2 = 24(t-1)\,\underset{\sim}{i} + [V(t-1) - 5(t-1)^2]\underset{\sim}{j}$. *(3 marks)* **Medium**

b It is known that these two objects collide. Find when they collide. *(2 marks)* **Medium**

c Hence find the value of V, the initial vertical velocity of the second object. *(2 marks)* **Medium**

Bonus question

8 A particle is projected with velocity V m/s from a point O at an angle of elevation θ. The particle just clears two vertical chimneys of height h metres at horizontal distances of p metres and $3p$ metres from O.

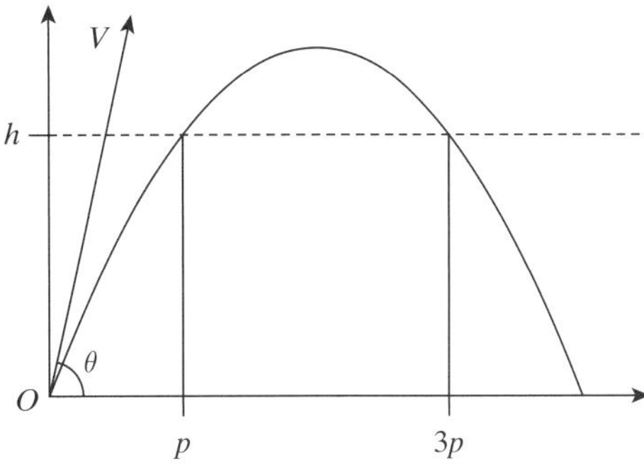

The position vector of the particle is given by:

$r(t) = Vt\cos\theta\underset{\sim}{i} + (Vt\sin\theta - 0.5gt^2)\underset{\sim}{j}$

where acceleration due to gravity is 10 m/s^2.

a Show that $V^2 = \dfrac{5p^2(1+\tan^2\theta)}{p\tan\theta - h}$. *(3 marks)* **Hard**

b Hence, show that $\tan\theta = \dfrac{4h}{3p}$. *(2 marks)* **Hard**

Bonus question

9 A projectile is fired from a point P, q metres above O with an initial velocity V m/s at an angle of elevation θ.

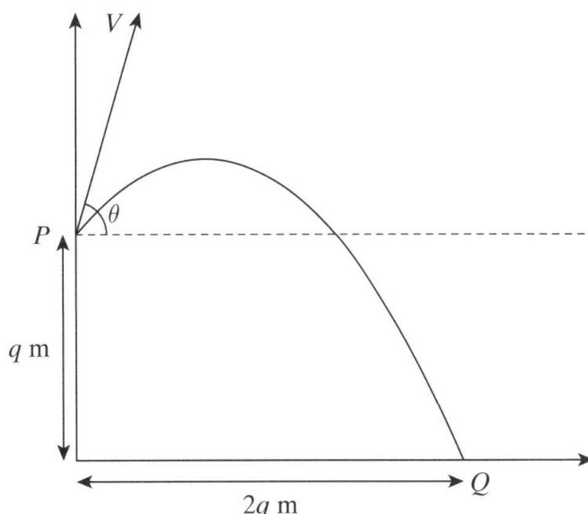

The position vector of the particle is given by:

$$r(t) = Vt\cos\theta\underset{\sim}{i} + (Vt\sin\theta + q - 0.5gt^2)\underset{\sim}{j}$$

where acceleration due to gravity is g m/s^2.

a Show that the time taken, in seconds, to reach Q, $2q$ metres from O in a horizontal direction is given by: $\dfrac{V(2\sin\theta + \cos\theta)}{g}$ *(3 marks)* **Hard**

b Hence show that $q = \dfrac{V^2(2\sin 2\theta + \cos 2\theta + 1)}{4g}$. *(3 marks)* **Hard**

Bonus question

10 The particles A and B have position vectors given as

$\underset{\sim}{r}_A(t) = (1 + 5t)\underset{\sim}{i} + (t^2 + 3)\underset{\sim}{j}$ and

$\underset{\sim}{r}_B(t) = (t^2 + 5)\underset{\sim}{i} + (7t - 9)\underset{\sim}{j}$.

Determine when and where the particles collide. *(3 marks)*

Bonus question Medium

11 A particle moves so that its position vector is given by $\underset{\sim}{r}(t) = \sin 2t\underset{\sim}{i} + 2\cos 2t\underset{\sim}{j}$.

Find its speed, in cms^{-1}, after $\frac{5\pi}{8}$ seconds. *(3 marks)*

Bonus question Medium

12 The point O is on a sloping plane that forms an angle of 45° to the horizontal. A particle is projected from the point O. The particle hits a point A on the sloping plane as shown in the diagram.

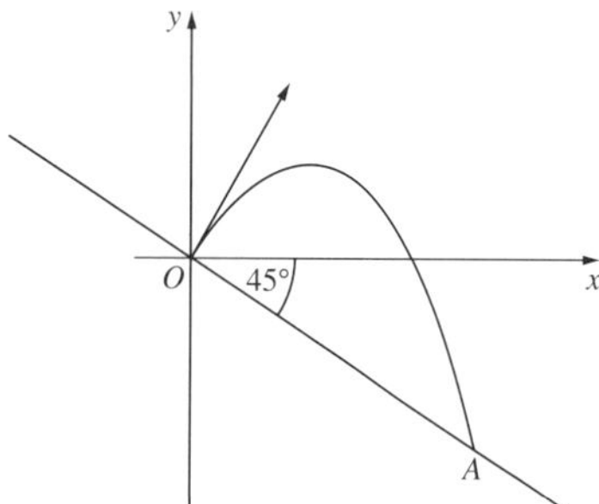

The equation of the line OA is $y = -x$. The equations of motion of the particle are

$$x = 18t$$
$$y = 18\sqrt{3}t - 5t^2,$$

where t is the time in seconds after projection. Do NOT prove these equations.

i Find the distance OA between the point of projection and the point where the particle hits the sloping plane. *(2 marks)* Medium

ii What is the size of the acute angle that the path of the particle makes with the sloping plane as the particle hits the point A? *(3 marks)* Hard

(Q13d, **2019 HSC**)

13 An object is projected from the origin with an initial velocity of V at an angle θ to the horizontal. The equations of motion of the object are

$$x(t) = Vt\cos\theta$$
$$y(t) = Vt\sin\theta - \frac{gt^2}{2}. \text{ (Do NOT prove these.)}$$

i Show that when the object is projected at an angle θ, the horizontal range is $\frac{V^2}{g}\sin 2\theta$. *(2 marks)* Medium

ii Show that when the object is projected at an angle $\frac{\pi}{2} - \theta$, the horizontal range is also $\frac{V^2}{g}\sin 2\theta$. *(1 mark)* Easy

iii The object is projected with initial velocity V to reach a horizontal distance d, which is less than the maximum possible horizontal range. There are two angles at which the object can be projected in order to travel that horizontal distance before landing.

Let these angles be α and β, where $\beta = \frac{\pi}{2} - \alpha$.

Let h_α be the maximum height reached by the object when projected at the angle α to the horizontal.

Let h_β be the maximum height reached by the object when projected at the angle β to the horizontal.

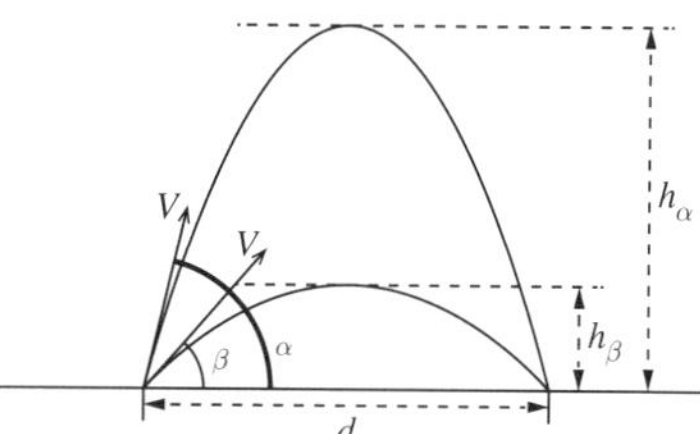

Show that the average of the two heights, $\frac{h_\alpha + h_\beta}{2}$, depends only on V and g. *(3 marks)* Medium

(Q13c, **2018 HSC**)

14 A golfer hits a golf ball with initial speed V ms^{-1} at an angle θ to the horizontal. The golf ball is hit from one side of a lake and must have a horizontal range of 100 m or more to avoid landing in the lake.

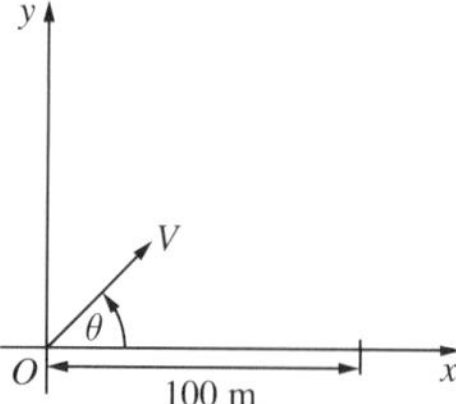

Neglecting the effects of air resistance, the equations describing the motion of the ball are

$$x = Vt\cos\theta$$
$$y = Vt\sin\theta - \frac{1}{2}gt^2,$$

where t is the time in seconds after the ball is hit and g is the acceleration due to gravity in ms^{-2}. Do NOT prove these equations.

i Show that the horizontal range of the golf ball is $\frac{V^2 \sin 2\theta}{g}$ metres. *(2 marks)* Easy

ii Show that if $V^2 < 100g$ then the horizontal range of the ball is less than 100 m. *(1 mark)* Medium

It is now given that $V^2 = 200g$ and that the horizontal range of the ball is 100 m or more.

iii Show that $\frac{\pi}{12} \le \theta \le \frac{5\pi}{12}$. *(2 marks)* Easy

iv Find the greatest height the ball can achieve. *(2 marks)* Hard

(Q13c, **2017 HSC**)

15 The trajectory of a projectile fired with speed u ms^{-1} at an angle θ to the horizontal is represented by the parametric equations

$$x = ut\cos\theta \quad \text{and} \quad y = ut\sin\theta - 5t^2,$$

where t is the time in seconds.

i Prove that the greatest height reached by the projectile is $\frac{u^2\sin^2\theta}{20}$. *(2 marks)* Easy

A ball is thrown from a point 20 m above the horizontal ground. It is thrown with speed 30 ms^{-1} at an angle of 30° to the horizontal. At its highest point the ball hits a wall, as shown in the diagram.

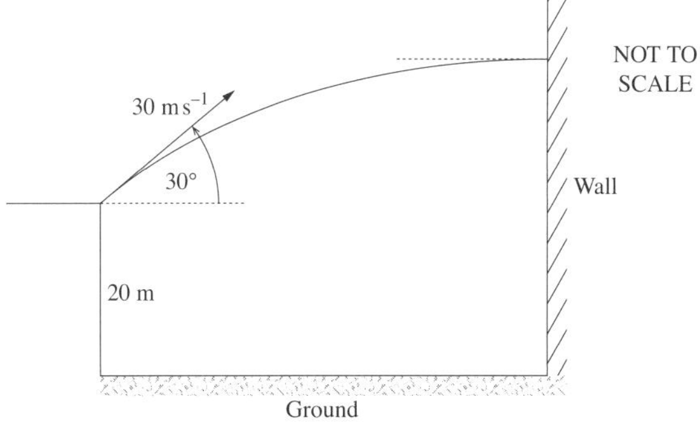

ii Show that the ball hits the wall at a height of $\frac{125}{4}$ m above the ground. *(2 marks)* Easy

The ball then rebounds horizontally from the wall with speed 10 ms^{-1}. You may assume that the acceleration due to gravity is 10 ms^{-2}.

iii How long does it take the ball to reach the ground after it rebounds from the wall? *(2 marks)* Hard

iv How far from the wall is the ball when it hits the ground? *(1 mark)* Medium

(Q13b, **2016 HSC**)

16 A projectile is fired from the origin O with initial velocity V ms^{-1} at an angle θ to the horizontal. The equations of motion are given by

$$x = Vt\cos\theta,\ y = Vt\sin\theta - \frac{1}{2}gt^2.$$

(Do NOT prove this.)

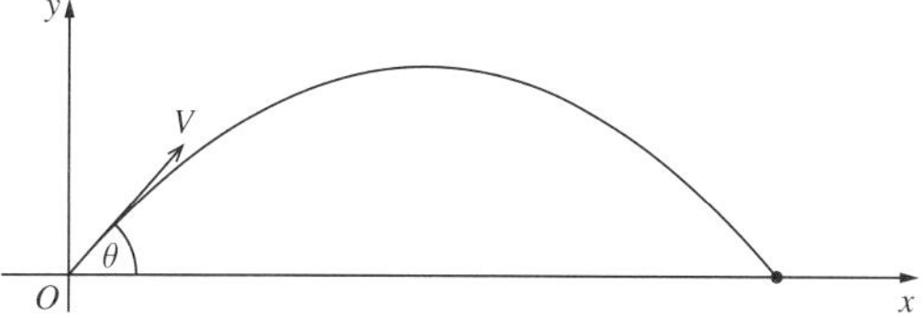

i Show that the horizontal range of the projectile is $\frac{V^2\sin 2\theta}{g}$. *(2 marks)* Easy

A particular projectile is fired so that $\theta = \frac{\pi}{3}$.

ii Find the angle that this projectile makes with the horizontal when $t = \frac{2V}{\sqrt{3}g}$. *(2 marks)* Hard

iii State whether this projectile is travelling upwards or downwards when $t = \frac{2V}{\sqrt{3}g}$. Justify your answer. *(1 mark)* Hard

(Q14a, **2015 HSC**)

17 The take-off point O on a ski jump is located at the top of a downslope. The angle between the downslope and the horizontal is $\frac{\pi}{4}$. A skier takes off from O with velocity V ms^{-1} at an angle θ to the horizontal, where $0 \le \theta < \frac{\pi}{2}$. The skier lands on the downslope at some point P, a distance D metres from O.

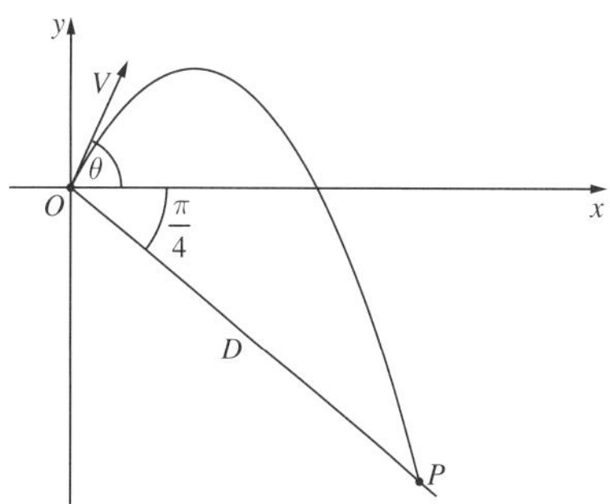

The flight path of the skier is given by

$$x = Vt\cos\theta, \quad y = -\frac{1}{2}gt^2 + Vt\sin\theta,$$

(Do NOT prove this.)

where t is the time in seconds after take-off.

i Show that the cartesian equation of the flight path of the skier is given by

$$y = x\tan\theta - \frac{gx^2}{2V^2}\sec^2\theta.$$ *(2 marks)* Easy

ii Show that $D = 2\sqrt{2}\,\frac{V^2}{g}\cos\theta(\cos\theta + \sin\theta)$. *(3 marks)* Hard

iii Show that $\frac{dD}{d\theta} = 2\sqrt{2}\frac{V^2}{g}\left(\cos 2\theta - \sin 2\theta\right)$.

(2 marks) **Medium**

iv Show that D has a maximum value and find the value of θ for which this occurs.

(3 marks) **Medium**

(Q14a, **2014 HSC**)

18 Points A and B are located d metres apart on a horizontal plane. A projectile is fired from A towards B with initial velocity u ms^{-1} at angle α to the horizontal.

At the same time, another projectile is fired from B towards A with initial velocity w ms^{-1} at angle β to the horizontal, as shown on the diagram.

The projectiles collide when they both reach their maximum height.

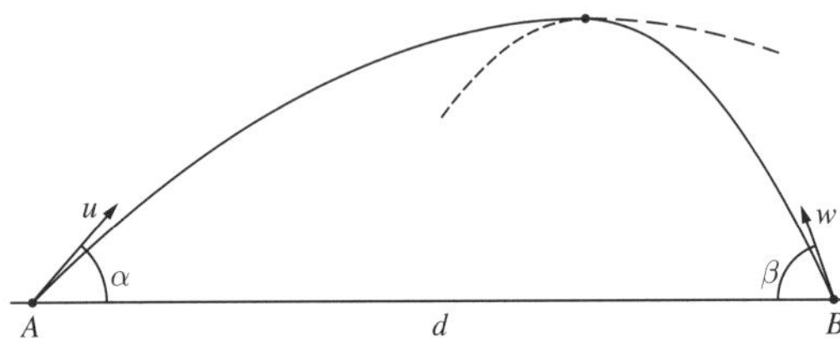

The equations of motion of a projectile fired from the origin with initial velocity V ms^{-1} at angle θ to the horizontal are

$$x = Vt\cos\theta \text{ and } y = Vt\sin\theta - \frac{g}{2}t^2.$$

(Do NOT prove this.)

i How long does the projectile fired from A take to reach its maximum height?

(2 marks) **Easy**

ii Show that $u\sin\alpha = w\sin\beta$.

(1 mark) **Medium**

iii Show that $d = \frac{uw}{g}\sin(\alpha + \beta)$.

(2 marks) **Hard**

(Q13c, **2013 HSC**)

19 A firework is fired from O, on level ground, with velocity 70 metres per second at an angle of inclination θ. The equations of motion of the firework are

$$x = 70t\cos\theta \text{ and } y = 70t\sin\theta - 4.9t^2.$$

(Do NOT prove this.)

The firework explodes when it reaches its maximum height.

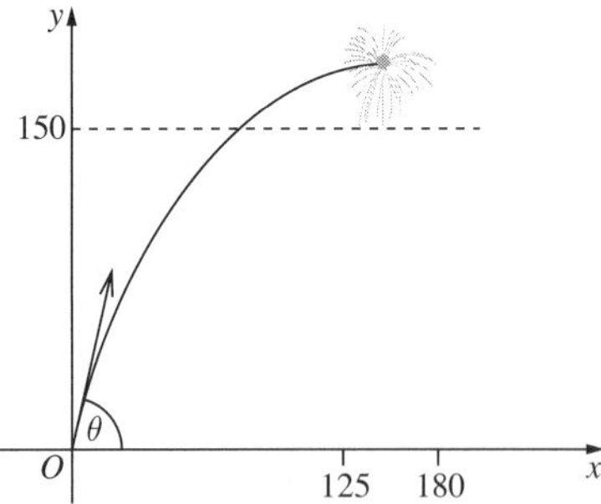

i Show that the firework explodes at a height of $250\sin^2\theta$ metres. *(2 marks)* **Medium**

ii Show that the firework explodes at a horizontal distance of $250\sin 2\theta$ metres from O. *(1 mark)* **Easy**

iii For best viewing, the firework must explode at a horizontal distance between 125 m and 180 m from O, and at least 150 m above the ground. For what values of θ will this occur?

(3 marks) **Hard**

(Q14b, **2012 HSC**)

20 The diagram shows the trajectory of a ball thrown horizontally, at speed v ms^{-1}, from the top of a tower h metres above ground level.

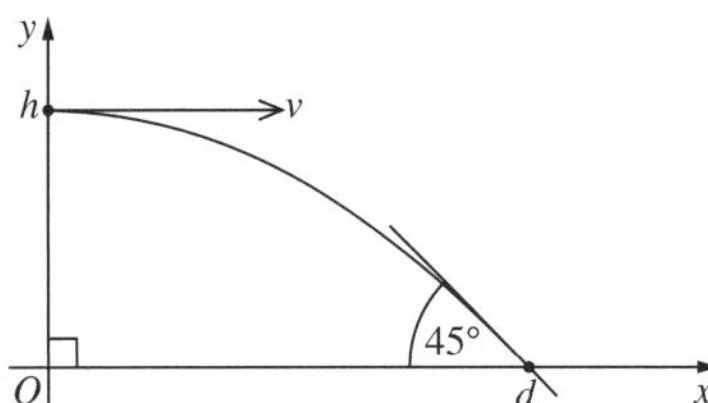

The ball strikes the ground at an angle of 45°, d metres from the base of the tower, as shown in the diagram. The equations describing the trajectory of the ball are

$$x = vt \text{ and } y = h - \frac{1}{2}gt^2, \text{ (Do NOT prove this.)}$$

where g is the acceleration due to gravity, and t is time in seconds.

i Prove that the ball strikes the ground at time $t = \sqrt{\frac{2h}{g}}$ seconds. *(1 mark)* **Easy**

ii Hence, or otherwise, show that $d = 2h$.

(2 marks) **Hard**

(Q6b, **2011 HSC**)

21 A basketball player throws a ball with an initial velocity v ms^{-1} at an angle θ to the horizontal. At the time the ball is released its centre is at $(0, 0)$, and the player is aiming for the point (d, h) as shown on the diagram.

The line joining $(0, 0)$ and (d, h) makes an angle α with the horizontal, where $0 < \alpha < \theta < \frac{\pi}{2}$.

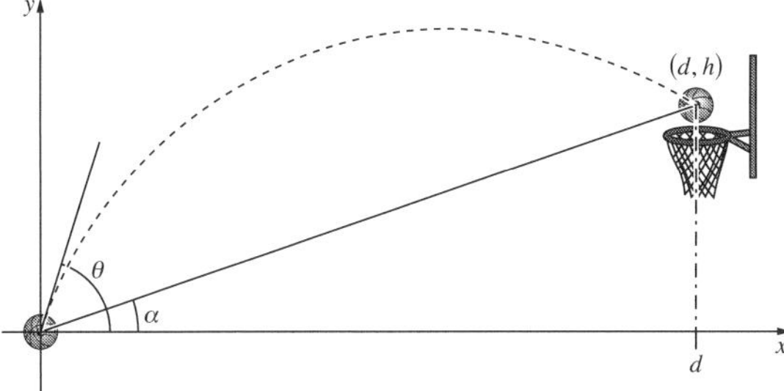

Assume that at time t seconds after the ball is thrown its centre is at the point (x, y), where

$$x = vt\cos\theta$$
$$y = vt\sin\theta - 5t^2.$$

(You are NOT required to prove these equations.)

i If the centre of the ball passes through (d, h) show that

$$v^2 = \frac{5d}{\cos\theta\sin\theta - \cos^2\theta\tan\alpha}.$$

(3 marks) **Medium**

ii **1** What happens to v as $\theta \to \alpha$?

(1 mark) **Easy**

2 What happens to v as $\theta \to \frac{\pi}{2}$?

(1 mark) **Easy**

iii For a fixed value of α, let $F(\theta) = \cos\theta\sin\theta - \cos^2\theta\tan\alpha$.
Show that $F'(\theta) = 0$ when $\tan 2\theta \tan\alpha = -1$.

(2 marks) **Hard**

iv Using part (a)(ii) [the result that if $\tan A \tan B = -1$, then $A - B = \frac{\pi}{2}$] or otherwise show that

$F'(\theta) = 0$ when $\theta = \frac{\alpha}{2} + \frac{\pi}{4}$. *(1 mark)* **Hard**

v Explain why v^2 is a minimum when $\theta = \frac{\alpha}{2} + \frac{\pi}{4}$.

(2 marks) **Hard**

(Q6b, **2010 HSC**)

22 Two points, A and B, are on cliff tops on either side of a deep valley. Let h and R be the vertical and horizontal distances between A and B as shown in the diagram. The angle of elevation of B from A is θ, so that $\theta = \tan^{-1}\left(\frac{h}{R}\right)$.

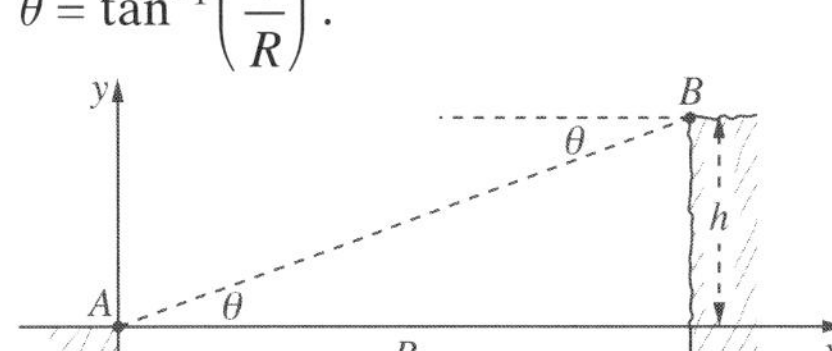

At time $t = 0$, projectiles are fired simultaneously from A and B. The projectile from A is aimed at B, and has initial speed U at an angle θ above the horizontal. The projectile from B is aimed at A and has initial speed V at an angle θ below the horizontal. The equations for the motion of the projectile from A are $x_1 = Ut\cos\theta$ and $y_1 = Ut\sin\theta - \frac{1}{2}gt^2$, and the equations for the motion of the projectile from B are $x_2 = R - Vt\cos\theta$ and $y_2 = h - Vt\sin\theta - \frac{1}{2}gt^2$.
(Do NOT prove these equations.)

i Let T be the time at which $x_1 = x_2$.
Show that $T = \frac{R}{(U + V)\cos\theta}$.

(1 mark) **Easy**

ii Show that the projectiles collide.

(2 marks) **Hard**

iii If the projectiles collide on the line $x = \lambda R$, where $0 < \lambda < 1$, show that $V = \left(\frac{1}{\lambda} - 1\right)U$.

(1 mark) **Medium**

(Q6a, **2009 HSC**)

23 A projectile is fired from O with velocity V at an angle of inclination θ across level ground. The projectile passes through the points L and M, which are both h metres above the ground, at times t_1 and t_2 respectively. The projectile returns to the ground at N.

The equations of motion of the projectile are

$$x\ Vt\cos\theta = \text{ and } y = Vt\sin\theta - \frac{1}{2}gt^2.$$

(Do NOT prove this.)

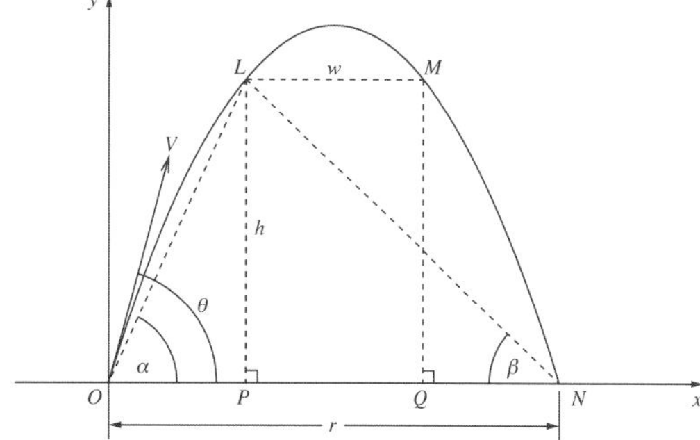

a Show that $t_1 + t_2 = \frac{2V}{g}\sin\theta$ AND $t_1 t_2 = \frac{2h}{g}$.

(2 marks) **Medium**

Let $\angle LON = \alpha$ and $\angle LNO = \beta$. It can be shown that

$$\tan\alpha = \frac{h}{Vt_1\cos\theta} \text{ and } \tan\beta = \frac{h}{Vt_2\cos\theta}.$$

(Do NOT prove this.)

b Show that $\tan\alpha + \tan\beta = \tan\theta$.

(2 marks) **Medium**

c Show that $\tan\alpha\tan\beta = \frac{gh}{2V^2\cos^2\theta}$.

(1 mark) **Medium**

Let $ON = r$ and $LM = w$.

d Show that $r = h(\cot\alpha + \cot\beta)$ and $w = h(\cot\beta - \cot\alpha)$. *(2 marks)* **Hard**

Let the gradient of the parabola at L be $\tan\phi$.

e Show that $\tan\phi = \tan\alpha - \tan\beta$. *(3 marks)* **Hard**

f Show that $\dfrac{w}{\tan\phi} = \dfrac{r}{\tan\theta}$. *(2 marks)* **Hard**

(Q7, **2008 HSC**)

24 A small paintball is fired from the origin with initial velocity 14 metres per second towards an eight-metre high barrier. The origin is at ground level, 10 metres from the base of the barrier.

The equations of motion are

$$x = 14t\cos\theta$$
$$y = 14t\sin\theta - 4.9t^2$$

where θ is the angle to the horizontal at which the paintball is fired and t is the time in seconds. (Do NOT prove these equations of motion.)

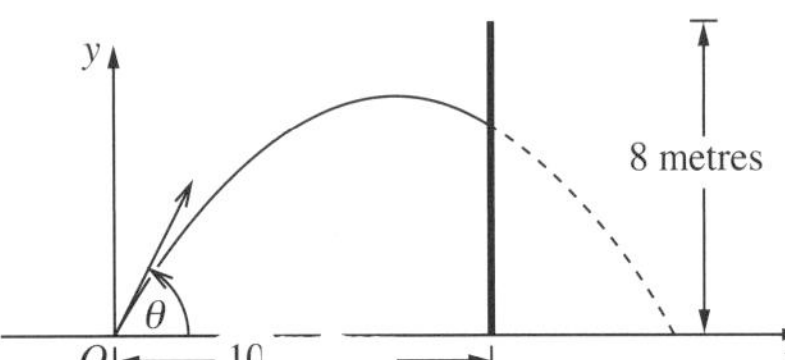

i Show that the equation of trajectory of the paintball is

$$y = mx - \left(\frac{1+m^2}{40}\right)x^2, \quad \text{where } m = \tan\theta.$$

(2 marks) **Easy**

ii Show that the paintball hits the barrier at height h metres when $m = 2 \pm \sqrt{3 - 0.4h}$. Hence determine the maximum value of h. *(2 marks)* **Medium**

iii There is a large hole in the barrier. The bottom of the hole is 3.9 metres above the ground and the top of the hole is 5.9 metres above the ground. The paintball passes through the hole if m is in one of two intervals.
One interval is $2.8 \le m \le 3.2$.
Find the other interval. *(2 marks)* **Medium**

iv Show that, if the paintball passes through the hole, the range is $\dfrac{40m}{1+m^2}$ metres.
Hence find the widths of the two intervals in which the paintball can land at ground level on the other side of the barrier. *(3 marks)* **Hard**

(Q7b, **2007 HSC**)

25 Two particles are fired simultaneously from the ground at time $t = 0$.

Particle 1 is projected from the origin at an angle θ, $0 < \theta < \dfrac{\pi}{2}$, with an initial velocity V.

Particle 2 is projected vertically upward from the point A, at a distance a to the right of the origin, also with an initial velocity of V.

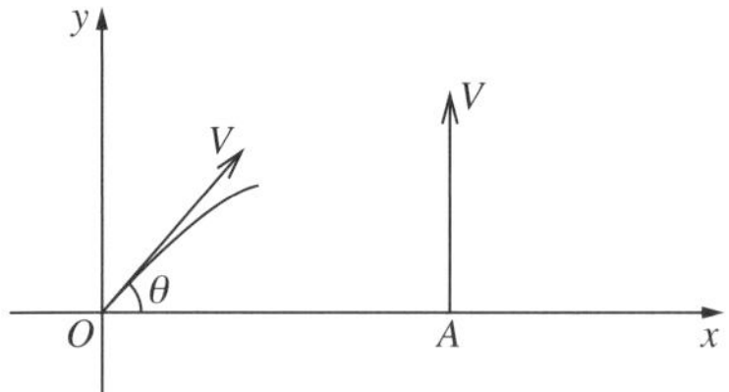

It can be shown that while both particles are in flight, Particle 1 has equations of motion:

$$x = Vt\cos\theta$$
$$y = Vt\sin\theta - \frac{1}{2}gt^2,$$

and Particle 2 has equations of motion:

$$x = a$$
$$y = Vt - \frac{1}{2}gt^2.$$

Do NOT prove these equations of motion.

Let L be the distance between the particles at time t.

i Show that, while both particles are in flight, $L^2 = 2V^2t^2(1 - \sin\theta) - 2aVt\cos\theta + a^2$. *(2 marks)* **Medium**

ii An observer notices that the distance between the particles in flight first decreases, then increases.
Show that the distance between the particles in flight is smallest when $t = \dfrac{a\cos\theta}{2V(1-\sin\theta)}$ and that this smallest distance is $a\sqrt{\dfrac{1-\sin\theta}{2}}$. *(3 marks)* **Hard**

iii Show that the smallest distance between the two particles in flight occurs while Particle 1 is ascending if

$$V > \sqrt{\frac{ag\cos\theta}{2\sin\theta(1-\sin\theta)}}.$$

(1 mark) **Hard**

(Q6a, **2006 HSC**)

26 An experimental rocket is at a height of 5000 m, ascending with a velocity of $200\sqrt{2}$ ms^{-1} at an angle of 45° to the horizontal, when its engine stops.

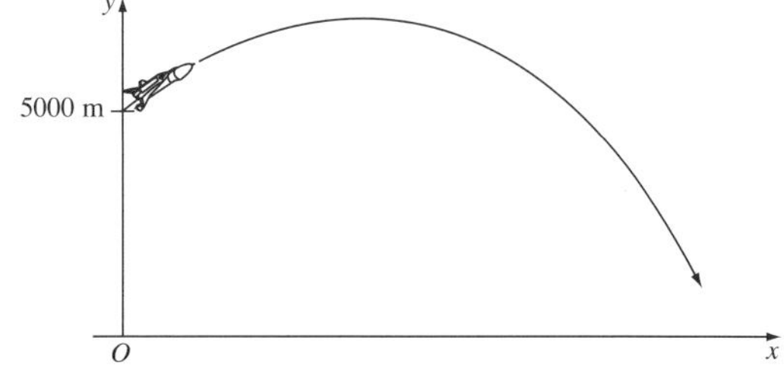

After this time, the equations of motion of the rocket are:
$x = 200t$
$y = -4.9t^2 + 200t + 5000$,
where t is measured in seconds after the engine stops. (Do NOT show this.)

i What is the maximum height the rocket will reach, and when will it reach this height? *(2 marks)* **Easy**

ii The pilot can only operate the ejection seat when the rocket is descending at an angle between 45° and 60° to the horizontal. What are the earliest and latest times that the pilot can operate the ejection seat? *(3 marks)* **Hard**

iii For the parachute to open safely, the pilot must eject when the speed of the rocket is no more than 350 ms^{-1}. What is the latest time at which the pilot can eject safely? *(2 marks)* **Hard**

(Q6b, **2005 HSC**)

27 A fire hose is at ground level on a horizontal plane. Water is projected from the hose. The angle of projection, θ, is allowed to vary. The speed of the water as it leaves the hose, v metres per second, remains constant. You may assume that if the origin is taken to be the point of projection, the path of the water is given by the parametric equations

$$x = vt\cos\theta$$
$$y = vt\sin\theta - \frac{1}{2}gt^2$$

where g ms^{-2} is the acceleration due to gravity. (Do NOT prove this.)

i Show that the water returns to ground level at a distance $\frac{v^2 \sin 2\theta}{g}$ metres from the point of projection. *(2 marks)* **Easy**

This fire hose is now aimed at a 20 metre high thin wall from a point of projection at ground level 40 metres from the base of the wall. It is known that when the angle θ is 15°, the water just reaches the base of the wall.

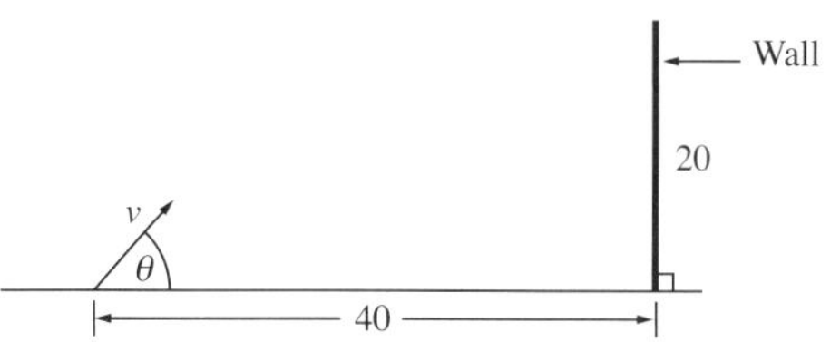

ii Show that $v^2 = 80g$. *(1 mark)* **Easy**

iii Show that the cartesian equation of the path of the water is given by

$$y = x\tan\theta - \frac{x^2 \sec^2\theta}{160}.$$ *(2 marks)* **Medium**

iv Show that the water just clears the top of the wall if $\tan^2\theta - 4\tan\theta + 3 = 0$. *(2 marks)* **Medium**

v Find all values of θ for which the water hits the front of the wall. *(2 marks)* **Hard**

(Q6b, **2004 HSC**)

28 A particle is projected from the origin with velocity v ms^{-1} at an angle α to the horizontal. The position of the particle at time t seconds is given by the parametric equations

$$x = vt\cos\alpha$$
$$y = vt\sin\alpha - \frac{1}{2}gt^2,$$

where g ms^{-2} is the acceleration due to gravity. (You are NOT required to derive these.)

i Show that the maximum height reached, h metres, is given by $h = \frac{v^2\sin^2\alpha}{2g}$. *(2 marks)* **Easy**

ii Show that it returns to the initial height at $x = \frac{v^2}{g}\sin 2\alpha$. *(2 marks)* **Easy**

iii Chris and Sandy are tossing a ball to each other in a long hallway. The ceiling height is H metres and the ball is thrown and caught at shoulder height, which is S metres for both Chris and Sandy.

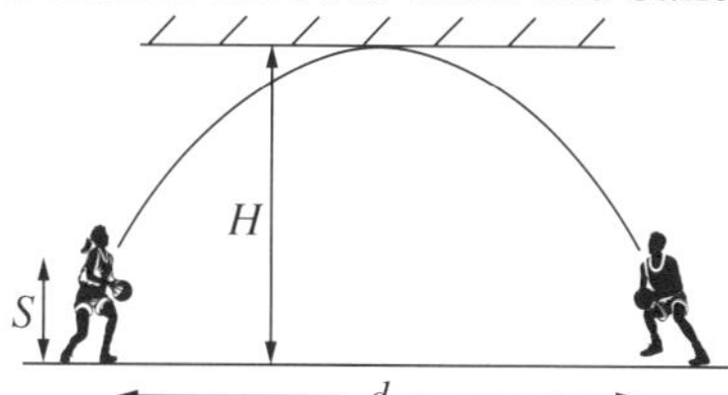

The ball is thrown with a velocity v ms^{-1}. Show that the maximum separation, d metres, that Chris and Sandy can have and still catch the ball is given by

$$d = 4 \times \sqrt{(H-S)\left(\frac{v^2}{2g}\right) - (H-S)^2},$$

if $v^2 \geq 4g(H-S)$, and

$$d = \frac{v^2}{g},$$

if $v^2 \leq 4g(H-S)$. *(4 marks)* **Hard**

(Q7b, **2003 HSC**)

29 An angler casts a fishing line so that the sinker is projected with a speed V ms^{-1} from a point 5 metres above a flat sea. The angle of projection to the horizontal is θ, as shown.

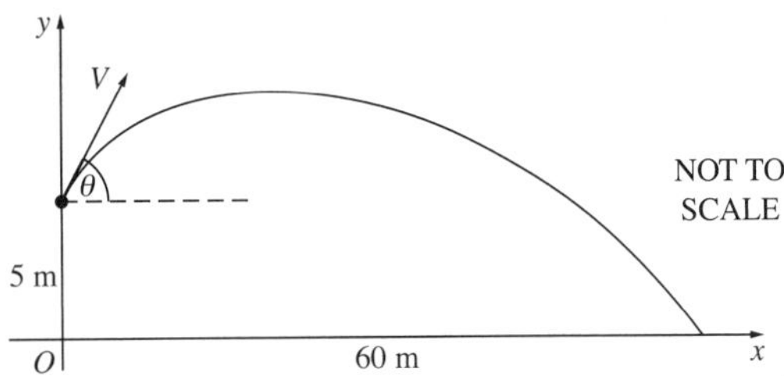

Assume that the equations of motion of the sinker are

$$\ddot{x} = 0 \text{ and } \ddot{y} = -10,$$

referred to the coordinate axes shown.

i Let (x, y) be the position of the sinker at time t seconds after the cast, and before the sinker hits the water.
It is known that $x = Vt\cos\theta$.
Show that $y = Vt\sin\theta - 5t^2 + 5$.
(2 marks) **Easy**

ii Suppose the sinker hits the sea 60 metres away as shown in the diagram.
Find the value of V if $\theta = \tan^{-1}\frac{3}{4}$.
(3 marks) **Medium**

iii For the cast described in part **ii**, find the maximum height above sea level that the sinker achieved. *(2 marks)* **Medium**

(Q6a, **2002 HSC**)

30 An aircraft flying horizontally at V ms^{-1} releases a bomb that hits the ground 4000 m away, measured horizontally. The bomb hits the ground at an angle of 45° to the vertical.

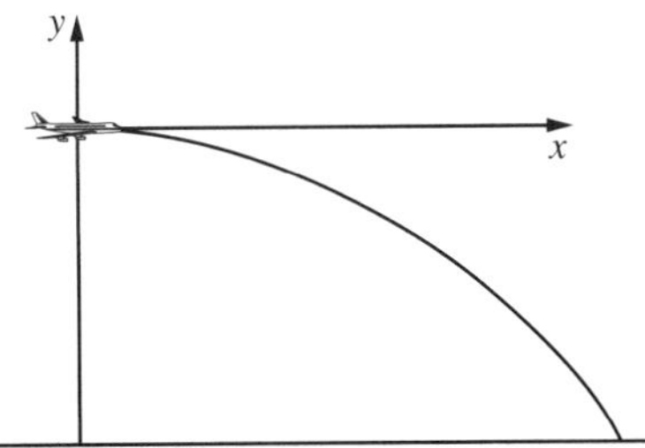

Assume that, t seconds after release, the position of the bomb is given by

$$x = Vt, y = -5t^2.$$

Find the speed V of the aircraft.
(4 marks) **Medium**

(Q4b, **2001 HSC**)

31 The diagram shows an inclined plane that makes an angle of α radians with the horizontal. A projectile is fired from O, at the bottom of the incline, with a speed of V ms^{-1} at an angle of elevation θ to the horizontal, as shown.

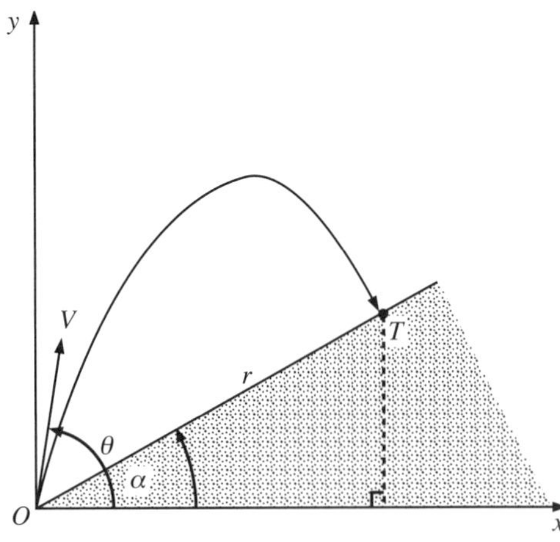

With the above axes, you may assume that the position of the projectile is given by

$$x = Vt\cos\theta$$
$$y = Vt\sin\theta - \tfrac{1}{2}gt^2,$$

where t is the time, in seconds, after firing, and g is the acceleration due to gravity.

For simplicity we assume that the unit of length has been chosen so that

$$\frac{2V^2}{g} = 1.$$

i Show that the path of the trajectory of the projectile is
$y = x\tan\theta - x^2\sec^2\theta$. **Medium**

ii Show that the range of the projectile, $r = OT$ metres, up the inclined plane is given by
$$r = \frac{\sin(\theta - \alpha)\cos\theta}{\cos^2\alpha}.$$ **Hard**

iii Hence, or otherwise, deduce that the maximum range, R metres, up the incline is
$$R = \frac{1}{2(1 + \sin\alpha)}.$$ **Hard**

(You may assume that
$2\sin A\cos B = \sin(A + B) + \sin(A - B)$.)

iv Consider the trajectory of the projectile for which the maximum range R is achieved. Show that, for this trajectory, the initial direction is perpendicular to the direction at which the projectile hits the inclined plane. **Hard**

(8 marks) (Q7b, **2000 HSC**)

32 A cricket ball leaves the bowler's hand 2 metres above the ground with a velocity of 30 ms^{-1} at an angle of 5° below the horizontal. The equations of motion for the ball are

$$\ddot{x} = 0 \text{ and } \ddot{y} = -10.$$

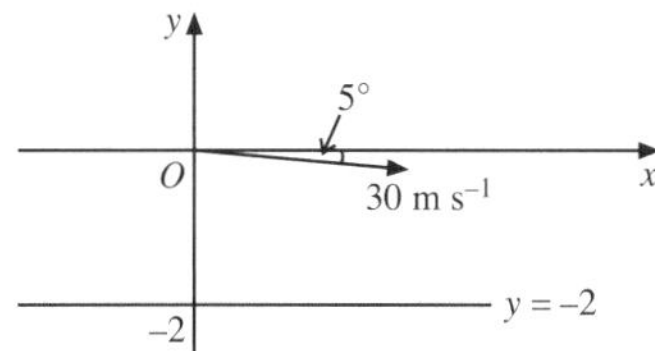

Take the origin to be the point where the ball leaves the bowler's hand.

i Using calculus, prove that the coordinates of the ball at time t are given by
$x = 30t\cos(5°)$, and
$y = -30t\sin(5°) - 5t^2$. Easy

ii Find the time at which the ball strikes the ground. Medium

iii Calculate the angle at which the ball strikes the ground. Medium

(8 marks) (Q7a, **1999 HSC**)

33 A particle is projected from the point (0, 1) at an angle of 45° with a velocity of V metres per second.

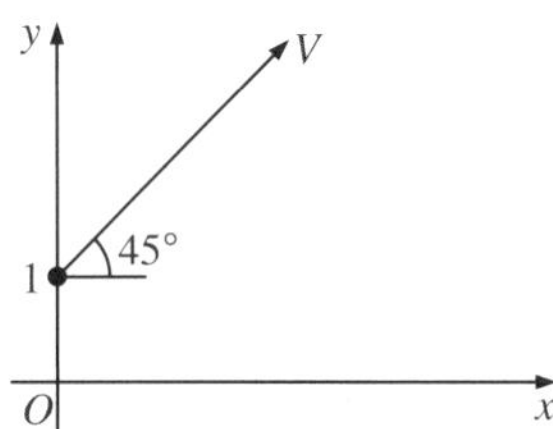

The equations of motion of the particle are

$$\ddot{x} = 0 \text{ and } \ddot{y} = -g,$$

i Using calculus, derive the expressions for the position of the particle at time t. Hence show that the path of the particle is given by

$$y = 1 + x - g\frac{x^2}{V^2}.$$ Medium

A volleyball player serves a ball with initial speed V metres per second and angle of projection 45°. At that moment the bottom of the ball is 1 metre above the ground and its horizontal distance from the net is 9·3 metres. The ball just clears the net, which is 2·3 metres high.

ii Show that the initial speed of the ball is approximately 10·3 metres per second. (Take $g = 9{\cdot}8\ \text{ms}^{-2}$.) Medium

iii What is the horizontal distance from the net to the point where the ball lands? Medium

(8 marks) (Q6a, **1998 HSC**)

34 A particle is projected horizontally from a point P, h metres above O, with a velocity of V metres per second. The equations of motion of the particle are

$\ddot{x} = 0$ and $\ddot{y} = -g$.

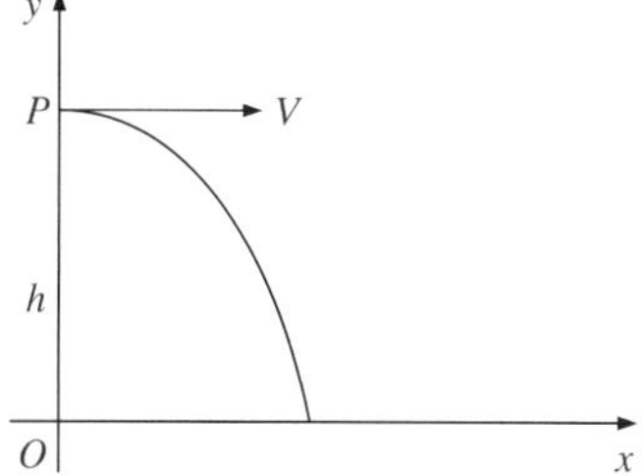

i Using calculus, show that the position of the particle at time t is given by

$$x = Vt,\quad y = h - \frac{1}{2}gt^2.$$ Easy

A canister containing a life raft is dropped from a plane to a stranded sailor. The plane is travelling at a constant velocity of 216 km/h, at a height of 120 metres above sea level, along a path that passes above the sailor.

ii How long will the canister take to hit the water? (Take $g = 10\ \text{m/s}^2$.) Medium

iii A current is causing the sailor to drift at a speed of 3·6 km/h in the same direction as the plane is travelling. The canister is dropped from the plane when the horizontal distance from the plane to the sailor is D metres. What values can D take if the canister lands at most 50 metres from the stranded sailor? Hard

(9 marks) (Q7a, **1997 HSC**)

35 A cap C is lying outside a softball field, r metres from the fence F, which is h metres high. The fence is R metres from the point O, and the point P is h metres above O. Axes are based at O, as shown.

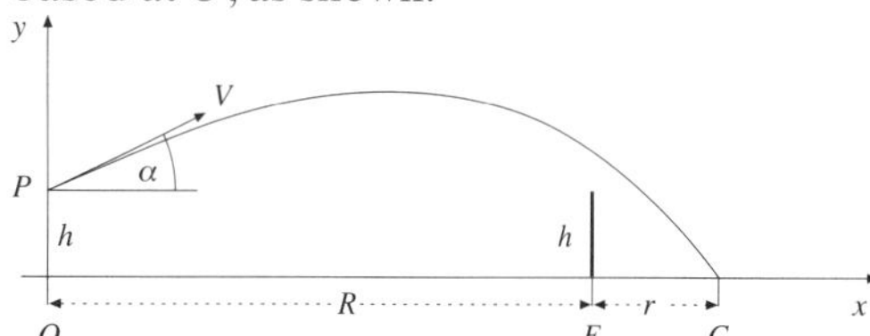

At time $t = 0$, a ball is hit from P at a speed V metres per second and at an angle α to the horizontal, towards the cap.

a The equations of motion of the ball are
$\ddot{x} = 0$ and $\ddot{y} = -g$.
Using calculus, show that the position of the ball at time t is given by

$$x = Vt\cos\alpha$$

$$y = Vt\sin\alpha - \tfrac{1}{2}gt^2 + h.$$ *(4 marks)* Easy

b Hence show that the trajectory of the ball is given by

$$y = h + x\tan\alpha - x^2\frac{g}{2V^2\cos^2\alpha}.$$ *(1 mark)* Medium

c The ball clears the fence. Show that

$$V^2 \geq \frac{gR}{2\sin\alpha\cos\alpha}.$$ *(2 marks)* Medium

d After clearing the fence, the ball hits the cap C. Show that

$\tan\alpha \geq \dfrac{Rh}{(R+r)r}$. *(3 marks)* Hard

e Suppose that the ball clears the fence, and that $V \leq 50$, $g = 10$, $R = 80$, and $h = 1$. What is the closest point to the fence where the ball can land? *(2 marks)* Hard

(Q7, **1995 HSC**)

36 A projectile is fired from the origin O with velocity V and with angle of elevation θ, where $\theta \neq \dfrac{\pi}{2}$. You may assume that

$x = Vt\cos\theta$ and $y = -\dfrac{1}{2}gt^2 + Vt\sin\theta$,

where x and y are the horizontal and vertical displacements of the projectile in metres from O at time t seconds after firing.

i Show that the equation of flight of the projectile can be written as

$y = x\tan\theta - \dfrac{1}{4h}x^2(1 + \tan^2\theta)$,

where $\dfrac{V^2}{2g} = h$. Medium

ii Show that the point (X, Y), where $X \neq 0$, can be hit by firing at two different angles θ_1 and θ_2 provided $X^2 < 4h(h - Y)$. Hard

iii Show that no point above the x axis can be hit by firing at two different angles θ_1 and θ_2, satisfying $\theta_1 < \dfrac{\pi}{4}$ and $\theta_2 < \dfrac{\pi}{4}$. Hard

(Q7b, **1993 HSC**)

37 A projectile, of initial speed V m/s, is fired at an angle of elevation α from the origin O towards a target T, which is moving away from O along the x axis.

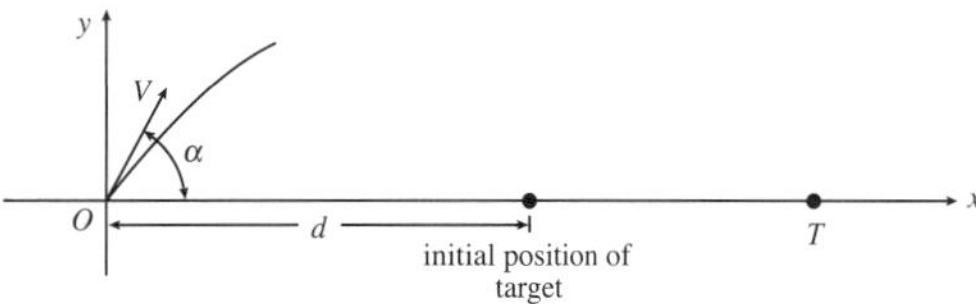

You may assume that the projectile's trajectory is defined by the equations

$$x = Vt\cos\alpha \quad \text{and} \quad y = -\frac{1}{2}gt^2 + Vt\sin\alpha,$$

where x and y are the horizontal and vertical displacements of the projectile in metres at time t seconds after firing, and where g is the acceleration due to gravity.

i Show that the projectile is above the x axis for a total of $\dfrac{2V\sin\alpha}{g}$ seconds. Easy

ii Show that the horizontal range of the projectile is $\dfrac{2V^2\sin\alpha\cos\alpha}{g}$ metres. Easy

iii At the instant the projectile is fired, the target T is d metres from O and it is moving away at a constant speed of u m/s. Suppose that the projectile hits the target when fired at an angle of elevation α. Show that

$u = V\cos\alpha - \dfrac{gd}{2V\sin\alpha}$. Hard

(Q7b, **1992 HSC**)

Year 12 Projectile motion—Worked Answers

1 Projectile will hit the ground when

$$2ut\sin\theta - \frac{g}{2}t^2 = 0.$$

$$t\left(2u\sin\theta - \frac{g}{2}t\right) = 0$$

$$2u\sin\theta - \frac{g}{2}t = 0$$

$$2u\sin\theta = \frac{g}{2}t$$

$$\therefore t = \frac{4u\sin\theta}{g} \quad ✓$$

$$x_P = 2ut\cos\theta$$

$$= 2u\left(\frac{4u\sin\theta}{g}\right)\cos\theta$$

$$= \frac{8u^2\sin\theta\cos\theta}{g}$$

$$= \frac{4u^2\sin 2\theta}{g}$$

The target's position is given by $x_T = d + ut$.

$$x_T = d + u\left(\frac{4u\sin\theta}{g}\right)$$

$$= d + \frac{4u^2\sin\theta}{g} \quad ✓$$

For the player to hit the target, $x_T \leq x_P$

$$d + \frac{4u^2\sin\theta}{g} \leq \frac{4u^2\sin 2\theta}{g}$$

$$d \leq \frac{4u^2\sin 2\theta}{g} - \frac{4u^2\sin\theta}{g}$$

$$d \leq \frac{4u^2}{g}(\sin 2\theta - \sin\theta)$$

Let $f(\theta) = \sin 2\theta - \sin\theta$

$$f'(\theta) = 2\cos 2\theta - \cos\theta$$

$$0 = 2\cos 2\theta - \cos\theta$$

$$0 = 2(2\cos^2\theta - 1) - \cos\theta$$

$$0 = 4\cos^2\theta - \cos\theta - 2 \quad ✓$$

$$\cos\theta = \frac{1 \pm \sqrt{(-1)^2 - 4\times 4 \times -2}}{8}$$

$$\cos\theta = \frac{1 \pm \sqrt{33}}{8}$$

$\cos\theta = 0.84$ or $\cos\theta = -0.59$

$\therefore \theta \approx 0.5678$

$\sin 2\theta - \sin\theta$

$= \sin(2(0.5678)) - \sin(0.5678)$

$= 0.369$ (3 d.p.)

$$\therefore d \leq 0.37\frac{4u^2}{g}$$

$\therefore d \leq 0.37\ \text{Max}(x_P)$ ✓

(4 marks)

2 The ball is projected at an angle of 30° with initial velocity of 12 m/s and initial height of 1 m.

So, $\theta = \frac{\pi}{6}$, $V = 12$ and $h = 1$.

Hence

$$\underset{\sim}{r}(t) = \left(12t\cos\frac{\pi}{6}\right)\underset{\sim}{i} + \left(-5t^2 + 12t\sin\frac{\pi}{6} + 1\right)\underset{\sim}{j}$$

So in parametric form: $x(t) = 6\sqrt{3}\,t$ and

$y(t) = -5t^2 + 6t + 1.$ ✓

To prove that the ball does not hit the roof, show that the max. height < 3:

$$\dot{y}(t) = -10t + 6$$

$$0 = -10t + 6$$

$$t = \frac{3}{5}$$

$\therefore$ Since $\ddot{y} = -10$, the maximum height occurs when

$t = \frac{3}{5}$ seconds. ✓

Substitute $t = \frac{3}{5}$ into $y(t)$ gives the maximum height.

$$y = -5\left(\frac{3}{5}\right)^2 + 6\left(\frac{3}{5}\right) + 1$$

$$y = \frac{14}{5}$$

$$< 3$$

$\therefore$ The ball does not touch the ceiling. ✓

To prove that the ball does not hit the floor, we can solve $x(t) = 10$ and substitute the solution into $y(t)$.

$$6\sqrt{3}\,t = 10$$

$$t = \frac{5}{3\sqrt{3}}$$

$\therefore$ the value of y when $x = 10$ is:

$$y\left(\frac{5}{3\sqrt{3}}\right) = -5\left(\frac{5}{3\sqrt{3}}\right)^2 + 6\left(\frac{5}{3\sqrt{3}}\right) + 1$$

$$\approx 2.144$$

$$> 0$$

$\therefore$ When $t = \frac{5}{3\sqrt{3}}$, $x = 10$

and $y \approx 2.144$, which shows the ball does not hit the floor. ✓

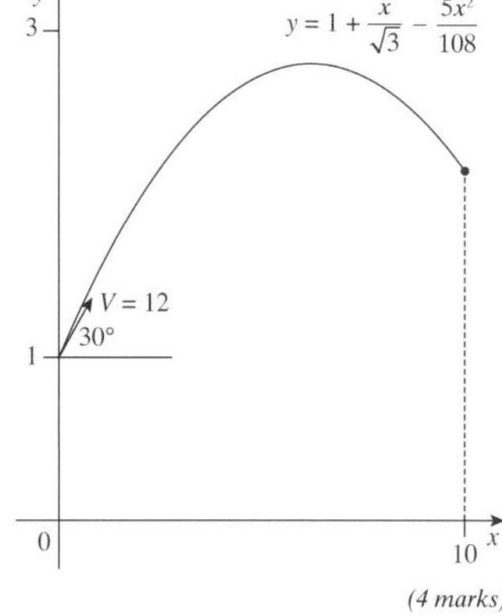

(4 marks)

3 **a** $$v(t) = \frac{d[r(t)]}{dt}$$

$$= 30\sqrt{3}\underset{\sim}{i} + (30 - 10t)\underset{\sim}{j} \quad ✓$$

(1 mark)

b At $t = 0$, $v(0) = 30\sqrt{3}\underset{\sim}{i} + 30\underset{\sim}{j}$

$$\therefore |v(0)| = \sqrt{(30\sqrt{3})^2 + 30^2}$$

$$= 60 \quad ✓$$

$$\theta = \tan^{-1}\left(\frac{30}{30\sqrt{3}}\right)$$

$$= 30° \quad ✓$$

$\therefore$ The particle is projected 30° above the horizontal at an initial velocity of 60 m/s.

(2 marks)

4

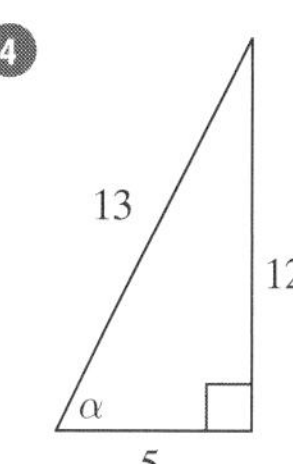

a 3rd side $= \sqrt{5^2 + 12^2}$

$= 13$

At $t = 0$

$$\dot{x} = V \times \cos\alpha$$

$$= \frac{5}{13}V$$

$$\dot{y} = V \times \sin\alpha$$

$$= \frac{12}{13}V \quad ✓$$

$$\ddot{x} = 0$$

$$\dot{x} = \int 0.dt$$

$$= 0 + c$$

At $t = 0$, $\dot{x} = \frac{5}{13}V$

$$\frac{5}{13}V = 0 + c$$

Hence, $\dot{x} = \frac{5}{13}V$

$$x = \int \frac{5}{13}V.dt$$

$$= \frac{5}{13}Vt + c$$

At $t = 0$, $x = 0$

$$\therefore 0 = \frac{5}{13}V \times 0 + c$$

So $c = 0$

Hence, $x = \frac{5}{13}Vt$ ✓

$$\ddot{y} = -10$$

$$\dot{y} = \int -10.dt$$

$$= -10t + c$$

At $t = 0$, $\dot{y} = \frac{12}{13}V$

$$\frac{12}{13}V = -10 \times 0 + c$$

Hence, $\dot{y} = \frac{12}{13}V - 10t$

$$y = \int \frac{12}{13}V - 10t.dt$$
$$= \frac{12}{13}Vt - 5t^2 + c$$

At $t = 0, y = 0$

$$\therefore 0 = \frac{12}{13}V \times 0 - 5 \times 0^2 + c$$

So $c = 0$

Hence, $y = \frac{12}{13}Vt - 5t^2$ ✓

Now $\underset{\sim}{r}(t) = x\underset{\sim}{i} + y\underset{\sim}{j}$

$$\therefore \underset{\sim}{r}(t) = \frac{5}{13}Vt\underset{\sim}{i} + \left(\frac{12}{13}Vt - 5t^2\right)\underset{\sim}{j}$$

$$\underset{\sim}{r}(t) = \frac{1}{13}[5Vt\underset{\sim}{i} + (12Vt - 65t^2)\underset{\sim}{j}]$$

(3 marks)

b Projectile passes through (10, 19),

$$\therefore 10 = \frac{5}{13}Vt \text{ or } t = \frac{26}{V}$$

And $19 = \frac{12}{13}Vt - 5t^2$ ✓

By substitution,

$$19 = \frac{12}{13}V \times \frac{26}{V} - 5\left(\frac{26}{V}\right)^2$$

$$19 = 24 - \frac{5 \times 26^2}{V^2}$$

Hence, $V^2 = 26^2$, so $V = 26$. ✓

(2 marks)

c At $x = 10$, $10 = \frac{5}{13} \times 26t$

Hence projectile at $x = 10$ at $t = 1$ second.

Now $\dot{x} = \frac{5}{13}V$ and $\dot{y} = \frac{12}{13}V - 10t$

At $t = 1$,

$\dot{x} = \frac{5}{13} \times 26 = 10$ and $\dot{y} = \frac{12}{13} \times 26 - 10 \times 1 = 14$

$\therefore \underset{\sim}{v}(t) = 10\underset{\sim}{i} + 14\underset{\sim}{j}$ ✓

Speed at 1 second $= \sqrt{10^2 + 14^2}$

$\cong 17.2$ m/s ✓

(2 marks)

5 Let $x = 24t\cos\theta$ and

$y = 24t\sin\theta - 5t^2$

Maximum range when $y = 0$

$\therefore 24t\sin\theta - 5t^2 = 0$

$t(24\sin\theta - 5t) = 0$

So $t = 0$ or $\frac{24\sin\theta}{5}$ ✓

Maximum height is $y = 10$ when $\dot{y} = 0$

$\therefore 0 = 24\sin\theta - 10t$

So maximum height at

$t = \frac{12\sin\theta}{5}$ ✓

$$\therefore 10 = 24\left(\frac{12\sin\theta}{5}\right)\sin\theta - 5\left(\frac{12\sin\theta}{5}\right)^2$$

$$50 = 288\sin^2\theta - 144\sin^2\theta$$

$$50 = 144\sin^2\theta$$

$$\sin^2\theta = \frac{50}{144}$$

$$\sin\theta = \pm\frac{5\sqrt{2}}{12}$$ ✓

But $0° \le \theta \le 90°$, $\therefore \sin\theta = \frac{5\sqrt{2}}{12}$

Maximum horizontal range at

$$x = 24\left(\frac{24\sin\theta}{5}\right)\cos\theta$$

$$x = 24 \times \frac{24}{5} \times \frac{5\sqrt{2}}{12} \times \frac{\sqrt{94}}{12}$$ ✓

$$x = 8\sqrt{47}$$

Therefore the maximum horizontal range along the tunnel is $8\sqrt{47}$ metres.

(4 marks)

6 a At the top of flight $\dot{y} = 0$

$\therefore 18 - 10t = 0$

$t = 1.8$ seconds ✓

So vertical height risen is

$= 18 \times 1.8 - 5 \times 1.8^2$

$= 16.2$ metres

Therefore maximum height is 24.2 metres. ✓

(2 marks)

b Object hits the ground when $y = -8$

$\therefore 18t - 5t^2 = -8$

$0 = 5t^2 - 18t - 8$

$0 = (5t + 2)(t - 4)$

$t = -\frac{2}{5}$ *or* 4 ✓

At $t = 4$, $x = 18\sqrt{3} \times 4$

$= 72\sqrt{3}$ metres. ✓

Therefore distance from window

$= \sqrt{8^2 + (72\sqrt{3})^2}$

$= \sqrt{15\,616}$

$\cong 125$ metres ✓

(3 marks)

7 a Let's consider the second object in isolation, such that at $t = 0$ it is projected from $x = 0$ and $y = 0$, having $\dot{x} = 24$ and $\dot{y} = V$.

$\ddot{x} = 0$

$\dot{x} = \int 0.dt$

$= 0 + c$

At $t = 0, \dot{x} = 24$

$24 = 0 + c$

Hence, $\dot{x} = 24$

$x = \int 24.dt$

$= 24t + c$

At $t = 0, x = 0$

$\therefore 0 = 24 \times 0 + c$

So $c = 0$

Hence, $x = 24t$ ✓

$\ddot{y} = -10$

$\dot{y} = \int -10.dt$

$= -10t + c$

At $t = 0, \dot{y} = V$

$V = -10 \times 0 + c$

Hence, $\dot{y} = V - 10t$

$y = \int V - 10t.dt$

$= Vt - 5t^2 + c$

At $t = 0, y = 0$

$\therefore 0 = V \times 0 - 5 \times 0^2 + c$

So $c = 0$

Hence, $y = Vt - 5t^2$ ✓

Now $r_2 = x\underset{\sim}{i} + y\underset{\sim}{j}$

$\therefore r_2 = 24t\underset{\sim}{i} + (Vt - 5t^2)\underset{\sim}{j}$

But since second object is projected at $t = 1$, we horizontally translate r_2 one unit to the right, hence at time of projection

$\therefore r_2 = 24(t-1)\underset{\sim}{i} + [V(t-1) - 5(t-1)^2]\underset{\sim}{j}$ ✓

(3 marks)

b Since objects collide their horizontal positions must be equal.

$\therefore 12t = 24(t-1)$ ✓

$12t = 24t - 24$

$12t = 24$

$\therefore t = 2$ seconds. ✓

(2 marks)

c $12t - 5t^2 = Vt - V - 5t^2 + 10t - 5$ ✓

$2t + 5 = Vt - V$

$2 \times 2 + 5 = 2V - V$

$\therefore V = 9$ ✓

(2 marks)

---

8 **a** Top of the first chimney is at (p, h).

Considering horizontal displacement,

$p = Vt\cos\theta$

$\therefore t = \dfrac{p}{V\cos\theta}$ ✓

Considering vertical displacement

$$h = V.\frac{p}{V\cos\theta}.\sin\theta - \frac{g}{2}\cdot\frac{p^2}{V^2\cos^2\theta}$$

$$= p\tan\theta - \frac{5p^2}{V^2\cos^2\theta}$$ ✓

$V^2h = V^2p\tan\theta - 5p^2(\tan^2\theta + 1)$

$5p^2(\tan^2\theta + 1) = V^2(p\tan\theta - h)$

$$\therefore V^2 = \frac{5p^2(\tan^2\theta+1)}{(p\tan\theta - h)}$$ ✓

(3 marks)

b Similarly,

$$V^2 = \frac{5(3p)^2(\tan^2\theta+1)}{(3p\tan\theta - h)}$$

$$\therefore \frac{5p^2(\tan^2\theta+1)}{(p\tan\theta-h)} = \frac{45p^2(\tan^2\theta+1)}{(3p\tan\theta-h)}$$ ✓

$(3p\tan\theta - h) = 9(p\tan\theta - h)$

$3p\tan\theta - h = 9p\tan\theta - 9h$

$8h = 6p\tan\theta$

$\therefore \tan\theta = \dfrac{4h}{3p}$ ✓

(2 marks)

---

9 **a** At $Q, x = 2q$, so

$2q = Vt\cos\theta$ or $t = \dfrac{2q}{V\cos\theta}$ ✓

At $Q, y = 0$.

$\therefore 0 = 2Vt\sin\theta + 2q - gt^2$ ✓

$0 = 2Vt\sin\theta + Vt\cos\theta - gt^2$

$0 = t(2V\sin\theta + V\cos\theta - gt)$

$gt = 2V\sin\theta + V\cos\theta$

$\therefore t = \dfrac{V(2\sin\theta + \cos\theta)}{g}$ ✓

Therefore time taken to reach Q is $\dfrac{V(2\sin\theta+\cos\theta)}{g}$.

(3 marks)

b By equating the equations, where t is the subject.

$$\frac{2q}{V\cos\theta} = \frac{V(2\sin\theta+\cos\theta)}{g}$$ ✓

$$2q = \frac{V(2\sin\theta+\cos\theta)}{g} \times \frac{V\cos\theta}{1}$$

$$q = \frac{V^2(2\sin\theta\cos\theta + \cos^2\theta)}{2g}$$

$$q = \frac{V^2\left(\sin 2\theta + \frac{1}{2}(1+\cos 2\theta)\right)}{2g}$$ ✓

$$\therefore q = \frac{V^2(2\sin 2\theta + \cos 2\theta + 1)}{4g}$$ ✓

(3 marks)

---

10 Equate the components of $\underset{\sim}{i}$: $1 + 5t = t^2 + 5$

$t^2 - 5t + 4 = 0$

$(t-1)(t-4) = 0$

$t = 1, 4$ ✓

Equate the components of $\underset{\sim}{j}$: $t^2 + 3 = 7t - 9$

$t^2 - 7t + 12 = 0$

$(t-4)(t-3) = 0$

$t = 4, 3$

$\therefore$ the common solution is $t = 4$.

The particles collide after 4 seconds. ✓

Substituting: $\underset{\sim}{r}_A(4) = (1 + 5(4))\underset{\sim}{i} + (4^2 + 3)\underset{\sim}{j}$

$= 21\underset{\sim}{i} + 19\underset{\sim}{j}$

$\therefore$ the particles collide at $21\underset{\sim}{i} + 19\underset{\sim}{j}$. ✓

(3 marks)

---

11 $\underset{\sim}{r}(t) = \sin 2t\underset{\sim}{i} + 2\cos 2t\underset{\sim}{j}$

$\underset{\sim}{v}(t) = 2\cos 2t\underset{\sim}{i} - 4\sin 2t\underset{\sim}{j}$

$$|\underset{\sim}{v}(t)| = \sqrt{(2\cos 2t)^2 + (-4\sin 2t)^2}$$

$$= \sqrt{4\cos^2 2t + 16\sin^2 2t}$$ ✓

$$\left|\underset{\sim}{v}\left(\frac{5\pi}{8}\right)\right| = \sqrt{4\cos^2 2\left(\frac{5\pi}{8}\right) + 16\sin^2 2\left(\frac{5\pi}{8}\right)}$$

$$= \sqrt{4\cos^2\frac{5\pi}{4} + 16\sin^2\frac{5\pi}{4}}$$ ✓

$$= \sqrt{4\left(-\frac{1}{\sqrt{2}}\right)^2 + 16\left(-\frac{1}{\sqrt{2}}\right)^2}$$

$= \sqrt{2+8}$

$= \sqrt{10}$

$\therefore$ a speed of $\sqrt{10}$ cms^{-1}. ✓

(3 marks)

---

12 **i** On OA, $y = -x$

$\therefore 18\sqrt{3}t - 5t^2 = -18t$

$5t^2 - (18 + 18\sqrt{3})t = 0$

$t(5t - (18 + 18\sqrt{3})) = 0$

$t = 0$ or $5t = 18 + 18\sqrt{3}$

$$t = \frac{18(1+\sqrt{3})}{5}$$

Now $t = 0$ when the particle is at O.

So at A, $t = \dfrac{18(1+\sqrt{3})}{5}$ ✓

$x = 18t$

$$= \frac{18^2(1+\sqrt{3})}{5}$$

$y = -x$

So $y = -\dfrac{18^2(1+\sqrt{3})}{5}$

$$x^2 + y^2 = 2\left(\frac{18^2(1+\sqrt{3})}{5}\right)^2$$

So $OA = \dfrac{18^2\sqrt{2}\left(1+\sqrt{3}\right)}{5}$

$= 250.367974\ldots$

The distance OA is about 250 units. ✓

(2 marks)

ii $x = 18t$

$\dot{x} = 18$

$y = 18\sqrt{3}t - 5t^2$

$\dot{y} = 18\sqrt{3} - 10t$ ✓

When $t = \dfrac{18\left(1+\sqrt{3}\right)}{5}$,

$\dot{y} = 18\sqrt{3} - 36(1 + \sqrt{3})$

$= 18\sqrt{3} - 36 - 36\sqrt{3}$

$= -36 - 18\sqrt{3}$

$= -18(2 + \sqrt{3})$ ✓

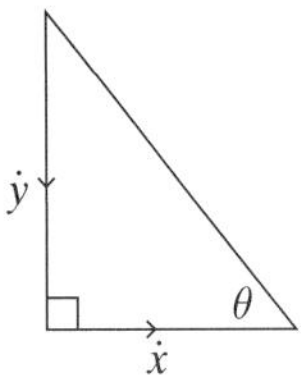

$\tan\theta = \left|\dfrac{\dot{y}}{\dot{x}}\right|$

$= \left|\dfrac{-18\left(2+\sqrt{3}\right)}{18}\right|$

$= 2 + \sqrt{3}$

$\theta = 75°$

So the required angle $= 75° - 45°$

$= 30°$ ✓

(3 marks)

13 **i** $y = Vt\sin\theta - \dfrac{gt^2}{2}$

When $y = 0$,

$Vt\sin\theta - \dfrac{gt^2}{2} = 0$

$t\left(V\sin\theta - \dfrac{gt}{2}\right) = 0$

$t = 0$ or $V\sin\theta - \dfrac{gt}{2} = 0$

$gt = 2V\sin\theta$

$t = \dfrac{2V\sin\theta}{g}$ ✓

Now $x = Vt\cos\theta$

When $t = \dfrac{2V\sin\theta}{g}$,

$x = \dfrac{2V^2\sin\theta\cos\theta}{g}$

$= \dfrac{V^2\sin 2\theta}{g}$

So the horizontal range of the projectile is $\dfrac{V^2}{g}\sin 2\theta$. ✓

(2 marks)

ii When the object is projected at $\dfrac{\pi}{2} - \theta$:

$x = \dfrac{V^2}{g}\sin 2\left(\dfrac{\pi}{2} - \theta\right)$

$= \dfrac{V^2}{g}\sin(\pi - 2\theta)$

$= \dfrac{V^2}{g}\sin 2\theta$

So the horizontal range of the projectile is also $\dfrac{V^2}{g}\sin 2\theta$. ✓

(1 mark)

iii When projected at angle α:

$y = Vt\sin\alpha - \dfrac{gt^2}{2}$

$\dot{y} = V\sin\alpha - gt$

Maximum height occurs when $\dot{y} = 0$

i.e $0 = V\sin\alpha - gt$

$gt = V\sin\alpha$

$t = \dfrac{V\sin\alpha}{g}$ ✓

When $t = \dfrac{V\sin\alpha}{g}$,

$h_\alpha = \dfrac{V^2\sin^2\alpha}{g} - \dfrac{g\left(\dfrac{V\sin\alpha}{g}\right)^2}{2}$

$= \dfrac{V^2\sin^2\alpha}{g} - \dfrac{V^2\sin^2\alpha}{2g}$

$= \dfrac{V^2\sin^2\alpha}{2g}$

Similarly, when projected at angle β:

$h_\beta = \dfrac{V^2\sin^2\beta}{2g}$ ✓

Now $\sin^2\beta = \sin^2\left(\dfrac{\pi}{2} - \alpha\right)$

$= \cos^2\alpha$

So $h_\beta = \dfrac{V^2\cos^2\alpha}{2g}$

$\dfrac{h_\alpha + h_\beta}{2} = \dfrac{\dfrac{V^2\sin^2\alpha}{2g} + \dfrac{V^2\cos^2\alpha}{2g}}{2}$

$= \dfrac{V^2\left(\sin^2\alpha + \cos^2\alpha\right)}{4g}$

$= \dfrac{V^2}{4g}$

So the average of the two heights depends only on V and g. ✓

(3 marks)

14 **i** $y = Vt\sin\theta - \dfrac{1}{2}gt^2$

When $y = 0$,

$Vt\sin\theta - \dfrac{1}{2}gt^2 = 0$

$t(V\sin\theta - \dfrac{1}{2}gt) = 0$

$t = 0$ or $V\sin\theta - \dfrac{1}{2}gt = 0$

$gt = 2V\sin\theta$

$t = \dfrac{2V\sin\theta}{g}$ ✓

Now $x = Vt\cos\theta$

When $t = \dfrac{2V\sin\theta}{g}$,

$$x = \frac{2V^2 \sin\theta\cos\theta}{g}$$
$$= \frac{V^2 \sin 2\theta}{g}$$

So the horizontal range of the projectile is $\frac{V^2 \sin 2\theta}{g}$. ✓

(2 marks)

ii If $V^2 < 100g$

the horizontal range is less than $100 \sin 2\theta$.

But $\sin 2\theta \leq 1$ for all values of θ.

So the horizontal range is less than 100 m. ✓

(1 mark)

iii Now horizontal range > 100 m.

So $\frac{V^2 \sin 2\theta}{g} > 100$

But $V^2 = 200g$

$\therefore \frac{200g \sin 2\theta}{g} > 100$

$200 \sin 2\theta > 100$

$\sin 2\theta > 0.5$ ✓

Now $0 < \theta < \frac{\pi}{2}$

$\therefore 0 < 2\theta < \pi$

If $\sin 2\theta = 0.5$

$2\theta = \frac{\pi}{6}$ or $\pi - \frac{\pi}{6}$

$= \frac{\pi}{6}$ or $\frac{5\pi}{6}$

$\theta = \frac{\pi}{12}$ or $\frac{5\pi}{12}$

$\sin 2\theta \geq 0.5$ when $\frac{\pi}{12} \leq \theta \leq \frac{5\pi}{12}$ ✓

(2 marks)

iv $y = Vt \sin\theta - \frac{1}{2}gt^2$

$\dot{y} = V \sin\theta - gt$

Maximum height occurs when $\dot{y} = 0$

i.e $0 = V \sin\theta - gt$

$gt = V \sin\theta$

$t = \frac{V \sin\theta}{g}$

When $t = \frac{V \sin\theta}{g}$,

$$y = \frac{V^2 \sin^2\theta}{g} - \frac{1}{2}g\left(\frac{V\sin\theta}{g}\right)^2$$
$$= \frac{V^2 \sin^2\theta}{2g}$$ ✓

But $V^2 = 200g$

$\therefore y = 100 \sin^2\theta$

For $\frac{\pi}{12} \leq \theta \leq \frac{5\pi}{12}$, the greatest value of $\sin\theta$ occurs when $\theta = \frac{5\pi}{12}$.

When $\theta = \frac{5\pi}{12}$,

$$y = 100 \sin^2\left(\frac{5\pi}{12}\right)$$
$= 93.301\,2701\ldots$

$= 93.3$ (1 d.p.)

So the greatest possible height that can be reached is 93.3 metres. ✓

(2 marks)

15 i $y = ut \sin\theta - 5t^2$

$\dot{y} = u \sin\theta - 10t$

The maximum height occurs when $\dot{y} = 0$

i.e. $0 = u \sin\theta - 10t$

$10t = u \sin\theta$

$t = \frac{u\sin\theta}{10}$

Now $\ddot{y} = -10$ (maximum)

So the maximum height occurs when

$t = \frac{u\sin\theta}{10}$ ✓

When $t = \frac{u\sin\theta}{10}$,

$$y = \frac{u^2 \sin^2\theta}{10} - \frac{5u^2 \sin^2\theta}{100}$$
$$= \frac{u^2 \sin^2\theta}{20}$$

The greatest height is $\frac{u^2 \sin^2\theta}{20}$ m. ✓

(2 marks)

ii $u = 30, \theta = 30°$

greatest height $= \frac{u^2 \sin^2\theta}{20}$

$= \frac{30^2 \sin^2 30°}{20}$

$= 11.25$ ✓

So the ball hits the wall at a point 11.25 m above the starting point.

Height above the ground $= (20 + 11\frac{1}{4})$ m

$= \frac{125}{4}$ m ✓

(2 marks)

iii $u = 10, \theta = 0°$

$y = ut \sin\theta - 5t^2$

Ball hits the ground when $y = -\frac{125}{4}$

$-\frac{125}{4} = 10t \times 0 - 5t^2$

$5t^2 = \frac{125}{4}$

$t^2 = \frac{25}{4}$ ✓

$t = \pm\frac{5}{2}$

But $t > 0$,

So the ball reaches the ground 2.5 seconds after it rebounds from the wall. ✓

(2 marks)

iv $x = ut \cos\theta$

When $u = 10, t = 2.5$ and $\theta = 0°$

$x = 10 \times 2.5 \times \cos 0°$

$= 25$

So the ball is 25 m from the wall when it hits the ground. ✓

(1 mark)

16 i $y = Vt \sin\theta - \frac{1}{2}gt^2$

When $y = 0$,

$Vt \sin\theta - \frac{1}{2}gt^2 = 0$

$t(V \sin\theta - \frac{1}{2}gt) = 0$

$t = 0$ or $V \sin\theta - \frac{1}{2}gt = 0$

$$gt = 2V\sin\theta$$
$$t = \frac{2V\sin\theta}{g}$$ ✓

Now $x = Vt\cos\theta$

When $t = \frac{2V\sin\theta}{g}$,

$$x = \frac{2V^2\sin\theta\cos\theta}{g}$$
$$= \frac{V^2\sin 2\theta}{g}$$

So the horizontal range of the projectile is $\frac{V^2\sin 2\theta}{g}$. ✓

(2 marks)

ii $\theta = \frac{\pi}{3}$

So $x = Vt\cos\frac{\pi}{3}$
$$= \frac{Vt}{2}$$
$$\dot{x} = \frac{V}{2}$$
$$y = Vt\sin\frac{\pi}{3} - \frac{1}{2}gt^2$$
$$= \frac{\sqrt{3}}{2}Vt - \frac{1}{2}gt^2$$
$$\dot{y} = \frac{\sqrt{3}}{2}V - gt$$

When $t = \frac{2V}{\sqrt{3}g}$,
$$\dot{y} = \frac{\sqrt{3}}{2}V - \frac{2V}{\sqrt{3}}$$
$$= V\left(\frac{\sqrt{3}}{2} - \frac{2\sqrt{3}}{3}\right)$$
$$= \frac{-\sqrt{3}V}{6}$$ ✓

Let α be the acute angle the projectile makes with the horizontal.
$$\tan\alpha = \left|\frac{\dot{y}}{\dot{x}}\right|$$
$$= \left|\frac{-\frac{\sqrt{3}V}{6}}{\frac{V}{2}}\right|$$
$$= \frac{\sqrt{3}V}{6} \times \frac{2}{V}$$
$$= \frac{\sqrt{3}}{3}$$
$$= \frac{1}{\sqrt{3}}$$

So $\alpha = \frac{\pi}{6}$ ✓ *(2 marks)*

iii The projectile is travelling downwards because $\dot{y}$ is negative. ✓

(1 mark)

17 i $x = Vt\cos\theta$
$$\therefore t = \frac{x}{V\cos\theta}$$
$$y = -\frac{1}{2}gt^2 + Vt\sin\theta$$
$$= -\frac{1}{2}g\left(\frac{x}{V\cos\theta}\right)^2 + V\frac{x}{V\cos\theta}\sin\theta$$
$$= -\frac{1}{2}g\frac{x^2}{V^2\cos^2\theta} + \frac{x\sin\theta}{\cos\theta}$$ ✓
$$= x\tan\theta - \frac{gx^2}{2V^2\cos^2\theta}$$
$$= x\tan\theta - \frac{gx^2}{2V^2}\sec^2\theta$$ ✓

(2 marks)

ii The angle between the slope and the horizontal is $\frac{\pi}{4}$.

$\therefore$ the slope has gradient -1.

Equation of OP is $y = -x$

At points of intersection:
$$-x = x\tan\theta - \frac{gx^2}{2V^2}\sec^2\theta$$
$$0 = x\tan\theta + x - \frac{gx^2}{2V^2}\sec^2\theta$$
$$= x(\tan\theta + 1 - \frac{gx}{2V^2}\sec^2\theta)$$
$$x = 0 \text{ or } \tan\theta + 1 - \frac{gx}{2V^2}\sec^2\theta = 0$$

At O, $x = 0$

At P, $\tan\theta + 1 - \frac{gx}{2V^2}\sec^2\theta = 0$ ✓
$$\frac{gx}{2V^2}\sec^2\theta = \tan\theta + 1$$
$$x = \frac{2V^2}{g\sec^2\theta}(\tan\theta + 1)$$
$$= \frac{2V^2}{g}\cos^2\theta\,(\tan\theta + 1)$$
$$= \frac{2V^2}{g}\cos\theta(\sin\theta + \cos\theta)$$ ✓

$y = -x$

So $y = -\frac{2V^2}{g}\cos\theta(\sin\theta + \cos\theta)$
$$D^2 = x^2 + y^2$$
$$= x^2 + (-x)^2$$
$$= 2x^2$$
$$D = \sqrt{2x^2}$$
$$= \sqrt{2}x \quad (x > 0)$$
$$= \sqrt{2} \times \frac{2V^2}{g}\cos\theta(\sin\theta + \cos\theta)$$
$$= 2\sqrt{2}\frac{V^2}{g}\cos\theta(\cos\theta + \sin\theta)$$ ✓

(3 marks)

iii
$$\frac{dD}{d\theta} = 2\sqrt{2}\frac{V^2}{g}(\cos\theta(-\sin\theta + \cos\theta) + (\cos\theta + \sin\theta)(-\sin\theta))$$
$$= 2\sqrt{2}\frac{V^2}{g}(-\sin\theta\cos\theta + \cos^2\theta - \sin\theta\cos\theta - \sin^2\theta)$$ ✓
$$= 2\sqrt{2}\frac{V^2}{g}(\cos^2\theta - \sin^2\theta - 2\sin\theta\cos\theta)$$
$$= 2\sqrt{2}\frac{V^2}{g}(\cos 2\theta - \sin 2\theta)$$ ✓

(2 marks)

iv $\frac{dD}{d\theta} = 2\sqrt{2}\frac{V^2}{g}(\cos 2\theta - \sin 2\theta)$

Stationary points occur when $\frac{dD}{d\theta} = 0$

i.e. $2\sqrt{2}\frac{V^2}{g}(\cos 2\theta - \sin 2\theta) = 0$

$\cos 2\theta - \sin 2\theta = 0$

$\sin 2\theta = \cos 2\theta$

$\tan 2\theta = 1 \qquad (\cos 2\theta \neq 0)$

$2\theta = \frac{\pi}{4} \qquad \left(0 \leq \theta < \frac{\pi}{2}\right)$

$\theta = \frac{\pi}{8}$ ✓

$\frac{d^2D}{d\theta^2} = 2\sqrt{2}\frac{V^2}{g}(-2\sin 2\theta - 2\cos 2\theta)$

$= -4\sqrt{2}\frac{V^2}{g}(\sin 2\theta + \cos 2\theta)$ ✓

When $\theta = \frac{\pi}{8}$,

$\frac{d^2D}{d\theta^2} = -4\sqrt{2}\frac{V^2}{g}\left(\frac{1}{\sqrt{2}} + \frac{1}{\sqrt{2}}\right) \quad (< 0)$

$\therefore$ D has a maximum value when $\theta = \frac{\pi}{8}$ ✓

(3 marks)

18 i Projectile fired from A:

$y = ut\sin\alpha - \frac{g}{2}t^2$

$\dot{y} = u\sin\alpha - gt$

$\ddot{y} = -g$ ✓

The maximum ($\ddot{y} < 0$) height will occur when $\dot{y} = 0$

i.e. $u\sin\alpha - gt = 0$

$gt = u\sin\alpha$

$t = \frac{u}{g}\sin\alpha$

The projectile fired from A will reach its maximum height after $\frac{u}{g}\sin\alpha$ seconds. ✓

(2 marks)

ii Similarly, the projectile fired from B will reach its maximum height after $\frac{w}{g}\sin\beta$ seconds.

But both reach their maximum height at the same time, when they collide.

So $\frac{u}{g}\sin\alpha = \frac{w}{g}\sin\beta$

$u\sin\alpha = w\sin\beta$ ✓

(1 mark)

iii For the projectile fired from A the horizontal distance from A to the point of impact is given by:

$x = ut\cos\alpha$

When $t = \frac{u}{g}\sin\alpha$,

$x = \frac{u^2}{g}\sin\alpha\cos\alpha$

Similarly for the projectile fired from B:

$x = \frac{w^2}{g}\sin\beta\cos\beta$ ✓

$d = \frac{u^2}{g}\sin\alpha\cos\alpha + \frac{w^2}{g}\sin\beta\cos\beta$

$= \frac{1}{g}(u^2\sin\alpha\cos\alpha + w^2\sin\beta\cos\beta)$

$= \frac{1}{g}(u\sin\alpha\,(u\cos\alpha) + w\sin\beta\,(w\cos\beta))$

But $u\sin\alpha = w\sin\beta$

$d = \frac{1}{g}(w\sin\beta\,(u\cos\alpha) + u\sin\alpha\,(w\cos\beta))$

$= \frac{uw}{g}(\sin\beta\cos\alpha + \sin\alpha\cos\beta)$

$= \frac{uw}{g}(\sin\alpha\cos\beta + \cos\alpha\sin\beta)$

$= \frac{uw}{g}\sin(\alpha + \beta)$ ✓

(2 marks)

19 i $y = 70t\sin\theta - 4.9t^2$

$\frac{dy}{dt} = 70\sin\theta - 9.8t$

Stationary point occurs when $\frac{dy}{dt} = 0$

i.e. $9.8t = 70\sin\theta$

$t = \frac{50}{7}\sin\theta$

$\frac{d^2y}{dt^2} = -9.8 \quad (< 0)$

$\therefore$ the maximum height is reached when

$t = \frac{50}{7}\sin\theta$ ✓

When $t = \frac{50}{7}\sin\theta$,

$y = 70 \times \frac{50}{7}\sin\theta \times \sin\theta - 4.9 \times (\frac{50}{7}\sin\theta)^2$

$= 500\sin^2\theta - 250\sin^2\theta$

$= 250\sin^2\theta$

$\therefore$ the firework explodes at a height of $250\sin^2\theta$ metres. ✓ *(2 marks)*

ii $x = 70t\cos\theta$

When $t = \frac{50}{7}\sin\theta$,

$x = 70 \times \frac{50}{7}\sin\theta \times \cos\theta$

$= 500\sin\theta\cos\theta$

$= 250 \times 2\sin\theta\cos\theta$

$= 250\sin 2\theta$

$\therefore$ the firework explodes at a horizontal distance of $250\sin 2\theta$ metres from O. ✓

(1 mark)

iii $0° < \theta < 90°$

Height ≥ 150 m

$250\sin^2\theta \geq 150$

$\sin^2\theta \geq 0.6$

$\sin\theta \geq 0.7745966\ldots \quad (\sin\theta > 0)$

$\theta \geq 50.768479\ldots°$ ✓

$0° < 2\theta < 180°$

125 m < horizontal distance < 180 m

So $250\sin 2\theta > 125$

$\sin 2\theta > 0.5$

$30° < 2\theta < 150°$

$15° < \theta < 75°$ ✓

And $250\sin 2\theta < 180$

$\sin 2\theta < 0.72$

$2\theta < 46.054...°$ or $2\theta > 133.945...°$

$\theta < 23.027...°$ or $\theta > 66.972...°$

$\therefore$ the firework will be at the best position for viewing when $67° \le \theta < 75°$. ✓ *(3 marks)*

20 **i** $y = h - \frac{1}{2}gt^2$

The ball hits the ground when $y = 0$

i.e. $h - \frac{1}{2}gt^2 = 0$

$$\frac{1}{2}gt^2 = h$$

$$gt^2 = 2h$$

$$t^2 = \frac{2h}{g}$$

$$t = \sqrt{\frac{2h}{g}} \quad (t \ge 0)$$

The ball strikes the ground at $t = \sqrt{\frac{2h}{g}}$ s. ✓

(1 mark)

ii $x = vt$

When $t = \sqrt{\frac{2h}{g}}, x = d$

$\therefore d = v\sqrt{\frac{2h}{g}}$

Now $\frac{dx}{dt} = v$

and $\frac{dy}{dt} = -gt$

When $t = \sqrt{\frac{2h}{g}}$, $\frac{dy}{dt} = -g\sqrt{\frac{2h}{g}}$

$$= -\sqrt{\frac{2hg^2}{g}}$$

$$= -\sqrt{2hg}$$ ✓

Now the ball strikes the ground at an angle of 45°.

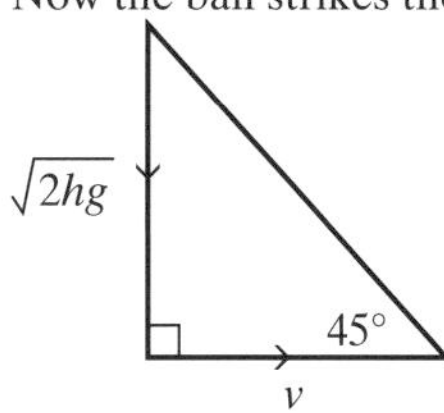

So $v = \sqrt{2hg}$

$\therefore d = \sqrt{2hg} \times \sqrt{\frac{2h}{g}}$

$= \sqrt{(2h)^2}$

$= 2h \quad (d > 0)$ ✓

(2 marks)

21 **i** $x = vt\cos\theta$

$\therefore t = \frac{x}{v\cos\theta}$

$y = vt\sin\theta - 5t^2$

$= v\left(\frac{x}{v\cos\theta}\right)\sin\theta - 5\left(\frac{x}{v\cos\theta}\right)^2$

$= x\tan\theta - \frac{5x^2}{v^2\cos^2\theta}$ ✓

The ball passes through the point (d, h).

Now $\tan\alpha = \frac{h}{d}$

$\therefore h = d\tan\alpha$

So when $x = d, y = d\tan\alpha$

But $y = x\tan\theta - \frac{5x^2}{v^2\cos^2\theta}$

$\therefore d\tan\alpha = d\tan\theta - \frac{5d^2}{v^2\cos^2\theta}$

$d\tan\alpha - d\tan\theta = -\frac{5d^2}{v^2\cos^2\theta}$

$\tan\theta - \tan\alpha = \frac{5d}{v^2\cos^2\theta}$ ✓

$$v^2 = \frac{5d}{\cos^2\theta(\tan\theta - \tan\alpha)}$$

$$= \frac{5d}{\cos^2\theta\left(\frac{\sin\theta}{\cos\theta}\right) - \cos^2\theta\tan\alpha}$$

$$= \frac{5d}{\cos\theta\sin\theta - \cos^2\theta\tan\alpha}$$ ✓ *(3 marks)*

ii **1** As $\theta \to \alpha$,

$\cos\theta\sin\theta - \cos^2\theta\tan\alpha \to 0$

$\therefore v \to \infty$ ✓ *(1 mark)*

2 As $\theta \to \frac{\pi}{2}, \cos\theta \to 0$

$\therefore \cos\theta\sin\theta - \cos^2\theta\tan\alpha \to 0$

$\therefore v \to \infty$ ✓ *(1 mark)*

iii $F(\theta) = \cos\theta\sin\theta - \cos^2\theta\tan\alpha$

$= \frac{1}{2}\sin 2\theta - \cos^2\theta\tan\alpha$

$F'(\theta) = \frac{1}{2} \times 2\cos 2\theta - 2\cos\theta(-\sin\theta)\tan\alpha$

$= \cos 2\theta + 2\cos\theta\sin\theta\tan\alpha$

$= \cos 2\theta + \sin 2\theta\tan\alpha$

$= \cos 2\theta + \cos 2\theta\tan 2\theta\tan\alpha$ ✓

If $\tan 2\theta\tan\alpha = -1$,

$F'(\theta) = \cos 2\theta - \cos 2\theta$

$= 0$ ✓

(2 marks)

iv $0 < \alpha < \frac{\pi}{2}$ and $\alpha < \theta < \pi$

If as given $\tan 2\theta\tan\alpha = -1$,

then $2\theta - \alpha = \frac{\pi}{2}$

$2\theta = \frac{\pi}{2} + \alpha$

$\theta = \frac{\pi}{4} + \frac{\alpha}{2}$

$\therefore$ if $\theta = \frac{\alpha}{2} + \frac{\pi}{4}, F'(\theta) = 0$ ✓

(1 mark)

v $F'\left(\frac{\alpha}{2}+\frac{\pi}{4}\right)=0$

$F'(\theta)=\cos 2\theta+\sin 2\theta\tan\alpha$

$F''(\theta)=-2\sin 2\theta+2\cos 2\theta\tan\alpha$

$$F''\left(\frac{\alpha}{2}+\frac{\pi}{4}\right)=-2\sin 2\left(\frac{\alpha}{2}+\frac{\pi}{4}\right)+2\cos 2\left(\frac{\alpha}{2}+\frac{\pi}{4}\right)\tan\alpha$$

$$=-2\sin\left(\alpha+\frac{\pi}{2}\right)+2\cos\left(\alpha+\frac{\pi}{2}\right)\tan\alpha \quad ✓$$

Now $0<\alpha<\frac{\pi}{2}$,

$\therefore \frac{\pi}{2}<\alpha+\frac{\pi}{2}<\pi$

$\therefore \sin\left(\alpha+\frac{\pi}{2}\right)>0$ and $\cos\left(\alpha+\frac{\pi}{2}\right)<0$

Also $\tan\alpha>0$

$\therefore F''\left(\frac{\alpha}{2}+\frac{\pi}{4}\right)<0$

$\therefore \cos\theta\sin\theta-\cos^2\theta\tan\alpha$ is a maximum when $\theta=\frac{\alpha}{2}+\frac{\pi}{4}$

Now $v^2=\frac{5d}{F(\theta)}$

$\therefore v^2$ is a minimum when $\theta=\frac{\alpha}{2}+\frac{\pi}{4}$. ✓

(2 marks)

22 Projectile from A:

$x_1=Ut\cos\theta$
$y_1=Ut\sin\theta-\frac{1}{2}gt^2$

Projectile from B:
$x_2=R-Vt\cos\theta$
$y_2=h-Vt\sin\theta-\frac{1}{2}gt^2$

i When $t=T$, $x_1=x_2$.

$$\therefore UT\cos\theta=R-VT\cos\theta$$
$$UT\cos\theta+VT\cos\theta=R$$
$$T(U\cos\theta+V\cos\theta)=R$$
$$T\cos\theta(U+V)=R$$
$$\therefore T=\frac{R}{(U+V)\cos\theta}. \quad ✓$$

(1 mark)

ii If the projectiles collide, then $x_1=x_2$ and $y_1=y_2$ at time, $t=T$ from **i**.

$y_1=Ut\sin\theta-\frac{1}{2}gt^2$

When $t=T=\frac{R}{(U+V)\cos\theta}$,

$$y_1=U\times\frac{R}{(U+V)\cos\theta}\times\sin\theta-\frac{1}{2}gT^2$$
$$=\frac{UR\tan\theta}{(U+V)}-\frac{1}{2}gT^2$$
$$=\frac{URh}{(U+V)R}-\frac{1}{2}gT^2$$
$\left(\text{since } \theta=\tan^{-1}\left(\frac{h}{R}\right), \text{ i.e. } \tan\theta=\frac{h}{R}\right)$
$$=\frac{Uh}{(U+V)}-\frac{1}{2}gT^2. \quad ✓$$

$y_2=h-Vt\sin\theta-\frac{1}{2}gt^2$

When $t=T=\frac{R}{(U+V)\cos\theta}$,

$$\therefore y_2=h-V\times\frac{R}{(U+V)\cos\theta}\times\sin\theta-\frac{1}{2}gT^2$$
$$=h-\frac{VR\tan\theta}{(U+V)}-\frac{1}{2}gT^2$$
$$=h-\frac{VRh}{(U+V).R}-\frac{1}{2}gT^2$$
$\left(\text{since } \theta=\tan^{-1}\left(\frac{h}{R}\right), \text{ i.e. } \tan\theta=\frac{h}{R}\right)$
$$=h-\frac{Vh}{(U+V)}-\frac{1}{2}gT^2$$
$$=\frac{h(U+V)-Vh}{(U+V)}-\frac{1}{2}gT^2$$
$$=\frac{Uh}{(U+V)}-\frac{1}{2}gT^2$$
$$=y_1$$

$\therefore$ The projectiles collide when $x_1=x_2$ and $y_1=y_2$ at $T=\frac{R}{(U+V)\cos\theta}$. ✓

(2 marks)

iii If the projectiles collide on $x=\lambda R$ where $0<\lambda<1$, and

$x_1=Ut\cos\theta$ when $t=\frac{R}{(U+V)\cos\theta}$,

then $\lambda R=U\times\frac{R}{(U+V)\cos\theta}\times\cos\theta$

$$\lambda R=\frac{UR}{U+V}$$
$$\lambda=\frac{U}{U+V}$$
$$U\lambda+V\lambda=U$$
$$V\lambda=U-U\lambda$$
$$V=\frac{U-U\lambda}{\lambda}$$
$$\therefore V=\left(\frac{1}{\lambda}-1\right)U. \quad ✓$$

(1 mark)

23 **a** At L and M, $y=h$ when $t=t_1$ and t_2 respectively.

Now $y=Vt\sin\theta-\frac{1}{2}gt^2$

$\therefore h=Vt\sin\theta-\frac{1}{2}gt^2$

(which has 2 solutions: t_1 and t_2)

$\therefore \frac{1}{2}gt^2-Vt\sin\theta+h=0$

Sum of the roots $= \frac{-b}{a}$

$$= \frac{V\sin\theta}{\frac{1}{2}g}$$

$$= \frac{2V\sin\theta}{g}$$

$$\therefore t_1 + t_2 = \frac{2V}{g}\sin\theta. \quad ✓$$

Product of the roots $= \frac{c}{a}$

$$= \frac{h}{\frac{1}{2}g}$$

$$= \frac{2h}{g}$$

$$\therefore t_1 t_2 = \frac{2h}{g}. \quad ✓$$

(2 marks)

b $\tan\alpha + \tan\beta$

$$= \frac{h}{Vt_1\cos\theta} + \frac{h}{Vt_2\cos\theta}$$

$$= \frac{ht_2 + ht_1}{Vt_1t_2\cos\theta}$$

$$= \frac{h(t_1+t_2)}{V\cos\theta(t_1t_2)} \quad ✓$$

$$= h\left(\frac{2V\sin\theta}{g}\right) \div V\cos\theta\left(\frac{2h}{g}\right) \quad (\text{using } \mathbf{a})$$

$$= \frac{2Vh\sin\theta}{g} \times \frac{g}{2Vh\cos\theta}$$

$$= \frac{\sin\theta}{\cos\theta}$$

$$= \tan\theta \quad ✓$$

(2 marks)

c $\tan\alpha\tan\beta = \frac{h}{Vt_1\cos\theta} \times \frac{h}{Vt_2\cos\theta}$

$$= \frac{h^2}{V^2\cos^2\theta\, t_1t_2}$$

$$= \frac{h^2}{V^2\cos^2\theta\left(\frac{2h}{g}\right)} \quad (\text{using } \mathbf{a})$$

$$= \frac{gh}{2V^2\cos^2\theta} \quad ✓$$

(1 mark)

d $ON = r = OP + PN$

Now in $\triangle OLP$,

$$\tan\alpha = \frac{h}{OP}$$

$$\therefore OP = \frac{h}{\tan\alpha}$$

Also in $\triangle LPN$

$$\tan\beta = \frac{h}{PN}$$

$$\therefore PN = \frac{h}{\tan\beta}$$

$ON = OP + PN$

$$\therefore r = \frac{h}{\tan\alpha} + \frac{h}{\tan\beta}$$

$$= h\left(\frac{1}{\tan\alpha} + \frac{1}{\tan\beta}\right)$$

$$= h(\cot\alpha + \cot\beta) \quad ✓$$

$LM = ON - 2 \times OP$

$$\therefore w = h(\cot\alpha + \cot\beta) - \frac{2h}{\tan\alpha}$$

$$= h(\cot\alpha + \cot\beta) - 2h\cot\alpha$$

$$= h\cot\alpha + h\cot\beta - 2h\cot\alpha$$

$$= h\cot\beta - h\cot\alpha$$

$$= h(\cot\beta - \cot\alpha) \quad ✓$$

(2 marks)

e At L, $\tan\phi = \frac{dy}{dx}$ when $t = t_1$

$$\frac{dy}{dx} = \frac{dy}{dt} \times \frac{dt}{dx}$$

$$y = Vt\sin\theta - \frac{1}{2}gt^2$$

$$\therefore \frac{dy}{dt} = V\sin\theta - gt$$

and $x = Vt\cos\theta$

$$\therefore \frac{dx}{dt} = V\cos\theta$$

$$\therefore \frac{dy}{dx} = (V\sin\theta - gt) \times \frac{1}{V\cos\theta}$$

$$= \frac{V\sin\theta - gt}{V\cos\theta} \quad ✓$$

$$\therefore \tan\phi = \frac{V\sin\theta}{V\cos\theta} - \frac{gt_1}{V\cos\theta} \text{ when } t = t_1.$$

Multiplying each term by $\frac{2}{g}$:

$$\tan\phi = \frac{\frac{2V\sin\theta}{g}}{\frac{2V\cos\theta}{g}} - \frac{2t_1}{\frac{2V\cos\theta}{g}}$$

$$= \frac{t_1+t_2}{\frac{2V\cos\theta}{g}} - \frac{2t_1}{\frac{2V\cos\theta}{g}} \quad (\text{from } \mathbf{a})$$

$$= \frac{t_2 - t_1}{\frac{2V\cos\theta}{g}} \quad ✓$$

$$= \frac{g(t_2-t_1)}{2V\cos\theta}$$

$$= \frac{gt_2}{2V\cos\theta} - \frac{gt_1}{2V\cos\theta}$$

$$= \frac{h}{Vt_1\cos\theta} - \frac{h}{Vt_2\cos\theta}$$

$$\left(\text{on multiplying each term by } \frac{2h}{gt_1t_2}\right)$$

$$= \tan\alpha - \tan\beta. \quad ✓$$

(3 marks)

f $\frac{w}{\tan\phi} = \frac{h(\cot\beta - \cot\alpha)}{\tan\alpha - \tan\beta}$ (from **d** and **e**)

$$= \left(\frac{h}{\tan\beta} - \frac{h}{\tan\alpha}\right) \div (\tan\alpha - \tan\beta)$$

$$= \left(\frac{h}{\tan\beta} - \frac{h}{\tan\alpha}\right) \times \frac{1}{\tan\alpha - \tan\beta}$$

$$= \frac{h\tan\alpha - h\tan\beta}{\tan\beta\tan\alpha} \times \frac{1}{\tan\alpha - \tan\beta}$$

$$= \frac{h(\tan\alpha - \tan\beta)}{\tan\beta \tan\alpha(\tan\alpha - \tan\beta)}$$

$$= \frac{h}{\tan\beta \tan\alpha}$$ ✓

$$\frac{r}{\tan\theta} = \frac{h(\cot\alpha + \cot\beta)}{\tan\alpha + \tan\beta} \quad \text{(from **b** and **d**)}$$

$$= \left(\frac{h}{\tan\alpha} + \frac{h}{\tan\beta}\right) \times \frac{1}{\tan\alpha + \tan\beta}$$

$$= \frac{h\tan\beta + h\tan\alpha}{\tan\alpha \tan\beta} \times \frac{1}{\tan\alpha + \tan\beta}$$

$$= \frac{h(\tan\beta + \tan\alpha)}{\tan\alpha \tan\beta(\tan\alpha + \tan\beta)}$$

$$= \frac{h}{\tan\alpha \tan\beta}$$

$$= \frac{w}{\tan\phi}$$ ✓

(2 marks)

24 $x = 14t\cos\theta$ ... (1)

$y = 14t\sin\theta - 4.9t^2$ ... (2)

When $t = 0, x = 0, v = 14$ and $y = 0$.

i From (1):

$$t = \frac{x}{14\cos\theta}$$

Substituting into (2):

$$y = 14\left(\frac{x}{14\cos\theta}\right)\sin\theta - 4.9\left(\frac{x}{14\cos\theta}\right)^2$$ ✓

$$= x\tan\theta - 4.9 . \frac{x^2}{196} . \sec^2\theta$$

$$= x\tan\theta - \frac{x^2}{40}(1 + \tan^2\theta)$$

$$= mx - \left(\frac{1+m^2}{40}\right)x^2 \quad \text{where } m = \tan\theta$$ ✓

(2 marks)

ii The paintball hits the barrier at height h metres when $x = 10$ metres.

i.e. $x = 10,\ y = h$

$$\therefore h = 10m - \left(\frac{1+m^2}{40}\right).100$$

$$= 10m - \frac{5}{2}(1 + m^2)$$

$$2h = 20m - 5 - 5m^2$$

$$5m^2 - 20m + 2h + 5 = 0$$ ✓

$$\therefore m = \frac{20 \pm \sqrt{(-20)^2 - 4(5)(2h+5)}}{2(5)}$$

$$= \frac{20 \pm \sqrt{400 - 20(2h+5)}}{10}$$

$$= \frac{20 \pm \sqrt{300 - 40h}}{10}$$

$$= \frac{20 \pm \sqrt{100(3 - 0.4h)}}{10}$$

$$= \frac{20 \pm 10\sqrt{3 - 0.4h}}{10}$$

$$= 2 \pm \sqrt{3 - 0.4h}$$

Now $3 - 0.4h \geqslant 0$

$0.4h \leqslant 3$

$h \leqslant 7.5$

$\therefore$ Maximum value of $h = 7.5$ m. ✓

(2 marks)

iii

7.5

5.9

hole

3.9

ground

The paintball passes through the hole if $3.9 \leqslant h \leqslant 5.9$

Now $m = 2 + \sqrt{3 - 0.4h}$ or $m = 2 - \sqrt{3 - 0.4h}$ ✓

If

$$\begin{aligned} 3.9 &\leqslant h \leqslant 5.9 \\ 1.56 &\leqslant 0.4h \leqslant 2.36 \\ -2.36 &\leqslant -0.4h \leqslant -1.56 \\ 0.64 &\leqslant 3 - 0.4h \leqslant 1.44 \\ 0.8 &\leqslant \sqrt{3 - 0.4h} \leqslant 1.2 \\ 2.8 &\leqslant 2 + \sqrt{3 - 0.4h} \leqslant 3.2 \end{aligned}$$

$$\begin{aligned} 3.9 &\leqslant h \leqslant 5.9 \\ 1.56 &\leqslant 0.4h \leqslant 2.36 \\ -2.36 &\leqslant -0.4h \leqslant -1.56 \\ 0.64 &\leqslant 3 - 0.4h \leqslant 1.44 \\ 0.8 &\leqslant \sqrt{3 - 0.4h} \leqslant 1.2 \\ -1.2 &\leqslant -\sqrt{3 - 0.4h} \leqslant -0.8 \\ 0.8 &\leqslant 2 - \sqrt{3 - 0.4h} \leqslant 1.2 \end{aligned}$$

$\therefore$ If $m = 2 + \sqrt{3 - 0.4h}$, $2.8 \leqslant m \leqslant 3.2$

If $m = 2 - \sqrt{3 - 0.4h}$, $0.8 \leqslant m \leqslant 1.2$ ✓

(2 marks)

iv The range occurs when

$$y = 0$$

$$mx - \left(\frac{1+m^2}{40}\right)x^2 = 0$$

$$x\left\{m - \left(\frac{1+m^2}{40}\right)x\right\} = 0$$

$$m - \left(\frac{1+m^2}{40}\right)x = 0$$

$$\left(\frac{1+m^2}{40}\right)x = m$$

$$x = \frac{40m}{1+m^2}$$

i.e. The range is $\frac{40m}{1+m^2}$ metres. ✓

If $m = 2.8$, $x_1 = \frac{40 \times 2.8}{1+(2.8)^2}$
$= 12.6696\ldots$

If $m = 3.2$, $x_2 = \frac{40 \times 3.2}{1+(3.2)^2}$
$= 11.3879\ldots$

∴ Width of the first interval
$= x_1 - x_2$
$= 1.3$ m to 1 decimal place.

If $m = 0.8$, $x_1 = \frac{40 \times 0.8}{1+0.8^2}$
$= 19.5121\ldots$

If $m = 1.2$, $x_2 = \frac{40 \times 1.2}{1+1.2^2}$
$= 19.6721\ldots$ ✓

However, maximum range occurs when
$\theta = 45°$
$\therefore m = \tan 45°$
$= 1$

If $m = 1$, $x_3 = \frac{40}{2}$
$= 20$ m

∴ Width of the second interval
$= 20 - 19.5121\ldots$
$= 0.4878\ldots$ m
$= 0.5$ m to 1 decimal place.

Hence, the widths of the two intervals are 1.3 m and 0.5 m. ✓

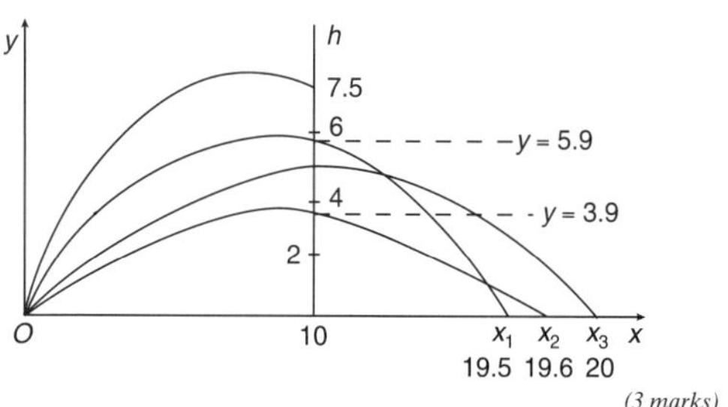

(3 marks)

25 i *Particle 1:*

$$x = Vt\cos\theta$$

$$y = Vt\sin\theta - \frac{1}{2}gt^2$$

Particle 2:

$$x = a$$

$$y = Vt - \frac{1}{2}gt^2$$

Let (x_1, y_1) and (x_2, y_2) be the positions of Particles 1 and 2 at time t.

Let L = distance between Particles 1 and 2 at time t.

$$\therefore L = \sqrt{(x_2 - x_1)^2 + (y_2 - y_1)^2}$$

$$= \sqrt{(a - Vt\cos\theta)^2 + \left[Vt - \frac{1}{2}gt^2 - \left(Vt\sin\theta - \frac{1}{2}gt^2\right)\right]^2}$$

$$= \sqrt{(a - Vt\cos\theta)^2 + [Vt - Vt\sin\theta]^2} \quad ✓$$

$$= \sqrt{a^2 - 2aVt\cos\theta + V^2t^2\cos^2\theta + V^2t^2 - 2V^2t^2\sin\theta + V^2t^2\sin^2\theta}$$

$$= \sqrt{a^2 - 2aVt\cos\theta + V^2t^2(\cos^2\theta + \sin^2\theta + 1 - 2\sin\theta)}$$

$$= \sqrt{a^2 - 2aVt\cos\theta + V^2t^2(2 - 2\sin\theta)}$$

$$= \sqrt{a^2 - 2aVt\cos\theta + 2V^2t^2(1 - \sin\theta)}$$

$$\therefore L^2 = 2V^2t^2(1 - \sin\theta) - 2aVt\cos\theta + a^2 \quad ✓$$

(2 marks)

ii Let $L^2 = D$

D is smallest when $\frac{dD}{dt} = 0$ and $\frac{d^2D}{dt^2} > 0$

Now $D = 2V^2t^2(1 - \sin\theta) - 2aVt\cos\theta + a^2$

$$\frac{dD}{dt} = 4V^2t(1 - \sin\theta) - 2aV\cos\theta$$

If $\frac{dD}{dt} = 0$

$\therefore 4V^2t(1 - \sin\theta) - 2aV\cos\theta = 0$

$4V^2t(1 - \sin\theta) = 2aV\cos\theta$ ✓

$$t = \frac{2aV\cos\theta}{4V^2(1 - \sin\theta)}$$

$$= \frac{a\cos\theta}{2V(1 - \sin\theta)}$$

$$\frac{d^2D}{dt^2} = 4V^2(1 - \sin\theta)$$

> 0 since $V > 0$ and $0 < \theta < \frac{\pi}{2}$

∴ The smallest distance occurs when $t = \frac{a\cos\theta}{2V(1 - \sin\theta)}$. ✓

$$\therefore L^2 = 2V^2\left(\frac{a^2\cos^2\theta}{4V^2(1 - \sin\theta)^2}\right)(1 - \sin\theta)$$

$$- 2aV\left(\frac{a\cos\theta}{2V(1 - \sin\theta)}\right).\cos\theta + a^2$$

$$= \frac{a^2\cos^2\theta}{2(1 - \sin\theta)} - \frac{a^2\cos^2\theta}{(1 - \sin\theta)} + a^2$$

$$= a^2\left[\frac{\cos^2\theta - 2\cos^2\theta + 2(1-\sin\theta)}{2(1-\sin\theta)}\right]$$

$$= a^2\left[\frac{-\cos^2\theta + 2 - 2\sin\theta}{2(1-\sin\theta)}\right]$$

$$= a^2\left[\frac{-(1-\sin^2\theta) + 2 - 2\sin\theta}{2(1-\sin\theta)}\right]$$

$$= a^2\left[\frac{\sin^2\theta - 2\sin\theta + 1}{2(1-\sin\theta)}\right]$$

$$= a^2\left[\frac{(1-\sin\theta)^2}{2(1-\sin\theta)}\right]$$

$$= a^2\left[\frac{1-\sin\theta}{2}\right]$$

$$\therefore L = a\sqrt{\frac{1-\sin\theta}{2}}.$$ ✓

(3 marks)

iii The smallest distance occurs when

$$t = \frac{a\cos\theta}{2V(1-\sin\theta)}$$

If Particle 1 is ascending, $\dot{y} > 0$.

Now $y = Vt\sin\theta - \frac{1}{2}gt^2$

$$\therefore \dot{y} = V\sin\theta - gt$$

$$= V\sin\theta - g\left(\frac{a\cos\theta}{2V(1-\sin\theta)}\right)$$

∴ The smallest distance occurs when Particle 1 is ascending if:

$$V\sin\theta - \frac{ag\cos\theta}{2V(1-\sin\theta)} > 0$$

$$\frac{2V^2\sin\theta(1-\sin\theta) - ag\cos\theta}{2V(1-\sin\theta)} > 0$$

$$2V^2\sin\theta(1-\sin\theta) - ag\cos\theta > 0$$

(N.B.: $2V(1-\sin\theta) > 0$, since $V > 0$ and $0 < \theta < \frac{\pi}{2}$)

$$2V^2\sin\theta(1-\sin\theta) > ag\cos\theta$$

$$V^2 > \frac{ag\cos\theta}{2\sin\theta(1-\sin\theta)}$$

$$\therefore V > \sqrt{\frac{ag\cos\theta}{2\sin\theta(1-\sin\theta)}}.$$ ✓

(1 mark)

26

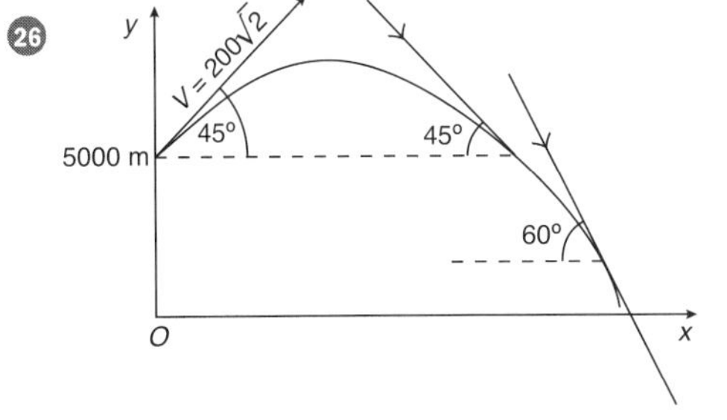

i

$$y = -4.9t^2 + 200t + 5000$$
$$\dot{y} = -9.8t + 200$$

Maximum height occurs when

$$\dot{y} = 0$$
$$\therefore -9.8t + 200 = 0$$
$$-9.8t = -200$$
$$t = 20.4081\ldots$$
$$= 20.4 \text{ seconds.}$$ ✓

When $t = 20.4081\ldots$

$$y = -4.9(20.4081\ldots)^2 + 200(20.4081\ldots) + 5000$$
$$= 7040.8163\ldots$$
$$\doteqdot 7041 \text{ m}$$

The rocket will reach a maximum height of approximately 7041 m at a time of approximately 20.4 seconds. ✓

(2 marks)

ii Since the path of the rocket is parabolic and symmetrical, the rocket will again be descending at an angle of 45° when t = 2 × 20.4 = 40.8 seconds.

∴ Earliest time pilot can eject is 40.8 seconds.

The latest time for ejection is when

$$\frac{dy}{dx} = \tan(-60°)$$

i.e. $\frac{dy}{dx} = -\sqrt{3}$ ✓

Now $x = 200t$

$$\therefore \frac{dx}{dt} = 200$$

and $y = -4.9t^2 + 200t + 5000$

$$\therefore \frac{dy}{dt} = -9.8t + 200$$

Also, $\frac{dy}{dx} = \frac{dy}{dt} \times \frac{dt}{dx}$

$$= -9.8t + 200 \times \frac{1}{200}$$

$$= \frac{-9.8t + 200}{200}$$ ✓

If $\frac{dy}{dx} = -\sqrt{3}$

$$\frac{-9.8t + 200}{200} = -\sqrt{3}$$

$$-9.8t = -200\sqrt{3} - 200$$

$$t = \frac{-200\sqrt{3} - 200}{-9.8}$$

$$= 55.7561\ldots$$

∴ The latest time the pilot can eject is approximately 55.76 seconds.

∴ Pilot can eject between $t = 40.8$ s and $t = 55.76$ s. ✓

(3 marks)

iii

$$v^2 = (\dot{x})^2 + (\dot{y})^2$$

Now $\dot{x} = 200$

and $\dot{y} = -9.8t + 200$

If $v = 350$,

$$\therefore 350^2 = 200^2 + (-9.8t + 200)^2$$
$$122\,500 = 40\,000 + (-9.8t + 200)^2$$
$$82\,500 = (-9.8t + 200)^2$$ ✓
$$-9.8t + 200 = \pm\sqrt{82\,500}$$

$$-9.8t = -200 \pm \sqrt{82\,500}$$
$$t = \frac{-200 \pm \sqrt{82\,500}}{-9.8}$$
$$= 49.7171\ldots$$
(as $t > 0$)

$\therefore$ The latest time to eject is 49.7 seconds after the engine stops. (Check if $t = 50.7 > 350 \text{ ms}^{-1}$.) ✓

(2 marks)

27 **i** $x = vt\cos\theta$

$y = vt\sin\theta - \frac{1}{2}gt^2$

Water returns to ground level when $y = 0$

i.e. $vt\sin\theta - \frac{1}{2}gt^2 = 0$

$t(v\sin\theta - \frac{1}{2}gt) = 0$

$\therefore t = 0$

or $v\sin\theta - \frac{1}{2}gt = 0$

$v\sin\theta = \frac{1}{2}gt$

$\therefore t = \frac{2v\sin\theta}{g}$ ✓

$\therefore$ Range $= x = v\left(\frac{2v\sin\theta}{g}\right)\cos\theta$

$= \frac{2v^2\sin\theta\cos\theta}{g}$

$= \frac{v^2\sin 2\theta}{g}$

(since $\sin 2\theta = 2\sin\theta\cos\theta$)

Hence water returns to ground level at $\frac{v^2\sin 2\theta}{g}$ metres from point of projection. ✓

(2 marks)

ii $x = \frac{v^2\sin 2\theta}{g}$ (from **i**)

When $\theta = 15°, x = 40$

$\therefore 40 = \frac{v^2\sin 30°}{g}$

$40g = \frac{1}{2}v^2$

$\therefore v^2 = 80g$ ✓

(1 mark)

iii Now $x = vt\cos\theta$

$\therefore t = \frac{x}{v\cos\theta}$

and $y = vt\sin\theta - \frac{1}{2}gt^2$

$\therefore y = v\left(\frac{x}{v\cos\theta}\right)\sin\theta - \frac{1}{2}g\left(\frac{x}{v\cos\theta}\right)^2$

$= \frac{x\sin\theta}{\cos\theta} - \frac{1}{2}g\frac{x^2}{v^2\cos^2\theta}$ ✓

$= x\tan\theta - \frac{1}{2}\cdot\frac{gx^2}{80g\cos^2\theta}$

(since $v^2 = 80g$)

$= x\tan\theta - \frac{x^2\sec^2\theta}{160}$ ✓

(2 marks)

iv The water just clears the top of the wall if $x = 40$ and $y = 20$. Substituting into equation in **iii**:

$20 = 40\tan\theta - \frac{40^2\sec^2\theta}{160}$

$20 = 40\tan\theta - 10\sec^2\theta$ ✓

$20 = 40\tan\theta - 10(1 + \tan^2\theta)$

(since $\sec^2\theta = 1 + \tan^2\theta$)

$20 = 40\tan\theta - 10 - 10\tan^2\theta$

$10\tan^2\theta - 40\tan\theta + 30 = 0$

$\therefore \tan^2\theta - 4\tan\theta + 3 = 0$ ✓

(2 marks)

v Now $y = x\tan\theta - \frac{x^2\sec^2\theta}{160}$

The water hits the front of the wall if $x = 40$ and $0 \leqslant y \leqslant 20$

$\therefore 0 \leqslant 40\tan\theta - \frac{40^2\sec^2\theta}{160} \leqslant 20$

$0 \leqslant 40\tan\theta - 10\sec^2\theta \leqslant 20$

$0 \leqslant 40\tan\theta - 10(1 + \tan^2\theta) \leqslant 20$

$0 \leqslant 40\tan\theta - 10 - 10\tan^2\theta \leqslant 20$ ✓

$\therefore 10\tan^2\theta - 40\tan\theta + 10 \leqslant 0$ *AND* $10\tan^2\theta - 40\tan\theta + 10 \geqslant -20$

$\tan^2\theta - 4\tan\theta + 1 \leqslant 0$

Consider:

$\tan^2\theta - 4\tan\theta + 1 = 0$

$\tan\theta = \frac{4 \pm \sqrt{16 - 4}}{2}$

$= \frac{4 \pm \sqrt{12}}{2}$

$= \frac{4 \pm 2\sqrt{3}}{2}$

$= 2 \pm \sqrt{3}$

$10\tan^2\theta - 40\tan\theta + 30 \geqslant 0$

$\tan^2\theta - 4\tan\theta + 3 \geqslant 0$

$(\tan\theta - 3)(\tan\theta - 1) \geqslant 0$

Consider:

$(\tan\theta - 3)(\tan\theta - 1) = 0$

$\tan\theta - 3 = 0$ and $\tan\theta - 1 = 0$

$\therefore \tan\theta = 3$ and $\tan\theta = 1$

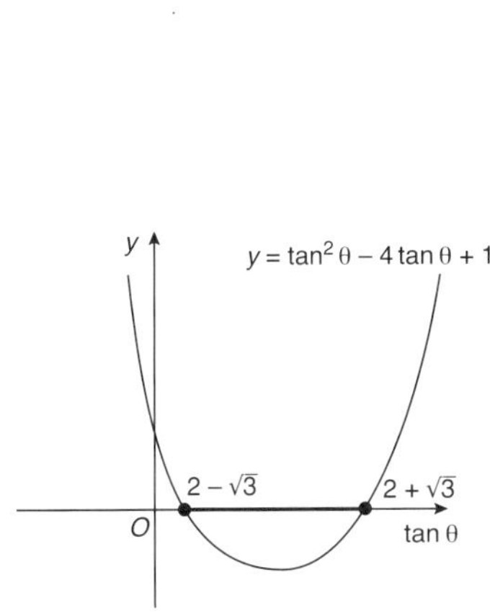

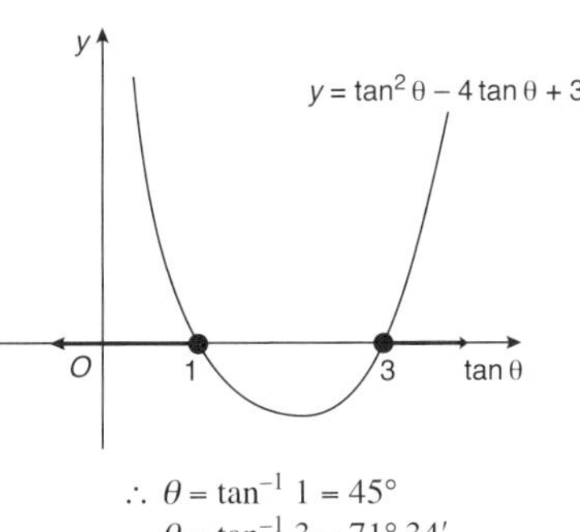

$\therefore\ \theta = \tan^{-1} 1 = 45°$
$\theta = \tan^{-1} 3 = 71° 34'$

$\therefore\ \theta = \tan^{-1}(2 \pm \sqrt{3})$
$= 15°, 75°$

$\therefore$ The water hits the front of the wall
if $15° \leqslant \theta \leqslant 75°$ *AND* $\theta \geqslant 71° 34', \theta \leqslant 45°$

$\therefore\ 15° \leqslant \theta \leqslant 45°$ *OR* $\tan^{-1} 3 \leqslant \theta \leqslant 75°$ ✓

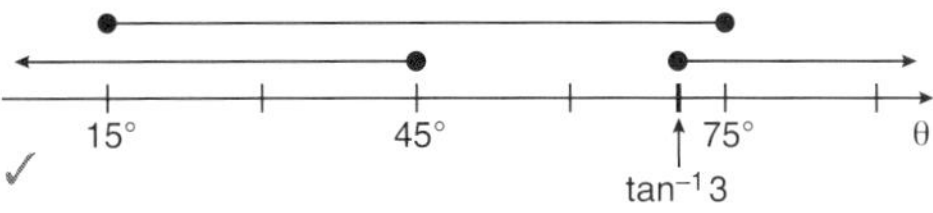

(2 marks)

28 i

$$y = vt \sin\alpha - \frac{1}{2}gt^2$$

$$\therefore\ \frac{dy}{dt} = v\sin\alpha - gt$$

The maximum height is reached when

$$\frac{dy}{dt} = 0$$

$$v\sin\alpha - gt = 0$$
$$v\sin\alpha = gt$$
$$\therefore\ t = \frac{v\sin\alpha}{g} \quad ✓$$

When $t = \dfrac{v\sin\alpha}{g}$,

$$h = v\left(\frac{v\sin\alpha}{g}\right)\sin\alpha - \frac{1}{2}g\left(\frac{v\sin\alpha}{g}\right)^2$$

$$= \frac{v^2\sin^2\alpha}{g} - \frac{v^2\sin^2\alpha}{2g}$$

$$= \frac{2v^2\sin^2\alpha}{2g} - \frac{v^2\sin^2\alpha}{2g}$$

$$= \frac{v^2\sin^2\alpha}{2g}. \quad ✓$$

(2 marks)

ii On returning to the initial height,

$$y = 0$$

$$\therefore\ vt\sin\alpha - \frac{1}{2}gt^2 = 0$$

$$t\left(v\sin\alpha - \frac{1}{2}gt\right) = 0$$

$$\therefore\ v\sin\alpha - \frac{1}{2}gt = 0 \text{ or } t = 0 \text{ (initial height)}$$

$$\therefore\ -\frac{1}{2}gt = -v\sin\alpha$$

$$\therefore\ t = \frac{2v\sin\alpha}{g} \quad ✓$$

Substituting this value of t in
$x = vt\cos\alpha$

$$\therefore\ x = v\left(\frac{2v\sin\alpha}{g}\right)\cos\alpha$$

$$\therefore\ x = \frac{2v^2\sin\alpha\cos\alpha}{g}$$

$$\therefore\ x = \frac{v^2}{g}\sin 2\alpha. \quad ✓$$

(since $2\sin\alpha\cos\alpha = \sin 2\alpha$)

(2 marks)

iii By positioning the co-ordinate axes such that the origin is at the shoulder height of the first player and the x-axis is parallel to the floor (and S metres above it), we will have the situation in parts **i** and **ii** above:

In general, the maximum separation,

$d = \dfrac{v^2}{g}\sin 2\alpha$ will occur when $\sin 2\alpha = 1$

(the maximum value of a sine function), i.e. when $2\alpha = 90°$, $\therefore\ \alpha = 45°$. This applies to all cases where $h \leqslant H - S$, otherwise the ball will hit the ceiling.

$\therefore$ When $\alpha = 45°$,

$$d = \frac{v^2}{g}\sin(2 \times 45°)$$

$$\therefore\ d = \frac{v^2}{g} \quad ✓$$

and $h = \dfrac{v^2\sin^2 45°}{2g}$

$$h = \frac{v^2\left(\frac{1}{\sqrt{2}}\right)^2}{2g}$$

$$h = \frac{v^2}{4g}$$

Now if $h \leqslant H - S$

$$\therefore \frac{v^2}{4g} \leqslant H - S$$

$$\therefore v^2 \leqslant 4g(H - S)$$ ✓

Hence, if $v^2 \leqslant 4g(H - S)$ then $d = \dfrac{v^2}{g}$.

Now if $h \geqslant H - S$, i.e. $v^2 \geqslant 4g(H - S)$, α may have other values, and the maximum separation will occur when

$$h = H - S$$

$$\therefore H - S = \frac{v^2 \sin^2 \alpha}{2g}$$

$$\therefore \sin^2 \alpha = \frac{2g(H - S)}{v^2} \quad \ldots (1)$$

$$\therefore \cos^2 \alpha = 1 - \frac{2g(H - S)}{v^2}$$

(since $\cos^2 \alpha = 1 - \sin^2 \alpha$)

$$= \frac{v^2 - 2g(H - S)}{v^2} \quad \ldots (2)$$

$$\therefore \sin \alpha = \frac{\sqrt{2g(H - S)}}{v} \quad \text{from (1)}$$

$$\text{and } \cos \alpha = \frac{\sqrt{v^2 - 2g(H - S)}}{v} \quad \text{from (2)}$$ ✓

Now the maximum separation,

$$d = \frac{v^2}{g} \sin 2\alpha$$

$$= \frac{v^2}{g} 2 \sin \alpha \cos \alpha$$

$$= \frac{2v^2}{g} \sin \alpha \cos \alpha$$

$$= \frac{2v^2}{g} \times \frac{\sqrt{2g(H - S)}}{v} \times \frac{\sqrt{v^2 - 2g(H - S)}}{v}$$

$$= \frac{2}{g} \times \frac{v^2}{v^2} \times \sqrt{2(H - S)g[v^2 - 2(H - S)g]}$$

$$= \frac{2}{g} \times \sqrt{2(H - S)gv^2 - 4(H - S)^2 g^2}$$

$$= 2 \times \sqrt{2(H - S)\left(\frac{gv^2}{g^2}\right) - 4(H - S)^2 \left(\frac{g^2}{g^2}\right)}$$

$$= 2 \times \sqrt{4(H - S)\left(\frac{v^2}{2g}\right) - 4(H - S)^2}$$

$$= 4 \times \sqrt{(H - S)\left(\frac{v^2}{2g}\right) - (H - S)^2}.$$ ✓

Hence, if $v^2 \geqslant 4g(H - S)$,

$$\boldsymbol{d = 4 \times \sqrt{(H - S)\left(\frac{v^2}{2g}\right) - (H - S)^2}}.$$

(4 marks)

29 **i** Given: $\ddot{x} = 0$ and $\ddot{y} = -10$

At $t = 0$, $x = 0$, $\dot{x} = V \cos \theta$
At $t = 0$, $y = 5$, $\dot{y} = V \sin \theta$

Now $\ddot{y} = -10$

$$\therefore \dot{y} = \int -10 \, dt$$

$$= -10t + C$$

At $t = 0$, $\dot{y} = V \sin \theta$ $\therefore C = V \sin \theta$

$$\therefore \dot{y} = -10t + V \sin \theta$$ ✓

$$\therefore y = \int (-10t + V \sin \theta) \, dt$$

$$= -5t^2 + Vt \sin \theta + C_1$$

At $t = 0, y = 5$ $\therefore C_1 = 5$

$$\therefore y = Vt \sin \theta - 5t^2 + 5 \quad \ldots (1)$$ ✓ *(2 marks)*

ii Given:

- $x = Vt \cos \theta$ (from **i**)
- At $x = 60, y = 0$
- $\tan \theta = \dfrac{3}{4}$

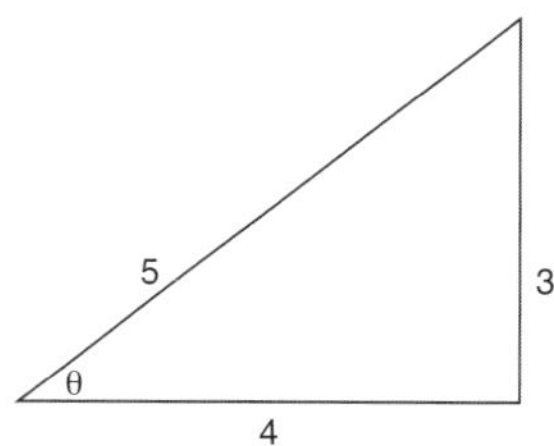

$$\therefore \sin \theta = \frac{3}{5} \text{ and } \cos \theta = \frac{4}{5}$$

To find time, t, required to reach $x = 60$:

$$\therefore 60 = Vt \cos \theta$$

$$\therefore 60 = Vt \times \frac{4}{5}$$

$$75 = Vt$$

$$t = \frac{75}{V}$$ ✓

Substituting this value of t in (1)

$$y = V\left(\frac{75}{V}\right) \sin \theta - 5\left(\frac{75}{V}\right)^2 + 5$$

$$= 75 \sin \theta - 5\left(\frac{75}{V}\right)^2 + 5$$ ✓

If $y = 0$ and $\sin \theta = \frac{3}{5}$:

$$0 = 75\left(\frac{3}{5}\right) - 5\left(\frac{75^2}{V^2}\right) + 5$$

$$0 = 45 - 5\left(\frac{75^2}{V^2}\right) + 5$$

$$5\left(\frac{75^2}{V^2}\right) = 50$$

$$\frac{75^2}{V^2} = 10$$

$$V^2 = \frac{75^2}{10}$$

$$V = \frac{75}{\sqrt{10}}.$$

$$\doteqdot 23.7$$ ✓ *(3 marks)*

iii At maximum height, $\dot{y} = 0$.

Also $V = \dfrac{75}{\sqrt{10}}$ and $\sin\theta = \dfrac{3}{5}$ (from above)

Now $\dot{y} = -10t + V\sin\theta$

$\therefore 0 = -10t + \dfrac{75}{\sqrt{10}} \times \dfrac{3}{5}$

$10t = \dfrac{15 \times 3}{\sqrt{10}}$

$t = \dfrac{45}{10\sqrt{10}}$

$= \dfrac{9}{2\sqrt{10}}$ ✓

Now $y = Vt\sin\theta - 5t^2 + 5$ (from **i**)

At $t = \dfrac{9}{2\sqrt{10}}$,

$y = \dfrac{75}{\sqrt{10}}\left(\dfrac{9}{2\sqrt{10}}\right)\left(\dfrac{3}{5}\right) - 5\left(\dfrac{9}{2\sqrt{10}}\right)^2 + 5$

$= \dfrac{81}{4} - \dfrac{81}{8} + 5$

$= \dfrac{121}{8}$

$\therefore$ The maximum height achieved is $\dfrac{121}{8}$ m ($\doteqdot$ 15.1) metres. ✓

(2 marks)

30 Now $x = Vt$ and $y = -5t^2$

$\therefore \dfrac{dx}{dt} = V$ and $\dfrac{dy}{dt} = -10t$

When the bomb hits the ground at an angle of 45° to the vertical, at $t = T$:

$\dfrac{dy}{dx} = \tan(180° - 45°) = -1$ ✓

Now $\dfrac{dy}{dx} = \dfrac{\frac{dy}{dt}}{\frac{dx}{dt}}$

$\therefore \dfrac{dy}{dx} = \dfrac{-10t}{V}$

$\therefore -1 = \dfrac{-10T}{V}$ (at $t = T$)

$\therefore V = 10T$ ...(1) ✓

Also, at $t = T$, $x = 4000$

$\therefore 4000 = VT$ (from $x = Vt$)

$\therefore T = \dfrac{4000}{V}$ ✓

Substituting this value of T in (1):

$V = \dfrac{40\,000}{V}$

$V^2 = 40\,000$

$\therefore V = 200$ m s^{-1} ✓

(4 marks)

31 i $x = Vt\cos\theta$ ...(1)

$y = Vt\sin\theta - \frac{1}{2}gt^2$ ...(2)

$\dfrac{2V^2}{g} = 1$ ...(3)

Now $t = \dfrac{x}{V\cos\theta}$ (from (1))

Substituting into (2):

$y = V\left(\dfrac{x}{V\cos\theta}\right)\sin\theta - \dfrac{1}{2}g\left(\dfrac{x}{V\cos\theta}\right)^2$

$= \dfrac{x\sin\theta}{\cos\theta} - \dfrac{gx^2}{2V^2\cos^2\theta}$

$= x\tan\theta - \dfrac{g}{2V^2} \cdot \dfrac{x^2}{\cos^2\theta}$

$= x\tan\theta - 1 \times x^2\sec^2\theta$ (using (3))

$= x\tan\theta - x^2\sec^2\theta$. ✓

ii Point T is the intersection of the path of the projectile with the plane.

(Find x-coordinage of point T, and then r using $\cos\alpha = \dfrac{x}{r}$)

Plane is represented by

$y = mx + b$

$= (\tan\alpha)x + 0$

$= x\tan\alpha$ ...(1)

Path of projectile:

$y = x\tan\theta - x^2\sec^2\theta$ ...(2)

Equating (1) and (2):

$x\tan\alpha = x\tan\theta - x^2\sec^2\theta$ ✓

$\therefore \tan\alpha = \tan\theta - x\sec^2\theta$ ($x \neq 0$)

$\therefore x\sec^2\theta = \tan\theta - \tan\alpha$

$= \dfrac{\sin\theta}{\cos\theta} - \dfrac{\sin\alpha}{\cos\alpha}$

$= \dfrac{\sin\theta\cos\alpha - \cos\theta\sin\alpha}{\cos\theta\cos\alpha}$

$= \dfrac{\sin(\theta - \alpha)}{\cos\theta\cos\alpha}$

$\therefore x = \dfrac{\sin(\theta-\alpha)}{\cos\theta\cos\alpha} \times \dfrac{1}{\sec^2\theta}$

$= \dfrac{\sin(\theta-\alpha)}{\cos\theta\cos\alpha} \times \cos^2\theta$

$= \dfrac{\sin(\theta-\alpha)\cos\theta}{\cos\alpha}$

Now $r = \dfrac{x}{\cos\alpha}$ (Since $\cos\alpha = \dfrac{x}{r}$)

$\therefore r = \dfrac{\sin(\theta-\alpha)\cos\theta}{\cos\alpha} \times \dfrac{1}{\cos\alpha}$

$= \dfrac{\sin(\theta-\alpha)\cos\theta}{\cos^2\alpha}$. ✓

iii $\dfrac{dr}{d\theta} = \dfrac{\sin(\theta-\alpha)\,.\,(-\sin\theta) + \cos\theta\,.\,\cos(\theta-\alpha))}{\cos^2\alpha}$

$= \dfrac{\cos\theta\cos(\theta-\alpha) - \sin\theta\sin(\theta-\alpha)}{\cos^2\alpha}$

$= \frac{\cos(\theta + \theta - \alpha)}{\cos^2\alpha}$

$= \frac{\cos(2\theta - \alpha)}{\cos^2\alpha}$.

Stationary points occur when

$\frac{dr}{d\theta} = 0$

$\frac{\cos(2\theta - \alpha)}{\cos^2\alpha} = 0$

$\therefore \cos(2\theta - \alpha) = 0$

$\therefore 2\theta - \alpha = \frac{\pi}{2} \quad \left(\text{or } \theta = \frac{\alpha}{2} + \frac{\pi}{4}\right)$

$\frac{d^2r}{d\theta^2} = \frac{-2\sin(2\theta - \alpha)}{\cos^2\alpha}$

< 0 when $2\theta - \alpha = \frac{\pi}{2}$

$\therefore r$ has a maximum value, R,

when $\theta = \frac{\alpha}{2} + \frac{\pi}{4}$. ✓

$\therefore R = \frac{\sin(\frac{\alpha}{2} + \frac{\pi}{4} - \alpha)\cos(\frac{\alpha}{2} + \frac{\pi}{4})}{\cos^2\alpha}$

$= \frac{\sin(\frac{\pi}{4} - \frac{\alpha}{2})\cos(\frac{\pi}{4} + \frac{\alpha}{2})}{1 - \sin^2\alpha}$

$= \frac{2\sin(\frac{\pi}{4} - \frac{\alpha}{2})\cos(\frac{\pi}{4} + \frac{\alpha}{2})}{2(1 - \sin^2\alpha)}$

$= \frac{\sin\left((\frac{\pi}{4} - \frac{\alpha}{2}) + (\frac{\pi}{4} + \frac{\alpha}{2})\right) + \sin\left((\frac{\pi}{4} - \frac{\alpha}{2}) - (\frac{\pi}{4} + \frac{\alpha}{2})\right)}{2(1 + \sin\alpha)(1 - \sin\alpha)}$

(using $2\sin A\cos B = \sin(A + B) + \sin(A - B)$)

$= \frac{\sin\frac{\pi}{2} + \sin(-\alpha)}{2(1 + \sin\alpha)(1 - \sin\alpha)}$

$= \frac{1 - \sin\alpha}{2(1 + \sin\alpha)(1 - \sin\alpha)}$

$= \frac{1}{2(1 + \sin\alpha)}$. ✓

iv The slope of the initial direction is $m = \tan\theta$. For the maximum range trajectory,

$\theta = \frac{\alpha}{2} + \frac{\pi}{4}$ (from **iii**)

$\therefore m = \tan(\frac{\alpha}{2} + \frac{\pi}{4})$

$= \frac{\tan\frac{\alpha}{2} + \tan\frac{\pi}{4}}{1 - \tan\frac{\alpha}{2}\tan\frac{\pi}{4}}$

$= \frac{1 + \tan\frac{\alpha}{2}}{1 - \tan\frac{\alpha}{2}} \quad \ldots(1)$ ✓

The slope of the direction at impact is given by $\frac{dy}{dx}$ when

$x = \frac{\sin(\theta - \alpha)\cos\theta}{\cos\alpha}$ (from **ii**)

Now $y = x\tan\theta - x^2\sec^2\theta$ (from **i**)

$\frac{dy}{dx} = \tan\theta - 2x\sec^2\theta$

$= \tan\theta - \frac{2\sin(\theta - \alpha)\cos\theta}{\cos\alpha} \cdot \frac{1}{\cos^2\theta}$

$= \tan\theta - \frac{2\sin(\theta - \alpha)}{\cos\alpha\cos\theta}$

$= \tan\theta - \frac{2\sin\theta\cos\alpha - 2\cos\theta\sin\alpha}{\cos\alpha\cos\theta}$

$= \tan\theta - \frac{2\sin\theta}{\cos\theta} + \frac{2\sin\alpha}{\cos\alpha}$

$= \tan\theta - 2\tan\theta + 2\tan\alpha$

$= 2\tan\alpha - \tan\theta$

$= \frac{4\tan\frac{\alpha}{2}}{1 - \tan^2\frac{\alpha}{2}}^{*} - \frac{1 + \tan\frac{\alpha}{2}}{1 - \tan\frac{\alpha}{2}}^{**}$

* using $\tan\alpha = \frac{2\tan\frac{\alpha}{2}}{1 - \tan^2\frac{\alpha}{2}}$

** from (1) above

$= \frac{4\tan\frac{\alpha}{2}}{1 - \tan^2\frac{\alpha}{2}} - \frac{(1 + \tan\frac{\alpha}{2})^2}{1 - \tan^2\frac{\alpha}{2}}$ ✓

$= \frac{4\tan\frac{\alpha}{2} - (1 + 2\tan\frac{\alpha}{2} + \tan^2\frac{\alpha}{2})}{1 - \tan^2\frac{\alpha}{2}}$

$= \frac{-\tan^2\frac{\alpha}{2} + 2\tan\frac{\alpha}{2} - 1}{1 - \tan^2\frac{\alpha}{2}}$

$= \frac{-(1 - 2\tan\frac{\alpha}{2} + \tan^2\frac{\alpha}{2})}{1 - \tan^2\frac{\alpha}{2}}$

$= -\frac{(1 - \tan\frac{\alpha}{2})^2}{1 - \tan^2\frac{\alpha}{2}}$

$= -\frac{1 - \tan\frac{\alpha}{2}}{1 + \tan\frac{\alpha}{2}}$

(n.b. $1 - \tan^2\frac{\alpha}{2} = (1 + \tan\frac{\alpha}{2})(1 - \tan\frac{\alpha}{2})$)

$= -\frac{1}{m}$.

$\therefore$ The initial direction is perpendicular to the direction at which the projectile hits the inclined plane. ✓

(8 marks)

(Total mark allocation only in exam)

32 **i** Given: $\ddot{x} = 0, \quad \ddot{y} = -10$

At $t = 0, \quad x = 0, \quad \dot{x} = 30\cos 5°$

At $t = 0, \quad y = 0, \quad \dot{y} = -30\sin 5°$

Horizontal motion:

$\ddot{x} = 0$

$\therefore \dot{x} = \int 0\, dt$

$= C_1$

At $t = 0, \dot{x} = 30\cos 5° \therefore C_1 = 30\cos 5°$

$\therefore \dot{x} = 30\cos 5°$

$\therefore x = \int 30\cos 5°\, dt$

$= 30t\cos 5° + C_2$

At $t = 0, x = 0 \therefore C_2 = 0$

$\therefore x = 30t\cos 5°$ ✓

Vertical motion:

$\ddot{y} = -10$

$\therefore \dot{y} = \int -10\, dt$

$= -10t + C_3$

At $t = 0, \dot{y} = -30 \sin 5° \therefore C_3 = -30 \sin 5°$

$\therefore \dot{y} = -30 \sin 5° - 10t$

$\therefore y = \int (-30 \sin 5° - 10t)\, dt$

$= -30t \sin 5° - 5t^2 + C_4$

At $t = 0, y = 0, \therefore C_4 = 0$

$\therefore y = -30t \sin 5° - 5t^2$ ✓

ii The ball strikes the ground when

$y = -2$

$\therefore -30t \sin 5° - 5t^2 = -2$

$\therefore 5t^2 + (30 \sin 5°)\, t - 2 = 0$ ✓

$$\therefore t = \frac{-30 \sin 5° \pm \sqrt{(30 \sin 5°)^2 - 4(5)(-2)}}{2 \times 5}$$

$$= \frac{-30 \sin 5° + \sqrt{900 \sin^2 5° + 40}}{10}$$ ✓

(since $t > 0$)

$= 0.4229\ ..$

$\doteqdot 0.423$ seconds. ✓

iii Let α be the acute angle at which the ball strikes the ground.

$\therefore \tan \alpha = \left|\frac{\dot{y}}{\dot{x}}\right|$ at $t \doteqdot 0.423$ seconds ✓

$$= \left|\frac{-30 \sin 5° - 10 \times 0.423}{30 \cos 5°}\right|$$

$= 0.229$ ✓

$\therefore \alpha = 13°$ to the nearest degree.

(The obtuse angle is 167°) ✓

(8 marks)

(Total mark allocation only in exam)

33 i Given: $\ddot{x} = 0, \ddot{y} = -g$

At $t = 0, x = 0, \dot{x} = V \cos 45° = \frac{\sqrt{2}}{2}V$

At $t = 0, y = 1, \dot{y} = V \sin 45° = \frac{\sqrt{2}}{2}V$

Horizontal motion:

$\ddot{x} = 0$

$\therefore \dot{x} = \int 0\, dt$

$= C$

At $t = 0, \dot{x} = \frac{\sqrt{2}}{2}V \therefore C = \frac{\sqrt{2}}{2}V$

$\therefore \dot{x} = \frac{\sqrt{2}}{2}V$

$\therefore x = \int \frac{\sqrt{2}}{2}V\, dt$

$= \frac{\sqrt{2}}{2}Vt + C_1$

At $t = 0, x = 0 \therefore C_1 = 0$

$\therefore x = \frac{\sqrt{2}}{2}Vt$

Vertical motion:

$\ddot{y} = -g$

$\therefore \dot{y} = \int -g\, dt$

$= -gt + C_2$

At $t = 0, \dot{y} = \frac{\sqrt{2}}{2}V \therefore C_2 = \frac{\sqrt{2}}{2}V$

$\therefore \dot{y} = -gt + \frac{\sqrt{2}}{2}V$

$\therefore y = \int (-gt + \frac{\sqrt{2}}{2}V)\, dt$

$= -\frac{1}{2}gt^2 + \frac{\sqrt{2}}{2}Vt + C_3$

At $t = 0, y = 1 \therefore C_3 = 1$

$\therefore y = -\frac{1}{2}gt^2 + \frac{2}{\sqrt{2}}Vt + 1$ ✓

Now $x = \frac{\sqrt{2}}{2}Vt$ (from above)

$\therefore t = \frac{x}{V} \times \frac{2}{\sqrt{2}}$

$= \frac{x}{V} \times \sqrt{2}$

$= \frac{\sqrt{2}x}{V}$ ✓

Substituting into the expression for y:

$$y = -\frac{1}{2}g\left(\frac{\sqrt{2}x}{V}\right)^2 + \frac{\sqrt{2}}{2}V\left(\frac{\sqrt{2}x}{V}\right) + 1$$

$$= -\frac{g2x^2}{2V^2} + \frac{2Vx}{2V} + 1$$

$$= -\frac{gx^2}{V^2} + x + 1$$

$$= 1 + x - \frac{gx^2}{V^2}$$

$$= 1 + x - g\frac{x^2}{V^2}$$ ✓

ii The ball passes through the point (9.3, 2.3).

Now $y = 1 + x - g\frac{x^2}{V^2}$

$\therefore 2.3 = 1 + 9.3 - g\left(\frac{9.3^2}{V^2}\right)$ ✓

$-8 = -g\left(\frac{9.3^2}{V^2}\right)$

$V^2 = \frac{9.8 \times 9.3^2}{8}$

$= \frac{847.602}{8}$

$\therefore V \doteqdot 10.3\ \text{ms}^{-1}$ ✓

iii When the ball lands, $y = 0$

$\therefore 0 = 1 + x - 9.8\frac{x^2}{10.3^2}$

$\therefore 0 = 1 + x - 9.8\frac{x^2}{106}$

$\therefore 106 + 106x - 9.8x^2 = 0$

$\therefore 4.9x^2 - 53x - 53 = 0$ ✓

$$\therefore x = \frac{53 \pm \sqrt{(-53)^2 - 4(4.9)(-53)}}{2 \times 4.9}$$

$$= \frac{53 \pm \sqrt{2809 + 1038.8}}{9.8}$$
$$= \frac{53 \pm 62}{9.8}$$
$$\doteqdot 11.7$$ ✓

(The negative value is irrelevant)

$\therefore$ Horizontal distance from the net
$= 11.7 - 9.3$
$= 2.4$ m ✓

(8 marks)
(Total mark allocation only in exam)

---

34 i Given: $\ddot{x} = 0$, $\ddot{y} = -g$

At $t = 0, x = 0, \dot{x} = V$
At $t = 0, y = h, \dot{y} = 0$

Horizontal motion:

$$\ddot{x} = 0$$
$$\therefore \dot{x} = \int 0\,dt$$
$$= C$$

At $t = 0, \dot{x} = V \quad \therefore C = V$
$\therefore \dot{x} = V$

$$\therefore x = \int V\,dt$$
$$= Vt + C_1$$

At $t = 0, x = 0, \therefore C_1 = 0$
$\therefore x = Vt$ ✓

Vertical motion:
$$\ddot{y} = -g$$
$$\therefore \dot{y} = \int -g\,dt$$
$$= -gt + C_2$$

At $t = 0, \dot{y} = 0 \quad \therefore C_2 = 0$
$\therefore \dot{y} = -gt$ ✓

$$\therefore y = \int -gt\,dt$$
$$= -\tfrac{1}{2}gt^2 + C_3$$

At $t = 0, y = h \quad \therefore C_3 = h$
$\therefore y = h - \tfrac{1}{2}gt^2$ ✓

ii Here $V = 216$ km/h
$$= \frac{216 \times 1000}{60 \times 60}\text{ m/s}$$
$$= 60\text{ m/s}$$ ✓

$h = 120$ m
$g = 10\text{ m/s}^2$

The canister hits the water when
$y = 0$
i.e. $h - \tfrac{1}{2}gt^2 = 0$ ✓
$\therefore 120 - 5t^2 = 0$
$t^2 = 24$
$t = \sqrt{24}$ (since $t > 0$)
$\doteqdot 4.9$ s ✓

iii In 4.9 s, the canister will travel in the horizontal direction:

$x = Vt$
$= 60 \times 4.9$
$= 294$ m ✓

Now the sailor is drifting at
$$3.6\text{ km/h} = \frac{3600\text{ m}}{60 \times 60\text{ s}} = 1\text{ m/s}$$

$\therefore$ In 4.9 s, she will drift 4.9 m and be at a distance from O of $(D + 4.9)$ m ✓

$\therefore$ To land at most 50 m from the sailor,
$294 - 50 \leqslant D + 4.9 \leqslant 294 + 50$
$244 \leqslant D + 4.9 \leqslant 344$
$239.1 \leqslant D \leqslant 339.1$ ✓

(9 marks)
(Total mark allocation only in exam)

---

35 a Given: $\ddot{x} = 0$, $\ddot{y} = -g$

At $t = 0$, $x = 0$, $\dot{x} = V\cos\alpha$
At $t = 0$, $y = h$, $\dot{y} = V\sin\alpha$

Horizontal motion:

$\ddot{x} = 0$
$\therefore \dot{x} = C_1$
At $t = 0, \dot{x} = V\cos\alpha \therefore C_1 = V\cos\alpha$
$\therefore \dot{x} = V\cos\alpha$ ✓
$$\therefore x = \int V\cos\alpha\,dt$$
$$= Vt\cos\alpha + C_2$$
At $t = 0, x = 0$
$\therefore 0 = 0 + C_2$
$\therefore C_2 = 0$
$\therefore x = Vt\cos\alpha$ ✓

Vertical motion:

$\ddot{y} = -g$
$$\therefore \dot{y} = \int -g\,dt$$
$$= -gt + C_3$$
At $t = 0, \dot{y} = V\sin\alpha \therefore C_3 = V\sin\alpha$
$\therefore \dot{y} = -gt + V\sin\alpha$
$= V\sin\alpha - gt$ ✓
$$\therefore y = \int (V\sin\alpha - gt)\,dt$$
$$= Vt\sin\alpha - \tfrac{1}{2}gt^2 + C_4$$
At $t = 0, y = h$
$\therefore h = 0 - 0 + C_4$
$\therefore C_4 = h$
$\therefore y = Vt\sin\alpha - \tfrac{1}{2}gt^2 + h$ ✓

(4 marks)

b Now $x = Vt\cos\alpha$ (from above)
$$\therefore t = \frac{x}{V\cos\alpha}$$
$$\therefore y = V\left(\frac{x}{V\cos\alpha}\right)\sin\alpha - \frac{1}{2}g\left(\frac{x}{V\cos\alpha}\right)^2 + h$$
$$= \frac{x\,V\sin\alpha}{V\cos\alpha} - \frac{gx^2}{2\,V^2\cos^2\alpha} + h$$
$$= x\tan\alpha - \frac{gx^2}{2\,V^2\cos^2\alpha} + h$$
$$= h + x\tan\alpha - x^2\frac{g}{2\,V^2\cos^2\alpha}.$$ ✓

(1 mark)

c The ball clears the fence if when $x = R, y \geqslant h$.

$$\therefore h + R\tan\alpha - \frac{R^2g}{2V^2\cos^2\alpha} \geqslant h$$
$$R\tan\alpha - \frac{gR^2}{2V^2\cos^2\alpha} \geqslant 0$$
$$\tan\alpha - \frac{gR}{2V^2\cos^2\alpha} \geqslant 0$$
$$\tan\alpha \geqslant \frac{gR}{2V^2\cos^2\alpha} \quad ✓$$
$$V^2\tan\alpha \geqslant \frac{gR}{2\cos^2\alpha}$$
$$V^2 \geqslant \frac{gR}{2\tan\alpha\cos^2\alpha}$$
(since $\tan\alpha > 0$)
$$V^2 \geqslant \frac{gR}{2.\frac{\sin\alpha}{\cos\alpha}.\cos^2\alpha}$$
$$V^2 \geqslant \frac{gR}{2\sin\alpha\cos\alpha}. \quad ✓$$

(2 marks)

d If the ball clears the fence and hits the cap C:

$y = 0$ at $x = R + r$ and $V^2 \geqslant \dfrac{gR}{2\sin\alpha\cos\alpha}$

Substituting into the equation in (b):

$$0 = h + (R + r)\tan\alpha - (R + r)^2\frac{g}{2V^2\cos^2\alpha}$$
$$\therefore 0 = \frac{h}{(R + r)^2} + \frac{\tan\alpha}{(R + r)} - \frac{g}{2V^2\cos^2\alpha} \quad ✓$$
$$\frac{g}{2V^2\cos^2\alpha} = \frac{h}{(R + r)^2} + \frac{\tan\alpha}{(R + r)}$$
$$\therefore \frac{g}{2\cos^2\alpha} = \left[\frac{h}{(R + r)^2} + \frac{\tan\alpha}{(R + r)}\right]V^2 \quad *$$
$$\text{But } V^2 \geqslant \frac{gR}{2\sin\alpha\cos\alpha} \quad \text{(From (c))}$$
$$\therefore \frac{g}{2\cos^2\alpha} \geqslant \left[\frac{h}{(R + r)^2} + \frac{\tan\alpha}{(R + r)}\right]\frac{gR}{2\sin\alpha\cos\alpha}$$
$$\therefore \frac{\sin\alpha\cos\alpha}{\cos^2\alpha} \geqslant \left[\frac{h}{(R + r)^2} + \frac{\tan\alpha}{(R + r)}\right]R$$
$$\therefore \frac{\sin\alpha\cos\alpha}{\cos^2\alpha} \geqslant \frac{Rh}{(R + r)^2} + \frac{R\tan\alpha}{R + r}$$
$$\therefore \tan\alpha \geqslant \frac{Rh}{(R + r)^2} + \frac{R\tan\alpha}{R + r} \quad ✓$$
$$\therefore (R + r)^2\tan\alpha \geqslant Rh + (R + r)R\tan\alpha$$
$$(R + r)^2\tan\alpha - (R + r)R\tan\alpha \geqslant Rh$$
$$\tan\alpha\,[(R + r)^2 - (R + r)R] \geqslant Rh$$
$$\tan\alpha\,[R^2 + 2Rr + r^2 - R^2 - Rr] \geqslant Rh$$
$$\tan\alpha(Rr + r^2) \geqslant Rh$$
$$\tan\alpha \geqslant \frac{Rh}{Rr + r^2}$$
$$\tan\alpha \geqslant \frac{Rh}{(R + r)r} \quad ✓$$

(3 marks)

e If $V \leqslant 50$

then $V^2 \leqslant 2500$

$$\text{Now } V^2 \geqslant \frac{gR}{2\sin\alpha\cos\alpha} \quad \text{(from (c))}$$
$$\therefore \frac{gR}{2\sin\alpha\cos\alpha} \leqslant V^2 \leqslant 2500$$
$$\therefore \frac{gR}{2\sin\alpha\cos\alpha} \leqslant 2500$$
$$\therefore 2500 \geqslant \frac{gR}{2\sin\alpha\cos\alpha}$$
$$\therefore 2\sin\alpha\cos\alpha \geqslant \frac{gR}{2500}$$
(since $2\sin\alpha\cos\alpha > 0$)

Now $g = 10$ and $R = 80$,

$$\therefore 2\sin\alpha\cos\alpha \geqslant \frac{10 \times 80}{2500}$$
$$\sin 2\alpha \geqslant \frac{8}{25} \quad (0° \leqslant 2\alpha \leqslant 180°)$$
$$\therefore 18.66° \leqslant 2\alpha \leqslant 161.34°$$
$$9.33° \leqslant \alpha \leqslant 80.67°. \quad ✓$$

Now the ball will land closest to the fence where α is a maximum:

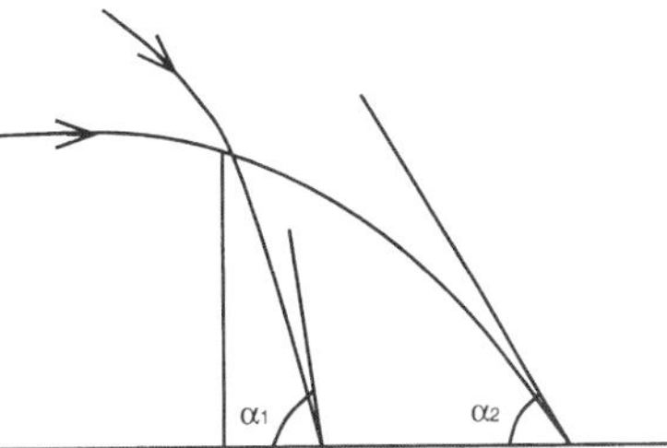

The ball will land closest to the fence where α is maximum, i.e. $\alpha \doteqdot 80.67°$.

$$\text{Now } \tan\alpha \geqslant \frac{Rh}{(R + r)r} \quad \text{(from (d))}$$
$R = 80, h = 1$ and $\alpha \doteqdot 80.67°$
$$\therefore \tan 80.67° \geqslant \frac{80(1)}{(80 + r)r}$$
$$6.09 \geqslant \frac{80}{(80 + r)r}$$
$$(80 + r)r \geqslant \frac{80}{6.09}$$
$$80r + r^2 \geqslant 13.14$$
$$r^2 + 80r - 13.14 \geqslant 0$$

The minimum **positive** value of r is

$$r = \frac{-80 + \sqrt{80^2 - 4(1)(-13.14)}}{2(1)}$$
$$= \frac{-80 + \sqrt{6452.56}}{2}$$
$$\doteqdot 0.16 \text{ m}$$

$\therefore$ The closest point to the fence where the ball can land is $80 + 0.16 = 80.16$ m from the point O. ✓

(2 marks)

36 i

$$x = Vt\cos\theta \quad \ldots(1)$$
$$y = -\tfrac{1}{2}gt^2 + VT\sin\theta \quad \ldots(2)$$
$$\left(\theta \neq \frac{\pi}{2}\right)$$

From (1),

$$t = \frac{x}{V\cos\theta}$$

Substituting this value of t into (2):

$$y = -\tfrac{1}{2}g\left(\frac{x}{V\cos\theta}\right)^2 + V\left(\frac{x}{V\cos\theta}\right)\sin$$
$$= -\tfrac{1}{2}g\left(\frac{x^2}{V^2\cos^2\theta}\right) + \frac{Vx\sin\theta}{V\cos\theta}$$

$$= -\frac{gx^2}{2V^2\cos\theta^2} + x\tan\theta$$
$$= -\frac{gx^2\sec^2\theta}{2V^2} + x\tan\theta$$
$$= -\frac{2gx^2\sec^2\theta}{4V^2} + x\tan\theta$$
$$= -\frac{x^2\sec^2\theta}{4h} + x\tan\theta$$
$$\left(\text{where } h = \frac{V^2}{2g}\right)$$
$$= x\tan\theta - \frac{1}{4h}x^2(1 + \tan^2\theta)$$
(since $\sec^2\theta = 1 + \tan^2\theta$).

ii

$$y = x\tan\theta - \frac{1}{4h}x^2(1 + \tan^2\theta)$$
$\therefore 4hy = 4hx\tan\theta - x^2(1 + \tan^2\theta)$
$\therefore 4hy = 4hx\tan\theta - x^2 - x^2\tan^2\theta$
$\therefore x^2\tan^2\theta - 4hx\tan\theta + x^2 + 4hy = 0$

Substituting (X, Y) into the above:

$X^2\tan^2\theta - 4hX\tan\theta + X^2 + 4hY = 0$

This quadratic equation in $\tan\theta$ has two distinct roots if $\Delta > 0$.
i.e. $(-4hX)^2 - 4(X^2)(X^2 + 4hY) > 0$
$16h^2X^2 - 4X^4 - 16hYX^2 > 0$
$4h^2 - X^2 - 4hY > 0$
(since $X^2 > 0$)
$4h^2 - 4hY > X^2$
$4h(h - Y) > X^2$
$X^2 < 4h(h - Y)$

$\therefore$ If $X_2 < 4h(h - Y)$, there are two solutions, $\tan\theta_1$ and $\tan\theta_2$, for the equation; i.e., two different angles, θ_1 and θ_2, can be used to hit the point (X, Y).

iii

Let $\tan\theta_1$ and $\tan\theta_2$ be the solutions of the quadratic equation:

$X^2\tan^2\theta - 4hX\tan\theta + (X^2 + 4hY) = 0$ (from (ii))
where X and Y are given fixed values with $Y > 0$.

$\therefore \tan\theta_1\tan\theta_2 = \dfrac{X^2 + 4hY}{X^2}$ (product of the roots)

$\therefore \tan\theta_1\tan\theta_2 = 1 + \dfrac{4hY}{X^2}$

$\therefore \tan\theta_1\tan\theta_2 > 1$ (since $Y > 0$)

But if both $0 < \theta_1 < \frac{\pi}{4}$ and $0 < \theta_2 < \frac{\pi}{4}$, then $0 < \tan\theta_1 < 1$ and $0 < \tan\theta_2 < 1$ and so $\tan\theta_1\tan\theta_2 < 1$.

$\therefore$ No point with $Y > 0$ can be hit from two different angles θ_1 and θ_2 satisfying $\theta_1 < \frac{\pi}{4}$ and $\theta_2 < \frac{\pi}{4}$.

(No mark allocation in exam)

37 **i** The projectile will come back to the x-axis at $y = 0$:

$$-\tfrac{1}{2}gt^2 + Vt\sin\alpha = 0$$
$$t(-\tfrac{1}{2}gt + V\sin\alpha) = 0$$
$$\therefore t = 0$$
$$\text{or } -\tfrac{1}{2}gt + V\sin\alpha = 0$$
$$\therefore t = \frac{2V\sin\alpha}{g}$$

$\therefore$ Particle is above x-axis for $\dfrac{2V\sin\alpha}{g}$ seconds.

ii The horizontal range is the value of x for the value of t found in (i).

$$x = V\left(\frac{2V\sin\alpha}{g}\right)\cos\alpha$$
$$= \frac{2V^2\sin\alpha\cos\alpha}{g}$$

$\therefore$ Horizontal range is $\dfrac{2V^2\sin\alpha\cos\alpha}{g}$ metres

iii Target T:

$$\frac{dx}{dt} = u, \text{ where } u \text{ is a constant}$$
$$\therefore x = \int u\,dt$$
$$= ut + C$$

At $t = 0$, $x = d$

$$\therefore d = 0 + C$$
$$\therefore C = d$$
$$\therefore x = ut + d \qquad \ldots(1)$$

If projectile hits target moving along x-axis, it will be for the values of t and x found in (i) and (ii):

$$t = \frac{2V\sin\alpha}{g}$$
$$x = \frac{2V^2\sin\alpha\cos\alpha}{g}$$

Substituting into (1):

$$\frac{2V^2\sin\alpha\cos\alpha}{g} = u \times \frac{2V\sin\alpha}{g} + d$$
$$2V^2\sin\alpha\cos\alpha = 2uV\sin\alpha + gd$$
$$2uV\sin\alpha = 2V^2\sin\alpha\cos\alpha - gd$$
$$u = \frac{2V^2\sin\alpha\cos\alpha}{2V\sin\alpha} - \frac{gd}{2V\sin\alpha}$$
$$= V\cos\alpha - \frac{gd}{2V\sin\alpha}.$$

(No mark allocation in exam)

1 Express $\sqrt{3}\sin(x) - 3\cos(x)$ in the form $R\sin(x + \alpha)$. *(3 marks)*

(Q11e, **2022 HSC**) Easy

2 Which curve best represents the graph of the function $f(x) = -a\sin x + b\cos x$ given that the constants a and b are both positive?

A

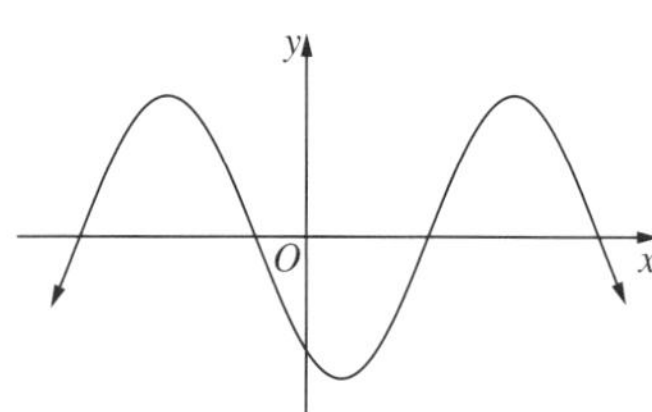

B

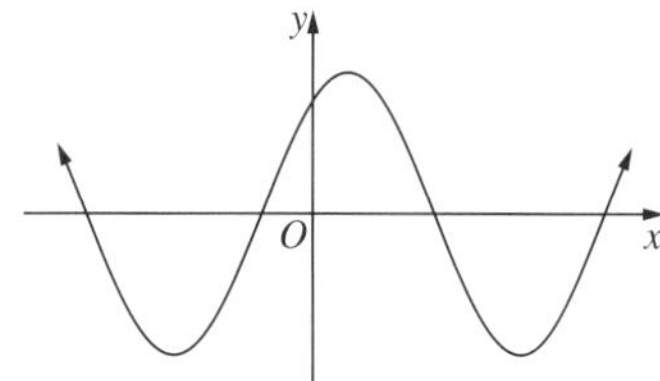

C

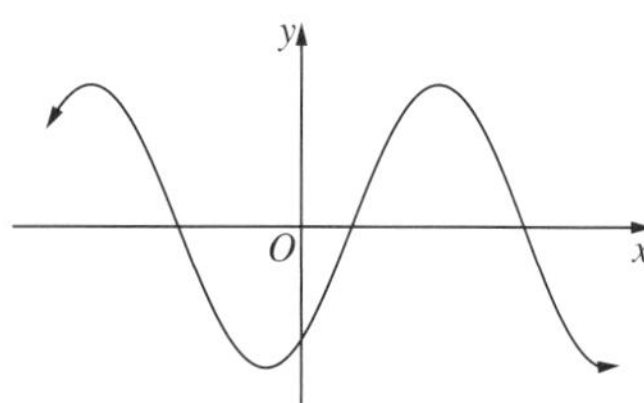

D

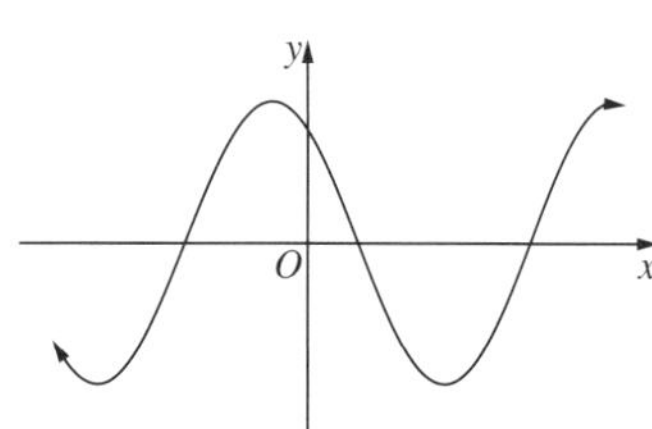

(1 mark)

(Q7, **2021 HSC**) Medium

3 By factorising, or otherwise, solve $2\sin^3 x + 2\sin^2 x - \sin x - 1 = 0$ for $0 \le x \le 2\pi$. *(3 marks)*

(Q11g, **2021 HSC**) Medium

4 **i** The numbers A, B and C are related by the equations $A = B - d$ and $C = B + d$, where d is a constant.

Show that $\dfrac{\sin A + \sin C}{\cos A + \cos C} = \tan B$. *(2 marks)* Medium

ii Hence, or otherwise, solve

$$\frac{\sin\frac{5\theta}{7} + \sin\frac{6\theta}{7}}{\cos\frac{5\theta}{7} + \cos\frac{6\theta}{7}} = \sqrt{3}, \text{ for } 0 \le \theta \le 2\pi.$$

(2 marks) Medium

(Q13d, **2021 HSC**)

5 Solve $\sin 5x = \sin 3x$, where $0 \le x \le \frac{\pi}{2}$. *(2 marks)*

Bonus question (see page iv) Medium

6 Show $\cos 3A \cos A = \cos^2 A - \sin^2 2A$. *(2 marks)*

Bonus question Easy

7 Solve $\sin \pi t + \sin 3\pi t = \cos \pi t$, where $0 \le t \le 1$. *(3 marks)*

Bonus question Medium

8 Show $\dfrac{\cos 3x + \cos x}{\cos 3x - \cos x} = -\cot 2x \cot x$. *(2 marks)*

Bonus question Medium

9 Solve $\cos 5\theta + \cos 3\theta + \cos\theta = 0$, where $0 \le \theta \le \pi$. *(3 marks)*

Bonus question Medium

10 By expressing $\sqrt{3}\sin x + 3\cos x$ in the form $A\sin(x + \alpha)$, solve $\sqrt{3}\sin x + 3\cos x = \sqrt{3}$, for $0 \le x \le 2\pi$. *(4 marks)*

(Q11d, **2020 HSC**) Easy

11 **i** Show that $\sin^3\theta - \frac{3}{4}\sin\theta + \frac{\sin(3\theta)}{4} = 0$.

(2 marks) Easy

ii By letting $x = 4\sin\theta$ in the cubic equation $x^3 - 12x + 8 = 0$. Show that $\sin(3\theta) = \frac{1}{2}$.

(2 marks) Medium

iii Prove that $\sin^2\frac{\pi}{18} + \sin^2\frac{5\pi}{18} + \sin^2\frac{25\pi}{18} = \frac{3}{2}$.

(3 marks) Hard

(Q14b, **2020 HSC**)

12 It is given that $\sin x = \frac{1}{4}$, where $\frac{\pi}{2} < x < \pi$.

What is the value of $\sin 2x$?

A $-\frac{7}{8}$

B $-\frac{\sqrt{15}}{8}$

C $\frac{\sqrt{15}}{8}$

D $\frac{7}{8}$

(1 mark)

(Q6, **2019 HSC**) Medium

13 The diagram shows the two curves $y = \sin x$ and $y = \sin(x - a) + k$, where $0 < \alpha < \pi$ and $k > 0$. The two curves have a common tangent at x_0 where $0 < x_0 < \frac{\pi}{2}$.

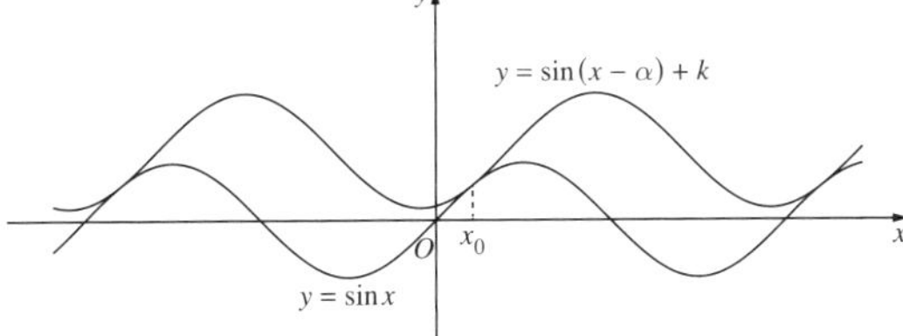

i Explain why $\cos x_0 = \cos(x_0 - \alpha)$. *(1 mark)* Medium

ii Show that $\sin x_0 = -\sin(x_0 - \alpha)$. *(2 marks)* Hard

iii Hence, or otherwise, find k in terms of α. *(2 marks)* Hard

(Q14c, **2019 HSC**)

14 Write $\sqrt{3}\sin x + \cos x$ in the form $R\sin(x + \alpha)$ where $R > 0$ and $0 \le \alpha \le \frac{\pi}{2}$. *(2 marks)*

(Q11c, **2018 HSC**) Easy

15 What is the value of $\tan\alpha$ when the expression $2\sin x - \cos x$ is written in the form $\sqrt{5}\sin(x - \alpha)$?

A -2 **B** $-\frac{1}{2}$

C $\frac{1}{2}$ **D** 2 *(1 mark)*

(Q4, **2017 HSC**) Easy

16 The displacement x of a particle at time t is given by

$$x = 5\sin 4t + 12\cos 4t.$$

What is the maximum velocity of the particle?

A 13 **B** 28

C 52 **D** 68 *(1 mark)*

(Q7, **2016 HSC**) Medium

17 The graph of the function $y = \cos\left(2t - \frac{\pi}{3}\right)$ is shown below.

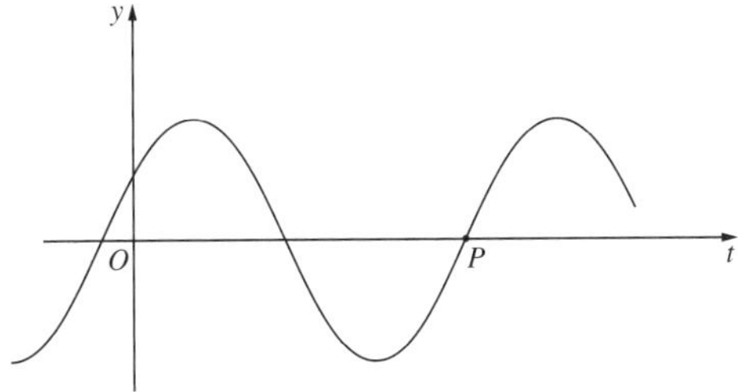

What are the coordinates of the point P?

A $\left(\frac{5\pi}{12}, 0\right)$ **B** $\left(\frac{2\pi}{3}, 0\right)$

C $\left(\frac{11\pi}{12}, 0\right)$ **D** $\left(\frac{7\pi}{6}, 0\right)$ *(1 mark)*

(Q10, **2015 HSC**) Medium

18 Express $5\cos x - 12\sin x$ in the form $A\cos(x + \alpha)$, where $0 \le \alpha \le \frac{\pi}{2}$. *(2 marks)*

(Q11d, **2015 HSC**) Easy

19 Which expression is equal to $\cos x - \sin x$?

A $\sqrt{2}\cos\left(x + \frac{\pi}{4}\right)$ **B** $\sqrt{2}\cos\left(x - \frac{\pi}{4}\right)$

C $2\cos\left(x + \frac{\pi}{4}\right)$ **D** $2\cos\left(x - \frac{\pi}{4}\right)$ *(1 mark)*

(Q2, **2014 HSC**) Easy

20 i Write $\sqrt{3}\cos x - \sin x$ in the form $2\cos(x + \alpha)$ where $0 < \alpha < \frac{\pi}{2}$. *(1 mark)* Easy

ii Hence, or otherwise, solve $\sqrt{3}\cos x = 1 + \sin x$, where $0 < x < 2\pi$. *(2 marks)* Easy

(Q12a, **2013 HSC**)

21 i Express $2\cos\theta + 2\cos\left(\theta + \frac{\pi}{3}\right)$ in the form $R\cos(\theta + \alpha)$, where $R > 0$ and $0 < \alpha < \frac{\pi}{2}$. *(3 marks)* Medium

ii Hence, or otherwise, solve $2\cos\theta + 2\cos\left(\theta + \frac{\pi}{3}\right) = 3$, for $0 < \theta < 2\pi$. *(2 marks)* Medium

(Q4b, **2010 HSC**)

22 i Express $3\sin x + 4\cos x$ in the form $A\sin(x + \alpha)$ where $0 \le \alpha \le \frac{\pi}{2}$. *(2 marks)* Easy

ii Hence, or otherwise, solve $3\sin x + 4\cos x = 5$ for $0 \le x \le 2\pi$. Give your answer, or answers, correct to two decimal places. *(2 marks)* Easy

(Q2b, **2009 HSC**)

23 It can be shown that $\sin 3\theta = 3\sin\theta - 4\sin^3\theta$ for all values of θ. (Do NOT prove this.)

Use this result to solve $\sin 3\theta + \sin 2\theta = \sin\theta$ for $0 \le \theta \le 2\pi$. *(3 marks)*

(Q6b, **2008 HSC**) Medium

24 By using the substitution $t = \tan\frac{\theta}{2}$, or otherwise, show that $\frac{1-\cos\theta}{\sin\theta} = \tan\frac{\theta}{2}$. *(2 marks)*

(Q2a, **2007 HSC**) Medium

---

25 i Write $8\cos x + 6\sin x$ in the form $A\cos(x-\alpha)$, where $A > 0$ and $0 \le \alpha \le \frac{\pi}{2}$. *(2 marks)* Easy

ii Hence, or otherwise, solve the equation $8\cos x + 6\sin x = 5$ for $0 \le x \le 2\pi$. Give your answers correct to three decimal places. *(2 marks)* Medium

(Q2d, **2004 HSC**)

---

26 i Express $\cos x - \sin x$ in the form $R\cos(x+\alpha)$, where α is in radians. *(2 marks)* Easy

ii Hence, or otherwise, sketch the graph of $y = \cos x - \sin x$ for $0 \le x \le 2\pi$. *(2 marks)* Easy

(Q2e, **2003 HSC**)

---

27 i Starting from the identity $\sin(\theta + 2\theta) = \sin\theta\cos 2\theta + \cos\theta\sin 2\theta$, and using the double angle formulae, prove the identity $\sin 3\theta = 3\sin\theta - 4\sin^3\theta$. *(2 marks)* Easy

ii Hence solve the equation $\sin 3\theta = 2\sin\theta$ for $0 \le \theta \le 2\pi$. *(3 marks)* Easy

(Q3c, **2001 HSC**)

---

28 Solve the equation $\cos 2\theta = \sin\theta$, $0 \le \theta \le 2\pi$. *(4 marks)*

(Q2c, **2000 HSC**) Easy

---

29 Find all values of θ in the range $0 \le \theta \le 2\pi$ for which $\cos\theta + \sqrt{3}\sin\theta = 1$. *(4 marks)*

(Q2b, **1999 HSC**) Medium

---

30 Find all angles θ, where $0 \le \theta \le 2\pi$, for which $\sqrt{3}\sin\theta - \cos\theta = 1$.

(Q5a, **1994 HSC**) Medium

---

31 i Sketch carefully on the same set of axes the graphs of $y = x^4$ and $y = \cos(\pi x)$ for $0 \le x \le 1{\cdot}5$. (Your diagram should be at least half a page in size.) Medium

ii On the same diagram, sketch the graph of $y = x^4 + \cos(\pi x)$. Label clearly the three curves on your diagram. Medium

iii Using the graph, determine the number of positive real roots of the equation $x^4 + \cos(\pi x) = 0$. Medium

(Q7a, **1994 HSC**)

---

32 Solve the equation $2\sin^2\theta = \sin 2\theta$ for $0 \le \theta \le 2\pi$.

(Q2a, **1992 HSC**) Medium

Year 12 Trigonometric equations—Worked Answers

1 $\sqrt{3}\sin(x) - 3\cos(x)$

$\equiv R\sin(x+\alpha)$

$R = \sqrt{(\sqrt{3})^2 + (3)^2}$

$= \sqrt{12}$

$= 2\sqrt{3}$ ✓

$\tan\alpha = \dfrac{-3}{\sqrt{3}}$

$\alpha = \tan^{-1}\left(\dfrac{-3}{\sqrt{3}}\right)$ ✓

$= -\dfrac{\pi}{3}\left(0 < |\alpha| < \dfrac{\pi}{2}\right)$

$\sqrt{3}\sin(x) - 3\cos(x)$

$= 2\sqrt{3}\sin\left(x - \dfrac{\pi}{3}\right)$ ✓

(3 marks)

2 $f(x) = -a\sin x + b\cos x$

$f(0) = -a\sin(0) + b\cos(0)$

$\therefore f(0) = b$ where $b > 0$

$\therefore f(x)$ has a positive y-intercept.

$f'(x) = -a\cos x - b\sin x$

$f'(0) = -a\cos(0) - b\sin(0)$

$f'(0) = -a$

$\therefore$ Graph of $f(x)$ is decreasing at $x = 0$ and has a positive y-intercept.

Answer D

3 Factorising $2\sin^3 x + 2\sin^2 x - \sin x - 1 = 0$ by grouping in pairs:

$2\sin^2 x(\sin x + 1) - 1(\sin x + 1) = 0$

$(2\sin^2 x - 1)(\sin x + 1) = 0$

$2\sin^2 x - 1 = 0$ and $\sin x + 1 = 0$

$\sin x = \pm\dfrac{1}{\sqrt{2}}$ ✓ and $\sin x = -1$ ✓

$\therefore x = \dfrac{\pi}{4}, \dfrac{3\pi}{4}, \dfrac{5\pi}{4}, \dfrac{3\pi}{2}, \dfrac{7\pi}{4}$ ✓ *(3 marks)*

4 **i** As $A = B - d$ and $C = B + d$ then,

$\dfrac{\sin A + \sin C}{\cos A + \cos C}$

$= \dfrac{\sin(B-d) + \sin(B+d)}{\cos(B-d) + \cos(B+d)}$

$= \dfrac{2\sin B\cos d}{2\cos B\cos d}$ ✓

(from the reference sheet)

$= \dfrac{\sin B}{\cos B}$

$= \tan B$ ✓ *(2 marks)*

ii Let $A = \dfrac{5\theta}{7}$ and $C = \dfrac{6\theta}{7}$.

Solving $\dfrac{5\theta}{7} = B - d$ and $\dfrac{6\theta}{7} = B + d$ simultaneously:

$\dfrac{5\theta}{7} + \dfrac{6\theta}{7} = 2B$

$B = \dfrac{11\theta}{14}$

Since $\dfrac{\sin A + \sin C}{\cos A + \cos C} = \tan B$, then

$\dfrac{\sin\dfrac{5\theta}{7} + \sin\dfrac{6\theta}{7}}{\cos\dfrac{5\theta}{7} + \cos\dfrac{6\theta}{7}} = \tan\dfrac{11\theta}{14}$ ✓

$\therefore \tan\dfrac{11\theta}{14} = \sqrt{3}$

$\dfrac{11\theta}{14} = \dfrac{\pi}{3}, \dfrac{4\pi}{3}$

$\therefore \theta = \dfrac{14\pi}{33}, \dfrac{56\pi}{33}$ ✓

(2 marks)

5 $\sin 5x = \sin 3x$

$\sin 5x - \sin 3x = 0$

$2\cos 4x\sin x = 0$ ✓

$\cos 4x = 0$ $\quad \sin x = 0$

$4x = \dfrac{\pi}{2}, \dfrac{3\pi}{2}$ $\quad x = 0$

$x = \dfrac{\pi}{8}, \dfrac{3\pi}{8}$

$\therefore x = 0, \dfrac{\pi}{8}, \dfrac{3\pi}{8}$ ✓ *(2 marks)*

6 LHS $= \cos 3A\cos A$

$= \dfrac{1}{2}[\cos 2A + \cos 4A]$ ✓

$= \dfrac{1}{2}[2\cos^2 A - 1 + 1 - 2\sin^2 2A]$

$= \dfrac{1}{2}[2\cos^2 A - 2\sin^2 2A]$

$= \cos^2 A - \sin^2 2A$

$=$ RHS ✓ *(2 marks)*

7 $\sin\pi t + \sin 3\pi t = \cos\pi t$

$2(\sin 2\pi t\cos(-\pi t)) = \cos\pi t$

$2(\sin 2\pi t\cos\pi t) = \cos\pi t$

$2\sin 2\pi t\cos\pi t - \cos\pi t = 0$ ✓

$\cos\pi t(2\sin 2\pi t - 1) = 0$

$\cos\pi t = 0 \quad 2\sin 2\pi t - 1 = 0$

$\cos\pi t = 0 \quad \sin 2\pi t = \dfrac{1}{2}$ ✓

$\pi t = \dfrac{\pi}{2} \quad 2\pi t = \dfrac{\pi}{6}, \dfrac{5\pi}{6}$

$t = \dfrac{1}{2} \quad t = \dfrac{1}{12}, \dfrac{5}{12}$

$\therefore t = \dfrac{1}{12}, \dfrac{5}{12}, \dfrac{1}{2}$ ✓ *(3 marks)*

8 LHS $= \dfrac{\cos 3x + \cos x}{\cos 3x - \cos x}$

$= \dfrac{2\cos 2x\cos x}{-2\sin 2x\sin x}$ ✓

$= \dfrac{\cos 2x\cos x}{-\sin 2x\sin x}$

$= -\cot 2x\cot x$

$=$ RHS ✓ *(2 marks)*

9 $\cos 5\theta + \cos 3\theta + \cos\theta = 0$

$2\cos 4\theta\cos\theta + \cos\theta = 0$ ✓

$\cos\theta(2\cos 4\theta + 1) = 0$

$\cos\theta = 0 \quad \cos 4\theta = -\dfrac{1}{2}$ ✓

$\theta = \dfrac{\pi}{2}$

$4\theta = \dfrac{2\pi}{3}, \dfrac{4\pi}{3}, \dfrac{8\pi}{3}, \dfrac{10\pi}{3}$

$\theta = \dfrac{\pi}{6}, \dfrac{\pi}{3}, \dfrac{2\pi}{3}, \dfrac{5\pi}{6}$

$\therefore x = \dfrac{\pi}{6}, \dfrac{\pi}{3}, \dfrac{\pi}{2}, \dfrac{2\pi}{3}, \dfrac{5\pi}{6}$ ✓ *(3 marks)*

10 $A\sin(x+\alpha)$

$= A\sin x\cos\alpha + A\cos x\sin\alpha$

Now $\sqrt{3}\sin x + 3\cos x \equiv A\sin(x+\alpha)$

$\therefore A\cos\alpha = \sqrt{3}$ and $A\sin\alpha = 3$

$\dfrac{A\sin\alpha}{A\cos\alpha} = \dfrac{3}{\sqrt{3}}$

$\tan\alpha = \sqrt{3}$

$\alpha = \dfrac{\pi}{3} \quad \left(0 < \alpha < \dfrac{\pi}{2}\right)$ ✓

$A\cos\dfrac{\pi}{3} = \sqrt{3}$

$A \times \dfrac{1}{2} = \sqrt{3}$

$A = 2\sqrt{3}$ ✓

So $\sqrt{3}\sin x + 3\cos x$

$\equiv 2\sqrt{3}\sin\left(x + \dfrac{\pi}{3}\right)$.

Now $0 \le x \le 2\pi$

$\dfrac{\pi}{3} \le x + \dfrac{\pi}{3} \le \dfrac{7\pi}{3}$

If $\sqrt{3}\sin x + 3\cos x = \sqrt{3}$

$2\sqrt{3}\sin\left(x + \dfrac{\pi}{3}\right) = \sqrt{3}$

$\sin\left(x + \dfrac{\pi}{3}\right) = \dfrac{1}{2}$ ✓

So $x + \dfrac{\pi}{3} = \dfrac{5\pi}{6}$ or $\dfrac{13\pi}{6}$

$x = \dfrac{\pi}{2}$ or $\dfrac{11\pi}{6}$ ✓ *(4 marks)*

11 **i** $\sin(3\theta) = \sin((2\theta) + \theta)$

$= \sin(2\theta)\cos\theta + \cos(2\theta)\sin\theta$

$= (2\sin\theta\cos\theta)\cos\theta + (\cos^2\theta - \sin^2\theta)\sin\theta$
$= 2\sin\theta\cos^2\theta + \cos^2\theta\sin\theta - \sin^3\theta$
$= 3\sin\theta\cos^2\theta - \sin^3\theta$
$= 3\sin\theta(1 - \sin^2\theta) - \sin^3\theta$
$= 3\sin\theta - 4\sin^3\theta$ ✓

Now $\sin^3\theta - \frac{3}{4}\sin\theta + \frac{\sin(3\theta)}{4}$

$= \sin^3\theta - \frac{3}{4}\sin\theta + \frac{3\sin\theta - 4\sin^3\theta}{4}$

$= \sin^3\theta - \frac{3}{4}\sin\theta + \frac{3}{4}\sin\theta - \sin^3\theta$

$= 0$ ✓

(2 marks)

ii $x^3 - 12x + 8 = 0$
Let $x = 4\sin\theta$.
$(4\sin\theta)^3 - 12(4\sin\theta) + 8 = 0$
$64\sin^3\theta - 48\sin\theta + 8 = 0$
Dividing both sides by 64:

$\sin^3\theta - \frac{3}{4}\sin\theta + \frac{1}{8} = 0$ ✓

But $\sin^3\theta - \frac{3}{4}\sin\theta + \frac{\sin(3\theta)}{4} = 0$ from part **i**

So $\frac{\sin(3\theta)}{4} = \frac{1}{8}$

$\sin(3\theta) = \frac{1}{2}$ ✓

(2 marks)

iii If $\sin(3\theta) = \frac{1}{2}$

$3\theta = \frac{\pi}{6}, \frac{5\pi}{6}, \frac{13\pi}{6}, \frac{17\pi}{6}, \frac{25\pi}{6}, \ldots \quad (\theta > 0)$

$\theta = \frac{\pi}{18}, \frac{5\pi}{18}, \frac{13\pi}{18}, \frac{17\pi}{18}, \frac{25\pi}{18}, \ldots \quad (\theta > 0)$

Now $4\sin\theta$ is a root of the equation
$x^3 - 12x + 8 = 0$.

$4\sin\frac{\pi}{18}$, $4\sin\frac{5\pi}{18}$ and $4\sin\frac{25\pi}{18}$ are distinct roots.

Now $\alpha + \beta + \gamma = -\frac{b}{a}$

So $4\sin\frac{\pi}{18} + 4\sin\frac{5\pi}{18} + 4\sin\frac{25\pi}{18} = 0$

$\sin\frac{\pi}{18} + \sin\frac{5\pi}{18} + \sin\frac{25\pi}{18} = 0$ ✓

$\alpha\beta + \alpha\gamma + \beta\gamma = \frac{c}{a}$

$$\left(4\sin\frac{\pi}{18}\right)\left(4\sin\frac{5\pi}{18}\right) + \left(4\sin\frac{\pi}{18}\right)\left(4\sin\frac{25\pi}{18}\right) + \left(4\sin\frac{5\pi}{18}\right)\left(4\sin\frac{25\pi}{18}\right) = -12$$

$$16\left[\left(\sin\frac{\pi}{18}\right)\left(\sin\frac{5\pi}{18}\right) + \left(\sin\frac{\pi}{18}\right)\left(\sin\frac{25\pi}{18}\right) + \left(\sin\frac{5\pi}{18}\right)\left(\sin\frac{25\pi}{18}\right)\right] = -12$$

$$\left(\sin\frac{\pi}{18}\right)\left(\sin\frac{5\pi}{18}\right) + \left(\sin\frac{\pi}{18}\right)\left(\sin\frac{25\pi}{18}\right) + \left(\sin\frac{5\pi}{18}\right)\left(\sin\frac{25\pi}{18}\right) = -\frac{3}{4}$$ ✓

$\sin^2\frac{\pi}{18} + \sin^2\frac{5\pi}{18} + \sin^2\frac{25\pi}{18}$

$$= \left(\sin\frac{\pi}{18} + \sin\frac{5\pi}{18} + \sin\frac{25\pi}{18}\right)^2 - 2\left(\left(\sin\frac{\pi}{18}\right)\left(\sin\frac{5\pi}{18}\right) + \left(\sin\frac{\pi}{18}\right)\left(\sin\frac{25\pi}{18}\right) + \left(\sin\frac{5\pi}{18}\right)\left(\sin\frac{25\pi}{18}\right)\right)$$

$= 0^2 - 2 \times -\frac{3}{4}$

$= \frac{3}{2}$ ✓

(3 marks)

12 $\sin x = \frac{1}{4}$

Now $\sin^2 x + \cos^2 x = 1$

So $\cos^2 x = 1 - \left(\frac{1}{4}\right)^2$

$= \frac{15}{16}$

$\cos x = -\frac{\sqrt{15}}{4} \quad \left(\frac{\pi}{2} < x < \pi\right)$

$\sin 2x = 2\sin x\cos x$

$= 2 \times \frac{1}{4} \times -\frac{\sqrt{15}}{4}$

$= -\frac{\sqrt{15}}{8}$

Answer B

13 **i** $y = \sin x$

$\frac{dy}{dx} = \cos x$

At $x = x_0$, $m = \cos x_0$.

$y = \sin(x - \alpha) + k$

$\frac{dy}{dx} = \cos(x - \alpha)$

At $x = x_0$, $m = \cos(x_0 - \alpha)$.

The gradients must be equal as there is a common tangent.

$\therefore \cos x_0 = \cos(x_0 - \alpha)$ ✓

(1 mark)

ii $\cos(-\theta) = \cos\theta$ and
$\sin(-\theta) = -\sin\theta$ for all values of θ.
Now $\cos x_0 = \cos(x_0 - \alpha)$ ✓
As $\alpha \neq 0$, $x_0 \neq x_0 - \alpha$
$\therefore x_0 = -(x_0 - \alpha)$
So $\sin x_0 = \sin(-(x_0 - \alpha))$
$= -\sin(x_0 - \alpha)$ ✓

(2 marks)

iii $x_0 = -(x_0 - \alpha)$
$2x_0 = \alpha$
$x_0 = \frac{\alpha}{2}$

The curves intersect when $x = x_0$.

$\therefore \sin(x_0 - \alpha) + k = \sin x_0$ ✓
$= -\sin(x_0 - \alpha)$
$k = -2\sin(x_0 - \alpha)$
$= -2\sin\left(\frac{\alpha}{2} - \alpha\right)$
$= -2\sin\left(-\frac{\alpha}{2}\right)$

So $k = 2\sin\left(\frac{\alpha}{2}\right)$ ✓

(2 marks)

14 $\sqrt{3}\sin x + \cos x \equiv R\sin(x+\alpha)$

Now $R\sin(x+\alpha)$

$= R\sin x\cos\alpha + R\cos x\sin\alpha$

$R\cos\alpha = \sqrt{3}$ and $R\sin\alpha = 1$

$\frac{R\sin\alpha}{R\cos\alpha} = \frac{1}{\sqrt{3}}$

$\tan\alpha = \frac{1}{\sqrt{3}}$

$\alpha = \frac{\pi}{6} \quad \left(0 \le \alpha \le \frac{\pi}{2}\right)$ ✓

$R\cos\frac{\pi}{6} = \sqrt{3}$

$R \times \frac{\sqrt{3}}{2} = \sqrt{3}$

$R = 2$

$\therefore \sqrt{3}\sin x + \cos x \equiv 2\sin\left(x + \frac{\pi}{6}\right)$ ✓

(2 marks)

15 $\sqrt{5}\sin(x-\alpha) \equiv 2\sin x - \cos x$

$\sin x\cos\alpha - \cos x\sin\alpha$

$\equiv \frac{2}{\sqrt{5}}\sin x - \frac{1}{\sqrt{5}}\cos x$

$\therefore \cos\alpha = \frac{2}{\sqrt{5}}$ and $\sin\alpha = \frac{1}{\sqrt{5}}$

Now $\tan\alpha = \frac{\sin\alpha}{\cos\alpha}$

$= \frac{1}{\sqrt{5}} \div \frac{2}{\sqrt{5}}$

$= \frac{1}{2}$

Answer C

16 $x = 5\sin 4t + 12\cos 4t$

$= 13\left(\frac{5}{13}\sin 4t + \frac{12}{13}\cos 4t\right)$

$= 13(\cos\alpha\sin 4t + \sin\alpha\cos 4t)$

$= 13\sin(4t+\alpha)$

$\dot{x} = 52\cos(4t+\alpha)$

(Right-angled triangle: hypotenuse 13, opposite 12, adjacent 5, angle α)

Now $-1 \le \cos(4t+\alpha) \le 1$

So the maximum velocity is 52.

Answer C

17 When $\cos\left(2t - \frac{\pi}{3}\right) = 0$

$2t - \frac{\pi}{3} = \frac{\pi}{2}$ or $\frac{3\pi}{2}$, or …

$2t = \frac{\pi}{2} + \frac{\pi}{3}$ or $\frac{3\pi}{2} + \frac{\pi}{3}$ or …

$= \frac{5\pi}{6}$ or $\frac{11\pi}{6}$ or …

$t = \frac{5\pi}{12}$ or $\frac{11\pi}{12}$ or…

So P is the point $\left(\frac{11\pi}{12}, 0\right)$.

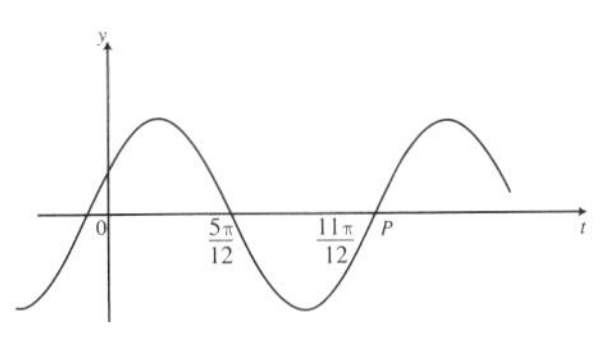

Answer C

18 $5\cos x - 12\sin x \equiv A\cos(x+\alpha)$

$\equiv A(\cos x\cos\alpha - \sin x\sin\alpha)$

$\therefore A\cos\alpha = 5$ and $A\sin\alpha = 12$

$A^2\cos^2\alpha + A^2\sin^2\alpha = 5^2 + 12^2$

$A^2 = 169$

$A = 13 \ (A > 0)$ ✓

Now $\frac{13\sin\alpha}{13\cos\alpha} = \frac{12}{5}$

$\tan\alpha = \frac{12}{5}$

$\alpha = 1.176$ (3 d.p.)

$\left(0 \le \alpha \le \frac{\pi}{2}\right)$

So $5\cos x - 12\sin x$

$\equiv 13\cos(x + 1.176)$ ✓

(2 marks)

19 $\cos\left(x + \frac{\pi}{4}\right)$

$= \cos x\cos\frac{\pi}{4} - \sin x\sin\frac{\pi}{4}$

$= \frac{1}{\sqrt{2}}(\cos x - \sin x)$

So $\sqrt{2}\cos\left(x + \frac{\pi}{4}\right)$

$= \cos x - \sin x$

Answer A

20 **i** $\sqrt{3}\cos x - \sin x$

$= 2\left(\frac{\sqrt{3}}{2}\cos x - \frac{1}{2}\sin x\right)$

$= 2\left(\cos\frac{\pi}{6}\cos x - \sin\frac{\pi}{6}\sin x\right)$

$= 2\cos\left(x + \frac{\pi}{6}\right)$ ✓

(1 mark)

ii $\sqrt{3}\cos x = 1 + \sin x$

$\sqrt{3}\cos x - \sin x = 1$

$2\cos\left(x + \frac{\pi}{6}\right) = 1$

$\cos\left(x + \frac{\pi}{6}\right) = \frac{1}{2}$ ✓

$\therefore x + \frac{\pi}{6} = \frac{\pi}{3}$ or $x + \frac{\pi}{6} = \frac{5\pi}{3}$

$(0 < x < 2\pi)$

$x = \frac{\pi}{6}$ or $x = \frac{3\pi}{2}$ ✓

(2 marks)

21 **i** $2\cos\theta + 2\cos\left(\theta + \frac{\pi}{3}\right)$

$\equiv R\cos(\theta + \alpha)$

$2\cos\theta + 2\cos\left(\theta + \frac{\pi}{3}\right)$

$= 2\cos\theta + 2\left(\cos\theta\cos\frac{\pi}{3} - \sin\theta\sin\frac{\pi}{3}\right)$

$= 2\cos\theta + 2\cos\theta \times \frac{1}{2} - 2\sin\theta \times \frac{\sqrt{3}}{2}$

$= 2\cos\theta + \cos\theta - \sqrt{3}\sin\theta$

$= 3\cos\theta - \sqrt{3}\sin\theta$ ✓

Now $R\cos(\theta + \alpha)$

$= R\cos\theta\cos\alpha - R\sin\theta\sin\alpha$

So $R\cos\alpha = 3$ and $R\sin\alpha = \sqrt{3}$

$\cos\alpha = \frac{3}{R}$

$\sin\alpha = \frac{\sqrt{3}}{R}$

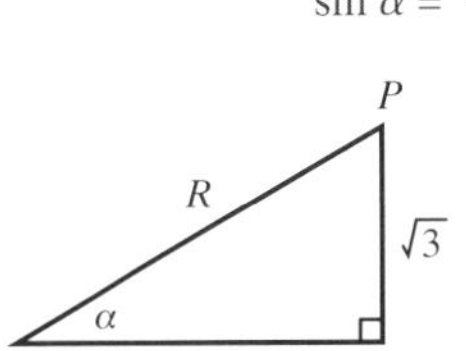

$R^2 = 3^2 + (\sqrt{3})^2$

$= 9 + 3$

$= 12$

$R = \sqrt{12} \quad (R > 0)$

$= 2\sqrt{3}$ ✓

$\tan\alpha = \frac{\sqrt{3}}{3}$

$= \frac{1}{\sqrt{3}}$

$\therefore \alpha = \frac{\pi}{6} \quad \left(0 < \alpha < \frac{\pi}{2}\right)$

$\therefore 2\cos\theta + 2\cos\left(\theta + \frac{\pi}{3}\right)$

$\equiv 2\sqrt{3}\cos\left(\theta + \frac{\pi}{6}\right)$ ✓

(3 marks)

ii $2\cos\theta + 2\cos\left(\theta + \frac{\pi}{3}\right) = 3$

$\therefore 2\sqrt{3}\cos\left(\theta + \frac{\pi}{6}\right) = 3$

$\cos\left(\theta + \frac{\pi}{6}\right) = \frac{3}{2\sqrt{3}}$

$= \frac{\sqrt{3}}{2}$ ✓

Now $0 < \theta < 2\pi$

$\therefore \frac{\pi}{6} < \theta + \frac{\pi}{6} < \frac{13\pi}{6}$

But $\cos\left(\theta + \frac{\pi}{6}\right) = \frac{\sqrt{3}}{2}$

$\therefore \theta + \frac{\pi}{6} = \frac{11\pi}{6}$

$\theta = \dfrac{10\pi}{6}$

$\theta = \dfrac{5\pi}{3}$ ✓

(2 marks)

22 **i** $A\sin(x+\alpha)$

$= A\sin x\cos\alpha + A\sin\alpha\cos x$

Let $3\sin x + 4\cos x$

$= A\sin x\cos\alpha + A\sin\alpha\cos x$

$\therefore A\cos\alpha = 3 \quad \ldots(1)$

and $A\sin\alpha = 4 \quad \ldots(2)$

Squaring (1) and (2):

$A^2\cos^2\alpha = 9 \quad \ldots(3)$

and $A^2\sin^2\alpha = 16 \quad \ldots(4)$

$\therefore A^2(\cos^2\alpha + \sin^2\alpha) = 25 \quad \ldots(3)+(4)$

$\therefore A^2 = 25$

$\therefore A = 5$ ✓

Dividing (2) by (1):

$\dfrac{A\sin\alpha}{A\cos\alpha} = \dfrac{4}{3}$

$\therefore \tan\alpha = \dfrac{4}{3}$

$\therefore \alpha = 0.927$

$\therefore 3\sin x + 4\cos x$

$= 5\sin(x + 0.927)$ ✓

(2 marks)

ii $3\sin x + 4\cos x = 5$

$\therefore 5\sin(x+0.927) = 5$

$\sin(x+0.927) = 1$

$\therefore x + 0.927 = \dfrac{\pi}{2}, \dfrac{5\pi}{2} \ldots$ ✓

$\therefore x = 0.64, 6.93\ldots$

Since $0 < x < 2\pi$, $x = 0.64$. ✓

(2 marks)

23 $\sin 3\theta + \sin 2\theta = \sin\theta$

Substituting for $\sin 3\theta$:

$(3\sin\theta - 4\sin^3\theta) + \sin 2\theta = \sin\theta$

$3\sin\theta - 4\sin^3\theta + 2\sin\theta\cos\theta - \sin\theta = 0$

$2\sin\theta - 4\sin^3\theta + 2\sin\theta\cos\theta = 0$ ✓

$2\sin\theta(1 - 2\sin^2\theta + \cos\theta) = 0$

$2\sin\theta[1 - 2(1-\cos^2\theta) + \cos\theta] = 0$

$2\sin\theta(1 - 2 + 2\cos^2\theta + \cos\theta) = 0$

$2\sin\theta(2\cos^2\theta + \cos\theta - 1) = 0$

$2\sin\theta(2\cos\theta - 1)(\cos\theta + 1) = 0$

$\therefore \sin\theta = 0$ *or* $\cos\theta = \dfrac{1}{2}$ *or* $\cos\theta = -1$ ✓

$\therefore \theta = 0, \pi, 2\pi$ *or* $\theta = \dfrac{\pi}{3}, \dfrac{5\pi}{3}$ or $\theta = \pi$

$\theta = 0, \dfrac{\pi}{3}, \pi, \dfrac{5\pi}{3}, 2\pi$

for $0 \le \theta \le 2\pi$. ✓

(3 marks)

24 If $t = \tan\dfrac{\theta}{2}$,

$\sin\theta = \dfrac{2t}{1+t^2}$

$\cos\theta = \dfrac{1-t^2}{1+t^2}$

$\therefore \dfrac{1-\cos\theta}{\sin\theta}$

$= \dfrac{1 - \dfrac{1-t^2}{1+t^2}}{\dfrac{2t}{1+t^2}}$ ✓

$= \dfrac{1+t^2-(1-t^2)}{1+t^2} \times \dfrac{1+t^2}{2t}$

$= \dfrac{2t^2}{2t}$

$= t$

$= \tan\dfrac{\theta}{2}$. ✓

(2 marks)

25 **i** $A\cos(x-\alpha)$

$= A\cos x\cos\alpha + A\sin x\sin\alpha$

If $8\cos x + 6\sin x$

$= A\cos x\cos\alpha + A\sin x\sin\alpha$

then $A\cos\alpha = 8 \quad \ldots(1)$

and $A\sin\alpha = 6 \quad \ldots(2)$

$\therefore \tan\alpha = \dfrac{3}{4} \quad \ldots(2)\div(1)$

$\therefore \alpha = 0.6435\ldots$

$\doteqdot 0.644$ radians ✓

Now squaring (1) and (2):

$A^2\cos^2\alpha = 64 \quad \ldots(3)$

$A^2\sin^2\alpha = 36 \quad \ldots(4)$

$\therefore A^2(\cos^2\alpha + \sin^2\alpha) = 100 \quad \ldots(3)+(4)$

$\therefore A^2 = 100$

$\therefore A = 10$

$\therefore 8\cos x + 6\sin x$

$= 10\cos(x - 0.644)$. ✓

(2 marks)

ii $8\cos x + 6\sin x = 5 \quad 0 \leqslant x \leqslant 2\pi$

$\therefore 10\cos(x - 0.64\ldots) = 5$

$\cos(x - 0.64\ldots) = \dfrac{1}{2}$

$\therefore x - 0.64\ldots = \dfrac{\pi}{3}, \dfrac{5\pi}{3}$ ✓

for $0 \leqslant x \leqslant 2\pi$

$\therefore x = \dfrac{\pi}{3} + 0.64\ldots$ or $\dfrac{5\pi}{3} + 0.64\ldots$

$= 1.691$ or 5.879

to 3 decimal places. ✓

(2 marks)

26 **i** $\cos x - \sin x$

$= \sqrt{1^2 + (-1)^2}\left(\dfrac{1}{\sqrt2}\cos x - \dfrac{1}{\sqrt2}\sin x\right)$

$= \sqrt2\left(\cos x \cdot \dfrac{1}{\sqrt2} - \sin x \cdot \dfrac{1}{\sqrt2}\right)$

$= \sqrt2(\cos x\cos\alpha - \sin x\sin\alpha)$

where $\cos\alpha = \dfrac{1}{\sqrt2}$

and $\sin\alpha = \dfrac{1}{\sqrt2}$ ✓

$\therefore \tan\alpha = \dfrac{\sin\alpha}{\cos\alpha}$

$= \dfrac{\frac{1}{\sqrt2}}{\frac{1}{\sqrt2}}$

$= 1$

$\therefore \alpha = \dfrac{\pi}{4}$.

$\therefore \cos x - \sin x$

$= \sqrt2\left(\cos x\cos\dfrac{\pi}{4} - \sin x\sin\dfrac{\pi}{4}\right)$

$= \sqrt2\cos\left(x + \dfrac{\pi}{4}\right)$ ✓

OR

Now $R\cos(x+\alpha)$

$= R(\cos x\cos\alpha - \sin x\sin\alpha)$

$= R\cos x\cos\alpha - R\sin x\sin\alpha$

Equating with $\cos x - \sin x$:

$R\cos\alpha = 1$ and $R\sin\alpha = 1$

$\cos\alpha = \dfrac{1}{R}$ and $\sin\alpha = \dfrac{1}{R}$

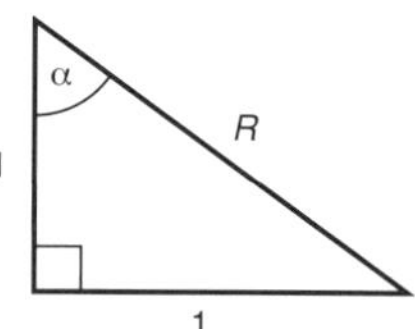

$\therefore R^2 = 1^2 + 1^2$

$\therefore R^2 = 2$

$\therefore R = \sqrt2$

and $\tan\alpha = \dfrac{1}{1}$

$= 1$

$\therefore \alpha = \dfrac{\pi}{4}$

$\therefore R\cos(x+\alpha)$ becomes

$\sqrt2\cos\left(x + \dfrac{\pi}{4}\right)$.

(2 marks)

ii $y = \cos x - \sin x$

$= \sqrt{2}\cos\left(x + \frac{\pi}{4}\right)$

Period $= 2\pi$

Range: $-\sqrt{2} \leq y \leq \sqrt{2}$

Curve intercepts x axis when

$\cos\left(x + \frac{\pi}{4}\right) = 0$

$\therefore x + \frac{\pi}{4} = \frac{\pi}{2}$ and

$x + \frac{\pi}{4} = \frac{3\pi}{2}$

for $0 \leq x \leq 2\pi$

$\therefore x + \frac{\pi}{4} = \frac{2\pi}{4}$ and $x + \frac{\pi}{4} = \frac{6\pi}{4}$

$\therefore x = \frac{\pi}{4}$ and $x = \frac{5\pi}{4}$ ✓

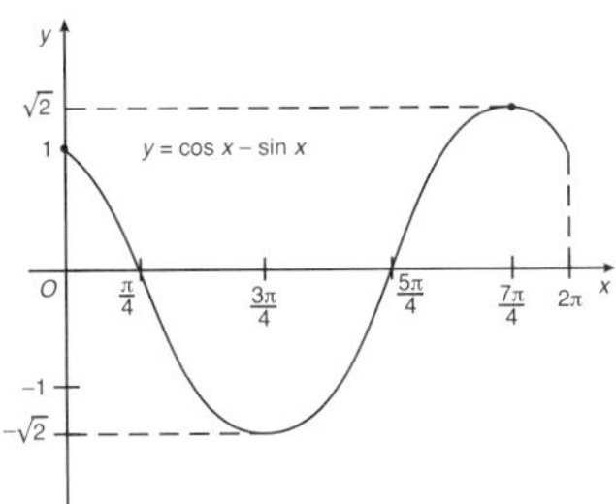

✓ *(2 marks)*

27 i $\sin(\theta + 2\theta)$

$= \sin\theta\cos 2\theta + \cos\theta\sin 2\theta$

$= \sin\theta(\cos^2\theta - \sin^2\theta) + \cos\theta(2\sin\theta\cos\theta)$ ✓

$= \sin\theta[(1 - \sin^2\theta) - \sin^2\theta] + 2\sin\theta\cos^2\theta$

$= \sin\theta[1 - 2\sin^2\theta] + 2\sin\theta(1 - \sin^2\theta)$

$= \sin\theta - 2\sin^3\theta + 2\sin\theta - 2\sin^3\theta$

$= 3\sin\theta - 4\sin^3\theta.$

$\therefore \sin 3\theta = 3\sin\theta - 4\sin^3\theta$ ✓ *(2 marks)*

ii $\sin 3\theta = 2\sin\theta$ for $0 \leq \theta \leq 2\pi$

$\therefore 3\sin\theta - 4\sin^3\theta = 2\sin\theta$

$\sin\theta - 4\sin^3\theta = 0$

$\sin\theta(1 - 4\sin^2\theta) = 0$

$\therefore \sin\theta = 0$ ✓

$\therefore \theta = 0, \pi$ or 2π.

or $1 - 4\sin^2\theta = 0$

$\sin^2\theta = \frac{1}{4}$

$\sin\theta = \pm\frac{1}{2}$ ✓

$\therefore \theta = \frac{\pi}{6}, \frac{5\pi}{6}, \frac{7\pi}{6}, \frac{11\pi}{6}$

$\therefore \theta = 0, \frac{\pi}{6}, \frac{5\pi}{6}, \pi, \frac{7\pi}{6}, \frac{11\pi}{6}$ or 2π. ✓ *(3 marks)*

28 $\cos 2\theta = \sin\theta,\ 0 \leq \theta \leq 2\pi$

Now $\cos 2\theta = 1 - 2\sin^2\theta$

$\therefore \sin\theta = 1 - 2\sin^2\theta$

$\therefore 2\sin^2\theta + \sin\theta - 1 = 0$ ✓

$\therefore (2\sin\theta - 1)(\sin\theta + 1) = 0$

$\therefore 2\sin\theta - 1 = 0$ or $\sin\theta + 1 = 0$

$\therefore 2\sin\theta = 1$ or $\sin\theta = -1$

$\therefore \sin\theta = \frac{1}{2}$ ✓ or $\sin\theta = -1$ ✓

$\therefore \theta = \frac{\pi}{6}, \frac{5\pi}{6}$ or $\frac{3\pi}{2}$ $0 \leq \theta \leq 2\pi$ ✓ *(4 marks)*

29 $\cos\theta + \sqrt{3}\sin\theta = 1 \quad 0 \leq \theta \leq 2\pi$

$\therefore \frac{1}{2}\cos\theta + \frac{\sqrt{3}}{2}\sin\theta = \frac{1}{2}$

(on dividing by $\sqrt{1^2 + (\sqrt{3})^2} = 2$)

$\therefore \sin\frac{\pi}{6}\cos\theta + \cos\frac{\pi}{6}\sin\theta = \frac{1}{2}$ ✓

(as $\sin\frac{\pi}{6} = \frac{1}{2}$ and $\cos\frac{\pi}{6} = \frac{\sqrt{3}}{2}$)

$\therefore \sin(\frac{\pi}{6} + \theta) = \frac{1}{2}$ ✓

$\therefore \frac{\pi}{6} + \theta = \frac{\pi}{6}, \pi - \frac{\pi}{6}, 2\pi + \frac{\pi}{6}, \ldots$ ✓

$\therefore \theta = 0,\ \pi - \frac{2\pi}{6},\ 2\pi, \ldots$

$= 0, \frac{2\pi}{3}, 2\pi\ (0 \leq \theta \leq 2\pi).$ ✓

(4 marks)

30 $\sqrt{3}\sin\theta - \cos\theta = 1 \quad 0 \leq \theta \leq 2\pi$

Now $\sqrt{(\sqrt{3})^2 + 1^2} = \sqrt{4}$

$= 2$

$\therefore 2\left(\frac{\sqrt{3}}{2}\sin\theta - \frac{1}{2}\cos\theta\right) = 1$

$\frac{\sqrt{3}}{2}\sin\theta - \frac{1}{2}\cos\theta = \frac{1}{2}$

$\therefore \sin\theta.\frac{\sqrt{3}}{2} - \cos\theta.\frac{1}{2} = \frac{1}{2}$

Let $\cos\alpha = \frac{\sqrt{3}}{2}$

$\therefore \sin\alpha = \frac{1}{2}$ and $\alpha = \frac{\pi}{6}$

$\therefore \sin\theta\cos\frac{\pi}{6} - \cos\theta\sin\frac{\pi}{6} = \frac{1}{2}$

$\therefore \sin\left(\theta - \frac{\pi}{6}\right) = \frac{1}{2}$

$\therefore \theta - \frac{\pi}{6} = \frac{\pi}{6}, \pi - \frac{\pi}{6}$

$\therefore \theta = \frac{\pi}{6} + \frac{\pi}{6}, \pi$

$= \frac{2\pi}{6}, \pi$

$= \frac{\pi}{3}, \pi$

(No mark allocation in exam)

31 **i and ii**

$$y = \cos(\pi x)$$

$$\text{Period} = \frac{2\pi}{\pi} = 2$$

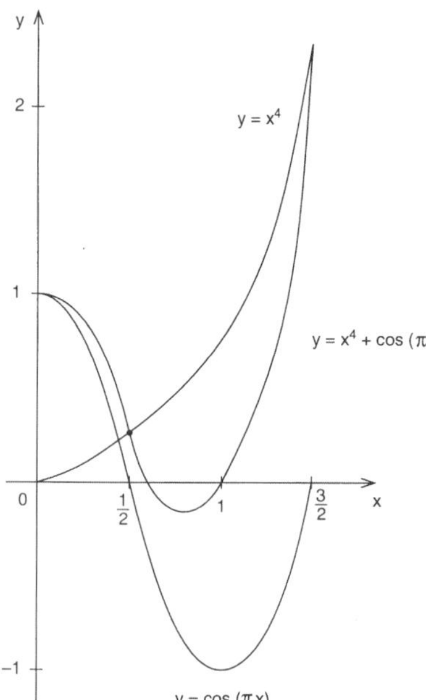

iii The number of positive real roots of $x^4 + \cos(\pi x) = 0$ is two.

(No mark allocation in exam)

---

32 $2\sin^2\theta = \sin 2\theta \qquad 0 \leqslant \theta \leqslant 2\pi$

$$2\sin^2\theta = 2\sin\theta\cos\theta$$
$$2\sin^2\theta - 2\sin\theta\cos\theta = 0$$
$$2\sin\theta(\sin\theta - \cos\theta) = 0$$
$$\therefore 2\sin\theta = 0$$
$$\sin\theta = 0$$
$$\theta = 0, \pi, 2\pi$$
$$\text{or } \sin\theta - \cos\theta = 0$$
$$\sin\theta = \cos\theta$$
$$\frac{\sin\theta}{\cos\theta} = 1$$
$$\therefore \tan\theta = 1$$
$$\theta = \frac{\pi}{4}, \frac{5\pi}{4}$$
$$\therefore \theta = 0, \frac{\pi}{4}, \pi, \frac{5\pi}{4}, 2\pi.$$

(No mark allocation in exam)

1 Find the exact value of $\int_0^1 \frac{x}{\sqrt{x^2+4}}\,dx$ using the substitution $u = x^2 + 4$. *(3 marks)*

(Q11b, **2022 HSC**) Easy

2 Which of the following integrals is equivalent to $\int \sin^2 3x\,dx$?

A $\int \frac{1+\cos 6x}{2}\,dx$ **B** $\int \frac{1-\cos 6x}{2}\,dx$

C $\int \frac{1+\sin 6x}{2}\,dx$ **D** $\int \frac{1-\sin 6x}{2}\,dx$ *(1 mark)*

(Q2, **2021 HSC**) Easy

3 Use the substitution $u = x + 1$ to find $\int x\sqrt{x+1}\,dx$. *(3 marks)*

(Q11c, **2021 HSC**) Easy

4 Evaluate $\int_0^{\sqrt{3}} \frac{1}{\sqrt{4-x^2}}\,dx$. *(2 marks)*

(Q11f, **2021 HSC**) Easy

5 The polynomial $g(x) = x^3 + 4x - 2$ passes through the point $(1, 3)$. Find the gradient of the tangent to $f(x) = xg^{-1}(x)$ at the point where $x = 3$. *(2 marks)*

(Q14e, **2021 HSC**) Hard

6 Which of the following is an anti-derivative of $\frac{1}{4x^2+1}$?

A $2\tan^{-1}\left(\frac{x}{2}\right) + c$ **B** $\frac{1}{2}\tan^{-1}\left(\frac{x}{2}\right) + c$

C $2\tan^{-1}(2x) + c$ **D** $\frac{1}{2}\tan^{-1}(2x) + c$ *(1 mark)*

(Q3, **2020 HSC**) Easy

7 Find $\int_0^{\frac{\pi}{2}} \cos 5x \sin 3x\,dx$. *(3 marks)*

(Q12d, **2020 HSC**) Medium

8 **i** Find $\frac{d}{d\theta}\left(\sin^3\theta\right)$. *(1 mark)* Easy

ii Use the substitution $x = \tan\theta$ to evaluate $\int_0^1 \frac{x^2}{\left(1+x^2\right)^{\frac{5}{2}}}\,dx$. *(4 marks)* Medium

(Q13a, **2020 HSC**)

9 Suppose $f(x) = \tan(\cos^{-1}(x))$ and $g(x) = \frac{\sqrt{1-x^2}}{x}$.

The graph of $y = g(x)$ is given.

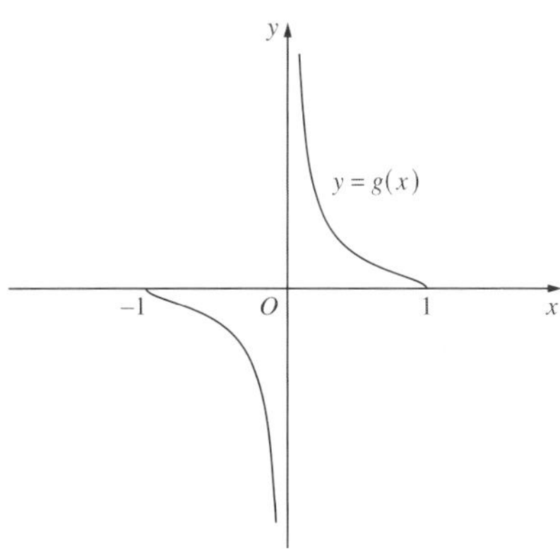

i Show that $f'(x) = g'(x)$. *(4 marks)* Medium

ii Using part **i**, or otherwise, show that $f(x) = g(x)$. *(3 marks)* Hard

(Q13c, **2020 HSC**)

10 Evaluate $\int_0^{\frac{\pi}{3}} \cos\frac{3x}{2}\cos\frac{x}{2}\,dx$. *(3 marks)*

Bonus question (see page iv) Medium

11 Find $\int \sin^2\left(\frac{x}{2}\right)dx$. *(1 mark)*

Bonus question Easy

12 Evaluate $\int_0^{\frac{\pi}{6}} \cos^2 2x \sin 2x\,dx$, using the substitution $u = \cos 2x$. *(4 marks)*

Bonus question Medium

13 Use the substitution $u = \cos 2x$ to evaluate $\int_0^{\frac{\pi}{8}} 2\cos^3 2x \sin 2x\,dx$. *(3 marks)*

Bonus question Medium

14 Use the substitution $u = \sin x$ to evaluate $\int_0^{\frac{\pi}{6}} \cos^3 x\,dx$. *(3 marks)*

Bonus question Medium

15 Using the substitution $u = \cos x$, evaluate $\int_{\frac{\pi}{6}}^{\frac{\pi}{4}} \sin x \cos^3 x\,dx$. *(3 marks)*

Bonus question Medium

16 Find the value of $\int_0^{\frac{\pi}{4}} 4\sin x \cos\frac{x}{3}\,dx$. *(3 marks)*

Bonus question Medium

17 Find the exact value of $\int_{\frac{\pi}{12}}^{\frac{\pi}{3}} \sin 3x \sin x \, dx$. *(3 marks)*

Bonus question **Medium**

18 Find the exact value of $\int_{0}^{\frac{\pi}{3}} \sin 3x \cos x \, dx$. *(3 marks)*

Bonus question **Medium**

19 What is the derivative of $\tan^{-1}\frac{x}{2}$?

A $\frac{1}{2(4+x^2)}$ **B** $\frac{1}{4+x^2}$

C $\frac{2}{4+x^2}$ **D** $\frac{4}{4+x^2}$ *(1 mark)*

(Q3, **2019 HSC**) **Easy**

20 Find $\int 2\sin^2 4x \, dx$. *(2 marks)*

(Q11e, **2019 HSC**) **Easy**

21 Use the substitution $u = \cos^2 x$ to evaluate

$\int_0^{\frac{\pi}{4}} \frac{\sin 2x}{4+\cos^2 x} dx$. *(3 marks)*

(Q13a, **2019 HSC**) **Medium**

22 Evaluate $\int_{-3}^{0} \frac{x}{\sqrt{1-x}} dx$,

using the substitution $u = 1 - x$. *(3 marks)*

(Q11f, **2018 HSC**) **Easy**

23 Find $\int \cos^2(3x) dx$. *(2 marks)*

(Q12a, **2018 HSC**) **Easy**

24 A ferris wheel has a radius of 20 metres and is rotating at a rate of 1.5 radians per minute. The top of a carriage is h metres above the horizontal diameter of the ferris wheel. The angle of elevation of the top of the carriage from the centre of the ferris wheel is θ.

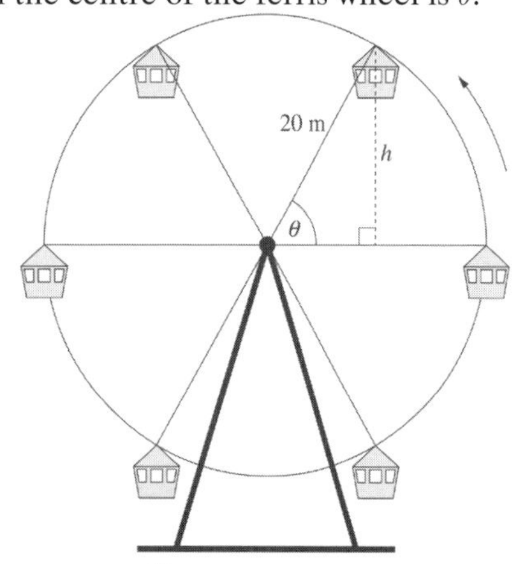

i Show that $\frac{dh}{d\theta} = 20\cos\theta$. *(1 mark)* **Easy**

ii At what speed is the top of the carriage rising when it is 15 metres higher than the horizontal diameter of the ferris wheel? Give your answer correct to one decimal place. *(2 marks)* **Medium**

(Q12b, **2018 HSC**)

25 Let $f(x) = \sin^{-1}x + \cos^{-1}x$.

i Show that $f'(x) = 0$. *(1 mark)* **Easy**

ii Hence, or otherwise, prove $\sin^{-1}x + \cos^{-1}x = \frac{\pi}{2}$. *(1 mark)* **Easy**

iii Hence, sketch $f(x) = \sin^{-1}x + \cos^{-1}x$. *(1 mark)* **Easy**

(Q12c, **2018 HSC**)

26 Differentiate $\tan^{-1}(x^3)$. *(2 marks)*

(Q11b, **2017 HSC**) **Easy**

27 Evaluate $\int_0^3 \frac{x}{\sqrt{x+1}} dx$, using the substitution $x = u^2 - 1$. *(3 marks)*

(Q11e, **2017 HSC**) **Easy**

28 Find $\int \sin^2 x \cos x \, dx$. *(1 mark)*

(Q11f, **2017 HSC**) **Medium**

29 The concentration of a drug in a body is $F(t)$, where t is the time in hours after the drug is taken.

Initially the concentration of the drug is zero. The rate of change of concentration of the drug is given by

$$F'(t)=50e^{-0.5t} - 0.4F(t).$$

i By differentiating the product $F(t)e^{0.4t}$ show that

$\frac{d}{dt}(F(t)e^{0.4t}) = 50e^{-0.1t}$. *(2 marks)* **Medium**

ii Hence, or otherwise, show that $F(t) = 500(e^{-0.4t} - e^{-0.5t})$. *(2 marks)* **Hard**

iii The concentration of the drug increases to a maximum. For what value of t does this maximum occur? *(2 marks)* **Hard**

(Q14c, **2017 HSC**)

30 Which expression is equal to $\int \sin^2 2x \, dx$?

A $\frac{1}{2}\left(x - \frac{1}{4}\sin 4x\right) + c$ **B** $\frac{1}{2}\left(x + \frac{1}{4}\sin 4x\right) + c$

C $\frac{\sin^3 2x}{6} + c$ **D** $\frac{-\cos^3 2x}{6} + c$ *(1 mark)*

(Q5, **2016 HSC**) **Easy**

31 Use the substitution $u = x - 4$ to find $\int x\sqrt{x-4}\,dx$. *(3 marks)*

(Q11b, **2016 HSC**) Easy

32 Differentiate $3\tan^{-1}(2x)$. *(2 marks)*

(Q11c, **2016 HSC**) Easy

33 What is the value of k such that $\int_0^k \frac{1}{\sqrt{4-x^2}}\,dx = \frac{\pi}{3}$?

A 1 **B** $\sqrt{3}$

C 2 **D** $2\sqrt{3}$ *(1 mark)*

(Q7, **2015 HSC**) Easy

34 Find $\int \sin^2 x\,dx$. *(2 marks)*

(Q11a, **2015 HSC**) Easy

35 Use the substitution $u = 2x - 1$ to evaluate $\int_1^2 \frac{x}{(2x-1)^2}\,dx$. *(3 marks)*

(Q11e, **2015 HSC**) Easy

36 Let $f(x) = \cos^{-1}(x) + \cos^{-1}(-x)$, where $-1 \le x \le 1$.

i By considering the derivative of $f(x)$, prove that $f(x)$ is constant. *(2 marks)* Easy

ii Hence deduce that $\cos^{-1}(-x) = \pi - \cos^{-1}(x)$. *(1 mark)* Hard

(Q13d, **2015 HSC**)

37 What is the derivative of $3\sin^{-1}\frac{x}{2}$?

A $\frac{6}{\sqrt{4-x^2}}$ **B** $\frac{3}{\sqrt{4-x^2}}$

C $\frac{3}{2\sqrt{4-x^2}}$ **D** $\frac{3}{4\sqrt{4-x^2}}$ *(1 mark)*

(Q6, **2014 HSC**) Easy

38 Evaluate $\int_2^5 \frac{x}{\sqrt{x-1}}\,dx$ using the substitution $x = u^2 + 1$. *(3 marks)*

(Q11d, **2014 HSC**) Medium

39 Which integral is obtained when the substitution $u = 1 + 2x$ is applied to $\int x\sqrt{1+2x}\,dx$?

A $\frac{1}{4}\int (u-1)\sqrt{u}\,du$ **B** $\frac{1}{2}\int (u-1)\sqrt{u}\,du$

C $\int (u-1)\sqrt{u}\,du$ **D** $2\int (u-1)\sqrt{u}\,du$ *(1 mark)*

(Q5, **2013 HSC**) Medium

40 Find $\int \frac{1}{\sqrt{49-4x^2}}\,dx$. *(2 marks)*

(Q11b, **2013 HSC**) Easy

41 Use the substitution $u = e^{3x}$ to evaluate $\int_0^{\frac{1}{3}} \frac{e^{3x}}{e^{6x}+1}\,dx$. *(3 marks)*

(Q11f, **2013 HSC**) Medium

42 Differentiate $x^2\sin^{-1}5x$. *(2 marks)*

(Q11g, **2013 HSC**) Easy

43 Which expression is equal to $\int \sin^2 3x\,dx$?

A $\frac{1}{2}\left(x - \frac{1}{3}\sin 3x\right) + C$ **B** $\frac{1}{2}\left(x + \frac{1}{3}\sin 3x\right) + C$

C $\frac{1}{2}\left(x - \frac{1}{6}\sin 6x\right) + C$ **D** $\frac{1}{2}\left(x + \frac{1}{6}\sin 6x\right) + C$ *(1 mark)*

(Q7, **2012 HSC**) Easy

44 What is the derivative of $\cos^{-1}(3x)$?

A $\frac{1}{3\sqrt{1-9x^2}}$ **B** $\frac{-1}{3\sqrt{1-9x^2}}$

C $\frac{3}{\sqrt{1-9x^2}}$ **D** $\frac{-3}{\sqrt{1-9x^2}}$ *(1 mark)*

(Q9, **2012 HSC**) Easy

45 Evaluate $\int_0^3 \frac{1}{9+x^2}\,dx$. *(3 marks)*

(Q11a, **2012 HSC**) Easy

46 Use the substitution $u = 2 - x$ to evaluate $\int_1^2 x(2-x)^5\,dx$. *(3 marks)*

(Q11d, **2012 HSC**) Easy

47 Using the substitution $u = \sqrt{x}$, evaluate $\int_1^4 \frac{e^{\sqrt{x}}}{\sqrt{x}}\,dx$. *(3 marks)*

(Q1d, **2011 HSC**) Easy

48 Use the substitution $u = 1 - x$ to evaluate

$$\int_0^1 x\sqrt{1-x}\,dx.$$ *(3 marks)*

(Q1e, **2010 HSC**) Easy

49 The derivative of a function $f(x)$ is given by

$$f'(x) = \sin^2 x.$$

Find $f(x)$, given that $f(0) = 2$. *(2 marks)*

(Q2a, **2010 HSC**) Easy

50 Let $f(x) = \tan^{-1}x + \tan^{-1}\left(\frac{1}{x}\right)$ for $x \neq 0$.

i By differentiating $f(x)$, or otherwise, show that $f(x) = \frac{\pi}{2}$ for $x > 0$. *(3 marks)* Medium

ii Given that $f(x)$ is an odd function, sketch the graph $y = f(x)$. *(1 mark)* Hard

(Q5b, **2010 HSC**)

51 Using the substitution $u = x^3 + 1$, or otherwise,

evaluate $\int_0^2 x^2 e^{x^3+1}\,dx.$ *(3 marks)*

(Q1f, **2009 HSC**) Easy

52 Differentiate $\cos^{-1}(3x)$ with respect to x. *(2 marks)*

(Q1b, **2008 HSC**) Easy

53 Evaluate $\int_{-1}^{1} \frac{1}{\sqrt{4-x^2}}\,dx.$ *(2 marks)*

(Q1c, **2008 HSC**) Easy

54 Evaluate $\int_0^{\frac{\pi}{4}} \cos\theta \sin^2\theta\, d\theta.$ *(2 marks)*

(Q1e, **2008 HSC**) Easy

55 Use the substitution $u = \log_e x$ to evaluate

$$\int_e^{e^2} \frac{1}{x(\log_e x)^2}\,dx.$$ *(3 marks)*

(Q2a, **2008 HSC**) Medium

56 Differentiate $\tan^{-1}(x^4)$ with respect to x. *(2 marks)*

(Q1c, **2007 HSC**) Easy

57 Use the substitution $u = 25 - x^2$ to evaluate

$$\int_3^4 \frac{2x}{\sqrt{25-x^2}}\,dx.$$ *(3 marks)*

(Q1e, **2007 HSC**) Medium

58 Find $\int \frac{dx}{49 + x^2}.$ *(2 marks)*

(Q1a, **2006 HSC**) Easy

59 Using the substitution $u = x^4 + 8$, or otherwise,

find $\int x^3\sqrt{x^4 + 8}\,dx.$ *(3 marks)*

(Q1b, **2006 HSC**) Easy

60 Let $f(x) = \sin^{-1}(x + 5)$.

i State the domain and range of the function $f(x)$. *(2 marks)* Easy

ii Find the gradient of the graph of $y = f(x)$ at the point where $x = -5$. *(2 marks)* Easy

iii Sketch the graph of $y = f(x)$. *(2 marks)* Easy

(Q2a, **2006 HSC**)

61 Find $\int_0^{\frac{\pi}{4}} \sin^2 x\,dx.$ *(2 marks)*

(Q3a, **2006 HSC**) Easy

62 Find $\int \frac{1}{x^2+49}\,dx.$ *(1 mark)*

(Q1a, **2005 HSC**) Easy

63 Using the substitution $u = 2x^2 + 1$, or otherwise, find $\int x\left(2x^2+1\right)^{\frac{5}{4}}dx.$ *(3 marks)*

(Q1d, **2005 HSC**) Easy

64 Find $\frac{d}{dx}\,2\sin^{-1}5x.$ *(2 marks)*

(Q2a, **2005 HSC**) Easy

65 i By expanding the left-hand side, show that $\sin(5x + 4x) + \sin(5x - 4x) = 2\sin 5x \cos 4x.$ *(1 mark)* Easy

ii Hence find $\int \sin 5x \cos 4x\,dx.$ *(2 marks)* Easy

(Q3b, **2005 HSC**)

66 Evaluate $\int_0^{\frac{\pi}{4}} \cos x \sin^2 x\,dx.$ *(2 marks)*

(Q4a, **2005 HSC**) Easy

67 Find $\int_0^1 \frac{dx}{\sqrt{4-x^2}}.$ *(2 marks)*

(Q1d, **2004 HSC**) Easy

68 Use the substitution $u = x - 3$ to evaluate

$$\int_3^4 x\sqrt{x-3}\,dx.$$ *(3 marks)*

(Q1e, **2004 HSC**) Medium

69 Find $\frac{d}{dx}\cos^{-1}\left(3x^2\right).$ *(2 marks)*

(Q2b, **2004 HSC**) Easy

70 Find $\int \cos^2 4x\,dx.$ *(2 marks)*

(Q3a, **2004 HSC**) Easy

71 Use the substitution $u = x^2 + 1$ to evaluate

$$\int_0^2 \frac{x}{\left(x^2+1\right)^3}\,dx.$$

(3 marks)

(Q1e, **2003 HSC**) Easy

72 Find $\frac{d}{dx}\left(x\tan^{-1}x\right)$. *(2 marks)*

(Q2b, **2003 HSC**) Easy

73 Evaluate $\int_0^1 \frac{1}{\sqrt{2-x^2}}\,dx$. *(2 marks)*

(Q2c, **2003 HSC**) Easy

74 Find $\int \cos^2 3x\,dx$. *(2 marks)*

(Q5a, **2003 HSC**) Easy

75 Use the substitution $u = 1 - x^2$ to evaluate

$$\int_2^3 \frac{2x}{\left(1-x^2\right)^2}\,dx.$$

(3 marks)

(Q1f, **2002 HSC**) Medium

76 Evaluate $2\int_0^{\frac{\pi}{4}} \sin^2 4x\,dx$. *(3 marks)*

(Q2d, **2002 HSC**) Easy

77 Consider the function

$f(x) = 2\sin^{-1}\sqrt{x} - \sin^{-1}(2x-1)$ for $0 \le x \le 1$.

i Show that $f'(x) = 0$ for $0 < x < 1$.

(3 marks) Medium

ii Sketch the graph of $y = f(x)$.

(2 marks) Medium

(Q5c, **2002 HSC**)

78 Use the table of standard integrals to find the exact value of

$$\int_0^2 \frac{dx}{\sqrt{16-x^2}}.$$

(2 marks)

(Q1a, **2001 HSC**) Easy

79 Use the substitution $u = 1 + x$ to evaluate

$$15\int_{-1}^{0} x\sqrt{1+x}\,dx.$$

(3 marks)

(Q1f, **2001 HSC**) Easy

80 Find

i $\int \frac{e^x}{1+e^x}\,dx$ *(1 mark)* Easy

ii $\int_0^{\pi} \cos^2 3x\,dx$ *(3 marks)* Easy

(Q2b, **2001 HSC**)

81 Differentiate $x\sin^{-1}x$. *(2 marks)*

(Q1a, **2000 HSC**) Easy

82 Evaluate $\int_0^{\sqrt{3}} \frac{4}{x^2+9}\,dx$. *(3 marks)*

(Q1d, **2000 HSC**) Easy

83 Use the substitution $u = 2 + x$ to find

$$\int \frac{x}{\sqrt{2+x}}\,dx.$$

(3 marks)

(Q2d, **2000 HSC**) Medium

84 Consider the function $f(x) = 3\tan^{-1}x$.

i State the range of the function $y = f(x)$.

Easy

ii Sketch the graph of $y = f(x)$.

Easy

iii Find the gradient of the tangent to the curve $y = f(x)$ at $x = \frac{1}{\sqrt{3}}$. Easy

(5 marks) (Q3b, **2000 HSC**)

85 Evaluate $\int_0^{\sqrt{3}} \frac{dx}{\sqrt{4-x^2}}$. *(2 marks)*

(Q1a, **1999 HSC**) Easy

86 Use the substitution $u = \tan x$ to evaluate

$$\int_0^{\frac{\pi}{3}} \tan^2 x\,\sec^2 x\,dx.$$

(3 marks)

(Q1f, **1999 HSC**) Medium

87 i By equating the coefficients of $\sin x$ and $\cos x$, or otherwise, find constants A and B satisfying the identity

$A(2\sin x + \cos x) + B(2\cos x - \sin x)$
$\equiv \sin x + 8\cos x$. Medium

ii Hence evaluate $\int \frac{\sin x + 8\cos x}{2\sin x + \cos x}\,dx$.

Medium

(4 marks) (Q3d, **1999 HSC**)

88 Differentiate $2x\tan^{-1}x$. *(2 marks)*

(Q1a, **1998 HSC**) Easy

89 Evaluate $\int_0^{\frac{\pi}{3}} \sin^2 x\,dx$. *(4 marks)*

(Q1f, **1998 HSC**) Easy

90 Evaluate $\int_0^2 \frac{dx}{4+x^2}$. *(3 marks)*

(Q1d, **1997 HSC**) Easy

91 Using the substitution $u = 2x + 1$, or otherwise, find $\int_0^1 \frac{4x}{2x+1}\,dx$. *(4 marks)*

(Q1e, **1997 HSC**) Easy

92 By using the substitution $x = \sin t$, or otherwise, evaluate $\int_0^{\frac{1}{2}} \sqrt{1-x^2}\, dx$. *(3 marks)*

(Q4b, **1997 HSC**) Medium

93 The function $f(x) = \sec x$ for $0 \le x < \frac{\pi}{2}$, and is not defined for other values of x.

- **i** State the domain of the inverse function $f^{-1}(x)$. Easy
- **ii** Show that $f^{-1}(x) = \cos^{-1}\left(\frac{1}{x}\right)$. Easy
- **iii** Hence find $\frac{d}{dx} f^{-1}(x)$. Medium

(4 marks) (Q6a, **1997 HSC**)

94 Using the substitution $u = e^x$, find $\int \frac{e^x}{1+e^{2x}}\, dx$. *(3 marks)*

(Q1f, **1996 HSC**) Easy

95 Show that $\int_{\frac{\pi}{4}}^{\frac{\pi}{2}} \cos^2 x\, dx = \frac{\pi}{8} - \frac{1}{4}$. *(2 marks)*

(Q3b, **1996 HSC**) Easy

96 The function $h(x)$ is given by

$$h(x) = \sin^{-1}x + \cos^{-1}x, 0 \le x \le 1.$$

- **i** Find $h'(x)$. Easy
- **ii** Sketch the graph of $y = h(x)$. Easy

(3 marks) (Q3d, **1996 HSC**)

97 Use the substitution $u = 9 - x^2$ to find $\int_0^1 6x\sqrt{9-x^2}\, dx$. *(4 marks)*

(Q1e, **1995 HSC**) Medium

98 Evaluate $\int_2^{10} \frac{x}{\sqrt{x-1}}\, dx$ using the substitution $x = t^2 + 1$.

(Q1b, **1994 HSC**) Medium

99 Evaluate $\int_0^{\frac{\pi}{3}} 3 \sin x \cos^2 x\, dx$.

(Q3b, **1994 HSC**) Medium

100 Consider the function $f(x) = 3x - x^3$.

- **i** Sketch $y = f(x)$, showing the x and y intercepts and the coordinates of the stationary points. Easy
- **ii** Find the largest domain containing the origin for which $f(x)$ has an inverse function, $f^{-1}(x)$. Easy
- **iii** State the domain of $f^{-1}(x)$. Easy
- **iv** Find the gradient of the inverse function at $x = 0$. Hard

(Q6a, **1994 HSC**

101 Evaluate $\int_{\frac{1}{2}}^{1} 4t(2t-1)^5\, dt$ by using the substitution $u = 2t - 1$.

(Q1c, **1993 HSC**) Easy

102 Find the exact value of $\int_1^{\sqrt{3}} \frac{1}{\sqrt{4-x^2}}\, dx$.

(Q1c, **1992 HSC**) Easy

103 Consider the function $f(x) = 2 \tan^{-1}x$.

- **i** Evaluate $f(\sqrt{3})$. Easy
- **ii** Draw the graph of $y = f(x)$, labelling any key features. Easy
- **iii** Find the slope of the curve at the point where it cuts the y axis. Medium

(Q3b, **1992 HSC**)

104 Evaluate $\int_0^1 \frac{2x}{(2x+1)^2}\, dx$ by using the substitution $u = 2x + 1$.

(Q4a, **1992 HSC**) Medium

Year 12 Further calculus skills—Worked Answers

1 Let $u = x^2 + 4$

$\frac{du}{dx} = 2x$

$du = 2x\,dx$

When $x = 1, u = 5$ and $x = 0, u = 4$ ✓

$$\int_0^1 \frac{x}{\sqrt{x^2+4}}\,dx = \frac{1}{2}\int_0^1 \frac{2x}{\sqrt{x^2+4}}\,dx$$

$$= \frac{1}{2}\int_4^5 \frac{1}{\sqrt{u}}\,du \quad ✓$$

$$= \frac{1}{2}\left[2\sqrt{u}\right]_4^5$$

$$= \left[\sqrt{5} - \sqrt{4}\right]$$

$$= \sqrt{5} - 2 \quad ✓$$

(3 marks)

2 From the formula sheet

$\sin^2 nx = \frac{1}{2}(1 - \cos 2nx)$, then

$$\sin^2 3x = \frac{1}{2}(1 - \cos 6x)$$

$$\int \sin^2 3x\,dx = \int \frac{1-\cos 6x}{2}\,dx$$

Answer B

3 If $u = x + 1$, then $x = u - 1$.

$\frac{du}{dx} = 1$

$du = dx$

$$\int x\sqrt{x+1}\,dx$$

$$= \int (u-1)\sqrt{u}\,du \quad ✓$$

$$= \int \left(u^{\frac{3}{2}} - u^{\frac{1}{2}}\right)du$$

$$= \frac{2u^{\frac{5}{2}}}{5} - \frac{2u^{\frac{3}{2}}}{3} + C \quad ✓$$

$$= \frac{2\sqrt{(x+1)^5}}{5} - \frac{2\sqrt{(x+1)^3}}{3} + C \quad ✓$$

(3 marks)

4 $$\int_0^{\sqrt{3}} \frac{1}{\sqrt{4-x^2}}\,dx$$

$$= \left[\sin^{-1}\left(\frac{x}{2}\right)\right]_0^{\sqrt{3}} \quad ✓$$

$$= \left[\sin^{-1}\left(\frac{\sqrt{3}}{2}\right) - \sin^{-1}\left(\frac{0}{2}\right)\right]$$

$$= \frac{\pi}{3} \quad ✓$$

(2 marks)

5 If $g(x) = x^3 + 4x - 2$ passes through the point $(1, 3)$, then $g^{-1}(3) = 1$

From the product rule:

$$f'(x) = x \cdot \frac{d}{dx}(g^{-1}(x)) + 1 \cdot g^{-1}(x)$$

Inverse functions have a gradient relationship where the gradient of $g(x)$ at (a, b) is the reciprocal of the gradient of $g^{-1}(x)$ at (b, a).

$\therefore$ The gradient of $g(x)$ at $x = 1$ is the reciprocal of the gradient of $g^{-1}(x)$ at $x = 3$

$g'(x) = 3x^2 + 4$

$g'(1) = 3(1)^2 + 4$

$g'(1) = 7$ ✓

$\therefore$ The gradient of $g^{-1}(x)$ at $x = 3$ is $\frac{1}{7}$

$$\therefore f'(3) = 3 \cdot \frac{1}{7} + 1 \cdot 1$$

$$\therefore f'(3) = \frac{10}{3} \quad ✓$$

(2 marks)

6 $$\int \frac{1}{4x^2+1}\,dx = \int \frac{1}{1+(2x)^2}\,dx$$

$$= \frac{1}{2}\int \frac{2}{1+(2x)^2}\,dx$$

$$= \frac{1}{2}\tan^{-1}(2x) + c$$

Answer D

7 $$\int_0^{\frac{\pi}{2}} \cos 5x \sin 3x\,dx$$

$$= \int_0^{\frac{\pi}{2}} \frac{1}{2}(\sin 8x - \sin 2x)\,dx \quad ✓$$

$$= \left[\frac{1}{2}\left(\frac{-\cos 8x}{8} - \frac{-\cos 2x}{2}\right)\right]_0^{\frac{\pi}{2}} \quad ✓$$

$$= \left[\frac{-\cos 8x}{16} + \frac{\cos 2x}{4}\right]_0^{\frac{\pi}{2}}$$

$$= \frac{-\cos 4\pi}{16} + \frac{\cos \pi}{4} - \left(\frac{-\cos 0}{16} + \frac{\cos 0}{4}\right)$$

$$= \frac{-1}{16} + \frac{-1}{4} + \frac{1}{16} - \frac{1}{4}$$

$$= -\frac{1}{2} \quad ✓$$

(3 marks)

8 **i** $\frac{d}{d\theta}\left(\sin^3\theta\right) = 3\sin^2\theta\cos\theta$ ✓

(1 mark)

ii $x = \tan\theta$

$dx = \sec^2\theta\,d\theta$

When $x = 0, \theta = 0$

When $x = 1, \theta = \frac{\pi}{4}$ ✓

$$\int_0^1 \frac{x^2}{(1+x^2)^{\frac{5}{2}}}\,dx$$

$$= \int_0^{\frac{\pi}{4}} \frac{\tan^2\theta}{(1+\tan^2\theta)^{\frac{5}{2}}}\sec^2\theta\,d\theta \quad ✓$$

$$= \int_0^{\frac{\pi}{4}} \frac{\tan^2\theta}{(\sec^2\theta)^{\frac{5}{2}}}\sec^2\theta\,d\theta$$

$$= \int_0^{\frac{\pi}{4}} \frac{\tan^2\theta}{\sec^3\theta}\,d\theta$$

$$= \int_0^{\frac{\pi}{4}} \left(\frac{\sin^2\theta}{\cos^2\theta} \times \cos^3\theta\right)d\theta$$

$$= \int_0^{\frac{\pi}{4}} \sin^2\theta\cos\theta\,d\theta \quad ✓$$

$$= \frac{1}{3}\left[\sin^3\theta\right]_0^{\frac{\pi}{4}} \quad \text{(from **i**)}$$

$$= \frac{1}{3}\left(\sin^3\frac{\pi}{4} - \sin^3 0\right)$$

$$= \frac{1}{3}\left(\left(\frac{1}{\sqrt{2}}\right)^3 - 0^3\right)$$

$$= \frac{1}{6\sqrt{2}} \quad ✓$$

(4 marks)

9 **i** $f(x) = \tan(\cos^{-1}(x))$

$f'(x)$

$$= \sec^2(\cos^{-1}(x)) \times \frac{-1}{\sqrt{1-x^2}}$$

$$= \frac{1}{(\cos(\cos^{-1}x))^2} \times \frac{-1}{\sqrt{1-x^2}}$$

$$= \frac{-1}{x^2\sqrt{1-x^2}} \quad ✓$$

Now $g(x) = \frac{\sqrt{1-x^2}}{x}$

$g'(x) =$

$$\frac{x \times \frac{1}{2}(1-x^2)^{-\frac{1}{2}} \times -2x - \sqrt{1-x^2} \times 1}{x^2} \quad ✓$$

$$= \frac{\frac{-x^2}{\sqrt{1-x^2}} - \sqrt{1-x^2}}{x^2}$$

$$= \frac{\frac{-x^2}{\sqrt{1-x^2}} - \frac{1-x^2}{\sqrt{1-x^2}}}{x^2} \quad ✓$$

$$= \frac{\frac{-1}{\sqrt{1-x^2}}}{x^2}$$

$$= \frac{-1}{x^2\sqrt{1-x^2}}$$

$\therefore f'(x) = g'(x)$ ✓ *(4 marks)*

ii $\cos^{-1}x$ has domain $[-1, 1]$.

$\cos^{-1}0 = \frac{\pi}{2}$ and $\tan\frac{\pi}{2}$ is undefined.

So both $f(x)$ and $g(x)$ have domain $[-1, 0) \cup (0, 1]$. ✓

From **i** the curves have the same gradient function.

So they only, at most, differ by a constant.

$\therefore f(x) = g(x) + c$ ✓

From the graph $g(1) = 0$.

$f(x) = \tan(\cos^{-1}(x))$

$f(1) = \tan(\cos^{-1}1)$

$= \tan 0$

$= 0$

So $c = 0$

$\therefore f(x) = g(x)$ ✓

OR: Let $\theta = \cos^{-1}(x)$

So $\cos\theta = x$

$f(x) = \tan(\cos^{-1}(x))$

$= \tan\theta$

$$= \frac{\sin\theta}{\cos\theta} \quad (\cos\theta \neq 0)$$

Now $0 \le \theta \le \pi$.
So $\sin\theta \ge 0$.
Now $\sin^2\theta = 1 - \cos^2\theta$.
So $\sin\theta = \sqrt{1-\cos^2\theta}$
$= \sqrt{1-x^2}$

$\therefore f(x) = \dfrac{\sqrt{1-x^2}}{x} = g(x)$ *(3 marks)*

10 $\displaystyle\int_0^{\frac{\pi}{3}} \cos\frac{3x}{2}\cos\frac{x}{2}\,dx$

$= \dfrac{1}{2}\displaystyle\int_0^{\frac{\pi}{3}}(\cos 2x + \cos x)\,dx$ ✓

$= \dfrac{1}{2}\left[\dfrac{1}{2}\sin 2x + \sin x\right]_0^{\frac{\pi}{3}}$ ✓

$= \dfrac{1}{2}\left[\dfrac{1}{2}\sin 2\left(\dfrac{\pi}{3}\right) + \sin\dfrac{\pi}{3} - 0\right]$

$= \dfrac{1}{2}\left[\dfrac{1}{2}\left(\dfrac{\sqrt{3}}{2}\right) + \dfrac{\sqrt{3}}{2} - 0\right]$

$= \dfrac{1}{2}\left[\dfrac{\sqrt{3}}{4} + \dfrac{\sqrt{3}}{2}\right]$

$= \dfrac{1}{2}\left[\dfrac{3\sqrt{3}}{4}\right]$

$= \dfrac{3\sqrt{3}}{8}$ ✓ *(3 marks)*

11 $\displaystyle\int \sin^2\left(\frac{x}{2}\right)dx = \int \frac{1}{2}(1 - \cos x)\,dx$

$= \dfrac{1}{2}(x - \sin x) + C$ ✓ *(1 mark)*

12 $u = \cos 2x \qquad \dfrac{du}{dx} = -2\sin 2x$

$dx = \dfrac{du}{-2\sin 2x}$

Also, $x = \dfrac{\pi}{6}, u = \dfrac{1}{2}$;
$x = 0, u = 1$ ✓

$\displaystyle\int_0^{\frac{\pi}{6}} \cos^2 2x \sin 2x\,dx$

$= \displaystyle\int_1^{\frac{1}{2}} u^2 \sin 2x.\frac{du}{-2\sin 2x}$

$= \dfrac{1}{2}\displaystyle\int_{\frac{1}{2}}^{1} u^2\,du$ ✓

$= \dfrac{1}{2}\left[\dfrac{u^3}{3}\right]_{\frac{1}{2}}^{1}$

$= \dfrac{1}{2}\left[\dfrac{1^3}{3} - \dfrac{\left(\frac{1}{2}\right)^3}{3}\right]$ ✓

$= \dfrac{1}{2}\left[\dfrac{1}{3} - \dfrac{1}{24}\right]$

$= \dfrac{7}{48}$ ✓ *(4 marks)*

13 $u = \cos 2x$

$\dfrac{du}{dx} = -2\sin 2x$

$dx = \dfrac{du}{-2\sin 2x} \qquad x = \dfrac{\pi}{8}$,

then $u = \dfrac{1}{\sqrt{2}}$; $x = 0$, then $u = 1$ ✓

$\displaystyle\int_0^{\frac{\pi}{8}} 2\cos^3 2x \sin 2x\,dx.$

$= \displaystyle\int_1^{\frac{1}{\sqrt{2}}} 2u^3 \sin 2x.\frac{du}{-2\sin 2x}$

$= \displaystyle\int_{\frac{1}{\sqrt{2}}}^{1} u^3 du$ ✓

$= \left[\dfrac{u^4}{4}\right]_{\frac{1}{\sqrt{2}}}^{1}$

$= \dfrac{1}{4}\left[1^4 - \left(\dfrac{1}{\sqrt{2}}\right)^4\right]$

$= \dfrac{1}{4}\left[1 - \dfrac{1}{4}\right]$

$= \dfrac{3}{16}$ ✓ *(3 marks)*

14 $u = \sin x$

$x = \dfrac{\pi}{6}$, then $u = \dfrac{1}{2}$

$\dfrac{du}{dx} = \cos x$

$x = 0$, then $u = 0$

$dx = \dfrac{du}{\cos x}$ ✓

$\displaystyle\int_0^{\frac{\pi}{6}} \cos^3 x\,dx$

$= \displaystyle\int_0^{\frac{\pi}{6}} \cos x(1 - \sin^2 x)\,dx$

$= \displaystyle\int_0^{\frac{1}{2}} \cos x(1 - u^2).\frac{du}{\cos x}$

$= \displaystyle\int_0^{\frac{1}{2}} (1 - u^2)\,du$ ✓

$= \left[u - \dfrac{u^3}{3}\right]_0^{\frac{1}{2}}$

$= \dfrac{1}{2} - \dfrac{\left(\frac{1}{2}\right)^3}{3} - 0$

$= \dfrac{1}{2} - \dfrac{1}{24}$

$= \dfrac{11}{24}$ ✓ *(3 marks)*

15 $u = \cos x$

$\dfrac{du}{dx} = -\sin x \qquad dx = \dfrac{du}{-\sin x}$

As $x = \dfrac{\pi}{4}, u = \dfrac{1}{\sqrt{2}}$; $x = \dfrac{\pi}{6}, u = \dfrac{\sqrt{3}}{2}$ ✓

$\displaystyle\int_{\frac{\pi}{6}}^{\frac{\pi}{4}} \sin x \cos^3 dx$

$= \displaystyle\int_{\frac{\sqrt{3}}{2}}^{\frac{1}{\sqrt{2}}} \sin x \cdot u^3 \cdot \frac{du}{-\sin x}$

$= -\displaystyle\int_{\frac{\sqrt{3}}{2}}^{\frac{1}{\sqrt{2}}} u^3\,du$

$= \displaystyle\int_{\frac{1}{\sqrt{2}}}^{\frac{\sqrt{3}}{2}} u^3\,du$ ✓

$= \left[\dfrac{u^4}{4}\right]_{\frac{1}{\sqrt{2}}}^{\frac{\sqrt{3}}{2}}$

$= \dfrac{\left(\frac{\sqrt{3}}{2}\right)^4}{4} - \dfrac{\left(\frac{1}{\sqrt{2}}\right)^4}{4}$

$= \dfrac{9}{64} - \dfrac{1}{16}$

$= \dfrac{5}{64}$ ✓ *(3 marks)*

16 $\displaystyle\int_0^{\frac{\pi}{4}} 4\sin x \cos\frac{x}{3}\,dx.$

$= 2\displaystyle\int_0^{\frac{\pi}{4}} 2\sin x\cos\frac{x}{3}\,dx$

$= 2\displaystyle\int_0^{\frac{\pi}{4}}\left(\sin\frac{4x}{3} + \sin\frac{2x}{3}\right)dx$ ✓

$= 2\left[-\dfrac{3}{4}\cos\dfrac{4x}{3} - \dfrac{3}{2}\cos\dfrac{2x}{3}\right]_0^{\frac{\pi}{4}}$

$= -\dfrac{3}{2}\left[\cos\dfrac{4x}{3} + 2\cos\dfrac{2x}{3}\right]_0^{\frac{\pi}{4}}$ ✓

$= -\dfrac{3}{2}\left[\cos\dfrac{\pi}{3} + 2\cos\dfrac{\pi}{6} - (\cos 0 + 2\cos 0)\right]$

$= -\dfrac{3}{2}\left[\dfrac{1}{2} + \sqrt{3} - (1 + 2)\right]$

$= -\dfrac{3}{2}\left[-\dfrac{5}{2} + \sqrt{3}\right]$

$= \dfrac{3}{4}\left[5 - 2\sqrt{3}\right]$ ✓ *(3 marks)*

17 $\displaystyle\int_{\frac{\pi}{12}}^{\frac{\pi}{3}} \sin 3x \sin x\,dx$

$= \displaystyle\int_{\frac{\pi}{12}}^{\frac{\pi}{3}} \frac{1}{2}(\cos 2x - \cos 4x)\,dx$ ✓

$= \dfrac{1}{2}\left[\dfrac{1}{2}\sin 2x - \dfrac{1}{4}\sin 4x\right]_{\frac{\pi}{12}}^{\frac{\pi}{3}}$ ✓

$= \dfrac{1}{2}\left[\dfrac{1}{2}\sin 2\left(\dfrac{\pi}{3}\right) - \dfrac{1}{4}\sin 4\left(\dfrac{\pi}{3}\right) - \left(\dfrac{1}{2}\sin 2\left(\dfrac{\pi}{12}\right) - \dfrac{1}{4}\sin 4\left(\dfrac{\pi}{12}\right)\right)\right]$

$= \dfrac{1}{2}\left[\dfrac{1}{2}\left(\dfrac{\sqrt{3}}{2}\right) - \dfrac{1}{4}\left(-\dfrac{\sqrt{3}}{2}\right) - \left(\dfrac{1}{2}\left(\dfrac{1}{2}\right) - \dfrac{1}{4}\left(\dfrac{\sqrt{3}}{2}\right)\right)\right]$

$= \dfrac{1}{2}\left[\dfrac{\sqrt{3}}{4} + \dfrac{\sqrt{3}}{8} - \dfrac{1}{4} + \dfrac{\sqrt{3}}{8}\right]$

$= \dfrac{-1 + 2\sqrt{3}}{8}$ ✓ *(3 marks)*

18 $\displaystyle\int_0^{\frac{\pi}{3}} \sin 3x \cos x\,dx$

$= \displaystyle\int_0^{\frac{\pi}{3}} \frac{1}{2}(\sin 4x + \sin 2x)\,dx$

$= \dfrac{1}{2}\displaystyle\int_0^{\frac{\pi}{3}}(\sin 4x + \sin 2x)\,dx$ ✓

$= \dfrac{1}{2}\left[-\dfrac{1}{4}\cos 4x - \dfrac{1}{2}\cos 2x\right]_0^{\frac{\pi}{3}}$ ✓

$$= \frac{1}{2}\left[-\frac{1}{4}\cos\frac{4\pi}{3} + \frac{1}{2}\cos\frac{2\pi}{3} - \frac{1}{4}\cos 0 - \frac{1}{2}\cos 0\right]$$

$$= \frac{1}{2}\left[-\frac{1}{4}\left(-\frac{1}{2}\right) - \frac{1}{2}\left(-\frac{1}{2}\right) - \left(-\frac{1}{4} - \frac{1}{2}\right)\right]$$

$$= \frac{1}{2}\left[\frac{1}{8} + \frac{1}{4} + \frac{1}{4} + \frac{1}{2}\right]$$

$$= \frac{1}{2}\left(\frac{9}{8}\right)$$

$$= \frac{9}{16}$$ ✓

(3 marks)

19 $y = \tan^{-1}\frac{x}{2}$

$$\frac{dy}{dx} = \frac{1}{1+\left(\frac{x}{2}\right)^2} \times \frac{1}{2}$$

$$= \frac{4}{4+x^2} \times \frac{1}{2}$$

$$= \frac{2}{4+x^2}$$

Answer C

20

$$\cos 8x = \cos^2 4x - \sin^2 4x$$
$$= 1 - 2\sin^2 4x$$

$\therefore 2\sin^2 4x = 1 - \cos 8x$ ✓

$$\int 2\sin^2 4x\,dx = \int (1 - \cos 8x)\,dx$$

$$= x - \frac{1}{8}\sin 8x + C$$ ✓

(2 marks)

21 Let $u = \cos^2 x$

$du = -2\cos x \sin x\,dx$

$= -\sin 2x\,dx$

When $x = 0$, $u = 1$

When $x = \frac{\pi}{4}$, $u = \frac{1}{2}$ ✓

$$\text{So } \int_0^{\frac{\pi}{4}} \frac{\sin 2x}{4+\cos^2 x}\,dx = \int_1^{\frac{1}{2}} \frac{-1}{4+u}\,du$$

$$= \int_{\frac{1}{2}}^1 \frac{1}{4+u}\,du$$ ✓

$$= \left[\ln(4+u)\right]_{\frac{1}{2}}^1$$

$$= \ln(4+1) - \ln\left(4+\frac{1}{2}\right)$$

$$= \ln\frac{5}{4.5}$$

$$= \ln\frac{10}{9}$$ ✓

(3 marks)

22 Let $u = 1 - x$

$x = 1 - u$

$dx = -du$

When $x = 0$, $u = 1$

When $x = -3$, $u = 4$ ✓

$$\therefore \int_{-3}^0 \frac{x}{\sqrt{1-x}}\,dx = \int_4^1 \frac{1-u}{\sqrt{u}} \cdot -du$$

$$= \int_1^4 \left(u^{-\frac{1}{2}} - u^{\frac{1}{2}}\right) du$$ ✓

$$= \left[\frac{u^{\frac{1}{2}}}{\frac{1}{2}} - \frac{u^{\frac{3}{2}}}{\frac{3}{2}}\right]_1^4$$

$$= \left[2\sqrt{u} - \frac{2u\sqrt{u}}{3}\right]_1^4$$

$$= 2\times\sqrt{4} - \frac{2\times4\times\sqrt{4}}{3} - \left(2\times\sqrt{1} - \frac{2\times1\times\sqrt{1}}{3}\right)$$

$$= -2\frac{2}{3}$$ ✓

(3 marks)

23 $\cos 2\theta = \cos^2\theta - \sin^2\theta$

$= 2\cos^2\theta - 1$

So $\cos^2\theta = \frac{1}{2}(\cos 2\theta + 1)$

$$\int \cos^2(3x)\,dx = \int \frac{1}{2}(\cos 6x + 1)\,dx$$ ✓

$$= \frac{1}{12}\sin 6x + \frac{x}{2} + C$$ ✓

(2 marks)

24 **i**

$\sin\theta = \frac{h}{20}$

$h = 20\sin\theta$

$\frac{dh}{d\theta} = 20\cos\theta$ ✓

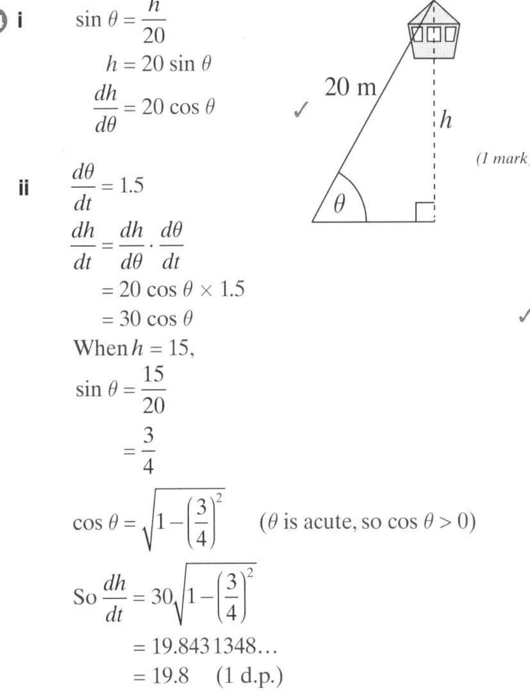

(1 mark)

ii

$\frac{d\theta}{dt} = 1.5$

$\frac{dh}{dt} = \frac{dh}{d\theta}\cdot\frac{d\theta}{dt}$

$= 20\cos\theta \times 1.5$

$= 30\cos\theta$ ✓

When $h = 15$,

$\sin\theta = \frac{15}{20}$

$= \frac{3}{4}$

$\cos\theta = \sqrt{1-\left(\frac{3}{4}\right)^2}$ (θ is acute, so $\cos\theta > 0$)

So $\frac{dh}{dt} = 30\sqrt{1-\left(\frac{3}{4}\right)^2}$

$= 19.843\,1348\ldots$

$= 19.8$ (1 d.p.)

The top of the carriage is rising at 19.8 m per minute, correct to one decimal place. ✓

(2 marks)

25 **i** $f(x) = \sin^{-1}x + \cos^{-1}x$

$$f'(x) = \frac{1}{\sqrt{1-x^2}} + \frac{-1}{\sqrt{1-x^2}}$$

$= 0$ ✓

(1 mark)

ii $f'(x) = 0$

$\therefore f(x)$ is constant.

When $x = 0$,

$f(x) = \sin^{-1}0 + \cos^{-1}0$

$= 0 + \frac{\pi}{2}$

$= \frac{\pi}{2}$

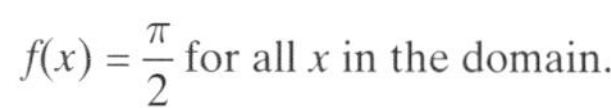

$f(x) = \frac{\pi}{2}$ for all x in the domain.

$\sin^{-1}x + \cos^{-1}x = \frac{\pi}{2}$ ✓

(1 mark)

iii $\sin^{-1}x$ and $\cos^{-1}x$ have domain $-1 \le x \le 1$.

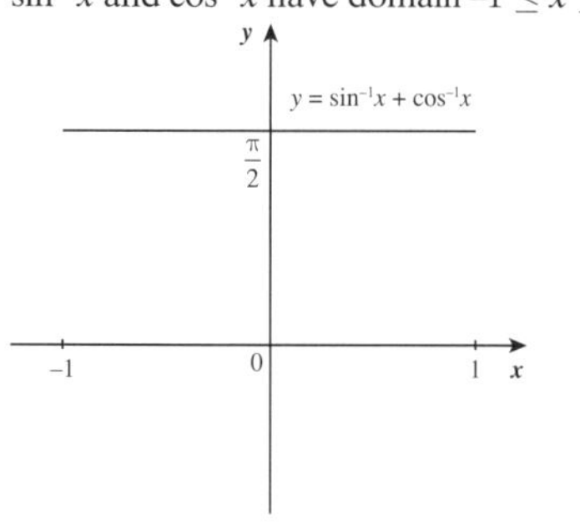

✓

(1 mark)

---

26 $y = \tan^{-1}(x^3)$

$$\frac{dy}{dx} = \frac{1}{1+(x^3)^2} \times 3x^2$$ ✓

$$= \frac{3x^2}{1+x^6}$$ ✓

(2 marks)

---

27 $\int_0^3 \frac{x}{\sqrt{x+1}}\,dx$

Let $x = u^2 - 1$

$dx = 2u\,du$

$u^2 = x + 1$

$u = \sqrt{x+1} \quad (u > 0)$

When $x = 0, u = 1$

When $x = 3, u = 2$ ✓

$$\therefore \int_0^3 \frac{x}{\sqrt{x+1}}\,dx = \int_1^2 \frac{u^2-1}{u}\cdot 2u\,du$$

$$= \int_1^2 (2u^2 - 2)\,du$$ ✓

$$= \left[\frac{2u^3}{3} - 2u\right]_1^2$$

$$= \frac{2\times 2^3}{3} - 2\times 2 - \left(\frac{2\times 1^3}{3} - 2\times 1\right)$$

$$= 2\frac{2}{3}$$ ✓

(3 marks)

---

28 Let $u = \sin x$

$du = \cos x\,dx$

$$\int \sin^2 x \cos x\,dx = \int u^2\,du$$

$$= \frac{u^3}{3} + C$$

$$= \frac{1}{3}\sin^3 x + C$$ ✓

(1 mark)

---

29 **i** $F'(t) = 50e^{-0.5t} - 0.4F(t)$

$\frac{d}{dt}(F(t)e^{0.4t})$

$= F(t) \times 0.4e^{0.4t} + e^{0.4t} \times F'(t)$

$= 0.4F(t)e^{0.4t} + e^{0.4t}(50e^{-0.5t} - 0.4F(t))$ ✓

$= 0.4F(t)e^{0.4t} + 50e^{-0.1t} - 0.4F(t)e^{0.4t}$

$= 50e^{-0.1t}$ ✓

(2 marks)

ii $F(t)e^{0.4t} = \int 50e^{-0.1t}\,dt$

$= \frac{50e^{-0.1t}}{-0.1} + C$

$= -500e^{-0.1t} + C$ ✓

Now $F(0) = 0$

When $t = 0$,

$0 = -500e^0 + C$

$C = 500$

So $F(t)e^{0.4t} = 500(1 - e^{-0.1t})$

$F(t) = \frac{500(1-e^{-0.1t})}{e^{0.4t}}$

$= 500(e^{-0.4t} - e^{-0.5t})$ ✓

(2 marks)

iii $F'(t) = 50e^{-0.5t} - 0.4F(t)$

$= 50e^{-0.5t} - 0.4 \times 500(e^{-0.4t} - e^{-0.5t})$

$= 50e^{-0.5t} - 200e^{-0.4t} + 200e^{-0.5t}$

$= 250e^{-0.5t} - 200e^{-0.4t}$ ✓

When $F'(t) = 0$,

$250e^{-0.5t} = 200e^{-0.4t}$

$\frac{250}{200} = \frac{e^{-0.4t}}{e^{-0.5t}}$

$1.25 = e^{0.1t}$

$\ln 1.25 = 0.1t$

$t = 10\ln 1.25$

$F''(t) = -125e^{-0.5t} + 80e^{-0.4t}$

$F''(10\ln 1.25) < 0$ (maximum)

The maximum concentration occurs when $t = 10\ln 1.25$ hours ($\approx 2\frac{1}{4}$ h). ✓

(2 marks)

---

30 $\cos 4x = \cos^2 2x - \sin^2 2x$

$= 1 - 2\sin^2 2x$

So $\sin^2 2x = \frac{1}{2}(1 - \cos 4x)$

$$\int \sin^2 2x\,dx = \frac{1}{2}\int (1 - \cos 4x)\,dx$$

$$= \frac{1}{2}(x - \frac{1}{4}\sin 4x) + c$$

Answer A

---

31 $u = x - 4$

$du = dx$ ✓

$$\int x\sqrt{x-4}\,dx = \int (u+4)\sqrt{u}\,du$$

$$= \int \left(u^{\frac{3}{2}} + 4u^{\frac{1}{2}}\right)du$$ ✓

$$= \frac{2u^{\frac{5}{2}}}{5} + \frac{8u^{\frac{3}{2}}}{3} + c$$

$$= \frac{2(x-4)^2\sqrt{x-4}}{5} + \frac{8(x-4)\sqrt{x-4}}{3} + c$$

$$= \frac{2(x-4)\sqrt{x-4}(3(x-4)+20)}{15} + c$$

$$= \frac{2(3x+8)(x-4)\sqrt{x-4}}{15} + c$$ ✓

(3 marks)

---

32 $y = 3\tan^{-1}(2x)$

$$\frac{dy}{dx} = 3 \times \frac{1}{1+(2x)^2} \times 2$$ ✓

$$= \frac{6}{1+4x^2}$$ ✓

(2 marks)

33 $\int_0^k \frac{1}{\sqrt{4-x^2}}\,dx = \left[\sin^{-1}\left(\frac{x}{2}\right)\right]_0^k$

$= \sin^{-1}\left(\frac{k}{2}\right) - \sin^{-1}0$

$= \sin^{-1}\left(\frac{k}{2}\right)$

So $\sin^{-1}\left(\frac{k}{2}\right) = \frac{\pi}{3}$

$\frac{k}{2} = \sin\frac{\pi}{3}$

$= \frac{\sqrt{3}}{2}$

$k = \sqrt{3}$

Answer B

34 $\int \sin^2 x\,dx = \int \frac{1}{2}(1 - \cos 2x)\,dx$ ✓

$= \frac{1}{2}(x - \frac{1}{2}\sin 2x) + C$

$= \frac{x}{2} - \frac{1}{4}\sin 2x + C$ ✓

(2 marks)

35 Let $u = 2x - 1$

$\frac{du}{dx} = 2$

$du = 2\,dx$

$2x = u + 1$

$x = \frac{1}{2}(u + 1)$

When $x = 1, u = 1$

When $x = 2, u = 3$ ✓

$\therefore \int_1^2 \frac{x}{(2x-1)^2}\,dx$

$= \int_1^3 \frac{\frac{1}{2}(u+1)}{u^2}\cdot\frac{1}{2}du$

$= \frac{1}{4}\int_1^3 \frac{u+1}{u^2}\,du$

$= \frac{1}{4}\int_1^3 \left(\frac{1}{u} + \frac{1}{u^2}\right)du$ ✓

$= \frac{1}{4}\left[\ln u - \frac{1}{u}\right]_1^3$

$= \frac{1}{4}\left(\left(\ln 3 - \frac{1}{3}\right) - (\ln 1 - 1)\right)$

$= \frac{1}{4}\left(\ln 3 + \frac{2}{3}\right)$

$= \frac{1}{12}(3\ln 3 + 2)$ ✓

(3 marks)

36 **i** $f(x) = \cos^{-1}(x) + \cos^{-1}(-x)$

$f'(x) = \frac{-1}{\sqrt{1-x^2}} + \frac{-1}{\sqrt{1-(-x)^2}} \times -1$ ✓

$= -\frac{1}{\sqrt{1-x^2}} + \frac{1}{\sqrt{1-x^2}}$

$= 0$

So, as the derivative is zero, $f(x)$ is constant. ✓

(2 marks)

ii $f(x) = \cos^{-1}(x) + \cos^{-1}(-x)$

$f(0) = \cos^{-1}(0) + \cos^{-1}(-0)$

$= \frac{\pi}{2} + \frac{\pi}{2}$

$= \pi$

But $f(x)$ is constant, so $f(x) = \pi$ for all values of x $(-1 \le x \le 1)$

$\cos^{-1}(x) + \cos^{-1}(-x) = \pi$

$\cos^{-1}(-x) = \pi - \cos^{-1}(x)$ ✓

(1 mark)

37 $y = 3\sin^{-1}\frac{x}{2}$

$\frac{dy}{dx} = 3 \times \frac{1}{\sqrt{2^2 - x^2}}$

$= \frac{3}{\sqrt{4 - x^2}}$

Answer B

38 Let $x = u^2 + 1$

$dx = 2u\,du$

If $u = 1, x = 2$

if $u = 2, x = 5$ ✓

$\int_2^5 \frac{x}{\sqrt{x-1}}dx = \int_1^2 \frac{u^2+1}{u}2u\,du$

$= 2\int_1^2 (u^2 + 1)du$ ✓

$= 2\left[\frac{u^3}{3} + u\right]_1^2$

$= 2\left(\frac{2^3}{3} + 2 - \left(\frac{1^3}{3} + 1\right)\right)$

$= 6\frac{2}{3}$ ✓

(3 marks)

39 Let $u = 1 + 2x$

$2x = u - 1$

$x = \frac{1}{2}(u - 1)$

$dx = \frac{1}{2}du$

$\int x\sqrt{1 + 2x}\,dx = \int\frac{1}{2}(u - 1)\sqrt{u}\,\frac{1}{2}du$

$= \frac{1}{4}\int(u - 1)\sqrt{u}\,du$

Answer A

40 $\int\frac{1}{\sqrt{49 - 4x^2}}dx = \frac{1}{2}\int\frac{1}{\sqrt{\frac{49}{4} - x^2}}dx$ ✓

$= \frac{1}{2}\sin^{-1}\frac{2x}{7} + C$ ✓

(2 marks)

41 $u = e^{3x}$

$u^2 = (e^{3x})^2 = e^{6x}$

$du = 3e^{3x}\,dx$

When $x = 0, u = 1$

When $x = \frac{1}{3}, u = e$ ✓

$$\int_0^{\frac{1}{3}} \frac{e^{3x}}{e^{6x}+1}\,dx = \frac{1}{3}\int_1^e \frac{du}{u^2+1}$$

$$= \frac{1}{3}[\tan^{-1}u]_1^e \quad ✓$$

$$= \frac{1}{3}(\tan^{-1}e - \tan^{-1}1)$$

$$= \frac{1}{3}(\tan^{-1}e - \frac{\pi}{4}) \quad ✓$$

$$= 0.144\,2949\ldots$$

$$= 0.144 \text{ [3 d.p.]}$$

(3 marks)

42 $y = x^2\sin^{-1}5x$

$$\frac{dy}{dx} = x^2 \times \frac{1}{\sqrt{1-(5x)^2}} \times 5 + (\sin^{-1}5x) \times 2x \quad ✓$$

$$= \frac{5x^2}{\sqrt{1-25x^2}} + 2x\sin^{-1}5x \quad ✓$$

(2 marks)

43

$$\cos 2x = \cos^2 x - \sin^2 x$$

$$= 1 - 2\sin^2 x$$

$$2\sin^2 x = 1 - \cos 2x$$

$$\sin^2 x = \frac{1}{2}(1 - \cos 2x)$$

$$\int \sin^2 3x\,dx = \int \frac{1}{2}(1 - \cos 6x)\,dx$$

$$= \frac{1}{2}\left(x - \frac{1}{6}\sin 6x\right) + C$$

Answer C

44 $y = \cos^{-1}(3x)$

$$\frac{dy}{dx} = \frac{-1}{\sqrt{1-(3x)^2}} \times 3$$

$$= \frac{-3}{\sqrt{1-9x^2}}$$

Answer D

45

$$\int_0^3 \frac{1}{9+x^2}\,dx = \left[\frac{1}{3}\tan^{-1}\frac{x}{3}\right]_0^3 \quad ✓$$

$$= \frac{1}{3}\tan^{-1}1 - \frac{1}{3}\tan^{-1}0$$

$$= \frac{1}{3} \times \frac{\pi}{4} - \frac{1}{3} \times 0 \quad ✓$$

$$= \frac{\pi}{12} \quad ✓$$

(3 marks)

46 Let $u = 2 - x$

$du = -dx$

$x = 2 - u$

When $x = 1, u = 1$

When $x = 2, u = 0$ ✓

$$\int_1^2 x(2-x)^5 dx = \int_1^0 -(2-u)u^5 du$$

$$= \int_0^1 (2u^5 - u^6)\,du \quad ✓$$

$$= \left[\frac{u^6}{3} - \frac{u^7}{7}\right]_0^1$$

$$= \frac{1}{3} - \frac{1}{7} - 0$$

$$= \frac{4}{21} \quad ✓$$

(3 marks)

47

$$u = \sqrt{x}$$

$$= x^{\frac{1}{2}}$$

$$du = \frac{1}{2}x^{-\frac{1}{2}}\,dx$$

$$= \frac{1}{2\sqrt{x}}\,dx$$

$$2\,du = \frac{1}{\sqrt{x}}\,dx$$

When $x = 1, u = 1$

When $x = 4, u = 2$ ✓

$$\int_1^4 \frac{e^{\sqrt{x}}}{\sqrt{x}}\,dx = \int_1^2 2e^u\,du \quad ✓$$

$$= \left[2e^u\right]_1^2$$

$$= 2e^2 - 2e^1$$

$$= 2e(e-1) \quad ✓$$

(3 marks)

48

$$u = 1 - x$$

$$du = -dx$$

$$x = 1 - u$$

When $x = 0,\ u = 1 - 0 = 1$

When $x = 1,\ u = 1 - 1 = 0$ ✓

$$\therefore \int_0^1 x\sqrt{1-x}\,dx$$

$$= -\int_1^0 (1-u)\sqrt{u}\,du$$

$$= \int_0^1 (1-u)u^{\frac{1}{2}}\,du$$

$$= \int_0^1 (u^{\frac{1}{2}} - u^{\frac{3}{2}})\,du \quad ✓$$

$$= \left[\frac{u^{\frac{3}{2}}}{\frac{3}{2}} - \frac{u^{\frac{5}{2}}}{\frac{5}{2}}\right]_0^1$$

$$= \frac{2(1^{\frac{3}{2}})}{3} - \frac{2(1^{\frac{5}{2}})}{5} - \left(\frac{2(0^{\frac{3}{2}})}{3} - \frac{2(0^{\frac{5}{2}})}{5}\right)$$

$= \frac{2}{3} - \frac{2}{5}$

$= \frac{4}{15}$ ✓

(3 marks)

49 $f'(x) = \sin^2 x$

$\therefore f(x) = \int \sin^2 x \, dx$

$= \int \left(\frac{1}{2} - \frac{1}{2}\cos 2x\right) dx$

$= \frac{x}{2} - \frac{1}{4}\sin 2x + C$ ✓

Now $f(0) = 2$

$\therefore 2 = \frac{0}{2} - \frac{1}{4}\sin 2(0) + C$

$2 = 0 + C$

$C = 2$

$\therefore f(x) = \frac{x}{2} - \frac{1}{4}\sin 2x + 2$ ✓

(2 marks)

50 $f(x) = \tan^{-1}x + \tan^{-1}\left(\frac{1}{x}\right) \quad (x \neq 0)$

i $f'(x) = \frac{1}{1 + x^2} + \frac{1}{1 + \left(\frac{1}{x}\right)^2} \times - x^{-2}$

$= \frac{1}{1 + x^2} - \frac{1}{x^2\left(1 + \frac{1}{x^2}\right)}$

$= \frac{1}{1 + x^2} - \frac{1}{x^2 + 1}$

$= 0$ ✓

$\therefore f(x)$ is constant, (but with a discontinuity at $x = 0$). ✓

Consider $x > 0$

Now $f(1) = \tan^{-1}1 + \tan^{-1}\left(\frac{1}{1}\right)$

$= 2\tan^{-1}1$

$= 2 \times \frac{\pi}{4}$

$= \frac{\pi}{2}$

$\therefore f(x) = \frac{\pi}{2} \quad (x > 0)$ ✓

(3 marks)

ii $f(x)$ is an odd function.

$\therefore f(-x) = -f(x)$

So $f(x) = -\frac{\pi}{2}$ if $x < 0$.

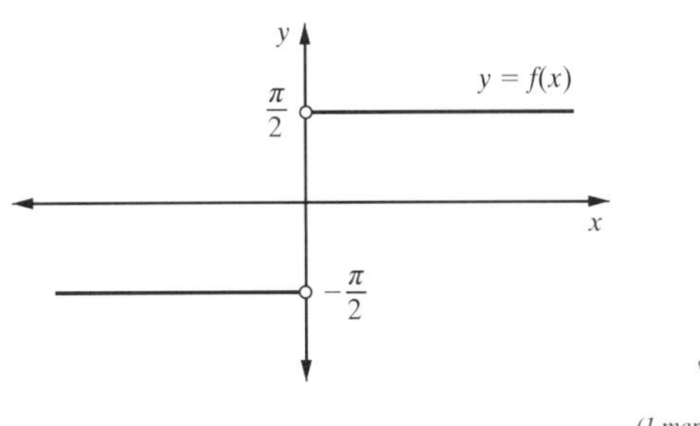

✓

(1 mark)

51 $\int_0^2 x^2 e^{x^3 + 1} dx$

Let $u = x^3 + 1$

$\therefore \frac{du}{dx} = 3x^2$

$\therefore x^2\,dx = \frac{1}{3}du$

When $x = 2,\ u = 9$

$x = 0,\ u = 1$ ✓

$\therefore \int_0^2 x^2 e^{x^3 + 1} dx$

$= \frac{1}{3}\int_1^9 e^u du$ ✓

$= \frac{1}{3}\left[e^u\right]_1^9$

$= \frac{1}{3}(e^9 - e^1)$

$= \frac{1}{3}(e^9 - e).$ ✓

(3 marks)

52 $\frac{d}{dx}\cos^{-1}(3x) = \frac{-1}{\sqrt{1 - (3x)^2}} \times 3$ ✓

$= \frac{-3}{\sqrt{1 - 9x^2}}.$ ✓ (2 marks)

53 $\int_{-1}^{1} \frac{1}{\sqrt{4 - x^2}}dx = \left[\sin^{-1}\frac{x}{2}\right]_{-1}^{1}$ ✓

(using standard integrals)

$= \sin^{-1}\frac{1}{2} - \sin^{-1}\left(-\frac{1}{2}\right)$

$= \frac{\pi}{6} - \left(-\frac{\pi}{6}\right)$

$= \frac{2\pi}{6}$

$= \frac{\pi}{3}.$ ✓

(2 marks)

54 $\int_0^{\frac{\pi}{4}} \cos\theta\sin^2\theta\,d\theta$

$= \int_0^{\frac{\pi}{4}} \frac{d}{d\theta}(\sin\theta)\,.\,(\sin\theta)^2\,d\theta$

$= \left[\frac{\sin^3\theta}{3}\right]_0^{\frac{\pi}{4}}$ ✓

$\left(\text{N.B.} \int f'(x)\,.\,(f(x))^n dx = \frac{(f(x))^{n+1}}{n+1}\right)$

$= \frac{1}{3}\left(\sin^3\left(\frac{\pi}{4}\right) - \sin^3 0\right)$

$= \frac{1}{3} \times \left(\frac{1}{\sqrt{2}}\right)^3$

$= \frac{1}{6\sqrt{2}}$

$= \frac{\sqrt{2}}{12}.$ ✓

(2 marks)

55 Let $u = \log_e x$

$\therefore \dfrac{du}{dx} = \dfrac{1}{x}$

$\therefore dx = x\,du$

At $x = e^2, u = \log_e e^2$
$= 2$

At $x = e, u = \log_e e$
$= 1$ ✓

$$\therefore \int_e^{e^2} \frac{1}{x(\log_e x)^2}\,dx$$
$$= \int_1^2 \frac{1}{xu^2}.\,x\,du$$
$$= \int_1^2 u^{-2}\,du \quad ✓$$
$$= -\left[u^{-1}\right]_1^2$$
$$= -(2^{-1} - 1^{-1})$$
$$= -\frac{1}{2} + 1$$
$$= \frac{1}{2}. \quad ✓$$

(3 marks)

56 $\dfrac{d}{dx}(\tan^{-1} x^4) = \dfrac{1}{1 + (x^4)^2}.4x^3$ ✓

$= \dfrac{4x^3}{1 + x^8}.$ ✓

or

Let $u = x^4$

$\therefore \dfrac{du}{dx} = 4x^3$

Now $y = \tan^{-1} x^4$

$\therefore y = \tan^{-1} u$

$\dfrac{dy}{du} = \dfrac{1}{1 + u^2}$

Now $\dfrac{dy}{dx} = \dfrac{dy}{du} \times \dfrac{du}{dx}$

$= \dfrac{1}{1 + u^2} \times 4x^3$

$= \dfrac{4x^3}{1 + (x^4)^2}$

$= \dfrac{4x^3}{1 + x^8}.$

(2 marks)

57 Let $u = 25 - x^2$

$\therefore \dfrac{du}{dx} = -2x$

$\therefore 2x dx = -du$

At $x = 4, u = 25 - 16$
$= 9$

At $x = 3, u = 25 - 9$
$= 16$ ✓

$$\therefore \int_3^4 \frac{2x}{\sqrt{25 - x^2}}\,dx$$
$$= \int_{16}^9 \frac{-du}{\sqrt{u}}$$
$$= -\int_{16}^9 u^{-\frac{1}{2}}\,du \quad ✓$$
$$= -2\left[u^{\frac{1}{2}}\right]_{16}^9$$
$$= -2\left(9^{\frac{1}{2}} - 16^{\frac{1}{2}}\right)$$
$$= -2(3 - 4)$$
$$= 2. \quad ✓$$

(3 marks)

58 $\displaystyle\int \frac{dx}{49 + x^2} = \int \frac{dx}{7^2 + x^2}$

$= \dfrac{1}{7}\tan^{-1}\dfrac{x}{7} + C.$ ✓✓

(using standard integrals)

(2 marks)

59 Let $u = x^4 + 8$

$\therefore \dfrac{du}{dx} = 4x^3$

$\therefore dx = \dfrac{du}{4x^3}$

$\therefore x^3 dx = \dfrac{du}{4}$ ✓

Now $\displaystyle\int x^3\sqrt{x^4 + 8}\,dx$

$$= \int \sqrt{x^4 + 8}\,.x^3 dx$$
$$= \int \sqrt{u}\,.\frac{du}{4}$$
$$= \frac{1}{4}\int u^{\frac{1}{2}}\,du \quad ✓$$
$$= \frac{1}{4}.\frac{2}{3}u^{\frac{3}{2}} + C$$
$$= \frac{1}{6}u^{\frac{3}{2}} + C$$
$$= \frac{1}{6}(x^4 + 8)^{\frac{3}{2}} + C$$
$$= \frac{1}{6}\sqrt{(x^4 + 8)^3} + C. \quad ✓$$

(3 marks)

60 **i** $f(x) = \sin^{-1}(x + 5)$

Domain:

$-1 \leqslant x + 5 \leqslant 1$

$\therefore -6 \leqslant x \leqslant -4$ ✓

Range:

$-\dfrac{\pi}{2} \leqslant y \leqslant \dfrac{\pi}{2}$ ✓

(2 marks)

ii $f'(x) = \dfrac{1}{\sqrt{1 - (x + 5)^2}}$ ✓

$\therefore f'(-5) = \dfrac{1}{\sqrt{1 - (-5 + 5)^2}}$

$= 1.$ ✓

(2 marks)

iii

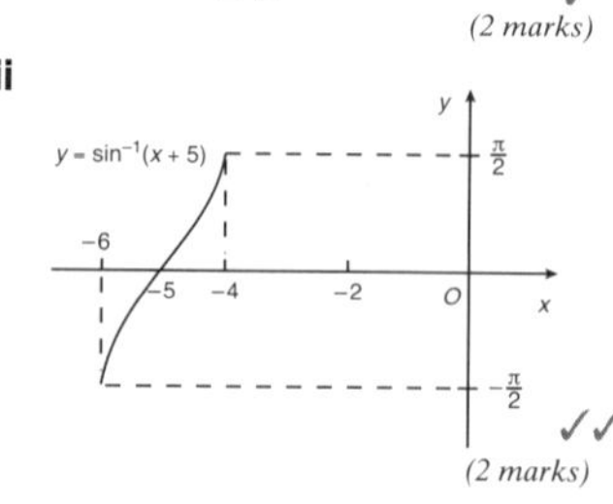

✓✓

(2 marks)

61 $\displaystyle\int_0^{\frac{\pi}{4}} \sin^2 x\,dx$

$$= \int_0^{\frac{\pi}{4}} \frac{1}{2}(1 - \cos 2x)\,dx$$
$$= \frac{1}{2}\int_0^{\frac{\pi}{4}} (1 - \cos 2x)\,dx$$
$$= \frac{1}{2}\left[x - \frac{1}{2}\sin 2x\right]_0^{\frac{\pi}{4}} \quad ✓$$
$$= \frac{1}{2}\left[\left(\frac{\pi}{4} - \frac{1}{2}\sin\frac{\pi}{2}\right) - \left(0 - \frac{1}{2}\sin 0\right)\right]$$
$$= \frac{1}{2}\left[\frac{\pi}{4} - \frac{1}{2} - 0\right]$$
$$= \frac{1}{2}\left[\frac{\pi}{4} - \frac{1}{2}\right]. \quad ✓$$

(2 marks)

62 $\displaystyle\int \frac{1}{x^2 + 49}\,dx = \int \frac{1}{x^2 + 7^2}\,dx$

$= \dfrac{1}{7}\tan^{-1}\dfrac{x}{7} + C.$ ✓

(using standard integrals)

(1 mark)

63 $u = 2x^2 + 1$

$\dfrac{du}{dx} = 4x$

$\therefore \dfrac{1}{4}du = x\,dx$ ✓

$$\therefore \int x(2x^2 + 1)^{\frac{5}{4}}\,dx$$
$$= \int (2x^2 + 1)^{\frac{5}{4}}\,x\,dx$$
$$= \int u^{\frac{5}{4}}.\frac{1}{4}\,du$$
$$= \frac{1}{4}\int u^{\frac{5}{4}}\,du \quad ✓$$
$$= \frac{1}{4} \times \frac{4}{9}u^{\frac{9}{4}} + C$$
$$= \frac{1}{9}u^{\frac{9}{4}} + C$$
$$= \frac{1}{9}(2x^2 + 1)^{\frac{9}{4}} + C \quad ✓$$

(3 marks)

64 Now $\frac{d}{dx}\sin^{-1} f(x) = \frac{f'(x)}{\sqrt{1-(f(x))^2}}$

$\therefore \frac{d}{dx}(2\sin^{-1} 5x) = 2 \times \frac{5}{\sqrt{1-25x^2}}$ ✓

$= \frac{10}{\sqrt{1-25x^2}}$. ✓

(2 marks)

65 **i** $\sin(5x+4x) + \sin(5x-4x)$
$= \sin 5x \cos 4x + \cos 5x \sin 4x$
$+ \sin 5x \cos 4x - \cos 5x \sin 4x$
$= 2\sin 5x \cos 4x$. ✓

(1 mark)

ii $\int \sin 5x \cos 4x \, dx$

Now $\sin(5x+4x) + \sin(5x-4x)$
$= 2\sin 5x \cos 4x$

$\therefore \frac{1}{2}\left[\sin(5x+4x) + \sin(5x-4x)\right]$

$= \sin 5x \cos 4x$.

$\therefore \int \sin 5x \cos 4x \, dx$

$= \frac{1}{2}\int \sin(5x+4x) + \sin(5x-4x)\, dx$ ✓

$= \frac{1}{2}\int \sin 9x + \sin x \, dx$

$= \frac{1}{2}\left[-\frac{1}{9}\cos 9x - \cos x\right] + C$

$= -\frac{1}{18}\cos 9x - \frac{1}{2}\cos x + C$. ✓

(2 marks)

66 $\int_0^{\frac{\pi}{4}} \cos x \sin^2 x \, dx$

$= \int_0^{\frac{\pi}{4}} \frac{d}{dx}(\sin x) \, . \, (\sin x)^2 \, dx$

$= \left[\frac{(\sin x)^3}{3}\right]_0^{\frac{\pi}{4}}$

$\left(\text{N.B.: } \int f'(x) \, . \, (f(x))^n dx = \frac{(f(x))^{n+1}}{n+1}\right)$

$= \frac{1}{3}\left[(\sin x)^3\right]_0^{\frac{\pi}{4}}$ ✓

$= \frac{1}{3}\left[\left(\sin\frac{\pi}{4}\right)^3 - (\sin 0)^3\right]$

$= \frac{1}{3}\left[\left(\frac{1}{\sqrt{2}}\right)^3 - 0\right]$

$= \frac{1}{3} \times \frac{1}{2\sqrt{2}}$

$= \frac{1}{6\sqrt{2}}$

$= \frac{\sqrt{2}}{12}$. ✓

OR

By substitution:

Let $u = \sin x$

$\therefore \frac{du}{dx} = \cos x$

$\therefore du = \cos x \, dx$

When $x = \frac{\pi}{4}$, $u = \frac{1}{\sqrt{2}}$

When $x = 0$, $u = 0$

$\therefore \int_0^{\frac{\pi}{4}} \cos x \sin^2 x \, dx$

$= \int_0^{\frac{1}{\sqrt{2}}} u^2 \, du$

$= \frac{1}{3}\left[u^3\right]_0^{\frac{1}{\sqrt{2}}}$

$= \frac{1}{3}\left(\left(\frac{1}{\sqrt{2}}\right)^3 - 0\right)$

$= \frac{1}{3} \times \frac{1}{2\sqrt{2}}$

$= \frac{1}{6\sqrt{2}}$

$= \frac{\sqrt{2}}{12}$.

(2 marks)

67 $\int_0^1 \frac{dx}{\sqrt{4-x^2}}$

$= \left[\sin^{-1}\frac{x}{2}\right]_0^1$ ✓

(using standard integrals)

$= \sin^{-1}\frac{1}{2} - \sin^{-1} 0$

$= \frac{\pi}{6}$. ✓

(2 marks)

68 $u = x - 3 \quad \therefore x = u + 3$

$\therefore \frac{du}{dx} = 1$

$\therefore du = dx$

At $x = 4$, $u = 1$
At $x = 3$, $u = 0$ ✓

$\therefore \int_3^4 x\sqrt{x-3} \, dx$

$= \int_0^1 (u+3)\sqrt{u} \, du$

$= \int_0^1 u^{\frac{3}{2}} + 3u^{\frac{1}{2}} du$ ✓

$= \left[\frac{2}{5}u^{\frac{5}{2}} + 3 . \frac{2}{3}u^{\frac{3}{2}}\right]_0^1$

$= \left[\frac{2}{5}u^{\frac{5}{2}} + 2u^{\frac{3}{2}}\right]_0^1$

$= \left(\frac{2}{5} + 2\right) - (0 + 0)$

$= 2\frac{2}{5}$. ✓

(3 marks)

69 $\frac{d}{dx}\cos^{-1}(3x^2)$

$= \frac{-1}{\sqrt{1-(3x^2)^2}} \, . \, 6x$ ✓

$= \frac{-6x}{\sqrt{1-9x^4}}$. ✓

(2 marks)

70 $\int \cos^2 4x \, dx$

Now $\cos^2 x = \frac{1}{2}(1 + \cos 2x)$

$\therefore \cos^2 4x = \frac{1}{2}(1 + \cos 8x)$

$\therefore \int \cos^2 4x \, dx$

$= \frac{1}{2}\int (1 + \cos 8x) dx$ ✓

$= \frac{1}{2}\left(x + \frac{1}{8}\sin 8x\right) + C$. ✓

(2 marks)

71 Let $u = x^2 + 1$

$\therefore \frac{du}{dx} = 2x$

$\therefore du = 2x \, dx$

At $x = 0$, $u = 1$
At $x = 2$, $u = 5$ ✓

Now $\int_0^2 \frac{x}{(x^2+1)^3} dx$

$= \frac{1}{2}\int_0^2 \frac{2x}{(x^2+1)^3} dx$

$= \frac{1}{2}\int_1^5 \frac{du}{u^3}$

$= \frac{1}{2}\int_1^5 u^{-3} \, du$ ✓

$= \frac{1}{2}\left[\frac{u^{-2}}{-2}\right]_1^5$

$= \frac{1}{2}\left[-\frac{1}{2u^2}\right]_1^5$

$= \frac{1}{2}\left[-\frac{1}{50} - \left(-\frac{1}{2}\right)\right]$

$= \frac{1}{2}\left(-\frac{1}{50} + \frac{1}{2}\right)$

$= \frac{1}{2}\left(-\frac{1}{50} + \frac{25}{50}\right)$

$= \frac{1}{2} \times \frac{24}{50}$

$= \frac{12}{50}$

$= \frac{6}{25}$. ✓

(3 marks)

72 $\frac{d}{dx}(x\tan^{-1}x)$

$= x \cdot \frac{d}{dx}(\tan^{-1}x) + \tan^{-1}x \cdot \frac{d}{dx}(x)$ ✓

$= x \cdot \frac{1}{1+x^2} + \tan^{-1}x \cdot 1$

$= \frac{x}{1+x^2} + \tan^{-1}x.$ ✓

(2 marks)

73 $\int_0^1 \frac{1}{\sqrt{2-x^2}}\,dx$

$= \int_0^1 \frac{1}{\sqrt{(\sqrt{2})^2 - x^2}}\,dx$

$= \left[\sin^{-1}\frac{x}{\sqrt{2}}\right]_0^1$ ✓

(using standard integrals)

$= \sin^{-1}\frac{1}{\sqrt{2}} - \sin^{-1}0$

$= \frac{\pi}{4}$. ✓

(2 marks)

74 $\int \cos^2 3x\,dx$

$= \frac{1}{2}\int (1+\cos 6x)\,dx$ ✓

$\left(\text{since } \cos^2 x = \frac{1}{2}(1+\cos 2x)\right)$

$= \frac{1}{2}\left(x + \frac{1}{6}\sin 6x\right) + C.$ ✓

(2 marks)

75 Let $u = 1 - x^2$

$\therefore \frac{du}{dx} = -2x$

$\therefore du = -2x\,dx$

At $x = 2, u = -3$

At $x = 3, u = -8$ ✓

Now $\int_2^3 \frac{2x}{(1-x^2)^2}\,dx$

$= \int_{-3}^{-8} \frac{-du}{u^2}$

$= -\int_{-3}^{-8} \frac{du}{u^2}$

$= -\int_{-3}^{-8} u^{-2}\,du$ ✓

$= -\left[\frac{u^{-1}}{-1}\right]_{-3}^{-8}$

$= \left[\frac{1}{u}\right]_{-3}^{-8}$

$= -\frac{1}{8} - \left(-\frac{1}{3}\right)$

$= -\frac{1}{8} + \frac{1}{3}$

$= -\frac{3}{24} + \frac{8}{24}$

$= \frac{5}{24}$. ✓

(3 marks)

76 $2\int_0^{\frac{\pi}{4}} \sin^2 4x\,dx$

$= 2\int_0^{\frac{\pi}{4}} \frac{1}{2}(1-\cos 8x)\,dx$ ✓

$\left(\text{n.b. } \sin^2\theta = \tfrac{1}{2}(1-\cos 2\theta)\right)$

$= \int_0^{\frac{\pi}{4}} (1-\cos 8x)\,dx$

$= \left[x - \frac{\sin 8x}{8}\right]_0^{\frac{\pi}{4}}$ ✓

$= \left(\frac{\pi}{4} - \frac{\sin 2\pi}{8}\right) - (0-0)$

$= \frac{\pi}{4} - \frac{0}{8}$

$= \frac{\pi}{4}$. ✓

(3 marks)

77 **i** $f(x) = 2\sin^{-1}\sqrt{x} - \sin^{-1}(2x-1)$ for $0 < x < 1$

$f'(x) = 2\frac{1}{\sqrt{1-(\sqrt{x})^2}} \cdot \frac{d}{dx}(x^{\frac{1}{2}})$

$\qquad - \frac{1}{\sqrt{1-(2x-1)^2}} \cdot \frac{d}{dx}(2x-1)$

$= \frac{2}{\sqrt{1-x}} \cdot \frac{1}{2}x^{-\frac{1}{2}}$

$\qquad - \frac{1}{\sqrt{1-(4x^2-4x+1)}} \cdot 2$ ✓

$= \frac{1}{\sqrt{1-x}} \cdot \frac{1}{\sqrt{x}} - \frac{2}{\sqrt{4x-4x^2}}$ ✓

(n.b. $x \neq 0$ and $x \neq 1$)

$= \frac{1}{\sqrt{x-x^2}} - \frac{2}{\sqrt{4(x-x^2)}}$

$= \frac{1}{\sqrt{x-x^2}} - \frac{1}{\sqrt{x-x^2}}$

$= 0$ for $0 < x < 1$. ✓

(3 marks)

ii Since $f'(x) = 0$, $f(x)$ is constant for $0 < x < 1$.

Now $f\left(\frac{1}{2}\right) = 2\sin^{-1}\sqrt{\frac{1}{2}} - \sin^{-1}0$

$= 2\sin^{-1}\frac{1}{\sqrt{2}} - \sin^{-1}0$

$= 2 \times \frac{\pi}{4} - 0$

$= \frac{\pi}{2} \quad \therefore f(x) = \frac{\pi}{2}$ for $0 < x < 1$ ✓

Also $f(0) = 2\sin^{-1}0 - \sin^{-1}(-1)$

$= 0 - \left(-\frac{\pi}{2}\right)$

$= \frac{\pi}{2}$

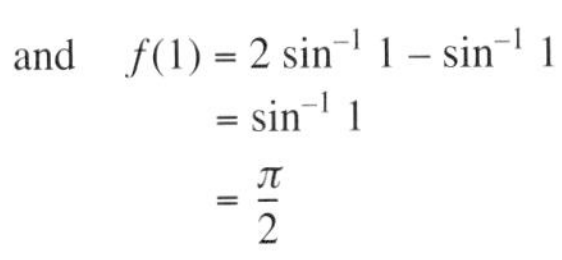

and $f(1) = 2\sin^{-1} 1 - \sin^{-1} 1$
$= \sin^{-1} 1$
$= \dfrac{\pi}{2}$

Hence $f(x) = \dfrac{\pi}{2}$ for $0 \leqslant x \leqslant 1$.

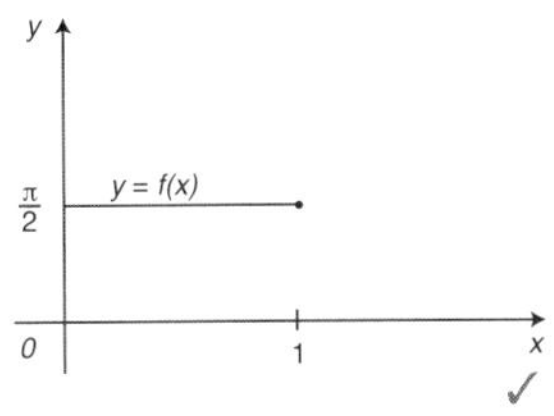

✓

(2 marks)

78 $\displaystyle\int_0^2 \frac{dx}{\sqrt{16 - x^2}} = \left[\sin^{-1}\frac{x}{4}\right]_0^2$ ✓

$= \sin^{-1}\dfrac{1}{2} - \sin^{-1} 0$

$= \dfrac{\pi}{6}$. ✓

(2 marks)

79 Let $u = 1 + x$

$\therefore \dfrac{du}{dx} = 1$

$\therefore du = dx$

and $x = u - 1$

At $x = -1$, $u = 0$
At $x = 0$, $u = 1$ ✓

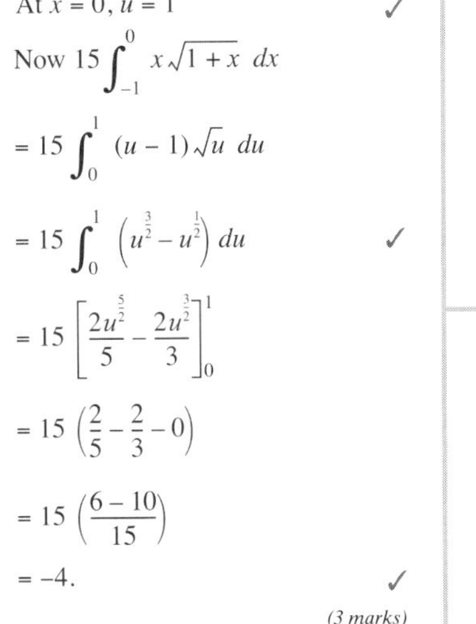

Now $15\displaystyle\int_{-1}^{0} x\sqrt{1 + x}\, dx$

$= 15\displaystyle\int_0^1 (u - 1)\sqrt{u}\, du$

$= 15\displaystyle\int_0^1 \left(u^{\frac{3}{2}} - u^{\frac{1}{2}}\right) du$ ✓

$= 15\left[\dfrac{2u^{\frac{5}{2}}}{5} - \dfrac{2u^{\frac{3}{2}}}{3}\right]_0^1$

$= 15\left(\dfrac{2}{5} - \dfrac{2}{3} - 0\right)$

$= 15\left(\dfrac{6 - 10}{15}\right)$

$= -4$. ✓

(3 marks)

80 **i** Let $u = 1 + e^x$

$\therefore \dfrac{du}{dx} = e^x$

$\therefore du = e^x\, dx$

Now $\displaystyle\int \frac{e^x}{1 + e^x}\, dx$

$= \displaystyle\int \frac{du}{u}$

$= \ln u + C$

$= \ln(1 + e^x) + C$ ✓

(1 mark)

ii $\displaystyle\int_0^{\pi} \cos^2 3x\, dx$

$= \displaystyle\int_0^{\pi} \frac{1}{2}(1 + \cos 6x)dx$ ✓

$= \dfrac{1}{2}\left[x + \dfrac{1}{6}\sin 6x\right]_0^{\pi}$ ✓

$= \dfrac{1}{2}\left(\pi + \dfrac{1}{6}\sin 6\pi - 0\right)$

$= \dfrac{1}{2} \times \pi$

$= \dfrac{\pi}{2}$. ✓

(3 marks)

81 $\dfrac{d}{dx}(x\sin^{-1} x)$

$= x \,.\, \dfrac{d}{dx}(\sin^{-1} x) + \sin^{-1} x \,.\, \dfrac{d}{dx}(x)$ ✓

$= x \,.\, \dfrac{1}{\sqrt{1 - x^2}} + \sin^{-1} x \,.\, 1$

$= \dfrac{x}{\sqrt{1 - x^2}} + \sin^{-1} x$. ✓

(2 marks)

82 $\displaystyle\int_0^{\sqrt{3}} \frac{4}{x^2 + 9}\, dx$

$= 4\displaystyle\int_0^{\sqrt{3}} \frac{1}{x^2 + 3^2}\, dx$

$= \dfrac{4}{3}\left[\tan^{-1}\dfrac{x}{3}\right]_0^{\sqrt{3}}$ ✓

$= \dfrac{4}{3}\left(\tan^{-1}\dfrac{\sqrt{3}}{3} - \tan^{-1} 0\right)$ ✓

$= \dfrac{4}{3} \times \dfrac{\pi}{6}$

$= \dfrac{2\pi}{9}$. ✓

(3 marks)

83 Let $u = 2 + x$

$\therefore \dfrac{du}{dx} = 1$

$\therefore du = dx$

and $x = u - 2$ ✓

Now $\displaystyle\int \frac{x}{\sqrt{2 + x}}\, dx$

$= \displaystyle\int \frac{u - 2}{\sqrt{u}}\, du$

$= \displaystyle\int \left(u^{\frac{1}{2}} - 2u^{-\frac{1}{2}}\right) du$ ✓

$= \dfrac{2u^{\frac{3}{2}}}{3} - 4u^{\frac{1}{2}} + C$

$= 2u^{\frac{1}{2}}\left(\dfrac{u}{3} - 2\right) + C$

$= \frac{2}{3}\sqrt{u}\,(u - 6) + C$

$= \frac{2}{3}\sqrt{2 + x}\,(2 + x - 6) + C$

$= \frac{2}{3}\sqrt{2 + x}\,(x - 4) + C$. ✓

(3 marks)

84 **i** Range: $-\dfrac{3\pi}{2} < y < \dfrac{3\pi}{2}$ ✓

ii

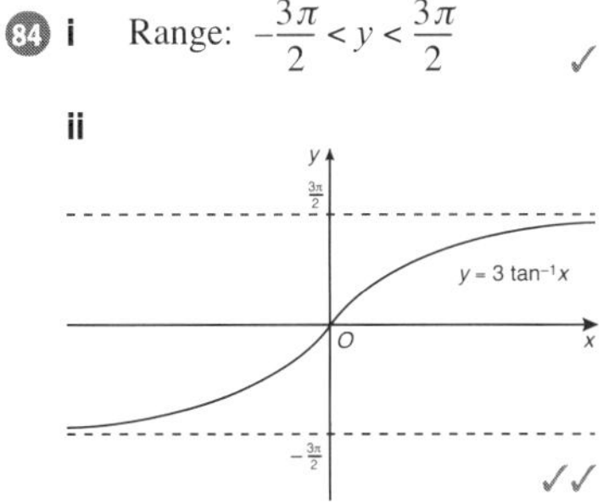

✓✓

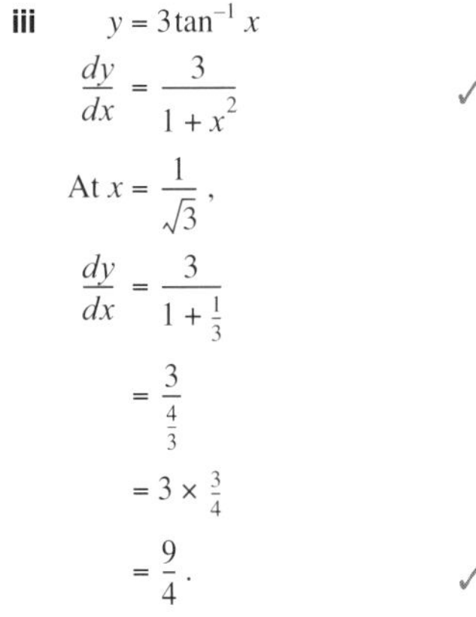

iii $y = 3\tan^{-1} x$

$\dfrac{dy}{dx} = \dfrac{3}{1 + x^2}$ ✓

At $x = \dfrac{1}{\sqrt{3}}$,

$\dfrac{dy}{dx} = \dfrac{3}{1 + \frac{1}{3}}$

$= \dfrac{3}{\frac{4}{3}}$

$= 3 \times \frac{3}{4}$

$= \dfrac{9}{4}$. ✓

(5 marks)
(Total mark allocation only in exam)

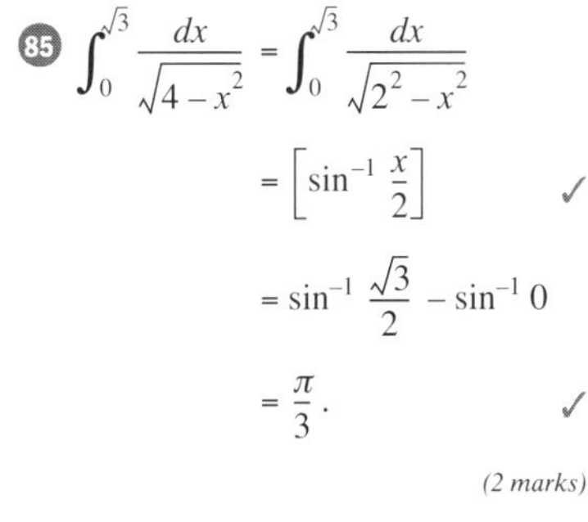

85 $\displaystyle\int_0^{\sqrt{3}} \frac{dx}{\sqrt{4 - x^2}} = \int_0^{\sqrt{3}} \frac{dx}{\sqrt{2^2 - x^2}}$

$= \left[\sin^{-1}\dfrac{x}{2}\right]$ ✓

$= \sin^{-1}\dfrac{\sqrt{3}}{2} - \sin^{-1} 0$

$= \dfrac{\pi}{3}$. ✓

(2 marks)

86 Let $u = \tan x$

$\therefore \dfrac{du}{dx} = \sec^2 x$

$\therefore du = \sec^2 x\, dx$

At $x = 0$, $u = 0$

At $x = \dfrac{\pi}{3}$, $u = \sqrt{3}$ ✓

Now $\displaystyle\int_0^{\frac{\pi}{3}} \tan^2 x \sec^2 x\, dx$

$= \displaystyle\int_0^{\sqrt{3}} u^2\, du$ ✓

$= \left[\dfrac{u^3}{3}\right]_0^{\sqrt{3}}$

$= \dfrac{(\sqrt{3})^3}{3} - 0$

$= \sqrt{3}$. ✓

(3 marks)

87 **i** $A(2\sin x + \cos x) + B(2\cos x - \sin x)$
$= 2A\sin x + A\cos x + 2B\cos x - B\sin x$
$= 2A\sin x - B\sin x + A\cos x + 2B\cos x$
$= (2A - B)\sin x + (A + 2B)\cos x$

If $(2A - B)\sin x + (A + 2B)\cos x \equiv \sin x + 8\cos x$

$\therefore 2A - B = 1 \quad \ldots(1)$
$A + 2B = 8 \quad \ldots(2)$ ✓

$(1) \times 2 \quad 4A - 2B = 2 \quad \ldots(3)$
$(2) + (3) \quad 5A = 10$
$\therefore A = 2$

Substituting $A = 2$ into (1)
$2(2) - B = 1$
$4 - B = 1$
$\therefore B = 3$ ✓

ii Now $2(2\sin x + \cos x) + 3(2\cos x - \sin x)$
$= \sin x + 8\cos x$ (from **ii**)

$$\therefore \int \frac{\sin x + 8\cos x}{2\sin x + \cos x}\,dx = \int \frac{2(2\sin x + \cos x) + 3(2\cos x - \sin x)}{2\sin x + \cos x}\,dx$$

$$= \int \left(2 + \frac{3(2\cos x - \sin x)}{2\sin x + \cos x}\right) dx$$

$$= 2x + \int \frac{3(2\cos x - \sin x)}{2\sin x + \cos x}\,dx$$ ✓

Let $u = 2\sin x + \cos x$

$\therefore \frac{du}{dx} = 2\cos x - \sin x$

$\therefore du = (2\cos x - \sin x)\,dx$

$$\therefore \int \frac{3(2\cos x - \sin x)}{2\sin x + \cos x}\,dx = \int \frac{3du}{u}$$

$= 3\ln u + C$
$= 3\ln(2\sin x + \cos x) + C$

$$\therefore \int \frac{\sin x + 8\cos x}{2\sin x + \cos x}\,dx = 2x + 3\ln(2\sin x + \cos x) + C$$ ✓

(4 marks)
(Total mark allocation only in exam)

88 $y = 2x\tan^{-1} x$

$$\frac{dy}{dx} = 2x \cdot \frac{1}{1 + x^2} + 2\tan^{-1} x$$ ✓

$$= \frac{2x}{1 + x^2} + 2\tan^{-1} x.$$ ✓

(2 marks)

89 $\int_0^{\frac{\pi}{3}} \sin^2 x\,dx$

$$= \frac{1}{2}\int_0^{\frac{\pi}{3}} (1 - \cos 2x)\,dx$$ ✓

$$= \frac{1}{2}\left[x - \frac{1}{2}\sin 2x\right]_0^{\frac{\pi}{3}}$$ ✓

$$= \frac{1}{2}\left(\frac{\pi}{3} - \frac{1}{2}\sin\frac{2\pi}{3}\right)$$ ✓

$$= \frac{1}{2}\left(\frac{\pi}{3} - \frac{1}{2} \times \frac{\sqrt{3}}{2}\right)$$

$$= \frac{1}{2}\left(\frac{\pi}{3} - \frac{\sqrt{3}}{4}\right)$$ ✓

(4 marks)

90 $\int_0^2 \frac{dx}{4 + x^2}$

$$= \int_0^2 \frac{dx}{2^2 + x^2}$$

$$= \left[\frac{1}{2}\tan^{-1}\frac{x}{2}\right]_0^2$$ ✓

$$= \tfrac{1}{2}(\tan^{-1} 1 - \tan^{-1} 0)$$ ✓

$$= \frac{1}{2} \times \frac{\pi}{4}$$

$$= \frac{\pi}{8}.$$ ✓

(3 marks)

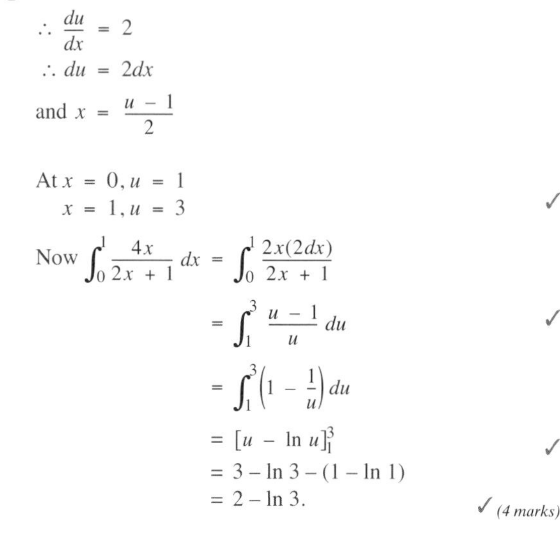

91 Let $u = 2x + 1$

$\therefore \frac{du}{dx} = 2$

$\therefore du = 2dx$

and $x = \frac{u - 1}{2}$

At $x = 0, u = 1$
$x = 1, u = 3$ ✓

$$\text{Now } \int_0^1 \frac{4x}{2x + 1}\,dx = \int_0^1 \frac{2x(2dx)}{2x + 1}$$

$$= \int_1^3 \frac{u - 1}{u}\,du$$ ✓

$$= \int_1^3 \left(1 - \frac{1}{u}\right) du$$

$$= [u - \ln u]_1^3$$ ✓

$= 3 - \ln 3 - (1 - \ln 1)$
$= 2 - \ln 3.$ ✓ *(4 marks)*

92 Let $x = \sin t$

$\therefore \cos t = \sqrt{1 - x^2}$ (from diagram)

$\therefore \frac{dx}{dt} = \cos t$

$\therefore dx = \cos t\,dt$

At $x = 0,\ t = 0$

At $x = \frac{1}{2},\ t = \frac{\pi}{6}$ ✓

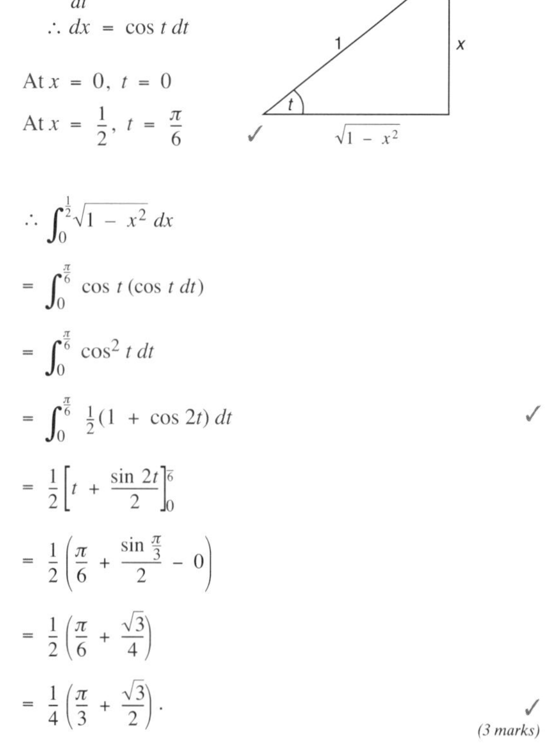

$$\therefore \int_0^{\frac{1}{2}} \sqrt{1 - x^2}\,dx$$

$$= \int_0^{\frac{\pi}{6}} \cos t(\cos t\,dt)$$

$$= \int_0^{\frac{\pi}{6}} \cos^2 t\,dt$$

$$= \int_0^{\frac{\pi}{6}} \tfrac{1}{2}(1 + \cos 2t)\,dt$$ ✓

$$= \frac{1}{2}\left[t + \frac{\sin 2t}{2}\right]_0^{\frac{\pi}{6}}$$

$$= \frac{1}{2}\left(\frac{\pi}{6} + \frac{\sin\frac{\pi}{3}}{2} - 0\right)$$

$$= \frac{1}{2}\left(\frac{\pi}{6} + \frac{\sqrt{3}}{4}\right)$$

$$= \frac{1}{4}\left(\frac{\pi}{3} + \frac{\sqrt{3}}{2}\right).$$ ✓

(3 marks)

93

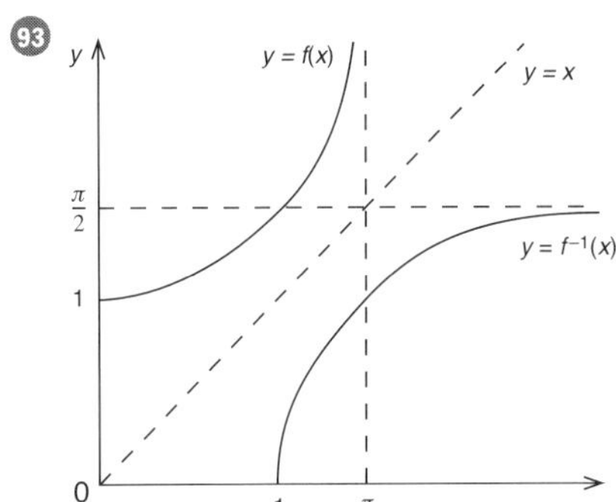

i $f(x) = \sec x$ for $0 \leqslant x < \frac{\pi}{2}$,

domain: $0 \leqslant x < \frac{\pi}{2}$

range: $y \geqslant 1$

The domain of the inverse function $f^{-1}(x)$ is $x \geqslant 1$. ✓

ii Let $y = f^{-1}(x)$

$\therefore \sec y = x$

$x = \frac{1}{\cos y}$

$\therefore \cos y = \frac{1}{x} \qquad 0 \leqslant y < \frac{\pi}{2}$

$\therefore y = \cos^{-1}\left(\frac{1}{x}\right)$

i.e. $f^{-1}(x) = \cos^{-1}\left(\frac{1}{x}\right)$ ✓

iii $\frac{d}{dx} f^{-1}(x) = \frac{d}{dx}\left(\cos^{-1}\left(\frac{1}{x}\right)\right)$

$= \frac{-1}{\sqrt{1 - \frac{1}{x^2}}} \times \left(-\frac{1}{x^2}\right)$ ✓

$= \frac{1}{x^2\sqrt{\frac{x^2 - 1}{x^2}}}$

$= \frac{1}{x\sqrt{x^2 - 1}}, \qquad x \neq 1$ ✓

(4 marks)
(Total mark allocation only in exam)

94

Let $u = e^x$

$\therefore \frac{du}{dx} = e^x$

$\therefore du = e^x\, dx$

and $e^{2x} = u^2$ ✓

$\therefore \int \frac{e^x}{1+e^{2x}}\, dx = \int \frac{du}{1+u^2}$ ✓

$= \tan^{-1} u + C$

$= \tan^{-1} e^x + C$ ✓ *(3 marks)*

95 $\int_{\frac{\pi}{4}}^{\frac{\pi}{2}} \cos^2 x\, dx$

$= \int_{\frac{\pi}{4}}^{\frac{\pi}{2}} \frac{1}{2}(1 + \cos 2x)\, dx$

$= \frac{1}{2}\int_{\frac{\pi}{4}}^{\frac{\pi}{2}} (1 + \cos 2x)\, dx$

$= \frac{1}{2}\left[x + \frac{1}{2}\sin 2x\right]_{\frac{\pi}{4}}^{\frac{\pi}{2}}$ ✓

$= \frac{1}{2}\left[\frac{\pi}{2} + \frac{1}{2}\sin \pi - \left(\frac{\pi}{4} + \frac{1}{2}\sin\frac{\pi}{2}\right)\right]$

$= \frac{1}{2}\left[\frac{\pi}{2} + 0 - \frac{\pi}{4} - \frac{1}{2}\right]$

$= \frac{1}{2}\left[\frac{\pi}{4} - \frac{1}{2}\right]$

$= \frac{\pi}{8} - \frac{1}{4}.$ ✓

(2 marks)

96 **i** $h(x) = \sin^{-1} x + \cos^{-1} x \qquad 0 \le x \le 1$

$h'(x) = \frac{1}{\sqrt{1-x^2}} - \frac{1}{\sqrt{1-x^2}}$

$= 0$ ✓

ii Since $h'(x) = 0$, $h(x)$ is constant.

Now $h(0) = 0 + \frac{\pi}{2}$

$= \frac{\pi}{2}$

$\therefore h(x) = \frac{\pi}{2}$ ✓

y
π/2
y = h (x)
0
1
x

✓

(3 marks)
(Total mark allocation only in exam)

97 Let $u = 9 - x^2$

$\therefore \frac{du}{dx} = -2x$

$\therefore du = -2x\, dx$

For $x = 0, u = 9$
For $x = 1, u = 8$ ✓

$\therefore \int_0^1 6x\sqrt{9 - x^2}\, dx$

$= -3\int_0^1 -2x\sqrt{9 - x^2}\, dx$

$= -3\int_9^8 \sqrt{u}\, du$ ✓

$= -3\int_9^8 u^{\frac{1}{2}}\, du$

$= -3\left[\frac{2u^{\frac{3}{2}}}{3}\right]_9^8$

$= -2\left[u\sqrt{u}\right]_9^8$ ✓

$= -2(8\sqrt{8} - 9\sqrt{9})$

$= -2(16\sqrt{2} - 27)$

$= 54 - 32\sqrt{2}.$ ✓ *(4 marks)*

98 Let $x = t^2 + 1$

$\therefore \frac{dx}{dt} = 2t$

$\therefore dx = 2t\, dt$

and $x - 1 = t^2$

$\therefore t = \sqrt{x - 1}$

At $x = 2$, $t = 1$
At $x = 10$, $t = 3$

$\therefore \int_2^{10} \frac{x}{\sqrt{x-1}}\,dx$

$= \int_1^3 \frac{t^2+1}{t} 2t\,dt$

$= 2\int_1^3 (t^2+1)\,dt$

$= 2\left[\frac{t^3}{3} + t\right]_1^3$

$= 2\left[(9+3) - \left(\frac{1}{3}+1\right)\right]$

$= 2\left[12 - \frac{4}{3}\right]$

$= 2\left[\frac{36}{3} - \frac{4}{3}\right]$

$= 2 \times \frac{32}{3}$

$= \frac{64}{3}$

$= 21\frac{1}{3}.$

(No mark allocation in exam)

99 Let $u = \cos x$

$\therefore \frac{du}{dx} = -\sin x$

$\therefore du = -\sin x\,dx$

At $x = 0, u = 1$

At $x = \frac{\pi}{3}, u = \frac{1}{2}$

$\therefore \int_0^{\frac{\pi}{3}} 3\sin x \cos^2 x\,dx$

$= \int_0^{\frac{\pi}{3}} -3\cos^2 x(-\sin x)\,dx$

$= \int_1^{\frac{1}{2}} -3u^2\,du$

$= \left[\frac{-3u^3}{3}\right]_1^{\frac{1}{2}}$

$= -\left[u^3\right]_1^{\frac{1}{2}}$

$= -\left(\frac{1}{8} - 1\right)$

$= 1 - \frac{1}{8}$

$= \frac{7}{8}.$

(No mark allocation in exam)

100 **i** $f(x) = 3x - x^3$

$= x(3 - x^2)$

$= x(\sqrt{3} + x)(\sqrt{3} - x)$

$\therefore$ x intercepts are 0 and $\pm\sqrt{3}$

$f'(x) = 3 - 3x^2$

Stationary points occur when

$f'(x) = 0$

$3 - 3x^2 = 0$

$1 - x^2 = 0$

$1 = x^2$

$x = \pm 1$

$f(1) = 3 - 1$

$= 2$

$f(-1) = -3 + 1$

$= -2$

$\therefore$ Stationary points are $(1, 2)$ and $(-1, -2)$.

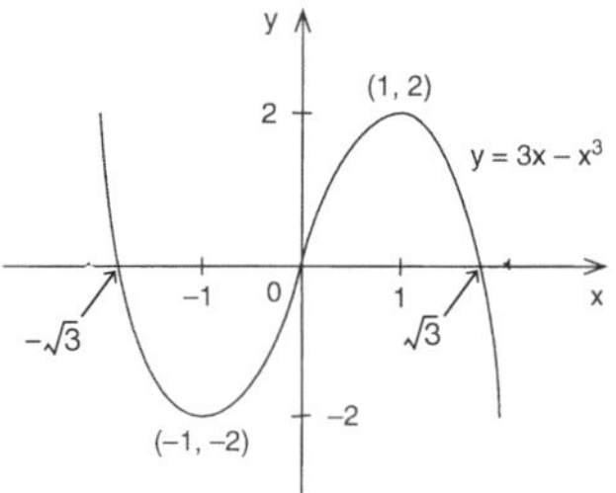

ii The required domain is

$-1 \leqslant x \leqslant 1$ (from graph.)

iii The range of $f(x)$ is

$-2 \leqslant y \leqslant 2$

$\therefore$ The domain of $f^{-1}(x)$ is

$-2 \leqslant x \leqslant 2$

iv The range of $f^{-1}(x)$ is

$-1 \leqslant y \leqslant 1$

The inverse function is

$x = 3y - y^3$

$= y(3 - y^2)$

Now $\frac{dy}{dx} = \frac{1}{\frac{dx}{dy}}$

and $\frac{dx}{dy} = 3 - 3y^2$

$\therefore \frac{dy}{dx} = \frac{1}{3 - 3y^2}$

At $x = 0, 0 = y(3 - y^2)$

$\therefore y = 0$

(since $-1 \leqslant y \leqslant 1$)

$\therefore \frac{dy}{dx} = \frac{1}{3 - 0}$ at $(0, 0)$

$= \frac{1}{3}.$

OR

By implicit differentiation:

$x = 3y - y^3$

$1 = 3\frac{dy}{dx} - 3y^2\frac{dy}{dx}$

At $(0, 0)$

$1 = 3\frac{dy}{dx}$

$\therefore \frac{dy}{dx} = \frac{1}{3}$

(No mark allocation in exam)

101 Let $u = 2t - 1$

$\frac{du}{dt} = 2$

$\therefore du = 2dt$

and $2t = u + 1$

When $t = \frac{1}{2}, u = 0$

When $t = 1, u = 1$

$\therefore \int_{\frac{1}{2}}^1 4t(2t-1)^5\,dt$

$= \int_{\frac{1}{2}}^1 2t(2t-1)^5\,2dt$

$= \int_0^1 (u+1)u^5\,du$

$= \int_0^1 (u^6 + u^5)\,du$

$= \left[\frac{u^7}{7} + \frac{u^6}{6}\right]_0^1$

$= \left[\left(\frac{1}{7} + \frac{1}{6}\right) - (0 + 0)\right]$

$= \frac{6}{42} + \frac{7}{42}$

$= \frac{13}{42}.$

(No mark allocation in exam)

102

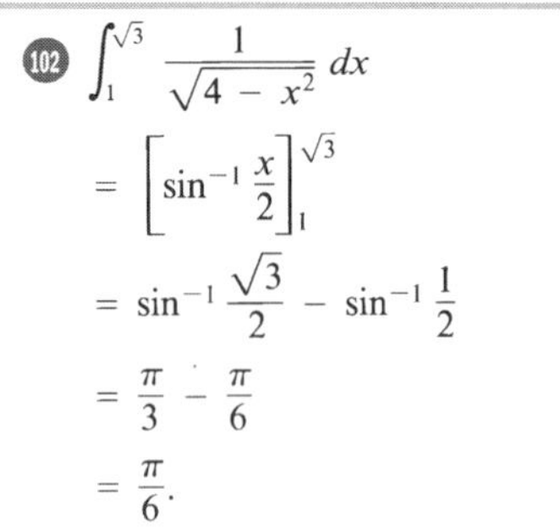

$\int_1^{\sqrt{3}} \frac{1}{\sqrt{4 - x^2}}\,dx$

$= \left[\sin^{-1}\frac{x}{2}\right]_1^{\sqrt{3}}$

$= \sin^{-1}\frac{\sqrt{3}}{2} - \sin^{-1}\frac{1}{2}$

$= \frac{\pi}{3} - \frac{\pi}{6}$

$= \frac{\pi}{6}.$

(No mark allocation in exam)

103 **i**

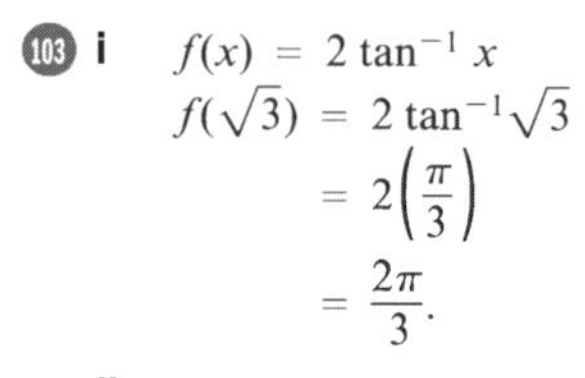

$f(x) = 2\tan^{-1} x$

$f(\sqrt{3}) = 2\tan^{-1}\sqrt{3}$

$= 2\left(\frac{\pi}{3}\right)$

$= \frac{2\pi}{3}.$

ii

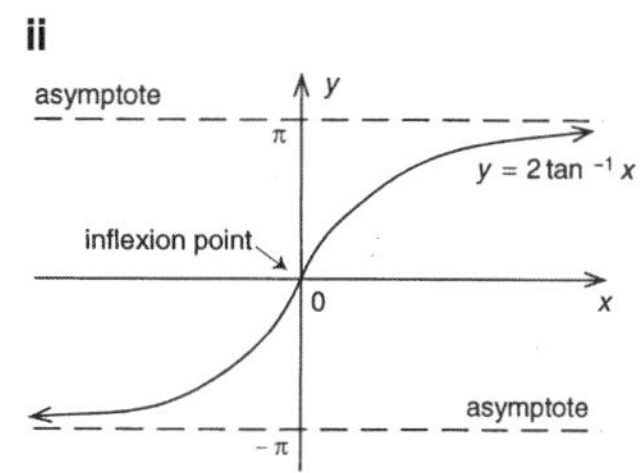

iii $f(x) = 2\tan^{-1} x$ cuts the y-axis at x = 0 (origin).

$f'(x) = \frac{2}{1 + x^2}$

$f'(0) = 2$

$\therefore$ Slope of curve where it cuts the y-axis is 2.

(No mark allocation in exam)

104 Let $u = 2x + 1$

$\frac{du}{dx} = 2$

$\therefore du = 2dx$

and $x = \frac{u - 1}{2}$

When $x = 0, u = 1$

When $x = 1, u = 3$

$\therefore \int_0^1 \frac{2x}{(2x+1)^2}\,dx$

$= \int_0^1 x.\frac{1}{(2x+1)^2}.2dx$

$= \int_1^3 \frac{u-1}{2}.\frac{1}{u^2}.du$

$= \int_1^3 \frac{u-1}{2u^2}\,du$

$= \frac{1}{2}\int_1^3 \left(\frac{u}{u^2} - \frac{1}{u^2}\right)du$

$= \frac{1}{2}\int_1^3 \left(\frac{1}{u} - u^{-2}\right)du$

$= \frac{1}{2}\left[\ln u + u^{-1}\right]_1^3$

$= \frac{1}{2}\left[(\ln 3 + \frac{1}{3}) - (\ln 1 + 1)\right]$
$= \frac{1}{2}\left[\ln 3 + \frac{1}{3} - 0 - 1\right]$
$= \frac{1}{2}\left[\ln 3 - \frac{2}{3}\right].$

(No mark allocation in exam)

1 A solid of revolution is to be found by rotating the region bounded by the x-axis and the curve $y = (k + 1)\sin(kx)$, where $k > 0$, between $x = 0$ and $x = \dfrac{\pi}{2k}$ about the x-axis.

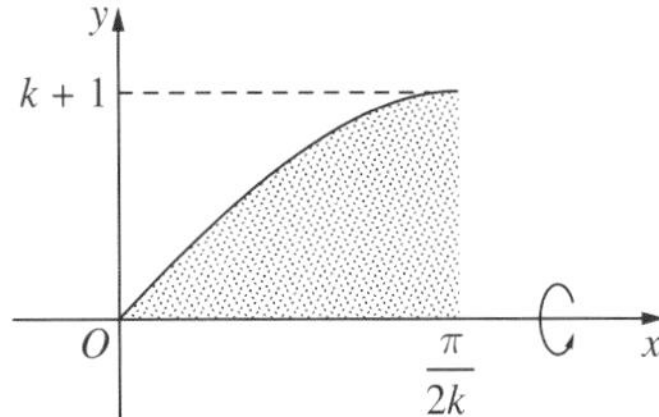

Find the value of k for which the volume is π^2.

(3 marks)

(Q13b, **2022 HSC**) Medium

2 A 2-metre-high sculpture is to be made out of concrete. The sculpture is formed by rotating the region between $y = x^2$, $y = x^2 + 1$ and $y = 2$ around the y-axis.

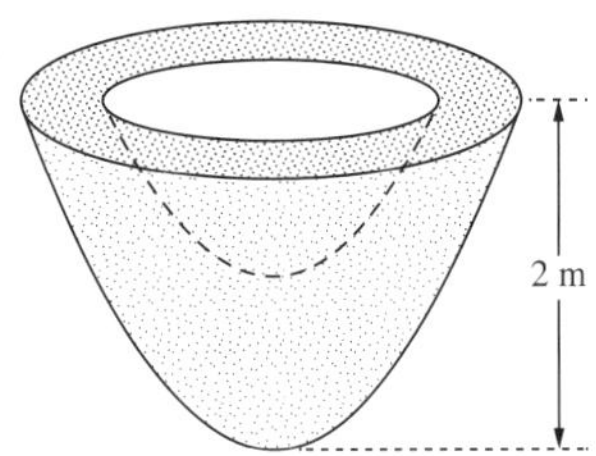

Find the volume of concrete needed to make the sculpture. *(3 marks)*

(Q13a, **2021 HSC**) Medium

3 The region enclosed by $y = 2 - |x|$ and $y = 1 - \dfrac{8}{4 + x^2}$ is shaded in the diagram.

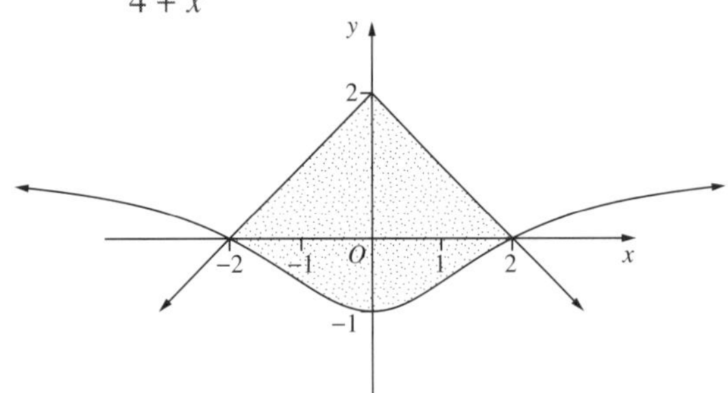

Find the exact value of the area of the shaded region. *(3 marks)*

(Q13c, **2021 HSC**) Medium

4 The region R is bounded by the y-axis, the graph of $y = \cos(2x)$ and the graph of $y = \sin x$, as shown in the diagram.

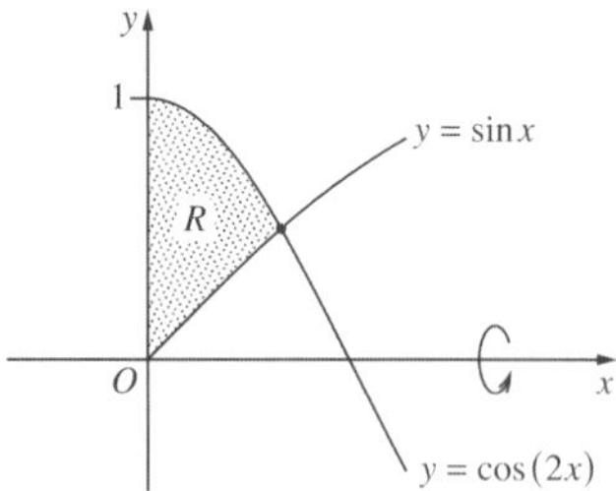

Find the volume of the solid of revolution formed when the region R is rotated about the x-axis. *(4 marks)*

(Q13b, **2020 HSC**) Medium

5 Find the area under the curve represented by the cartesian equations $x = 2t - 1$ and $y = t^2 + 3$ between the parameters $t = 1$ and $t = 2$. *(4 marks)*

Bonus question (see page iv) Medium

6 Find the volume of the solid formed when the curve $y = \sec x$ is rotated around the x-axis between $x = 0$ and $x = \dfrac{\pi}{4}$. *(2 marks)*

Bonus question Easy

7 The diagram shows part of the graph of $xy = 4$. A line passes through the points $A(1, 4)$ and $B(2, 2)$ which lie on the curve.

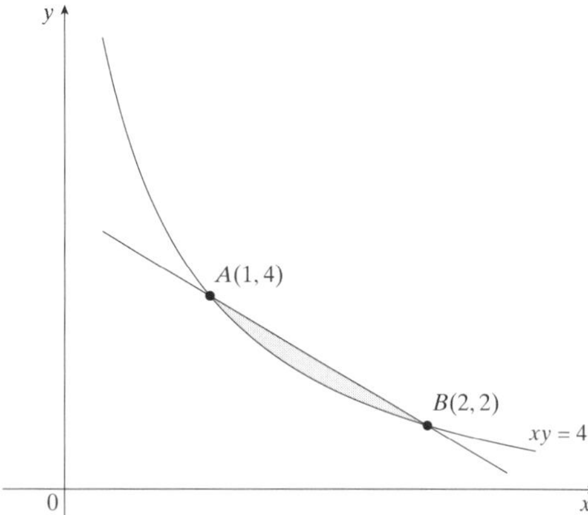

i Find the equation of the line AB.

(1 mark) Easy

ii The shaded area is rotated about the y-axis to form a solid.
Find the volume of the solid formed.

(4 marks) Medium

Bonus question

8 Find the volume of the solid of revolution when $y = \cos 2x \sin x$ between $x = 0$ and $x = \frac{\pi}{6}$ is rotated around the x-axis. *(4 marks)*

Bonus question Hard

9 **i** If $y = 2x \tan^{-1} 2x$, find $\frac{dy}{dx}$. *(1 mark)* Easy

ii Hence, find the area under the curve $y = \tan^{-1} 2x$ between $x = \frac{1}{2}$ and $x = \frac{\sqrt{3}}{2}$. *(3 marks)* Hard

Bonus question

10 The region in the first quadrant enclosed by the coordinate axes, the graph with equation $y = e^{-x}$ and the straight line $x = a$, where $a > 0$, is rotated about the x-axis to form a solid of revolution.

Find the exact value of a, if the volume of the solid is $\frac{3\pi}{8}$ units^3. *(3 marks)*

Bonus question Medium

11 The diagram shows the graph of a function in the form $y = a \cos^{-1} bx$.

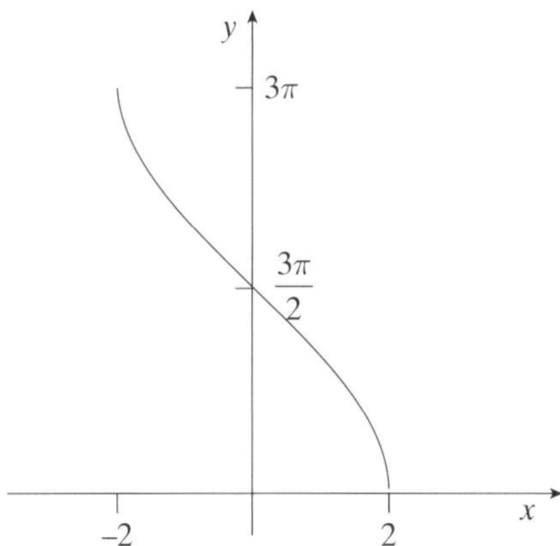

Show that the area between the curve and the y-axis between $y = 3\pi$ and $y = \frac{3\pi}{2}$ is 6 units^2. *(3 marks)*

Bonus question Medium

12 Show that the area bounded by the curve $y = -x^3 + 7x - 6$ and the x-axis is 32.75 units^2. *(4 marks)*

Bonus question Medium

13 A bowl is formed by rotating the curve $y = 4 \log_e(x + 1)$ about the y-axis for $0 \leq y \leq 4$.

Find the volume of the bowl.

Give your answer correct to 1 decimal place. *(4 marks)*

Bonus question Medium

14 **i** Show that $f(x) = \sqrt{\sin^2 x \cos x}$ is an even function. *(1 mark)* Easy

ii Use the substitution $u = \sin x$ to find the volume of the solid formed when the graph of $y = \sqrt{\sin^2 x \cos x}$ is rotated about the x-axis between $x = -\frac{\pi}{2}$ and $x = \frac{\pi}{2}$. *(3 marks)* Medium

Bonus question

15 The shaded region bounded by $y = 3 \sin 2x$, the x-axis and the line $x = \frac{\pi}{4}$ is rotated about the x-axis.

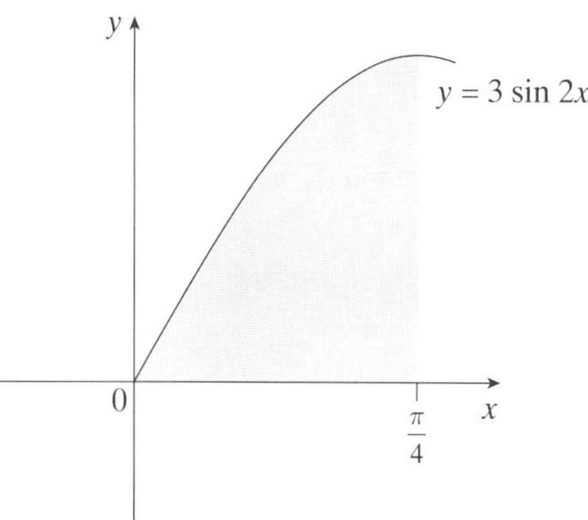

What is the exact volume of the solid formed? *(3 marks)*

Bonus question Medium

16 The diagram shows the region bounded by the curve $y = x - x^3$, and the x-axis between $x = 0$ and $x = 1$. The region is rotated about the x-axis to form a solid.

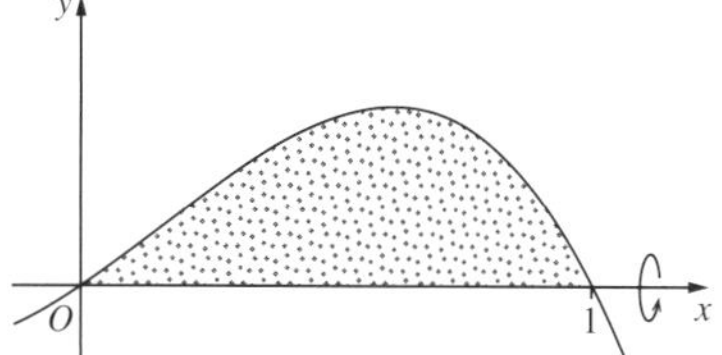

Find the exact value of the volume of the solid formed. *(3 marks)*

(Q13d, **2019 MA HSC**) Easy

17 The shaded region shown in the diagram is bounded by the curve $y = x^4 + 1$, the y-axis and the line $y = 10$.

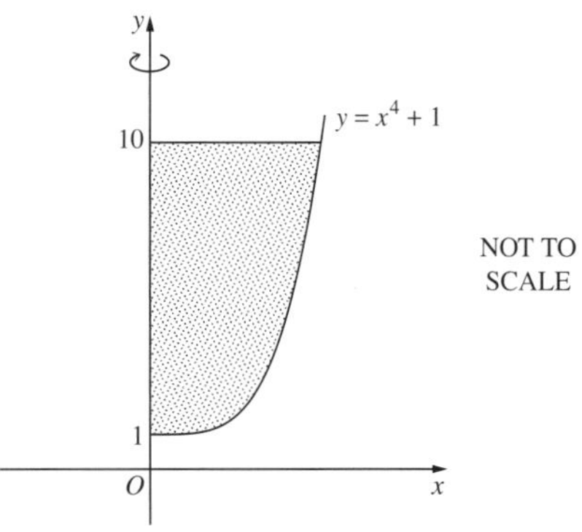

Find the volume of the solid of revolution formed when the shaded region is rotated about the y-axis. *(3 marks)*

(Q14b, **2018 MA HSC**) Easy

18 The diagram shows the region bounded by $y = \sqrt{16 - 4x^2}$ and the x-axis.

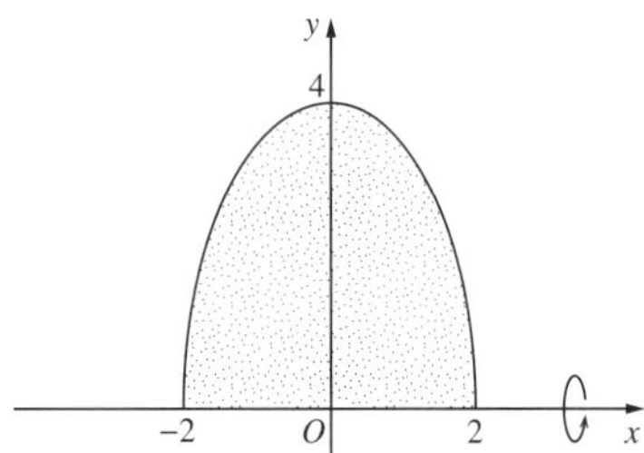

The region is rotated about the x-axis to form a solid.

Find the exact volume of the solid formed. *(3 marks)*

(Q12b, **2017 MA HSC**) Easy

19 The diagram shows two curves C_1 and C_2. The curve C_1 is the semicircle $x^2 + y^2 = 4$, $-2 \le x \le 0$. The curve C_2 has equation $\frac{x^2}{9} + \frac{y^2}{4} = 1, 0 \le x \le 3$.

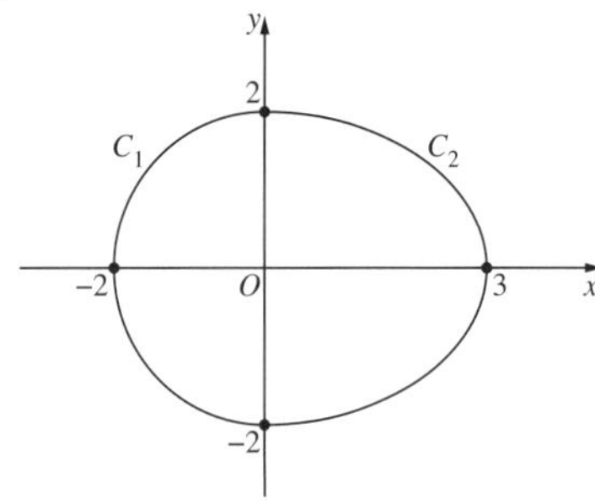

An egg is modelled by rotating the curves about the x-axis to form a solid of revolution.

Find the exact value of the volume of the solid of revolution. *(4 marks)*

(Q15a, **2016 MA HSC**) Medium

20 A bowl is formed by rotating the curve $y = 8\log_e(x - 1)$ about the y-axis for $0 \le y \le 6$.

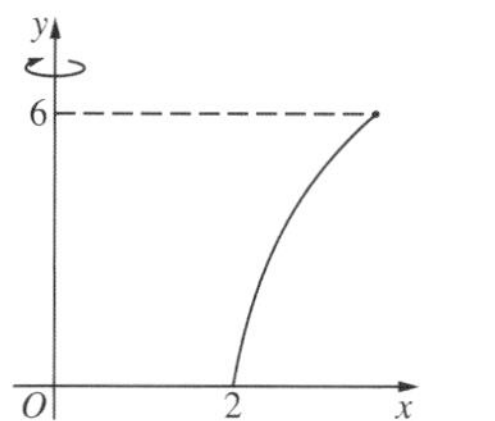

Find the volume of the bowl. Give your answer correct to 1 decimal place. *(3 marks)*

(Q16b, **2015 MA HSC**) Medium

21 The region bounded by $y = \cos 4x$ and the x-axis, between $x = 0$ and $x = \frac{\pi}{8}$, is rotated about the x-axis to form a solid.
Find the volume of the solid.

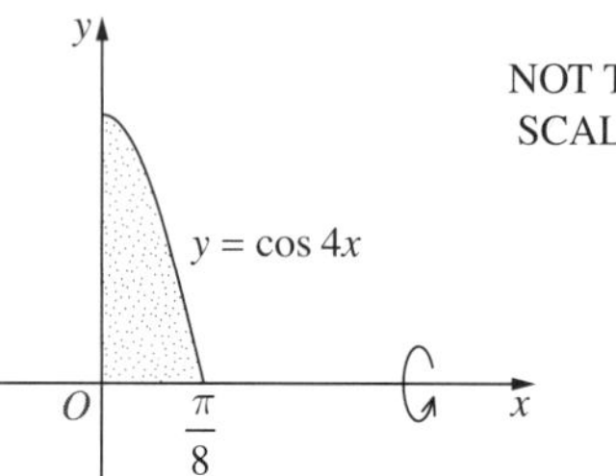

(3 marks)

(Q12b, **2014 HSC**) Easy

22 The region bounded by the curve $y = 1 + \sqrt{x}$ and the x-axis between $x = 0$ and $x = 4$ is rotated about the x-axis to form a solid.

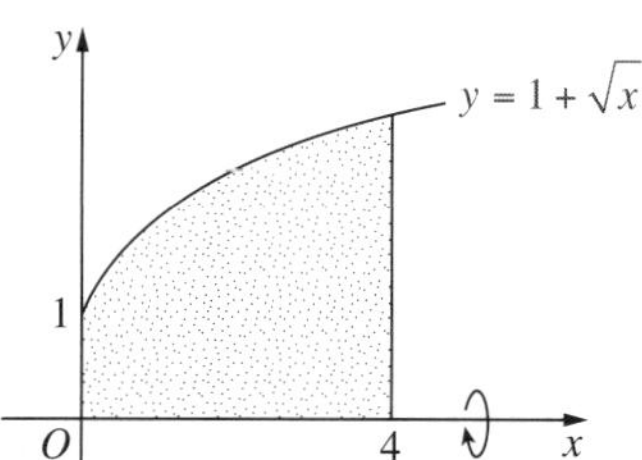

Find the volume of the solid. *(3 marks)*

(Q14c, **2014 MA HSC**) Medium

23 The region bounded by the graph $y = 3\sin\frac{x}{2}$ and the x-axis between $x = 0$ and $x = \frac{3\pi}{2}$ is rotated about the x-axis to form a solid.

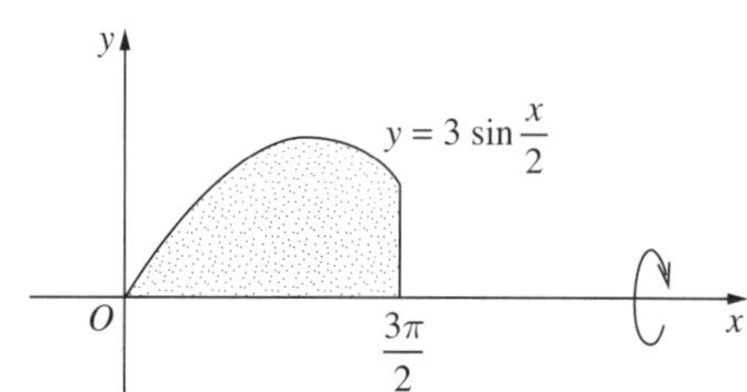

Find the exact volume of the solid. *(3 marks)*

(Q12b, **2013 HSC**) **Medium**

24 The region bounded by the x-axis, the y-axis and the parabola $y = (x - 2)^2$ is rotated about the y-axis to form a solid.

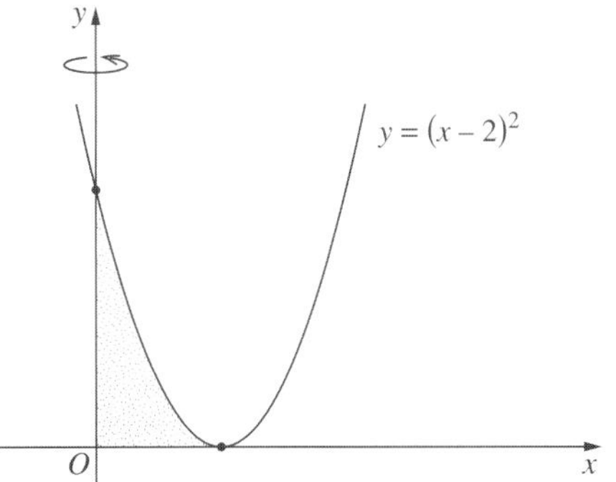

Find the volume of the solid. *(4 marks)*

(Q15b, **2013 MA HSC**) **Medium**

25 The diagram shows the region bounded by $y = \dfrac{3}{(x+2)^2}$, the x-axis, the y-axis, and the line $x = 1$.

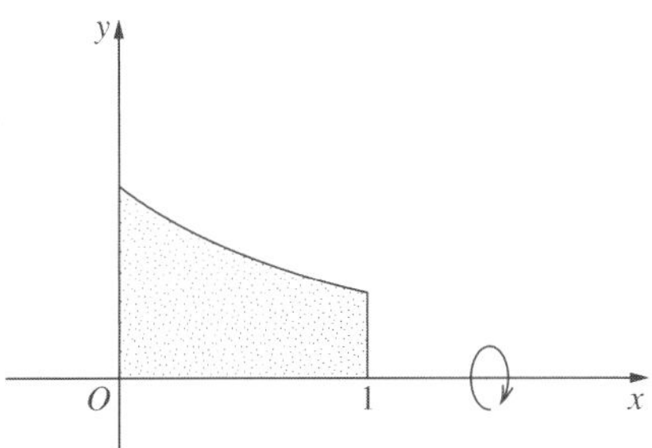

The region is rotated about the x-axis to form a solid. Find the volume of the solid. *(3 marks)*

(Q14b, **2012 MA HSC**) **Medium**

26 The diagram shows the region enclosed by the parabola $y = x^2$, the y-axis and the line $y = h$, where $h > 0$. This region is rotated about the y-axis to form a solid called a paraboloid. The point C is the intersection of $y = x^2$ and $y = h$. The point H has coordinates $(0, h)$.

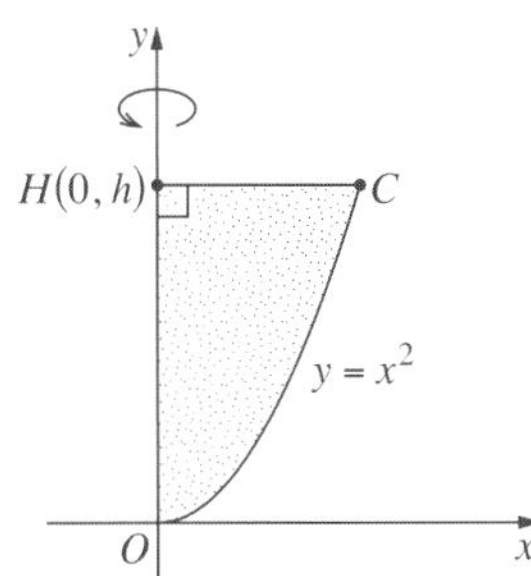

i Find the exact volume of the paraboloid in terms of h. *(2 marks)* **Easy**

ii A cylinder has radius HC and height h. What is the ratio of the volume of the paraboloid to the volume of the cylinder? *(1 mark)* **Medium**

(Q8b, **2011 MA HSC**)

27 The circle $x^2 + y^2 = r^2$ has radius r and centre O. The circle meets the positive x-axis at B. The point A is on the interval OB. A vertical line through A meets the circle at P. Let $\theta = \angle OPA$.

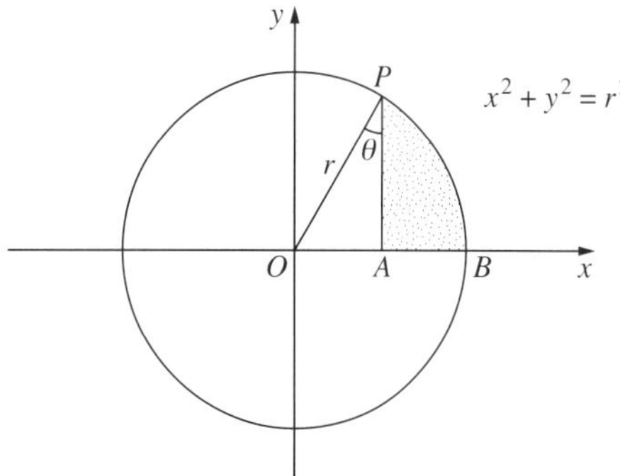

i The shaded region bounded by the arc PB and the intervals AB and AP is rotated about the x-axis. Show that the volume, V, formed is given by

$$V = \frac{\pi r^3}{3}\left(2 - 3\sin\theta + \sin^3\theta\right)$$

(3 marks) **Medium**

ii

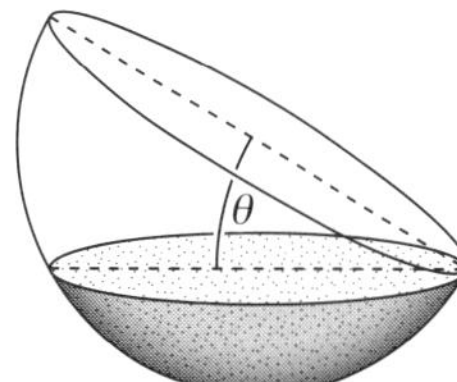

A container is in the shape of a hemisphere of radius r metres. The container is initially horizontal and full of water. The container is then tilted at an angle of θ to the horizontal so that some water spills out.

1 Find θ so that the depth of water remaining is one half of the original depth. *(1 mark)* **Hard**

2 What fraction of the original volume is left in the container? *(2 marks)* **Hard**

(Q10b, **2010 MA HSC**)

28 The diagram shows the region bounded by the curve $y = \sec x$, the lines $x = \dfrac{\pi}{3}$ and $x = -\dfrac{\pi}{3}$, and the x-axis.

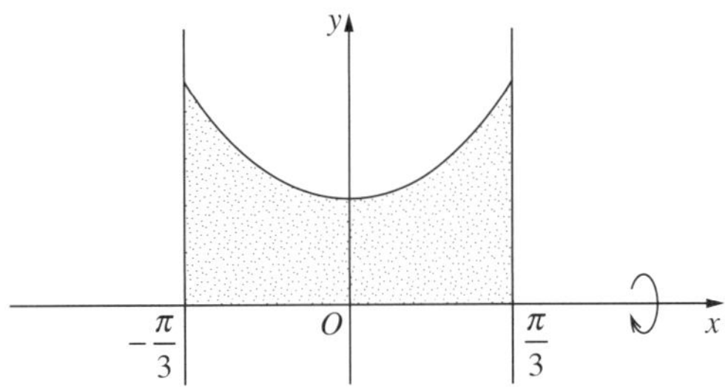

The region is rotated about the x-axis. Find the volume of the solid of revolution formed. *(3 marks)*

(Q6a, **2009 MA HSC**) Easy

29 The graph of $y = \dfrac{5}{x-2}$ is shown below.

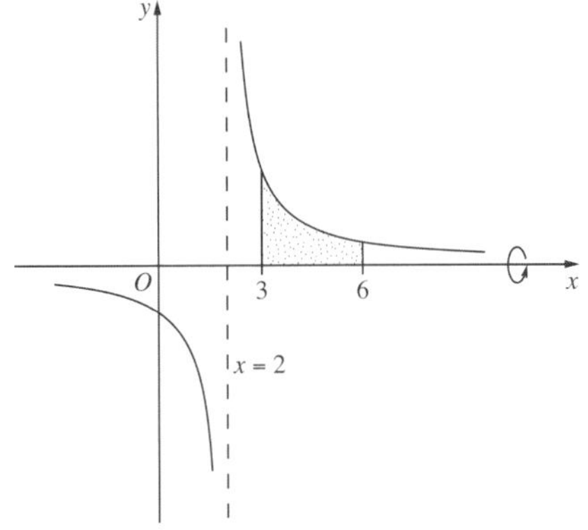

The shaded region in the diagram is bounded by the curve $y = \dfrac{5}{x-2}$, the x-axis and the lines $x = 3$ and $x = 6$.

Find the volume of the solid of revolution formed when the shaded region is rotated about the x-axis. *(3 marks)*

(Q6c, **2008 MA HSC**) Medium

30 Find the volume of the solid of revolution formed when the region bounded by the curve $y = \dfrac{1}{\sqrt{9+x^2}}$, the x-axis, the y-axis and the line $x = 3$, is rotated about the x-axis. *(3 marks)*

(Q3a, **2007 HSC**) Easy

31 The shaded region in the diagram is bounded by the curve $y = x^2 + 1$, the x-axis, and the lines $x = 0$ and $x = 1$.

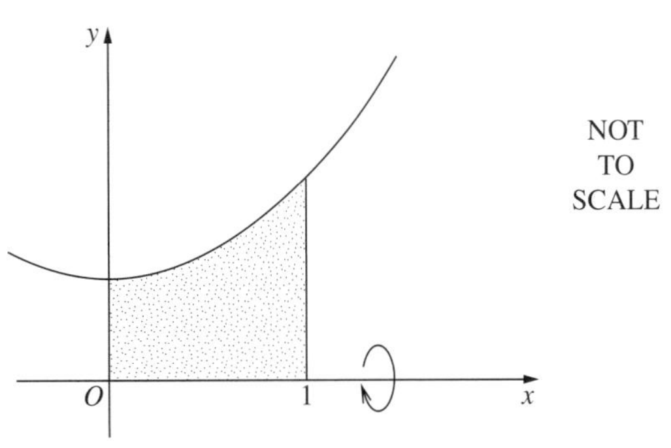

Find the volume of the solid of revolution formed when the shaded region is rotated about the x-axis. *(3 marks)*

(Q9a, **2007 MA HSC**) Easy

32 In the diagram, the shaded region is bounded by the parabola $y = x^2 + 1$, the y-axis and the line $y = 5$.

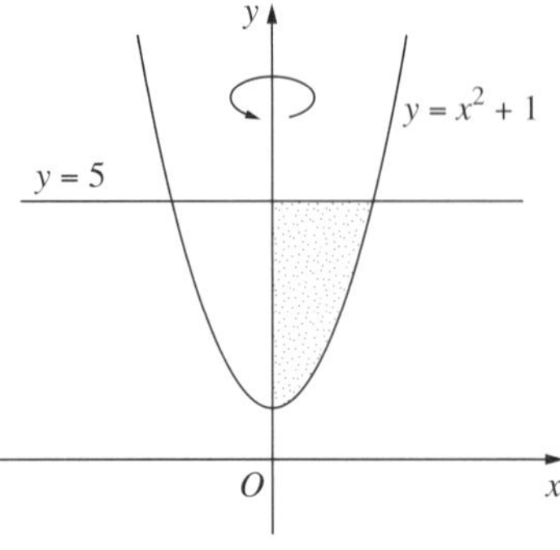

Find the volume of the solid formed when the shaded region is rotated about the y-axis. *(3 marks)*

(Q4b, **2006 MA HSC**) Easy

33 Find the exact value of the volume of the solid of revolution formed when the region bounded by the curve $y = \sin 2x$, the x-axis and the line $x = \dfrac{\pi}{8}$ is rotated about the x-axis. *(3 marks)*

(Q5a, **2005 HSC**) Medium

34 The graphs of the curves $y = x^2$ and $y = 12 - 2x^2$ are shown in the diagram.

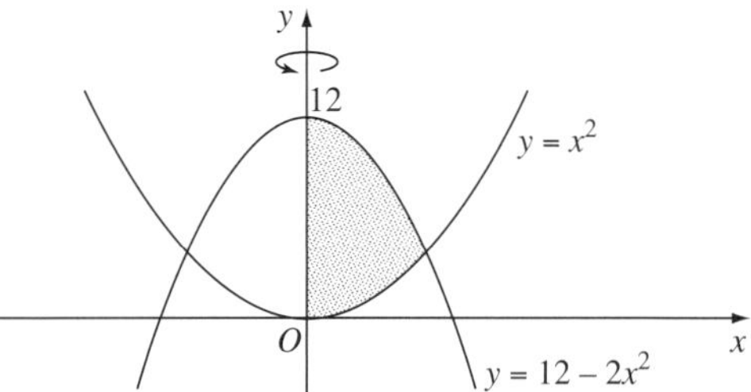

i Find the points of intersection of the two curves. *(1 mark)* Easy

ii The shaded region between the curves and the y-axis is rotated about the y-axis. By splitting the shaded region into two parts, or otherwise, find the volume of the solid formed. *(3 marks)* Medium

(Q6c, **2005 MA HSC**)

35 In the diagram, the shaded region is bounded by the curve $y = 2\sec x$, the coordinate axes and the line $x = \dfrac{\pi}{3}$. The shaded region is rotated about the x-axis.

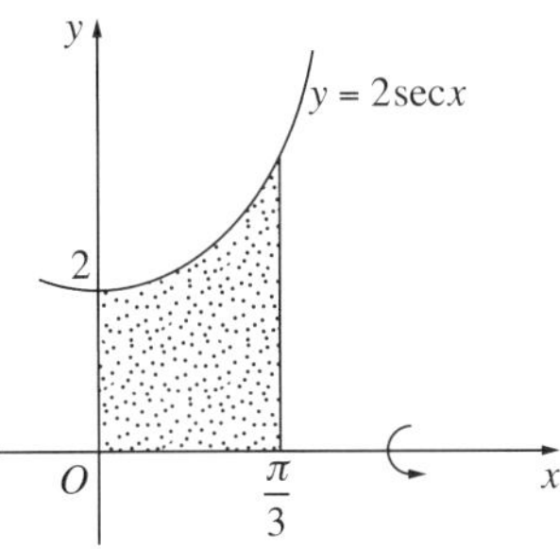

Calculate the exact volume of the solid of revolution formed. *(3 marks)*

(Q4c, **2004 MA HSC**) Easy

36 In the diagram, the shaded region is bounded by the y-axis, the curve $y = e^x$ and a horizontal line ℓ that cuts the curve at a point whose x coordinate is $\log_e 5$.

A solid is formed by rotating the shaded region about the y-axis.

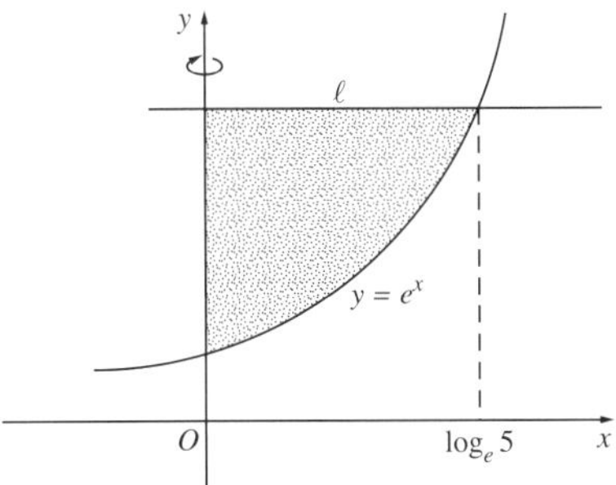

Write down a definite integral whose value is the volume of the solid. (Do NOT evaluate the integral.) *(3 marks)*

(Q8b, **2003 MA HSC**) Easy

37 A bowl is formed by rotating the part of the curve $y = \frac{x^4}{4}$ between $x = 0$ and $x = 2$ about the y axis.

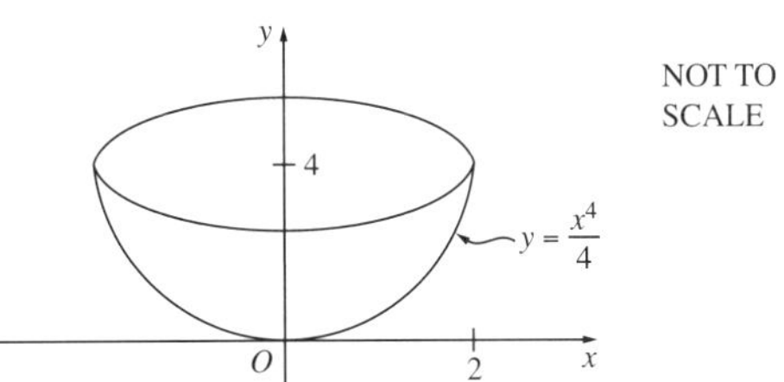

Find the volume of the bowl. *(4 marks)*

(Q6c, **2002 MA HSC**) Medium

38 The sketch shows the graph of the curve $y = f(x)$ where $f(x) = 2\cos^{-1}\frac{x}{3}$. The area under the curve for $0 \le x \le 3$ is shaded.

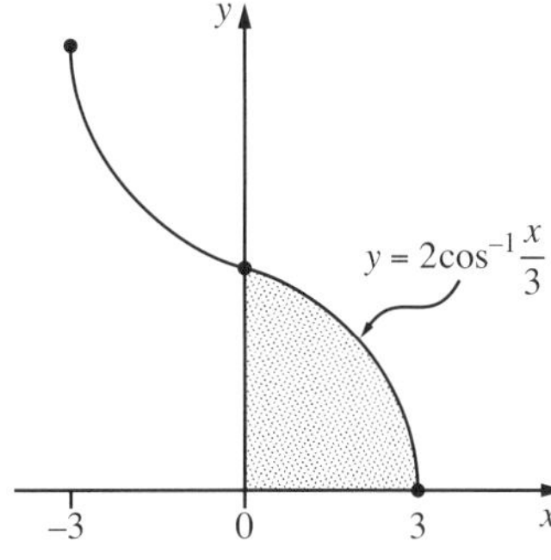

i Find the y intercept. *(1 mark)* Easy

ii Determine the inverse function $y = f^{-1}(x)$, and write down the domain D of this inverse function. *(2 marks)* Medium

iii Calculate the area of the shaded region. *(2 marks)* Medium

(Q5a, **2001 HSC**)

39 The part of the curve $\frac{x^2}{2} + y^2 = 8$ that lies in the first quadrant is rotated about the x axis.

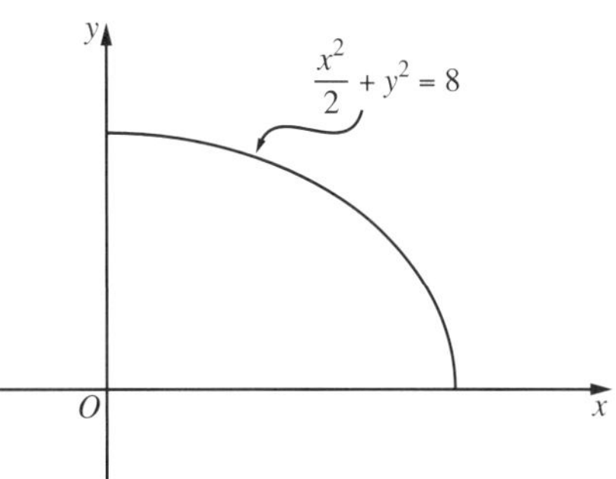

Find the volume of the solid of revolution. *(3 marks)*

(Q7a, **2001 MA HSC**) Medium

40 The area under the curve $y = \frac{1}{\sqrt{x}}$, for $1 \le x \le e^2$, is rotated about the x axis.

Find the exact volume of the solid of revolution. *(4 marks)*

(Q7a, **2000 MA HSC**) Medium

41 The shaded region bounded by $y = 3\sin x$, the x axis and the line $x = \frac{\pi}{2}$ is rotated about the x axis to form a solid. Calculate the volume of the solid.

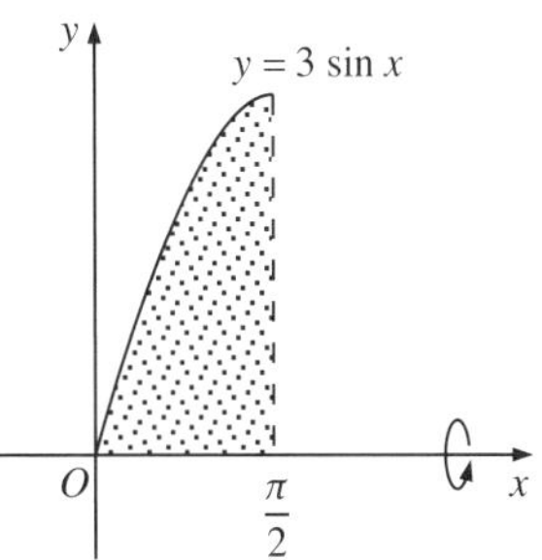

(4 marks)

(Q3a, **1999 HSC**) Medium

42 The shaded region bounded by $y = e^{x^2}$, $y = 7$ and the y axis is rotated around the y axis to form a solid.

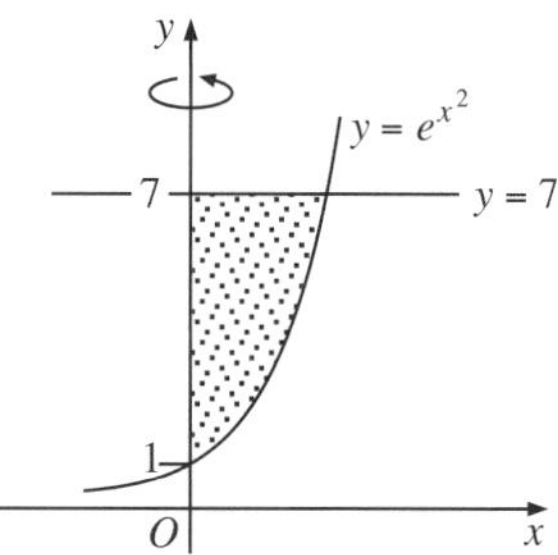

i Show that the volume of the solid is given by $V = \pi \int_1^7 \log_e y \, dy$. Easy

ii Copy and complete the table. Give your answers correct to 3 decimal places.

y	1	4	7
$\log_e y$			

Easy

iii Use the trapezoidal rule with 3 function values to approximate the volume, V. Easy

(5 marks) (Q8a, **1999 MA HSC**)

43 The left-hand diagram shows the lower half of the circle $x^2 + (y - 15)^2 = 15^2$.

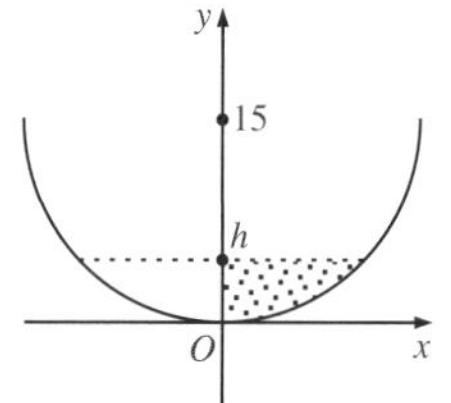

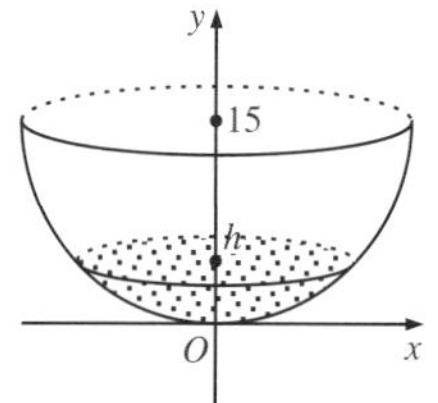

The shaded area in this diagram is bounded by the semicircle, the line $y = h$, and the y axis.

i Show that the volume V formed when the shaded area is rotated around the y axis is given by $V = 15\pi h^2 - \dfrac{\pi h^3}{3}$. Medium

A semicircle is rotated around the y axis to form a hemispherical bowl of radius 15 cm, as shown in the right-hand diagram.

ii The bowl is filled with water at a constant rate of 3 $cm^3 \, s^{-1}$. Find the rate at which the water level is rising when the water level is 6 cm. Medium

(5 marks) (Q4a, **1998 HSC**)

44 The diagram shows the graph of $y = \log_2 x$ between $x = 1$ and $x = 8$. The shaded region, bounded by $y = \log_2 x$, the line $y = 3$, and the x and y axes, is rotated about the y axis to form a solid.

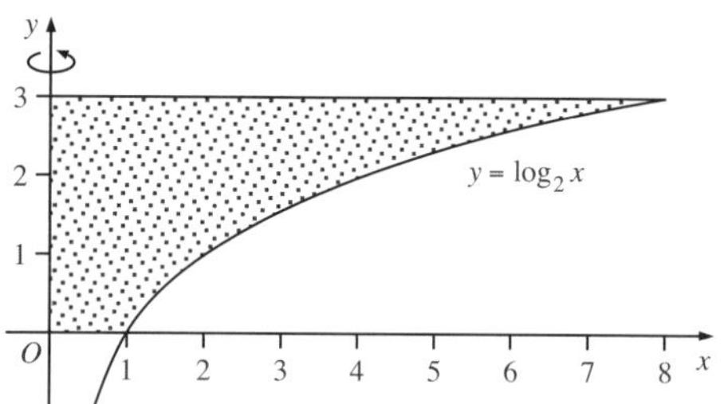

i Show that the volume of the solid is given by $V = \pi \int_0^3 e^{y \ln 4} \, dy$. Medium

ii Hence find the volume of the solid. Easy

(5 marks) (Q8b, **1998 MA HSC**)

45 The diagram shows part of the graph of the function $y = \tan 2x$. The shaded region is bounded by the curve, the x axis, and the line $x = \dfrac{\pi}{6}$. The region is rotated about the x axis to form a solid.

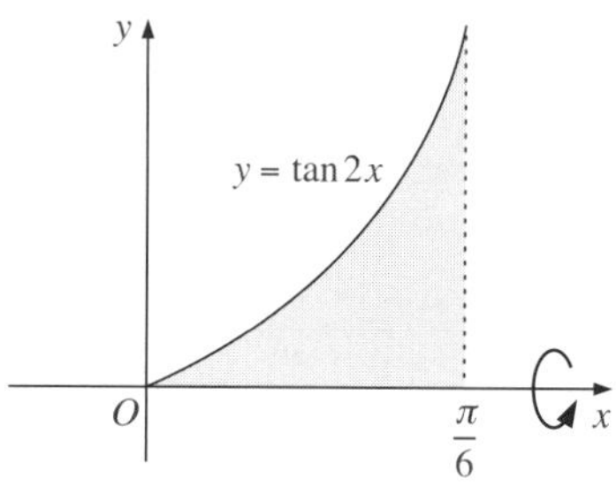

i Show that the volume of the solid is given by $V = \pi \int_0^{\frac{\pi}{6}} \left(\sec^2 2x - 1\right) dx$. Easy

(You may use the result of part (a).)
$[\sec^2\theta - \tan^2\theta = 1]$

ii Find the exact volume of the solid. Easy

(5 marks) (Q7b, **1997 MA HSC**)

46

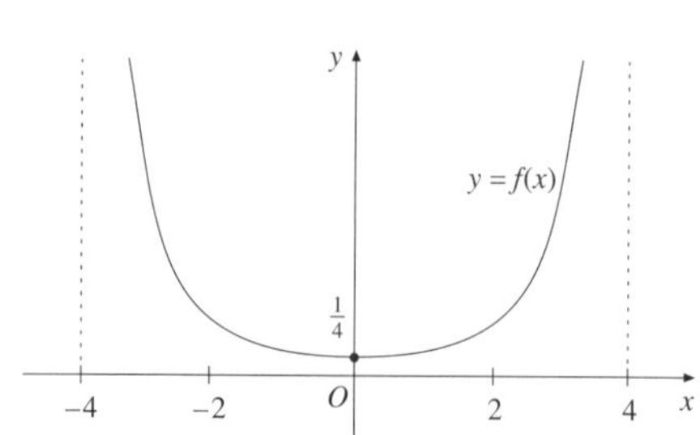

Let $f(x) = \dfrac{1}{\sqrt{16 - x^2}}$. The graph of $y = f(x)$ is sketched above.

i Show that $f(x)$ is an even function. Easy

ii Find the area enclosed by $y = f(x)$, the x axis, $x = -2$, and $x = 2$. Easy

(4 marks) (Q3a, **1996 HSC**)

47 The function $g(x)$ is given by $g(x) = 2 + \cos x$.

The graph $y = g(x)$ for $\frac{\pi}{4} \le x \le \frac{\pi}{2}$ is rotated about the x axis.

Find the volume of the solid generated. (You may use the result of part (b).

$\left[\int_{\frac{\pi}{4}}^{\frac{\pi}{2}} \cos^2 x\,dx = \frac{\pi}{8} - \frac{1}{4}\right]$) *(3 marks)*

(Q3c, **1996 HSC**) **Medium**

48 The region which lies between the x axis and the line $y = x + 1$ from $x = 0$ to $x = 3$ is rotated about the x axis to form a solid.

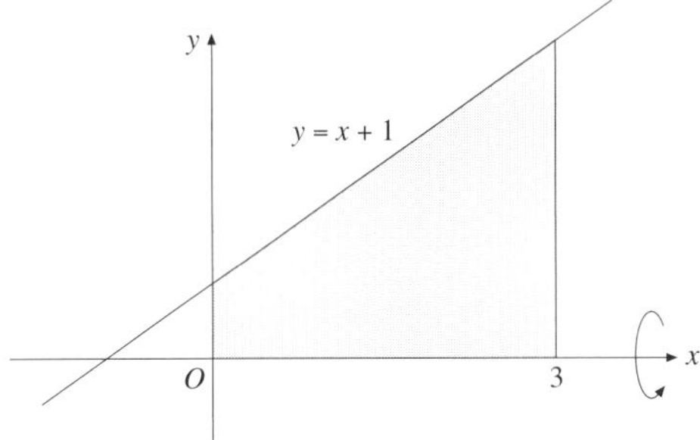

Find the volume of the solid. *(3 marks)*

(Q4a, **1996 MA HSC**) **Easy**

49 The shaded area is bounded by the curve $xy = 3$, the lines $x = 1$ and $y = 6$, and the two axes. A solid is formed by rotating the shaded area about the y axis.

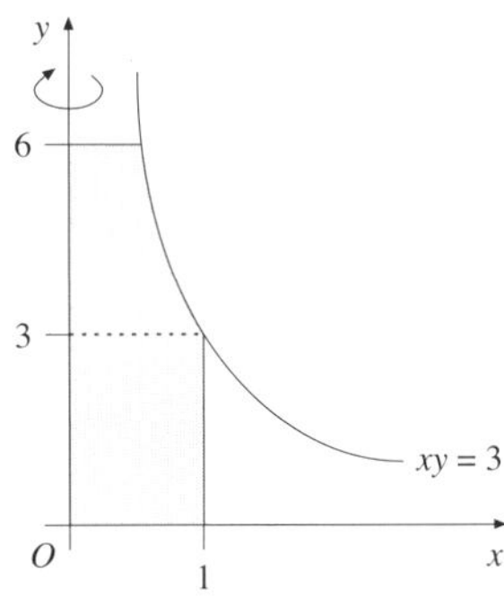

Find the volume of this solid by considering separately the regions above and below $y = 3$. *(4 marks)*

(Q2b, **1995 HSC**) **Medium**

50 **i** Solve the equation $\sin 2x = 2\sin^2 x$ for $0 < x < \pi$. **Medium**

ii Show that if $0 < x < \frac{\pi}{4}$, then $\sin 2x > 2\sin^2 x$. **Medium**

iii Find the area enclosed between the curves $y = \sin 2x$ and $y = 2\sin^2 x$ for $0 \le x \le \frac{\pi}{4}$. **Medium**

(6 marks) (Q5a, **1995 HSC**)

51 A glass has a shape obtained by rotating part of the parabola $x = \frac{y^2}{30}$ about the y axis as shown. The glass is 10 cm deep.

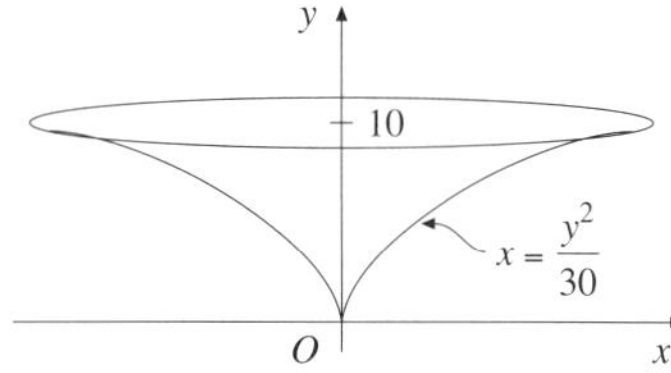

Find the volume of liquid which the glass will hold. *(3 marks)*

(Q6a, **1995 MA HSC**) **Medium**

52 The part of the curve $x^2 = 32y$ between $y = 0$ and $y = h$ is rotated about the y axis.

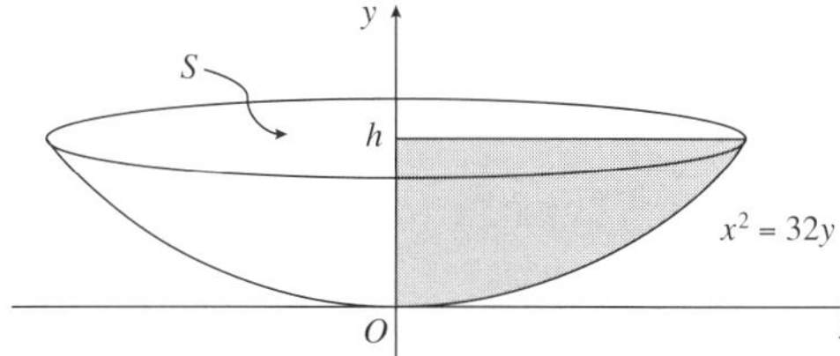

Show that the volume enclosed is given by $V = 16\pi h^2$.

(Q5b, **1994 HSC**) **Easy**

53 The region enclosed by the curve $y = 6\sqrt{x}$ and the x axis between $x = 0$ and $x = 9$ is rotated about the x axis as shown.

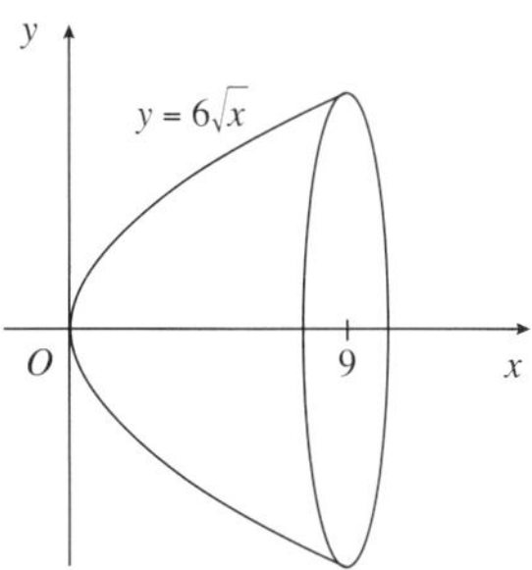

Find the volume of the solid of revolution.

(Q5d, **1994 MA HSC**) **Easy**

54 The region beneath the curve $y = e^{-x}$ which is above the x axis and between the lines $x = 0$ and $x = 1$ is rotated about the x axis.

i Sketch the region. **Easy**

ii Find the volume of the resulting solid of revolution. Easy

(Q6b, **1993 MA HSC**)

55 The shaded region in the diagram is bounded by the curve $y = x^4$, the y axis, and the line $y = 16$.

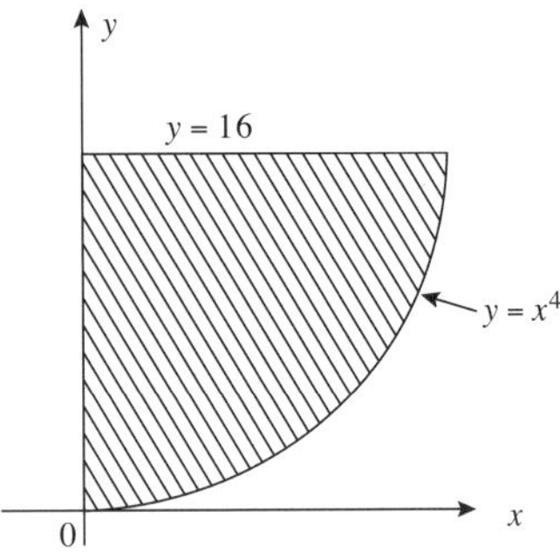

Calculate the volume of the solid of revolution when this region is rotated about the y axis.

(Q6b, **1992 MA HSC**) Easy

Year 12 Further area and volume of solids—Worked Answers

1 $y=(k+1)\sin(kx)$

$V=\pi\int_a^b y^2\,dx$

$\pi^2=\pi\int_0^{\frac{\pi}{2k}}(k+1)^2\sin^2(kx)\,dx$ ✓

$=\pi(k+1)^2\int_0^{\frac{\pi}{2k}}\frac{1-\cos(2kx)}{2}\,dx$

$\pi=\frac{(k+1)^2}{2}\left[x-\frac{\sin(2kx)}{2k}\right]_0^{\frac{\pi}{2k}}$ ✓

$=\frac{(k+1)^2}{2}\left[\left(\frac{\pi}{2k}-\frac{\sin(\pi)}{2k}\right)-\left(0-\frac{\sin(0)}{2k}\right)\right]$

$=\frac{(k+1)^2}{2}\left[\frac{\pi}{2k}-0\right]$

$=\frac{\pi(k+1)^2}{4k}$

$1=\frac{(k+1)^2}{4k}$

$4k=k^2+2k+1$

$0=k^2-2k+1$

$=(k-1)^2$

$\therefore k=1$ ✓

(3 marks)

2 $V=\pi\int_0^2 x^2\,dy-\pi\int_1^2 x^2\,dy$

$V=\pi\int_0^2 y\,dy-\pi\int_1^2(y-1)\,dy$ ✓

$V=\pi\left[\frac{y^2}{2}\right]_0^2-\pi\left[\frac{y^2}{2}-y\right]_1^2$

$V=\pi\left[\frac{2^2}{2}-\frac{0^2}{2}\right]-\pi\left[\left(\frac{2^2}{2}-2\right)-\left(\frac{1^2}{2}-1\right)\right]$ ✓

$V=2\pi-\pi\left[(0)-\left(-\frac{1}{2}\right)\right]$

$V=2\pi-\frac{\pi}{2}$

$V=\frac{3\pi}{2}$

$\therefore$ The volume is $\frac{3\pi}{2}$ units^3. ✓

(3 marks)

3 $A=\int_{-2}^2\left[(2-|x|)-\left(1-\frac{8}{4+x^2}\right)\right]dx$

$=2\int_0^2\left[(2-|x|)-\left(1-\frac{8}{4+x^2}\right)\right]dx$

$=2\int_0^2 2-x-\left(1-\frac{8}{4+x^2}\right)dx$

$=2\int_0^2 1-x+\frac{8}{4+x^2}\,dx$ ✓

$=2\left[x-\frac{x^2}{2}+4\tan^{-1}\frac{x}{2}\right]_0^2$ ✓

$=2\left[\left(2-\frac{2^2}{2}+4\tan^{-1}\frac{2}{2}\right)-\left(0-\frac{0^2}{2}+4\tan^{-1}\frac{0}{2}\right)\right]$

$=2[\pi-0]$

$=2\pi$

$\therefore$ The area is 2π units^2. ✓

(3 marks)

4 $\cos 2x=\cos^2x-\sin^2x$

$=1-2\sin^2x$

At the point of intersection:

$\sin x=1-2\sin^2x$

$2\sin^2x+\sin x-1=0$

$(2\sin x-1)(\sin x+1)=0$

$\sin x=\frac{1}{2}$ or $\sin x=-1$

From the diagram $0<x<\frac{\pi}{2}$ at the point of intersection.

So $\sin x=\frac{1}{2}$

$\therefore x=\frac{\pi}{6}$ ✓

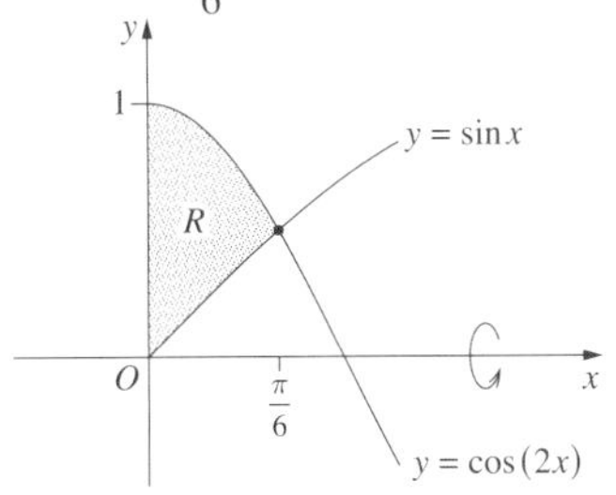

$V=\int_0^{\frac{\pi}{6}}\pi(\cos^2(2x)-\sin^2x)\,dx$

$=\pi\int_0^{\frac{\pi}{6}}\left(\frac{1}{2}(1+\cos(4x))-\frac{1}{2}(1-\cos(2x))\right)dx$

$=\frac{\pi}{2}\int_0^{\frac{\pi}{6}}(\cos(4x)+\cos(2x))\,dx$ ✓

$=\frac{\pi}{2}\left[\frac{\sin(4x)}{4}+\frac{\sin(2x)}{2}\right]_0^{\frac{\pi}{6}}$

$=\frac{\pi}{2}\left(\frac{\sin\frac{2\pi}{3}}{4}+\frac{\sin\frac{\pi}{3}}{2}\right)$ ✓

$=\frac{\pi}{2}\left(\frac{\sqrt{3}}{8}+\frac{\sqrt{3}}{4}\right)$

$=\frac{3\sqrt{3}\pi}{16}$ units^3 ✓

(4 marks)

5 $x=2t-1$

$\therefore\ t=\frac{x+1}{2}$

Substitute into y:

$y=\left(\frac{x+1}{2}\right)^2+3$

$=\frac{x^2+2x+1}{4}+3$

$=\frac{x^2+2x+1+12}{4}$

$=\frac{x^2+2x+13}{4}$ ✓

As $t=1$, then $x=1$; $t=2$, $x=3$.

Area $=\int_1^3\frac{x^2+2x+13}{4}\,dx$

$=\frac{1}{4}\int_1^3(x^2+2x+13)\,dx$ ✓

$=\frac{1}{4}\left[\frac{x^3}{3}+x^2+13x\right]_1^3$

$=\frac{1}{4}\left[\frac{3^3}{3}+3^2+13(3)-\left(\frac{1^3}{3}+1^2+13(1)\right)\right]$ ✓

$=\frac{1}{4}\left[9+9+39-\left(\frac{1}{3}+1+13\right)\right]$

$=\frac{32}{3}$

$\therefore$ area of $\frac{32}{3}$ units^2. ✓ *(4 marks)*

6 Volume $=\pi\int_0^{\frac{\pi}{4}}\sec^2x\,dx$

$=2\pi\int_0^{\frac{\pi}{4}}\sec^2x\,dx$

$=2\pi[\tan x]_0^{\frac{\pi}{4}}$ ✓

$=2\pi\left[\tan\frac{\pi}{4}-\tan 0\right]$

$=2\pi[1-0]$

$=2\pi$

$\therefore$ the volume is 2π units^3. ✓

(2 marks)

7 **i** $m=\frac{4-2}{1-2}$

$=-2$

Equation of the line is $y=-2x+c$.

Substitute $(1, 4)$:

$4=-2(1)+c$

$c=6$

$\therefore y=6-2x$ ✓

(1 mark)

ii As $y=6-2x$,

$x=\frac{6-y}{2}$

$\therefore x^2=\frac{(6-y)^2}{4}$

Also, $x=\frac{4}{y}$

$\therefore x^2=\frac{16}{y^2}$

Volume

$=\pi\int_2^4\frac{(6-y)^2}{4}\,dy-\pi\int_2^4\frac{16}{y^2}\,dy$ ✓

$$= \frac{\pi}{4}\int_2^4 (6-y)^2 dy - 16\pi\int_2^4 y^{-2} dy$$

$$= \frac{\pi}{4}\left[\frac{(6-y)^3}{-3}\right]_2^4 - 16\pi\left[\frac{y^{-1}}{-1}\right]_2^4 \quad ✓$$

$$= -\frac{\pi}{12}\left[(6-y)^3\right]_2^4 + 16\pi\left[\frac{1}{y}\right]_2^4$$

$$= -\frac{\pi}{12}\left[(6-4)^3 - (6-2)^3\right] + 16\pi\left[\frac{1}{4} - \frac{1}{2}\right] \quad ✓$$

$$= -\frac{\pi}{12}[8-64] + 16\pi\left[-\frac{1}{4}\right]$$

$$= \frac{56\pi}{12} - 4\pi$$

$$= \frac{2\pi}{3}$$

$\therefore$ volume of $\frac{2\pi}{3}$ units^3. ✓

(4 marks)

8 $y = \cos 2x \sin x$

$$= \frac{1}{2}(\sin 3x - \sin x) \quad ✓$$

$$\therefore y^2 = \frac{1}{4}(\sin^2 3x - 2\sin 3x \sin x + \sin^2 x)$$

$$= \frac{1}{4}\left(\frac{1}{2}(1-\cos 6x) + \cos 4x - \cos 2x + \frac{1}{2}(1-\cos 2x)\right)$$

$$= \frac{1}{4}\left(\frac{1}{2} - \frac{1}{2}\cos 6x + \cos 4x - \cos 2x + \frac{1}{2} - \frac{1}{2}\cos 2x\right)$$

$$= \frac{1}{4}\left(1 - \frac{1}{2}\cos 6x + \cos 4x - \frac{3}{2}\cos 2x\right) \quad ✓$$

$$\text{Volume} = \pi\int_a^b y^2\,dx$$

$$= \frac{\pi}{4}\int_0^{\frac{\pi}{6}}\left(1 - \frac{1}{2}\cos 6x + \cos 4x - \frac{3}{2}\cos 2x\right)dx$$

$$= \frac{\pi}{4}\left[x - \frac{1}{12}\sin 6x + \frac{1}{4}\sin 4x - \frac{3}{4}\sin 2x\right]_0^{\frac{\pi}{6}} \quad ✓$$

$$= \frac{\pi}{4}\left[\frac{\pi}{6} - \frac{1}{12}\sin 6\left(\frac{\pi}{6}\right) + \frac{1}{4}\sin 4\left(\frac{\pi}{6}\right) - \frac{3}{4}\sin 2\left(\frac{\pi}{6}\right) - 0\right]$$

$$= \frac{\pi}{4}\left[\frac{\pi}{6} - 0 + \frac{1}{4}\left(\frac{\sqrt{3}}{2}\right) - \frac{3}{4}\left(\frac{\sqrt{3}}{2}\right)\right]$$

$$= \frac{\pi}{4}\left[\frac{\pi}{6} + \frac{\sqrt{3}}{8} - \frac{3\sqrt{3}}{8}\right]$$

$$= \frac{\pi}{4}\left[\frac{\pi}{6} - \frac{\sqrt{3}}{4}\right]$$

$$= \frac{\pi}{48}\left[2\pi - 3\sqrt{3}\right] \text{ units}^3$$

✓ *(4 marks)*

9 **i** $y = 2x\tan^{-1} 2x$

$$\frac{dy}{dx} = 2\tan^{-1} 2x + 2x.\frac{1}{1+4x^2}.2$$

$$= 2\tan^{-1} 2x + \frac{4x}{1+4x^2} \quad ✓$$

(1 mark)

ii

$$\int_{\frac{1}{2}}^{\frac{\sqrt{3}}{2}}\left(2\tan^{-1} 2x + \frac{4x}{1+4x^2}\right)dx = \left[2x\tan^{-1} 2x\right]_{\frac{1}{2}}^{\frac{\sqrt{3}}{2}}$$

$$\int_{\frac{1}{2}}^{\frac{\sqrt{3}}{2}}\left(\tan^{-1} 2x + \frac{2x}{1+4x^2}\right)dx = \left[x\tan^{-1} 2x\right]_{\frac{1}{2}}^{\frac{\sqrt{3}}{2}}$$

$$\therefore \int_{\frac{1}{2}}^{\frac{\sqrt{3}}{2}} \tan^{-1} 2x\,dx$$

$$= \left[x\tan^{-1} 2x\right]_{\frac{1}{2}}^{\frac{\sqrt{3}}{2}} - \int_{\frac{1}{2}}^{\frac{\sqrt{3}}{2}}\frac{2x}{1+4x^2}dx \quad ✓$$

$$= \left[\frac{\sqrt{3}}{2}\tan^{-1} 2\left(\frac{\sqrt{3}}{2}\right) - \frac{1}{2}\tan^{-1} 2\left(\frac{1}{2}\right)\right] - \frac{1}{4}\left[\log_e(1+4x^2)\right]_{\frac{1}{2}}^{\frac{\sqrt{3}}{2}} \quad ✓$$

$$= \left[\frac{\sqrt{3}}{2}\tan^{-1}\sqrt{3} - \frac{1}{2}\tan^{-1} 1\right] - \frac{1}{4}\left[\log_e\left(1+4\left(\frac{\sqrt{3}}{2}\right)^2\right) - \log_e\left(1+4\left(\frac{1}{2}\right)^2\right)\right]$$

$$= \left[\frac{\sqrt{3}}{2}\left(\frac{\pi}{3}\right) - \frac{1}{2}\left(\frac{\pi}{4}\right)\right] - \frac{1}{4}\left[\log_e(1+3) - \log_e(1+1)\right]$$

$$= \frac{\pi\sqrt{3}}{6} - \frac{\pi}{8} - \frac{1}{4}\log\frac{4}{2}$$

$$= \frac{\pi\sqrt{3}}{6} - \frac{\pi}{8} - \frac{1}{4}\log_e 2$$

$\therefore$ an area of $\left(\frac{\pi\sqrt{3}}{6} - \frac{\pi}{8} - \frac{1}{4}\log_e 2\right)$ units^2. ✓ *(3 marks)*

10 Volume $= \pi\int_0^a (e^{-x})^2 dx = \frac{3\pi}{8}$

$$\pi\int_0^a e^{-2x} dx = \frac{3\pi}{8} \quad ✓$$

$$\frac{-\pi}{2}\left[e^{-2x}\right]_0^a = \frac{3\pi}{8}$$

$$\frac{-\pi}{2}\left[e^{-2a} - e^{-2(0)}\right] = \frac{3\pi}{8}$$

$$-\frac{1}{2}\left[e^{-2a} - 1\right] = \frac{3}{8} \quad ✓$$

$$e^{-2a} - 1 = -\frac{3}{4}$$

$$e^{-2a} = \frac{1}{4}$$

$$-2a = \log_e\frac{1}{4}$$

$$-2a = -\log_e 4$$

$$a = \frac{1}{2}\log_e 4$$

$$a = \log_e 2$$

✓ *(3 marks)*

11 Range is $0 \le y \le 3\pi$, then $a = 3$.

Domain is $-2 \le x \le 2$, and hence $b = \frac{1}{2}$.

$\therefore a = 3$ and $b = \frac{1}{2}$

$$y = 3\cos^{-1}\frac{x}{2}. \quad ✓$$

Now, $\frac{y}{3} = \cos^{-1}\frac{x}{2}$

$$\cos\frac{y}{3} = \frac{x}{2}$$

$$x = 2\cos\frac{y}{3}$$

$$\text{Area} = \left|\int_{\frac{3\pi}{2}}^{3\pi} 2\cos\frac{y}{3}dy\right| \quad ✓$$

$$= \int_0^{\frac{3\pi}{2}} 2\cos\frac{y}{3}dy$$

$$= 2\left[3\sin\frac{y}{3}\right]_0^{\frac{3\pi}{2}}$$

$$= 6\left[\sin\frac{\frac{3\pi}{2}}{3} - \sin 0\right]$$

$$= 6\left[\sin\frac{\pi}{2} - \sin 0\right]$$
$$= 6$$

$\therefore$ the area is 6 units^2. ✓ *(3 marks)*

12 Let $P(x) = -x^3 + 7x - 6$

Consider $P(1) = -1^3 + 7(1) - 6 = 0$

As $x = 1$ is root, then $(x - 1)$ is factor.

$$\begin{array}{r} -x^2 - x + 6 \\ x-1\overline{)-x^3 + 0x^2 + 7x - 6} \\ \underline{-x^3 + x^2} \\ -x^2 + 7x \\ \underline{-x^2 + x} \\ 6x - 6 \\ \underline{6x - 6} \\ 0 \end{array}$$

$$P(x) = (x - 1)(6 - x - x^2)$$
$$= (x - 1)(3 + x)(2 - x)$$

$\therefore y = (x - 1)(3 + x)(2 - x)$ has zeros of $1, -3, 2$ ✓

$$\text{Area} = \left|\int_{-3}^{1}\left(-x^3 + 7x - 6\right)\right| + \int_{1}^{2}\left(-x^3 + 7x - 6\right)dx \quad ✓$$

$$= \left|\left[-\frac{x^4}{4} + \frac{7x^2}{2} - 6x\right]_{-3}^{1}\right| + \left[-\frac{x^4}{4} + \frac{7x^2}{2} - 6x\right]_{1}^{2} \quad ✓$$

$$= \left|\left[-\frac{1^4}{4} + \frac{7(1)^2}{2} - 6(1) - \left(-\frac{(-3)^4}{4} + \frac{7(-3)^2}{2} - 6(-3)\right)\right]\right| + \left[-\frac{2^4}{4} + \frac{7(2)^2}{2} - 6(2) - \left(-\frac{1^4}{4} + \frac{7(1)^2}{2} - 6(1)\right)\right]$$

$$= 32 + 0.75$$
$$= 32.75$$

$\therefore$ area is 32.75 units^2. ✓ *(4 marks)*

13
$$y = 4\log_e(x + 1)$$
$$\frac{y}{4} = \log_e(x + 1)$$
$$x + 1 = e^{\frac{y}{4}}$$
$$x = e^{\frac{y}{4}} - 1 \quad ✓$$
$$x^2 = (e^{\frac{y}{4}} - 1)^2$$
$$= e^{\frac{y}{2}} - 2e^{\frac{y}{4}} + 1$$
$$V = \pi\int_0^4 x^2 dy$$
$$= \pi\int_0^4\left(e^{\frac{y}{2}} - 2e^{\frac{y}{4}} + 1\right)dy \quad ✓$$
$$= \pi\left[2e^{\frac{y}{2}} - 8e^{\frac{y}{4}} + y\right]_0^4 \quad ✓$$
$$= \pi\left[2e^{\frac{4}{2}} - 8e^{\frac{4}{4}} + 4 - (2e^0 - 8e^0 + 0)\right]$$
$$= 9.524\,861\,469...$$
$$= 9.5 \text{ (1 dec. pl.)}$$

$\therefore$ the volume is 9.5 units^3. ✓ *(4 marks)*

14 **i** $f(x) = \sqrt{\sin^2 x\cos x}$

$f(-x) = \sqrt{\sin^2(-x)\cos(-x)}$

As $\sin^2(-x) = \sin^2 x$ and $\cos(-x) = \cos x$, then $f(-x) = f(x)$.

$\therefore$ the function is even. ✓ *(1 mark)*

ii
$$\text{Volume} = \pi\int_{-\frac{\pi}{2}}^{\frac{\pi}{2}}\left(\sqrt{\sin^2 x\cos x}\right)^2 dx$$
$$= 2\pi\int_0^{\frac{\pi}{2}}\sin^2 x\cos x\,dx \quad ✓$$

$u = \sin x \qquad \frac{du}{dx} = \cos x \qquad dx = \frac{du}{\cos x}$

$$= 2\pi\int_0^1 u^2\cos x.\frac{du}{\cos x}$$

As $x = \frac{\pi}{2}$, then $u = 1$; as $x = 0$, then $u = 0$

$$= 2\pi\int_0^1 u^2 du \quad ✓$$
$$= 2\pi\left[\frac{u^3}{3}\right]_0^1$$
$$= 2\pi\left[\frac{1^3}{3} - 0\right]$$
$$= \frac{2\pi}{3}$$

$\therefore$ the volume is $\frac{2\pi}{3}$ units^3. ✓ *(3 marks)*

15
$$\text{Volume} = \pi\int_0^{\frac{\pi}{4}}(3\sin 2x)^2 dx \quad ✓$$
$$= 9\pi\int_0^{\frac{\pi}{4}}\sin^2 2x\,dx$$
$$= 9\pi\int_0^{\frac{\pi}{4}}\frac{1}{2}(1 - \cos 4x)\,dx$$
$$= \frac{9\pi}{2}\left[x - \frac{1}{4}\sin 4x\right]_0^{\frac{\pi}{4}} \quad ✓$$
$$= \frac{9\pi}{2}\left[\frac{\pi}{4} - \frac{1}{4}\sin 4\left(\frac{\pi}{4}\right) - 0 - \frac{1}{4}\sin 4(0)\right]$$
$$= \frac{9\pi}{2}\left[\frac{\pi}{4}\right]$$
$$= \frac{9\pi^2}{8}$$

$\therefore$ the volume is $\frac{9\pi^2}{8}$ units^3. ✓ *(3 marks)*

16

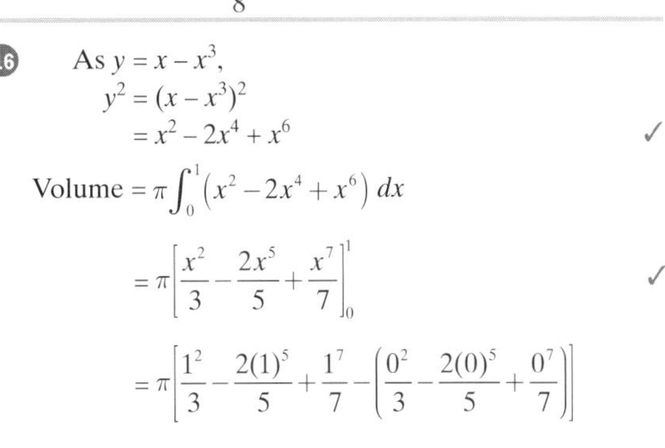

As $y = x - x^3$,
$$y^2 = (x - x^3)^2$$
$$= x^2 - 2x^4 + x^6 \quad ✓$$
$$\text{Volume} = \pi\int_0^1\left(x^2 - 2x^4 + x^6\right)dx$$
$$= \pi\left[\frac{x^2}{3} - \frac{2x^5}{5} + \frac{x^7}{7}\right]_0^1 \quad ✓$$
$$= \pi\left[\frac{1^2}{3} - \frac{2(1)^5}{5} + \frac{1^7}{7} - \left(\frac{0^2}{3} - \frac{2(0)^5}{5} + \frac{0^7}{7}\right)\right]$$

$= \dfrac{8\pi}{105}$

$\therefore$ the volume is $\dfrac{8\pi}{105}$ units^3. ✓ *(3 marks)*

17 $y = x^4 + 1$

$x^4 = y - 1$

$x^2 = \sqrt{y-1}$ ✓

$$\text{Volume} = \pi\int_1^{10} (y-1)^{\frac{1}{2}}dy$$

$$= \pi\left[\frac{2(y-1)^{\frac{3}{2}}}{3}\right]_1^{10}$$ ✓

$$= \frac{2\pi}{3}\left[(10-1)^{\frac{3}{2}} - 0\right]$$

$$= 18\pi$$

$\therefore$ the volume is 18π u^3. ✓ *(3 marks)*

18 $y = \sqrt{16 - 4x^2}$

$y^2 = 16 - 4x^2$

$$V = 2\pi\int_0^2 y^2 \; dx$$

$$= 2\pi\int_0^2 16 - 4x^2 \; dx$$ ✓

$$= 2\pi\left[16x - \frac{4x^3}{3}\right]_0^2$$ ✓

$$= 2\pi\left(16(2) - \frac{4(2)^3}{3} - 0\right)$$

$$= 2\pi\left(32 - \frac{32}{3} - 0\right)$$

$$= \frac{128\pi}{3}$$

$\therefore \dfrac{128\pi}{3}$ units^3 ✓ *(3 marks)*

19 Consider two volumes:

For C_1, rotated around x-axis, use $y^2 = 4 - x^2$. The solid is a hemisphere with radius 2.

For C_2, rotated around x-axis, use $\dfrac{y^2}{4} = 1 - \dfrac{x^2}{9}$, i.e. $y^2 = 4 - \dfrac{4x^2}{9}$. ✓

$$\text{Total volume} = \frac{1}{2} \times \frac{4}{3}\pi r^3 + \pi\int_a^b y^2 \, dx$$

$$= \frac{1}{2} \times \frac{4}{3} \times \pi \times 2^3 + \pi\int_0^3\left(4 - \frac{4x^2}{9}\right)dx$$ ✓

$$= \frac{16\pi}{3} + \pi\left[4x - \frac{4x^3}{27}\right]_0^3$$ ✓

$$= \frac{16\pi}{3} + \pi\left[4(3) - \frac{4(3)^3}{27} - (4(0) - \frac{4(0)^3}{27})\right]$$

$$= \frac{16\pi}{3} + \pi[12 - 4]$$

$$= \frac{16\pi}{3} + 8\pi$$

$$= \frac{40\pi}{3}$$

$\therefore$ the volume is $\dfrac{40\pi}{3}$ units^3. ✓ *(4 marks)*

20 $y = 8\log_e(x-1)$

$\dfrac{y}{8} = \log_e(x-1)$

$x - 1 = e^{\frac{y}{8}}$

$x = e^{\frac{y}{8}} + 1$

$x^2 = (e^{\frac{y}{8}} + 1)^2$

$= e^{\frac{y}{4}} + 2e^{\frac{y}{8}} + 1$ ✓

$$V = \pi\int_0^6 x^2dy$$

$$= \pi\int_0^6\left(e^{\frac{y}{4}} + 2e^{\frac{y}{8}} + 1\right)dy$$

$$= \pi\left[4e^{\frac{y}{4}} + 16e^{\frac{y}{8}} + y\right]_0^6$$ ✓

$= \pi(4e^{1.5} + 16e^{0.75} + 6 - (4e^0 + 16e^0 + 0))$

$= 118.748\,2959...$

$= 118.7$ (1 dec. pl.)

$\therefore$ the volume is 118.7 units^3. ✓ *(3 marks)*

21 $V = \displaystyle\int \pi y^2 dx$

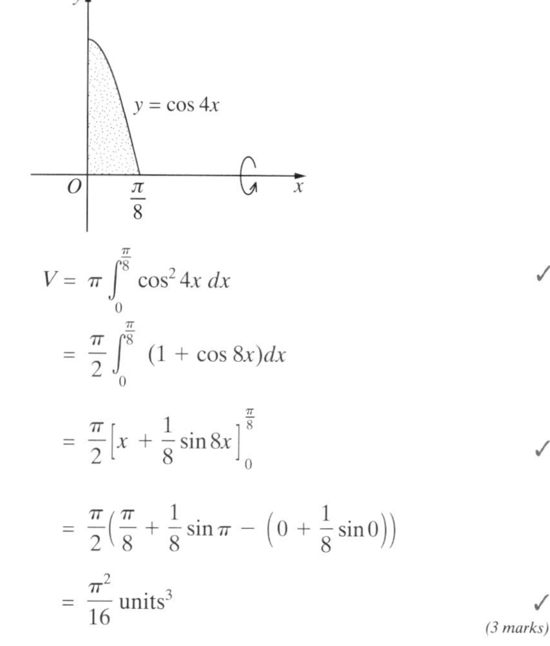

$$V = \pi\int_0^{\frac{\pi}{8}} \cos^2 4x \, dx$$ ✓

$$= \frac{\pi}{2}\int_0^{\frac{\pi}{8}} (1 + \cos 8x)dx$$

$$= \frac{\pi}{2}\left[x + \frac{1}{8}\sin 8x\right]_0^{\frac{\pi}{8}}$$ ✓

$$= \frac{\pi}{2}\left(\frac{\pi}{8} + \frac{1}{8}\sin\pi - \left(0 + \frac{1}{8}\sin 0\right)\right)$$

$$= \frac{\pi^2}{16} \text{ units}^3$$ ✓ *(3 marks)*

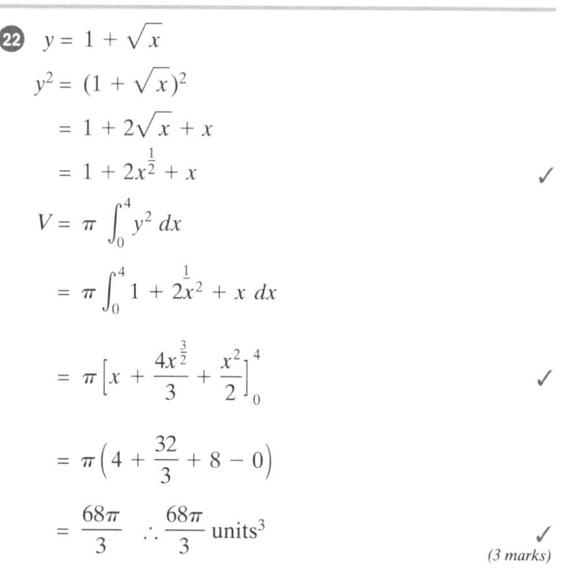

22 $y = 1 + \sqrt{x}$

$y^2 = (1 + \sqrt{x})^2$

$= 1 + 2\sqrt{x} + x$

$= 1 + 2x^{\frac{1}{2}} + x$ ✓

$$V = \pi\int_0^4 y^2 \, dx$$

$$= \pi\int_0^4 1 + 2x^{\frac{1}{2}} + x \, dx$$

$$= \pi\left[x + \frac{4x^{\frac{3}{2}}}{3} + \frac{x^2}{2}\right]_0^4$$ ✓

$$= \pi\left(4 + \frac{32}{3} + 8 - 0\right)$$

$= \dfrac{68\pi}{3} \quad \therefore \dfrac{68\pi}{3}$ units^3 ✓ *(3 marks)*

23 $V = \int_0^{\frac{3\pi}{2}} \pi\left(3\sin\frac{x}{2}\right)^2 dx$

$= 9\pi \int_0^{\frac{3\pi}{2}} \sin^2\left(\frac{x}{2}\right) dx$ ✓

$= 9\pi \int_0^{\frac{3\pi}{2}} \frac{1}{2}(1 - \cos x)\, dx$

$= \frac{9\pi}{2}[x - \sin x]_0^{\frac{3\pi}{2}}$ ✓

$= \frac{9\pi}{2}\left(\frac{3\pi}{2} - \sin\frac{3\pi}{2} - (0 - \sin 0)\right)$

$= \frac{9\pi}{2}\left(\frac{3\pi}{2} + 1\right)$

$= \frac{9\pi}{4}(3\pi + 2)$ units^3 ✓

(3 marks)

24 $y = (x - 2)^2$

$(x - 2)^2 = y$

$x - 2 = \pm\sqrt{y}$

$x = 2 \pm \sqrt{y}$ ✓

When $y = 4, x = 0$, so $x = 2 - \sqrt{y}$

$V = \pi \int_0^4 (2 - \sqrt{y})^2\, dy$ ✓

$= \pi \int_0^4 (4 - 4y^{\frac{1}{2}} + y)\, dy$

$= \pi\left[4y - \frac{8y^{\frac{3}{2}}}{3} + \frac{y^2}{2}\right]_0^4$ ✓

$= \pi\left(16 - \frac{64}{3} + 8 - 0\right)$

$= \frac{8\pi}{3}$ $\therefore \frac{8\pi}{3}$ units^3 ✓

(4 marks)

25 $V = \pi \int_0^1 y^2\, dx$

$= \pi \int_0^1 \left[\frac{3}{(x + 2)^2}\right]^2 dx$

$= \pi \int_0^1 \frac{9}{(x + 2)^4}\, dx$ ✓

$= 9\pi \int_0^1 (x + 2)^{-4}\, dx$

$= 9\pi\left[\frac{(x + 2)^{-3}}{-3}\right]_0^1$ ✓

$= -3\pi\left[\frac{1}{(x + 2)^3}\right]_0^1$

$= -3\pi\left[\frac{1}{27} - \frac{1}{8}\right]$

$= -3\pi\left[\frac{-19}{216}\right]$

$= \frac{19\pi}{72}$

$\therefore$ volume is $\frac{19\pi}{72}$ units^3 ✓

(3 marks)

26 **i** $V = \pi \int x^2\, dy$

$= \pi \int_0^h y\, dy$

$= \pi\left[\frac{y^2}{2}\right]_0^h$ ✓

$= \pi\left[\frac{h^2}{2} - 0\right]$

$= \frac{\pi h^2}{2}$

$\therefore$ the volume is $\frac{\pi h^2}{2}$ units^3. ✓

(2 marks)

ii If $y = h$, then $x^2 = h$

$x = \sqrt{h}$

$\therefore C(\sqrt{h}, h)$

V of cylinder $= \pi \times (\sqrt{h})^2 \times h$

$= \pi h^2$

Ratio $= \frac{\pi h^2}{2} : \pi h^2$

$= 1 : 2$ ✓ *(1 mark)*

27 **i**

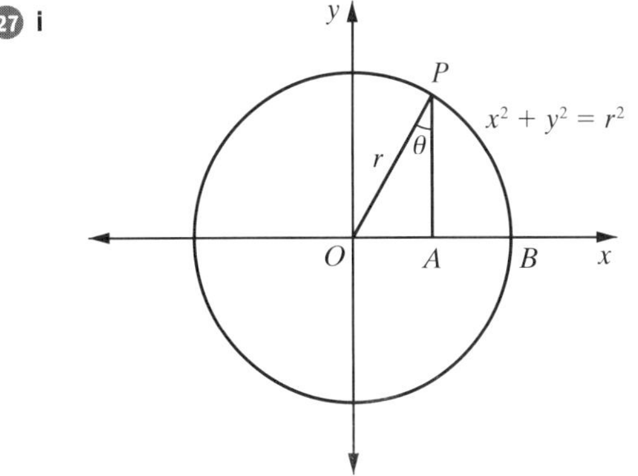

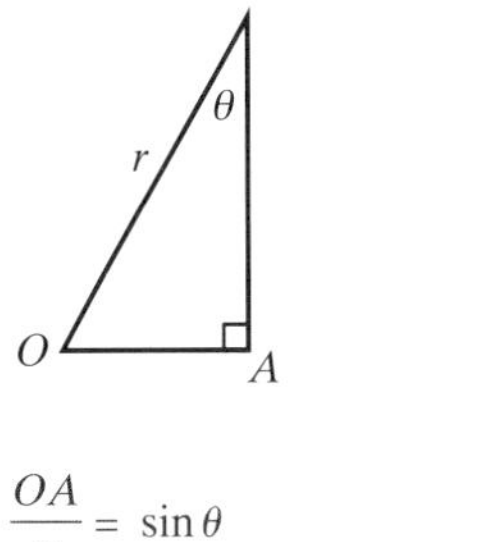

Using ΔPAO, $\frac{OA}{r} = \sin\theta$

$OA = r\sin\theta$

Also, as $x^2 + y^2 = r^2$

$\therefore y^2 = r^2 - x^2$ ✓

$$\text{Now } V = \pi \int_{r\sin\theta}^{r} r^2 - x^2 dx$$
$$= \pi \left[r^2x - \frac{x^3}{3}\right]_{r\sin\theta}^{r} \quad ✓$$
$$= \pi \left[r^3 - \frac{r^3}{3} - \left(r^3\sin\theta - \frac{r^3\sin^3\theta}{3}\right)\right]$$
$$= \pi \left[\frac{2r^3}{3} - r^3\sin\theta + \frac{r^3\sin^3\theta}{3}\right]$$
$$= \frac{\pi r^3}{3}[2 - 3\sin\theta + \sin^3\theta] \quad ✓$$

(3 marks)

ii 1

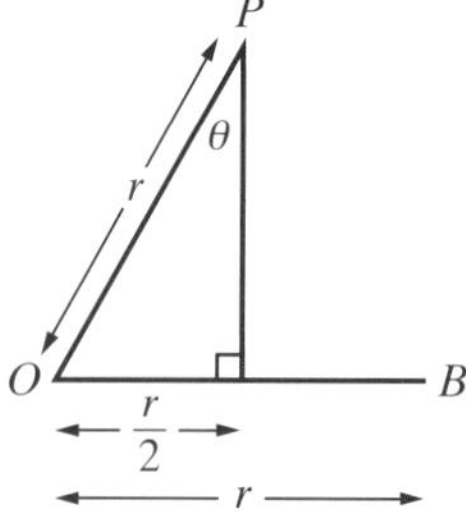

As $OB = r$, then $OA = \frac{r}{2}$ (half-depth)

$$\therefore \sin\theta = \frac{\frac{r}{2}}{r}$$
$$= \frac{r}{2} \div r$$
$$= \frac{1}{2}$$
$$\therefore \theta = 30°$$ ✓ *(1 mark)*

2 Volume of hemisphere $= \frac{1}{2} \times \frac{4}{3}\pi r^3$
$$= \frac{2}{3}\pi r^3$$

$\therefore$ Volume of original hemisphere is $\frac{2}{3}\pi r^3$ units^3.

When water is one half of the original depth:

subs $\theta = 30°$ in V:

$$V = \frac{\pi r^3}{3}[2 - 3\sin 30° + \sin^3 30°]$$
$$= \frac{\pi r^3}{3}\left[2 - \frac{3}{2} + \frac{1}{8}\right]$$
$$= \frac{5\pi r^3}{24}$$

$\therefore$ New volume is $\frac{5\pi r^3}{24}$ units^3. ✓

$$\therefore \text{Fraction} = \frac{5\pi r^3}{24} \div \frac{2}{3}\pi r^3$$
$$= \frac{5\pi r^3}{24} \times \frac{3}{2\pi r^3}$$
$$= \frac{5}{16} \quad ✓$$

(2 marks)

28 $y = \sec x$

$\therefore y^2 = \sec^2 x$

$$\text{Now } V = \pi\int_a^b y^2\,dx$$
$$\therefore V = 2 \times \pi\int_0^{\frac{\pi}{3}} \sec^2 x\,dx \quad ✓$$
$$= 2\pi\Big[\tan x\Big]_0^{\frac{\pi}{3}} \quad ✓$$
$$= 2\pi\left(\tan\frac{\pi}{3} - \tan 0\right)$$
$$= 2\pi\left(\sqrt{3} - 0\right)$$
$$= 2\sqrt{3}\,\pi \text{ units}^3. \quad ✓$$

(3 marks)

29 $y = \frac{5}{x-2}$

$$\therefore y^2 = \left(\frac{5}{x-2}\right)^2$$
$$= \frac{25}{(x-2)^2}$$
$$v = \pi\int_a^b y^2\,dx$$
$$= \pi\int_3^6 \frac{25}{(x-2)^2}\,dx \quad ✓$$
$$= 25\pi\int_3^6 (x-2)^{-2}\,dx$$
$$= 25\pi\left[\frac{(x-2)^{-1}}{-1}\right]_3^6 \quad ✓$$
$$= 25\pi\left[-\frac{1}{(x-2)}\right]_3^6$$
$$= 25\pi\left[\left(-\frac{1}{6-2}\right) - \left(-\frac{1}{3-2}\right)\right]$$
$$= 25\pi\left[-\frac{1}{4} + 1\right]$$
$$= 25\pi \times \frac{3}{4}$$
$$= \frac{75\pi}{4} \text{ units}^3. \quad ✓$$

(3 marks)

30 $$V = \pi\int_0^3 \left(\frac{1}{\sqrt{9+x^2}}\right)^2 dx$$

$$= \pi\int_0^3 \frac{1}{9+x^2}\,dx \quad ✓$$
$$= \pi\left[\frac{1}{3}\tan^{-1}\frac{x}{3}\right]_0^3 \quad \text{(using standard integrals)}$$
$$= \frac{\pi}{3}(\tan^{-1}1 - \tan^{-1}0) \quad ✓$$
$$= \frac{\pi}{3}\left(\frac{\pi}{4} - 0\right)$$
$$= \frac{\pi^2}{12} \text{ units}^3. \quad ✓$$

(3 marks)

31
$$y = x^2 + 1$$
$$\therefore y^2 = (x^2+1)^2$$
$$= x^4 + 2x^2 + 1$$

$$V = \pi\int_a^b y^2\,dx$$
$$= \pi\int_0^1 (x^4 + 2x^2 + 1)\,dx \quad ✓$$
$$= \pi\left[\frac{x^5}{5} + \frac{2x^3}{3} + x\right]_0^1 \quad ✓$$
$$= \pi\left\{\left(\frac{(1)^5}{5} + \frac{2(1)^3}{3} + 1\right) - \left(\frac{(0)^5}{5} + \frac{2(0)^3}{3} + 0\right)\right\}$$
$$= \pi\left\{\left(\frac{1}{5} + \frac{2}{3} + 1\right) - (0)\right\}$$
$$= \frac{28\pi}{15}\text{ units}^3. \quad ✓$$

(3 marks)

32 Now $y = x^2 + 1$
$$\therefore x^2 = y - 1$$

Now $V = \pi\int_a^b x^2\,dy$
$$\therefore V = \pi\int_1^5 (y-1)\,dy \quad ✓$$
$$= \pi\left[\frac{y^2}{2} - y\right]_1^5 \quad ✓$$
$$= \pi\left[\left(\frac{5^2}{2} - 5\right) - \left(\frac{1^2}{2} - 1\right)\right]$$
$$= \pi\left(7\tfrac{1}{2} + \tfrac{1}{2}\right)$$
$$= 8\pi\text{ units}^3. \quad ✓$$

(3 marks)

33

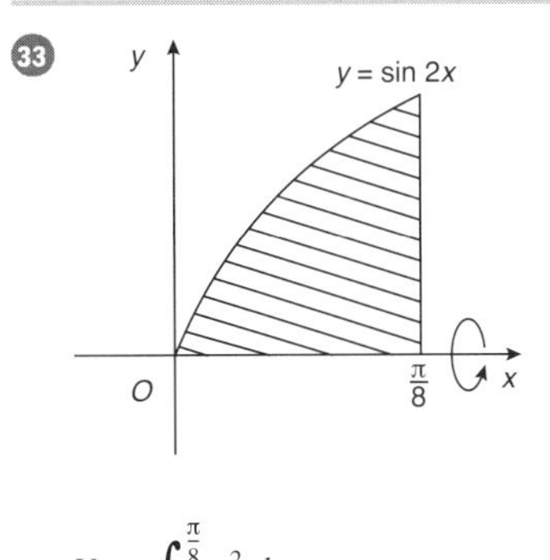

$$V = \pi\int_0^{\frac{\pi}{8}} y^2\,dx$$
$$= \pi\int_0^{\frac{\pi}{8}} (\sin 2x)^2\,dx$$

Now $\sin^2 x = \frac{1}{2}(1 - \cos 2x)$

$$\therefore \sin^2 2x = \frac{1}{2}(1 - \cos 4x)$$

$$\therefore V = \pi\int_0^{\frac{\pi}{8}} \frac{1}{2}(1 - \cos 4x)\,dx \quad ✓$$
$$= \frac{\pi}{2}\int_0^{\frac{\pi}{8}} (1 - \cos 4x)\,dx$$
$$= \frac{\pi}{2}\left[x - \frac{1}{4}\sin 4x\right]_0^{\frac{\pi}{8}} \quad ✓$$
$$= \frac{\pi}{2}\left[\left(\frac{\pi}{8} - \frac{1}{4}\sin\frac{\pi}{2}\right) - \left(0 - \frac{1}{4}\sin 0\right)\right]$$
$$= \frac{\pi}{2}\left(\frac{\pi}{8} - \frac{1}{4} - 0\right)$$
$$= \frac{\pi}{2}\left(\frac{\pi}{8} - \frac{2}{8}\right)$$
$$= \frac{\pi}{16}(\pi - 2)\text{ units}^3. \quad ✓$$

(3 marks)

34 **i** Solving simultaneously:

$y = 12 - 2x^2 \quad \ldots(1)$

$y = x^2 \quad \ldots(2)$

Equating (1) and (2):
$$x^2 = 12 - 2x^2$$
$$3x^2 - 12 = 0$$
$$3(x^2 - 4) = 0$$
$$3(x+2)(x-2) = 0$$
$$\therefore x = \pm 2$$

When $x = -2$, $y = 4$
When $x = 2$, $y = 4$

$\therefore$ Points of intersection are $(-2, 4)$ and $(2, 4)$. ✓

(1 mark)

ii

For $y = 12 - 2x^2$
$$2x^2 = 12 - y$$
$$\therefore x^2 = \frac{12-y}{2}$$

$$V = \pi\int_0^4 x^2\,dy + \pi\int_4^{12} x^2\,dy$$
$$= \pi\int_0^4 y\,dy + \pi\int_4^{12}\left(\frac{12-y}{2}\right)dy \quad ✓$$
$$= \pi\int_0^4 y\,dy + \frac{\pi}{2}\int_4^{12}(12-y)\,dy$$
$$= \pi\left[\frac{y^2}{2}\right]_0^4 + \frac{\pi}{2}\left[12y - \frac{y^2}{2}\right]_4^{12} \quad ✓$$
$$= \pi\left(\frac{16}{2} - 0\right) + \frac{\pi}{2}\left[\left(12(12) - \frac{12^2}{2}\right) - \left(12(4) - \frac{4^2}{2}\right)\right]$$
$$= 8\pi + \frac{\pi}{2}\left[(144 - 72) - (48 - 8)\right]$$
$$= 8\pi + \frac{\pi}{2}(72 - 40)$$
$$= 8\pi + \frac{\pi}{2}(32)$$
$$= 8\pi + 16\pi$$
$$= 24\pi\text{ units}^3. \quad ✓$$

(3 marks)

35 Now $y = 2\sec x$
$$\therefore y^2 = 4\sec^2 x$$

$$V = \pi\int_0^{\frac{\pi}{3}} y^2\,dx$$
$$= \pi\int_0^{\frac{\pi}{3}} 4\sec^2 x\,dx \quad ✓$$
$$= 4\pi\int_0^{\frac{\pi}{3}} \sec^2 x\,dx$$
$$= 4\pi\left[\tan x\right]_0^{\frac{\pi}{3}} \quad ✓$$
$$= 4\pi\left(\tan\frac{\pi}{3} - \tan 0\right)$$
$$= 4\pi(\sqrt{3} - 0)$$
$$= 4\sqrt{3}\,\pi\text{ units}^3. \quad ✓$$

(3 marks)

36
$$y = e^x$$
At $x = \log_e 5$, $y = e^{\log_e 5}$
$$= 5$$

At $x = 0$, $y = e^0 = 1$ ✓

Now $V = \pi\int_1^5 x^2\,dy$ ✓

and $y = e^x$
$$\therefore x = \log_e y$$
$$\therefore V = \pi\int_1^5 (\log_e y)^2\,dy \quad ✓$$

(3 marks)

37 $V = \pi\int_0^4 x^2\,dy$

Now $y = \frac{x^4}{4}$
$$x^4 = 4y$$
$$x^2 = 2\sqrt{y}$$
$$= 2y^{\frac{1}{2}} \quad ✓$$
$$\therefore V = \pi\int_0^4 2y^{\frac{1}{2}}\,dy$$
$$= 2\pi\int_0^4 y^{\frac{1}{2}}\,dy \quad ✓$$
$$= 2\pi\left[\frac{2y^{\frac{3}{2}}}{3}\right]_0^4 \quad ✓$$

$= 2\pi\left[\frac{2}{3}(4)^{\frac{3}{2}} - 0\right]$

$= 2\pi\left[\frac{2}{3} \times 8\right]$

$= \frac{32\pi}{3}$ units^3. ✓

(4 marks)

38 **i** The y-intercept occurs at $x = 0$

$\therefore y = 2\cos^{-1} 0$

$= 2 \times \frac{\pi}{2}$

$= \pi$ (note the range of $f(x)$ is $0 \le y \le 2\pi$) ✓

(1 mark)

ii The inverse function is given by

$x = 2\cos^{-1}\frac{y}{3}$

$\cos^{-1}\frac{y}{3} = \frac{x}{2}$

$\frac{y}{3} = \cos\frac{x}{2}$

$y = 3\cos\frac{x}{2}$ ✓

with domain: $0 \le x \le 2\pi$ ✓

(2 marks)

iii

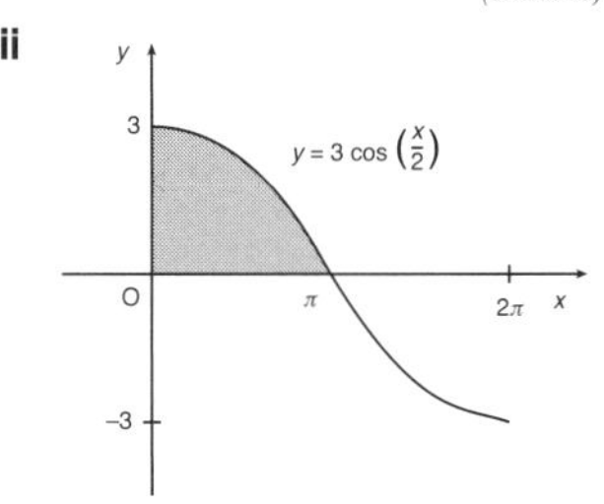

Area of shaded region

$= \int_0^3 2\cos^{-1}\frac{x}{3}\,dx$

$= \int_0^{\pi} 3\cos\frac{x}{2}\,dx$ ✓

$= \left[6\sin\frac{x}{2}\right]_0^{\pi}$

$= 6\sin\frac{\pi}{2} - 6\sin 0$

$= 6 \times 1 - 0$

$= 6$ units^2. ✓

(2 marks)

39

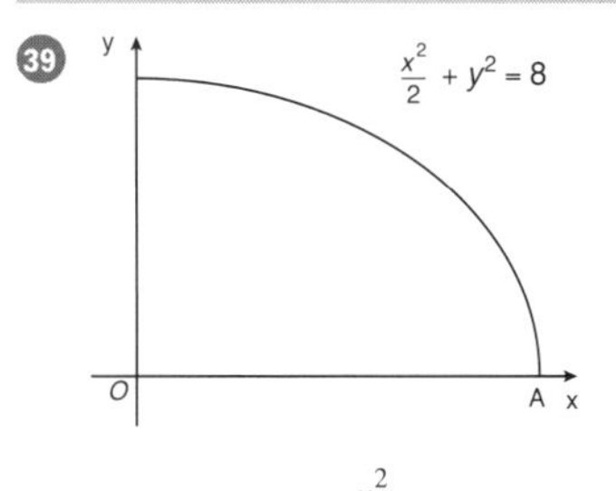

At A, $y = 0 \therefore \frac{x^2}{2} + 0 = 8$

$x^2 = 16$

$\therefore x = \pm 4$

$\therefore A \equiv (4, 0)$ ✓

Now $\frac{x^2}{2} + y^2 = 8$

$\therefore y^2 = 8 - \frac{x^2}{2}$

and $V = \pi\int_0^4 y^2\,dx$

$= \pi\int_0^4 \left(8 - \frac{x^2}{2}\right)dx$ ✓

$= \pi\left[8x - \frac{x^3}{6}\right]_0^4$

$= \pi\left[\left(8(4) - \frac{(4)^3}{6}\right) - 0\right]$

$= \pi\left(32 - \frac{64}{6}\right)$

$= \pi\left(32 - \frac{32}{3}\right)$

$= \pi\left(\frac{96}{3} - \frac{32}{3}\right)$

$= \frac{64\pi}{3}$ units^3. ✓

(3 marks)

40 $V = \pi\int_1^{e^2} y^2\,dx$

$= \pi\int_1^{e^2}\left(\frac{1}{\sqrt{x}}\right)^2 dx$ ✓

$= \pi\int_1^{e^2}\frac{1}{x}\,dx$

$= \pi[\ln x]_1^{e^2}$ ✓

$= \pi(\ln e^2 - \ln 1)$

$= \pi(2\ln e - 0)$ ✓

$= \pi \times 2$

$= 2\pi$ units^3. ✓

(4 marks)

41 $V = \pi\int_0^{\frac{\pi}{2}} y^2\,dx$

$= \pi\int_0^{\frac{\pi}{2}} 9\sin^2 x\,dx$ ✓

$= 9\pi\int_0^{\frac{\pi}{2}} \sin^2 x\,dx$

$= 9\pi\int_0^{\frac{\pi}{2}} \frac{1 - \cos 2x}{2}\,dx$

$= \frac{9\pi}{2}\int_0^{\frac{\pi}{2}} (1 - \cos 2x)\,dx$ ✓

$= \frac{9\pi}{2}\left[x - \frac{1}{2}\sin 2x\right]_0^{\frac{\pi}{2}}$ ✓

$= \frac{9\pi}{2}\left[\left(\frac{\pi}{2} - \frac{1}{2}\sin\pi\right) - (0 - 0)\right]$

$= \frac{9\pi}{2} \times \frac{\pi}{2}$

$= \frac{9\pi^2}{4}$ units^3. ✓

(4 marks)

42 **i** $V = \pi\int_1^7 x^2\,dy$

Now $y = e^{x^2}$

$\therefore \log_e y = \log_e e^{x^2}$

$\log_e y = x^2$

$\therefore V = \pi\int_1^7 \log_e y\,dy$ ✓

ii

y	1	4	7
$\log_e y$	0	1.386	1.946

✓✓

iii

$V = \pi\left[\frac{b-a}{2n}(y_0 + 2(y_1) + y_2)\right]$

$\doteqdot \pi\left[\frac{7-1}{2(2)}(0 + 2(1.386) + 1.946)\right]$ ✓

$\doteqdot \pi\left[\frac{6}{4} \times 4.718\right]$

$\doteqdot 22.233$ units^3. ✓

(5 marks)
(Total mark allocation only in exam)

43 **i** $V = \pi\int_0^h x^2\,dy$

Now $x^2 + (y - 15)^2 = 15^2$

$x^2 = 15^2 - (y - 15)^2$

$= 15^2 - (y^2 - 30y + 15^2)$

$= 30y - y^2$

$\therefore V = \pi\int_0^h (30y - y^2)\,dy$ ✓

$= \pi\left[15y^2 - \frac{y^3}{3}\right]_0^h$

$= \pi\left[15h^2 - \frac{h^3}{3} - 0\right]$

$= 15\pi h^2 - \frac{\pi h^3}{3}$. ✓

ii Now $\frac{dV}{dt} = 3$

and $V = 15\pi h^2 - \frac{\pi h^3}{3}$

$\therefore \frac{dV}{dh} = 30\pi h - \pi h^2$ ✓

Now $\frac{dV}{dt} = \frac{dV}{dh} \times \frac{dh}{dt}$

$\therefore \frac{dV}{dt} = (30\pi h - \pi h^2)\frac{dh}{dt}$ ✓

When $h = 6$,

$3 = (30 \times \pi \times 6 - 36\pi)\frac{dh}{dt}$

$3 = 144\pi\frac{dh}{dt}$

$\frac{dh}{dt} = \frac{1}{48\pi}$

$\therefore$ The rate at which the water level is rising when $h = 6$ cm is $\frac{1}{48\pi}$ cm s^{-1}. ✓

(5 marks)
(Total mark allocation only in exam)

44 **i**

$$y = \log_2 x$$

$$\therefore y = \frac{\log_e x}{\log_e 2} \quad \text{(change of base)}$$

$$\log_e x = y \log_e 2$$

$$x = e^{y \log_e 2}$$

$$= e^{y \ln 2} \quad ✓$$

$$V = \pi \int_0^3 x^2\, dy$$

$$= \pi \int_0^3 (e^{y \ln 2})^2\, dy$$

$$= \pi \int_0^3 e^{2y \ln 2}\, dy$$

$$= \pi \int_0^3 e^{y \ln 2^2}\, dy$$

$$= \pi \int_0^3 e^{y \ln 4}\, dy \quad ✓$$

ii

$$V = \pi \int_0^3 e^{y \ln 4}\, dy$$

$$= \pi \left[\frac{e^{y \ln 4}}{\ln 4}\right]_0^3 \quad ✓$$

$$= \pi \left(\frac{e^{3 \ln 4}}{\ln 4} - \frac{e^0}{\ln 4}\right) \quad ✓$$

$$= \pi \left(\frac{e^{\ln 64}}{\ln 4} - \frac{1}{\ln 4}\right)$$

$$= \pi \left(\frac{64}{\ln 4} - \frac{1}{\ln 4}\right)$$

$$= \pi \times \frac{63}{\ln 4}$$

$$\doteqdot 142.8$$

$\therefore$ Volume of the solid is approximately 142.8 units^3. ✓

(5 marks)
(Total mark allocation only in exam)

45 **i**

$$V = \pi \int_0^{\frac{\pi}{6}} y^2\, dx$$

$$= \pi \int_0^{\frac{\pi}{6}} (\tan 2x)^2\, dx \quad ✓$$

$$= \pi \int_0^{\frac{\pi}{6}} \tan^2 2x\, dx$$

$$= \pi \int_0^{\frac{\pi}{6}} (\sec^2 2x - 1)\, dx \quad ✓$$

ii

$$V = \pi \left[\frac{\tan 2x}{2} - x\right]_0^{\frac{\pi}{6}} \quad ✓$$

$$= \pi \left[\left(\frac{\tan \frac{\pi}{3}}{2} - \frac{\pi}{6}\right) - (0 - 0)\right]$$

$$= \pi \left[\frac{\sqrt{3}}{2} - \frac{\pi}{6}\right] \quad ✓$$

$$= \frac{\pi(3\sqrt{3} - \pi)}{6} \text{ units}^3. \quad ✓$$

(5 marks)
(Total mark allocation only in exam)

46

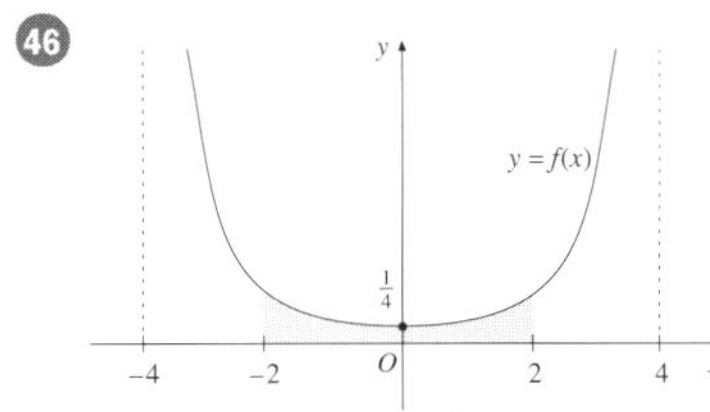

i

$$f(-x) = \frac{1}{\sqrt{16 - (-x)^2}}$$

$$= \frac{1}{\sqrt{16 - x^2}}$$

$$= f(x)$$

$\therefore f(x)$ is an even function. ✓

ii

$$A = \int_{-2}^{2} f(x)\, dx$$

$$= 2 \int_0^2 \frac{dx}{\sqrt{16 - x^2}} \quad \text{(as } f(x) \text{ is even)} ✓$$

$$= 2 \left[\sin^{-1} \frac{x}{4}\right]_0^2 \quad ✓$$

$$= 2 (\sin^{-1} \tfrac{1}{2} - \sin^{-1} 0)$$

$$= 2 (\tfrac{\pi}{6} - 0)$$

$$= \frac{\pi}{3} \text{ units}^2. \quad ✓$$

(4 marks)
(Total mark allocation only in exam)

47

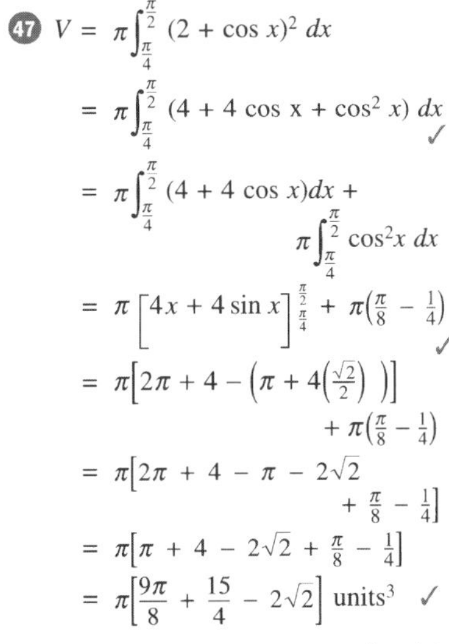

$$V = \pi \int_{\frac{\pi}{4}}^{\frac{\pi}{2}} (2 + \cos x)^2\, dx$$

$$= \pi \int_{\frac{\pi}{4}}^{\frac{\pi}{2}} (4 + 4 \cos x + \cos^2 x)\, dx \quad ✓$$

$$= \pi \int_{\frac{\pi}{4}}^{\frac{\pi}{2}} (4 + 4 \cos x)\,dx + \pi \int_{\frac{\pi}{4}}^{\frac{\pi}{2}} \cos^2 x\, dx$$

$$= \pi \Big[4x + 4 \sin x\Big]_{\frac{\pi}{4}}^{\frac{\pi}{2}} + \pi\left(\tfrac{\pi}{8} - \tfrac{1}{4}\right) \quad ✓$$

$$= \pi\left[2\pi + 4 - \left(\pi + 4\left(\tfrac{\sqrt{2}}{2}\right)\right)\right] + \pi\left(\tfrac{\pi}{8} - \tfrac{1}{4}\right)$$

$$= \pi\left[2\pi + 4 - \pi - 2\sqrt{2} + \tfrac{\pi}{8} - \tfrac{1}{4}\right]$$

$$= \pi\left[\pi + 4 - 2\sqrt{2} + \tfrac{\pi}{8} - \tfrac{1}{4}\right]$$

$$= \pi\left[\frac{9\pi}{8} + \frac{15}{4} - 2\sqrt{2}\right] \text{ units}^3 \quad ✓$$

(3 marks)

48

$$V = \pi \int_0^3 y^2\, dx$$

$$= \pi \int_0^3 (x + 1)^2\, dx \quad ✓$$

$$= \pi \left[\frac{(x + 1)^3}{3 \times 1}\right]_0^3$$

$$= \pi \left[\frac{(x + 1)^3}{3}\right]_0^3 \quad ✓$$

$$= \frac{\pi}{3} \left[(x + 1)^3\right]_0^3$$

$$= \frac{\pi}{3} (4^3 - 1^3)$$

$$= \frac{\pi}{3} (64 - 1)$$

$$= \frac{63\pi}{3}$$

$$= 21\pi \text{ units}^3 \quad ✓$$

(3 marks)

49 Let V_1 be the volume of the solid formed by rotating the shaded region below $y = 3$. Let V_2 be the volume of the solid formed by rotating the shaded region between $y = 3$ and $y = 6$.

$$V_1 = \pi r^2 h$$

$$= \pi(1)^2 \times 3$$

$$= 3\pi \text{ units}^3. \quad ✓$$

$$V_2 = \pi \int_3^6 x^2\, dy$$

Now $xy = 3$

$$\therefore x = \frac{3}{y}$$

$$\therefore x^2 = \frac{9}{y^2}$$

$$\therefore V_2 = \pi \int_3^6 \frac{9}{y^2}\, dy$$

$$= 9\pi \int_3^6 \frac{dy}{y^2}$$

$$= 9\pi \int_3^6 y^{-2}\, dy \quad ✓$$

$$= 9\pi \left[\frac{y^{-1}}{-1}\right]_3^6$$

$$= 9\pi \left[-\frac{1}{y}\right]_3^6$$

$$= 9\pi \left[-\tfrac{1}{6} - \left(-\tfrac{1}{3}\right)\right]$$

$$= 9\pi \left[-\tfrac{1}{6} + \tfrac{1}{3}\right]$$

$$= 9\pi \times \tfrac{1}{6}$$

$$= \frac{9\pi}{6}$$

$$= \frac{3\pi}{2} \text{ units}^3. \quad ✓$$

$$\therefore V_{\text{TOTAL}} = V_1 + V_2$$

$$= 3\pi + \frac{3\pi}{2}$$

$$= \frac{6\pi}{2} + \frac{3\pi}{2}$$

$$= \frac{9\pi}{2} \text{ units}^3. \quad ✓$$

(4 marks)

50 **i**

$$\begin{aligned} \sin 2x &= 2\sin^2 x \quad 0 < x < \pi \\ 2\sin x\cos x &= 2\sin^2 x \\ \sin x\cos x &= \sin^2 x \\ \sin x\cos x - \sin^2 x &= 0 \\ \sin x(\cos x - \sin x) &= 0 \quad ✓ \\ \therefore \cos x - \sin x &= 0 \\ \therefore \sin x &= \cos x \\ \frac{\sin x}{\cos x} &= 1 \\ \tan x &= 1 \\ \therefore x &= \frac{\pi}{4} \quad 0 < x < \pi. \quad ✓ \end{aligned}$$

(N.B. $\sin x = 0$ has no solutions for $0 < x < \pi$).

ii If $0 < x < \frac{\pi}{4}$, then $\tan x < 1$,

$\sin x > 0$ and $\cos x > 0$.

Now if $\tan x < 1$

$$\begin{aligned} &\therefore \frac{\sin x}{\cos x} < 1 \\ &\therefore \sin x < \cos x \quad ✓ \\ &\therefore \sin^2 x < \sin x\cos x \\ &\therefore 2\sin^2 x < 2\sin x\cos x \\ &\therefore 2\sin^2 x < \sin 2x \\ &\therefore \sin 2x > 2\sin^2 x \quad ✓ \end{aligned}$$

iii

$$\begin{aligned} A &= \int_0^{\frac{\pi}{4}} (\sin 2x - 2\sin^2 x)\,dx \\ &= \int_0^{\frac{\pi}{4}} \left(\sin 2x - 2.\tfrac{1}{2}(1 - \cos 2x)\right) dx \\ &= \int_0^{\frac{\pi}{4}} \sin 2x - (1 - \cos 2x)\,dx \\ &= \int_0^{\frac{\pi}{4}} (\sin 2x + \cos 2x - 1)\,dx \\ &= \left[-\frac{\cos 2x}{2} + \frac{\sin 2x}{2} - x\right]_0^{\frac{\pi}{4}} \quad ✓ \\ &= \left(-\frac{\cos\frac{\pi}{2}}{2} + \frac{\sin\frac{\pi}{2}}{2} - \frac{\pi}{4}\right) - \left(-\frac{\cos 0}{2} + \frac{\sin 0}{2} - 0\right) \\ &= \left(0 + \frac{1}{2} - \frac{\pi}{4}\right) - \left(-\frac{1}{2}\right) \\ &= \frac{1}{2} - \frac{\pi}{4} + \frac{1}{2} \\ &= 1 - \frac{\pi}{4} \\ &\doteqdot 0.21 \text{ units}^2. \quad ✓ \end{aligned}$$

(6 marks)
(Total mark allocation only in exam)

51

$$\begin{aligned} V &= \pi\int_0^{10} x^2\,dy \\ &= \pi\int_0^{10} \left(\frac{y^2}{30}\right)^2 dy \quad ✓ \\ &= \pi\int_0^{10} \frac{y^4}{900}\,dy \\ &= \frac{\pi}{900}\left[\frac{y^5}{5}\right]_0^{10} \quad ✓ \\ &= \frac{\pi}{4500}\left[y^5\right]_0^{10} \\ &= \frac{\pi}{4500}[10^5 - 0] \\ &= \frac{100\,000\pi}{4500} \\ &= \frac{1000}{45}\pi \\ &= \frac{200}{9}\pi \text{ cm}^3. \quad ✓ \end{aligned}$$

(3 marks)

52

$$\begin{aligned} V &= \pi\int_0^h x^2\,dy \\ &= \pi\int_0^h 32y\,dy \\ &= \pi\left[16y^2\right]_0^h \\ &= \pi(16h^2 - 0) \\ &= 16\pi h^2. \end{aligned}$$

(No mark allocation in exam)

53

$$\begin{aligned} V &= \pi\int_0^9 y^2\,dx \\ &= \pi\int_0^9 (6\sqrt{x})^2\,dx \\ &= \pi\int_0^9 (36x)\,dx \\ &= \pi\left[18x^2\right]_0^9 \\ &= \pi[18(9)^2 - 0] \\ &= 1458\pi \text{ units}^3. \end{aligned}$$

(No mark allocation in exam)

54 **i**

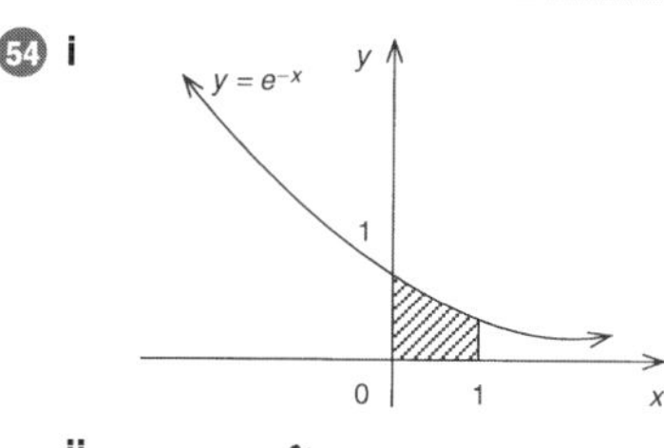

ii

$$\begin{aligned} V &= \pi\int_0^1 y^2\,dx \\ &= \pi\int_0^1 (e^{-x})^2\,dx \\ &= \pi\int_0^1 e^{-2x}\,dx \\ &= \pi\left[\frac{e^{-2x}}{-2}\right]_0^1 \\ &= -\frac{\pi}{2}\left[e^{-2x}\right]_0^1 \\ &= -\frac{\pi}{2}(e^{-2} - e^0) \\ &= -\frac{\pi}{2}\left(\frac{1}{e^2} - 1\right) \text{ units}^3 \\ &= \frac{\pi}{2}\left(1 - \frac{1}{e^2}\right) \text{ units}^3. \end{aligned}$$

(No mark allocation in exam)

55

$$V = \pi\int_0^{16} x^2\,dy$$

$$\begin{aligned} \text{Now } y &= x^4 \\ \therefore \sqrt{y} &= x^2 \\ \therefore x^2 &= y^{\frac{1}{2}} \end{aligned}$$

$$\begin{aligned} V &= \pi\int_0^{16} y^{\frac{1}{2}}\,dy \\ &= \pi\left[\frac{y^{\frac{3}{2}}}{\frac{3}{2}}\right]_0^{16} \end{aligned}$$

$= \pi\left[\frac{2}{3}y^{\frac{3}{2}}\right]_0^{16}$

$= \frac{2\pi}{3}[16^{\frac{3}{2}} - 0^{\frac{3}{2}}]$

$= \frac{2\pi}{3}[(\sqrt{16})^3 - 0]$

$= \frac{2\pi}{3}[4^3 - 0]$

$= \frac{2\pi}{3} \times 64$

$= \frac{128\pi}{3}$ units^3.

(No mark allocation in exam)

1 Which of the following could be the graph of a solution to the differential equation $\frac{dy}{dx} = \sin y + 1$?

A

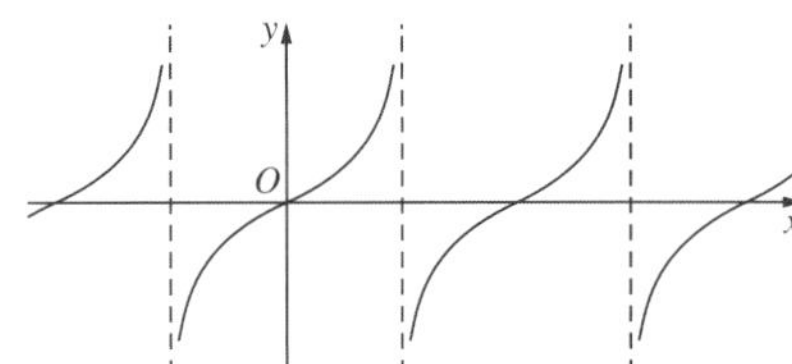

B

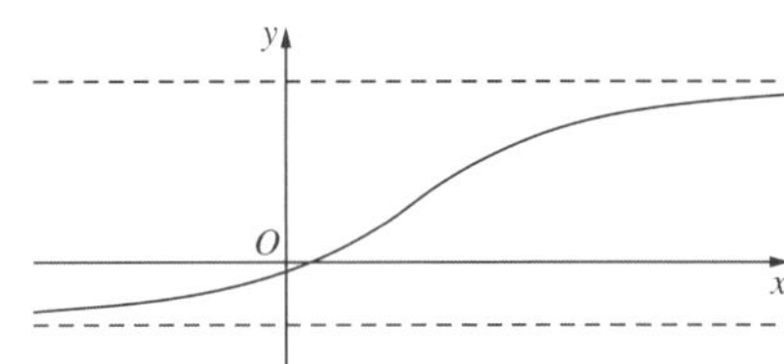

C

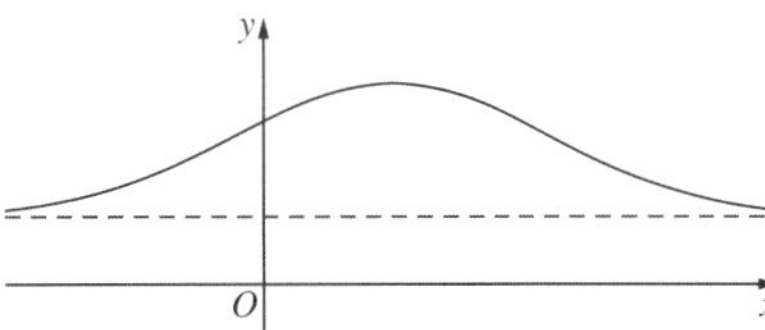

D

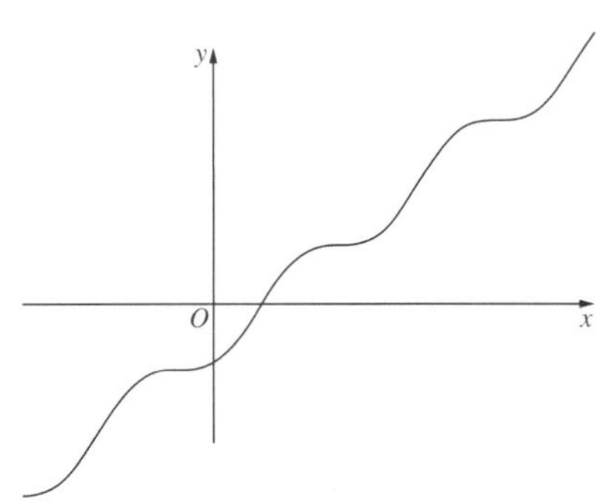

(1 mark)

(Q10, **2022 HSC**) Hard

2 A direction field is to be drawn for the differential equation $\frac{dy}{dx} = \frac{x - 2y}{x^2 + y^2}$.

On the diagram on page 1 of the Question 12 Writing Booklet [diagram provided below], clearly draw the correct slopes of the direction field at the points P, Q and R.

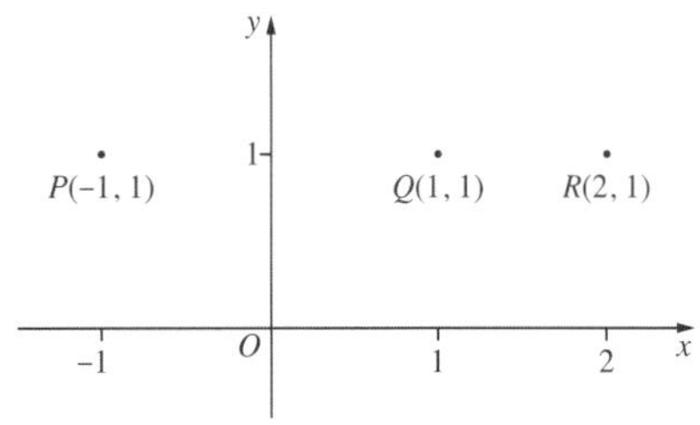

(2 marks)

(Q12a, **2022 HSC**) Easy

3 In a room with temperature 12 °C, coffee is poured into a cup. The temperature of the coffee when it is poured into the cup is 92 °C, and it is far too hot to drink.

The temperature, T, in degrees Celsius, of the coffee, t minutes after it is made, can be modelled using the differential equation $\frac{dT}{dt} = k(T - T_1)$, where k is the constant of proportionality and T_1 is a constant.

i It takes 5 minutes for the coffee to cool to a temperature of 76 °C.
Using separation of variables, solve the given differential equation to show that $T = 12 + 80e^{\frac{t}{5}\ln\left(\frac{4}{5}\right)}$.
(3 marks) Medium

ii The optimal drinking temperature for a hot beverage is 57 °C.
Find the value of t when the coffee reaches this temperature, giving your answer to the nearest minute. *(1 mark)* Easy

(Q12d, **2022 HSC**)

4 Find the particular solution to the differential equation $(x - 2)\frac{dy}{dx} = xy$ that passes through the point $(0, 1)$. *(4 marks)*

(Q14a, **2022 HSC**) Hard

5 Consider the differential equation $\frac{dy}{dx} = \frac{x}{y}$.

Which of the following equations best represents this relationship between x and y?

A $y^2 = x^2 + c$ **B** $y^2 = \frac{x^2}{2} + c$

C $y = x\ln|y| + c$ **D** $y = \frac{x^2}{2}\ln|y| + c$

(1 mark)

(Q4, **2021 HSC**) Easy

6 The direction field for a differential equation is given on page 1 of the Question 12 Writing Booklet.

[Direction field provided here]

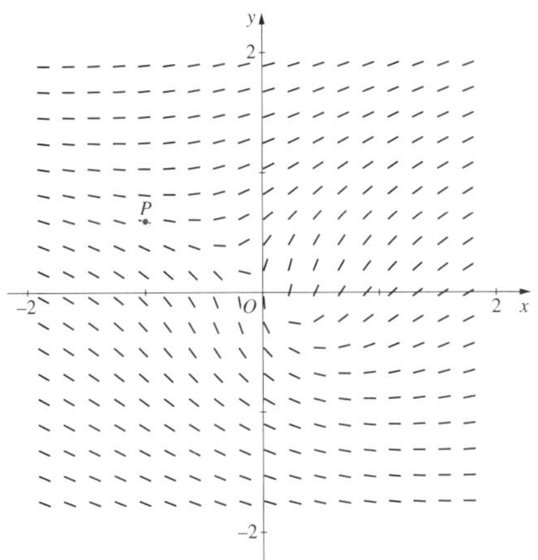

The graph of a particular solution to the differential equation passes through the point P.

On the diagram provided in the writing booklet, sketch the graph of this particular solution. *(1 mark)*

(Q12a, **2021 HSC**) **Hard**

7 In a certain country, the population of deer was estimated in 1980 to be 150 000.

The population growth is given by the logistic equation $\frac{dP}{dt} = 0.1P\left(\frac{C-P}{C}\right)$ where t is the number of years after 1980 and C is the carrying capacity.

In the year 2000, the population of deer was estimated to be 600 000.

Use the fact that $\frac{C}{P(C-P)} = \frac{1}{P} + \frac{1}{C-P}$ to show that the carrying capacity is approximately 1 130 000. *(4 marks)* **Medium**

(Q14b, **2021 HSC**

8 Which of the following best represents the direction field for the differential equation $\frac{dy}{dx} = -\frac{x}{4y}$?

A

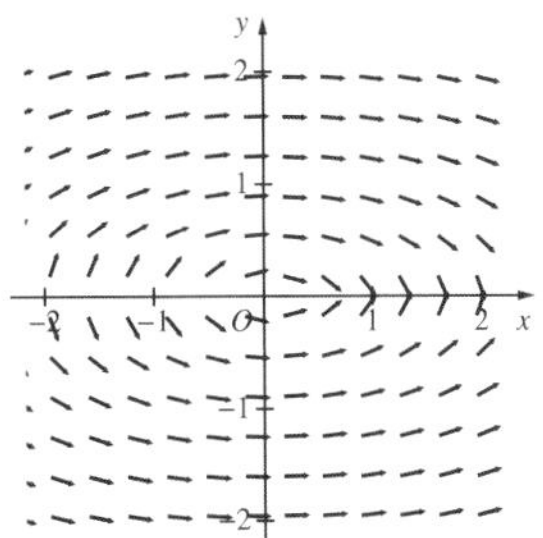

B

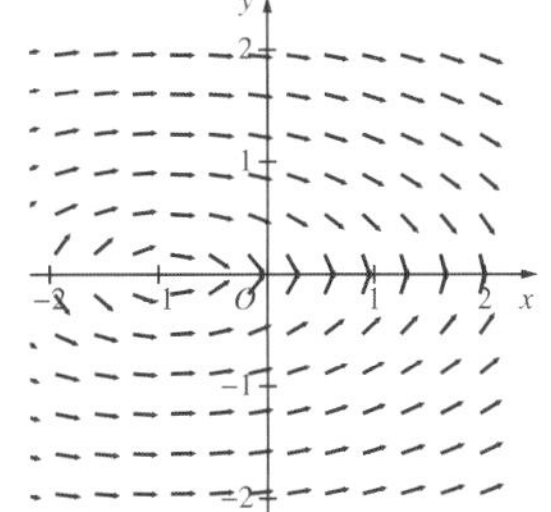

C

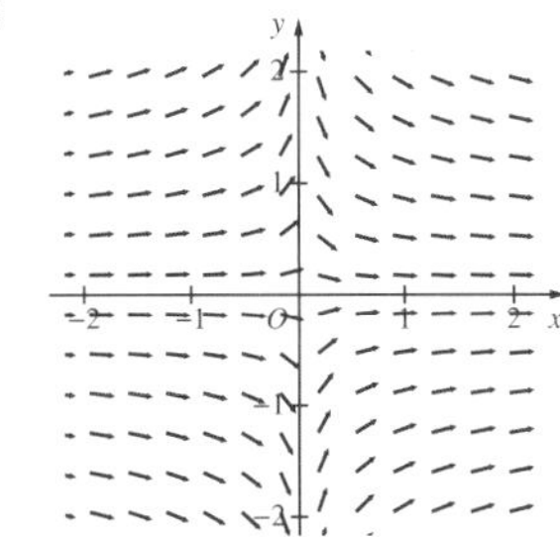

D

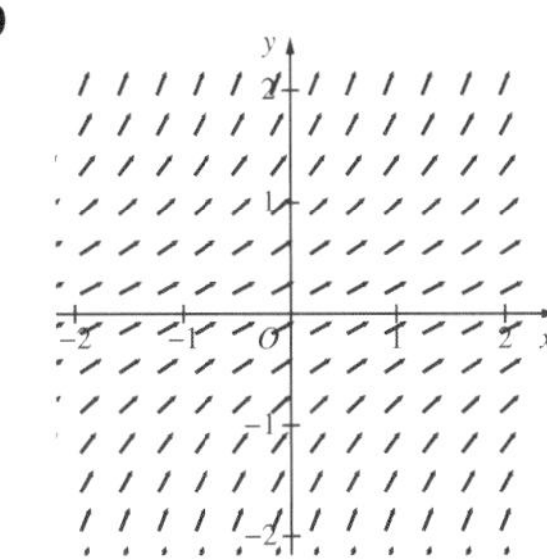

(1 mark)

(Q7, **2020 HSC**) **Medium**

9 Solve $\frac{dy}{dx} = e^{2y}$, finding x as a function of y. *(2 marks)*

(Q11e, **2020 HSC**) **Easy**

10 Find the curve which satisfies the differential equation $\frac{dy}{dx} = -\frac{x}{y}$ and passes through the point $(1, 0)$. *(3 marks)*

(Q12e, **2020 HSC**) **Easy**

11 Which of these is the general solution of the differential equation $2x^2\frac{dy}{dx} - 3x = 0$?

A $y = -\frac{3}{2}\log_e x + c$ **B** $y = -\frac{2}{3}\log_e x + c$

C $y = \frac{2}{3}\log_e x + c$ **D** $y = \frac{3}{2}\log_e x + c$

(1 mark)

Bonus question (see page iv) **Easy**

12 A population of wild dogs in a national park grows in a way described by the logistical differential equation $\frac{dP}{dt} = 0.1P\left(1 - \frac{P}{2500}\right)$, where P is the number of wild dogs after t months, and the initial population is $P_0 = 200$. Also, $0 < P < 2500$.

[Use the result $\frac{25000}{P(2500-P)} = \left(\frac{10}{P} + \frac{10}{2500-P}\right)$]

i By solving the differential equation, show that $P = \frac{5000}{2+23e^{-0.1t}}$. *(4 marks)* **Hard**

ii How many wild dogs are there after 5 years? *(1 mark)* **Easy**

Bonus question

13 Solve the differential equation $(x^2+5)\frac{dy}{dx} - xy = 0$, given that $y = 3$ when $x = 2$ and hence find the value of y when $x = 2\sqrt{5}$. *(4 marks)*

Bonus question **Medium**

14 Consider the differential equation $\frac{dy}{dx} = xy$

If $y = e^2$ when $x = \sqrt{2}$, solve the differential equation, leaving your answer with y in terms of x. *(3 marks)*

Bonus question **Medium**

15 Solve the differential equation $\frac{dy}{dx} = \frac{x}{1+y^2}$, given that $y = 3$ when $x = 4$ and find the positive value of x when $y = 2$. *(4 marks)*

Bonus question **Medium**

16 Which of these diagrams best represents the direction field of the differential equation $\frac{dy}{dx} = \frac{y-2x}{2y+x}$?

A

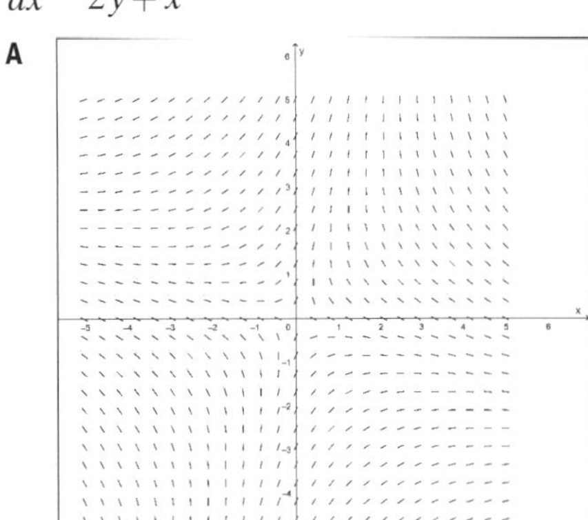

B

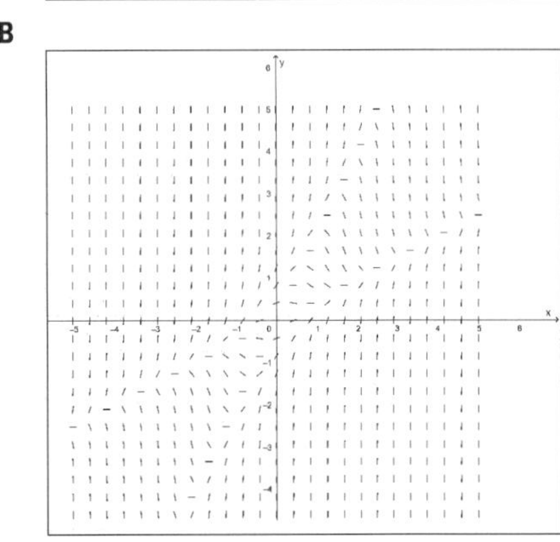

C

D

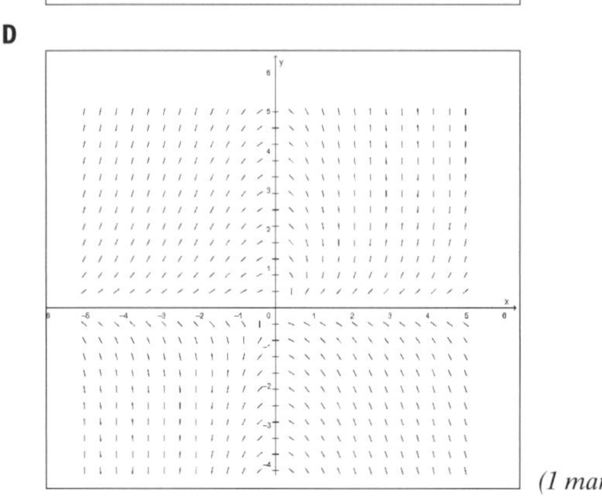

(1 mark)

Bonus question **Easy**

17 A population grows according to the logistical differential equation $\frac{dP}{dt} = 0.02P\left(1 - \frac{P}{1200}\right)$, where P is the population after t months, and the initial population is 800.
Also, $0 < P < 1200$.

[Use the result $\frac{60\,000}{P(1200-P)} = \frac{50}{P} + \frac{50}{1200-P}$]

Find the population P at time t. *(4 marks)*

Bonus question **Hard**

18 Express y in terms of x for the differential equation $(1+x^2)\frac{dy}{dx} = 4xy$, where $x = 1$ when $y = 2$. *(3 marks)*

Bonus question **Medium**

19 Given that $\frac{dy}{dx} = y$ and $y = 1$ when $x = 1$, then

A $y = e^x$ **B** $y = e^{x-1}$

C $y = e^{1-x}$ **D** $y = e^x - 1$ *(1 mark)*

Bonus question **Easy**

20 Which of these diagrams best represents the direction field of the differential equation $\frac{dy}{dx} = \frac{x}{y}$?

A

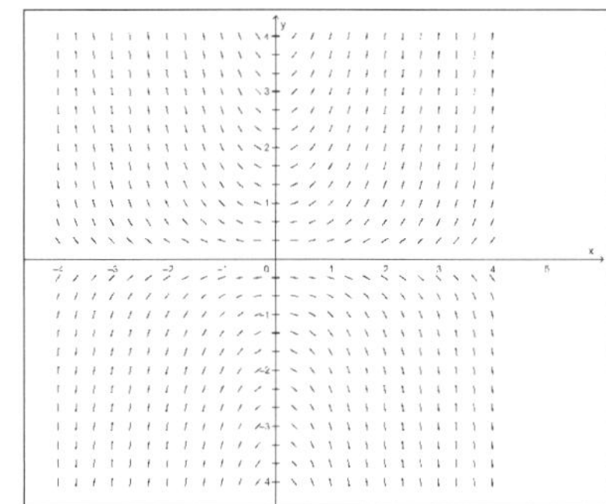

B

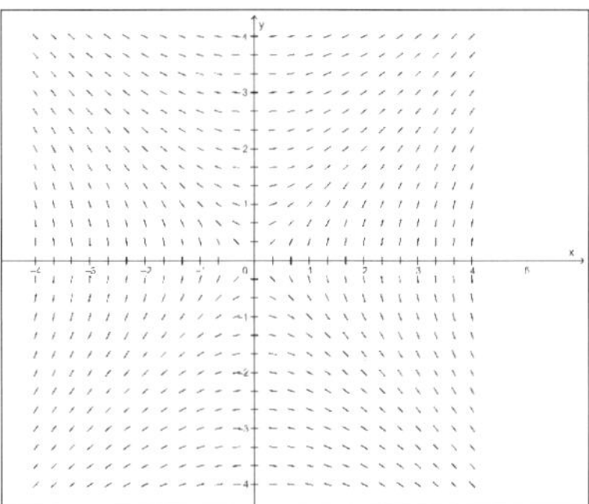

C

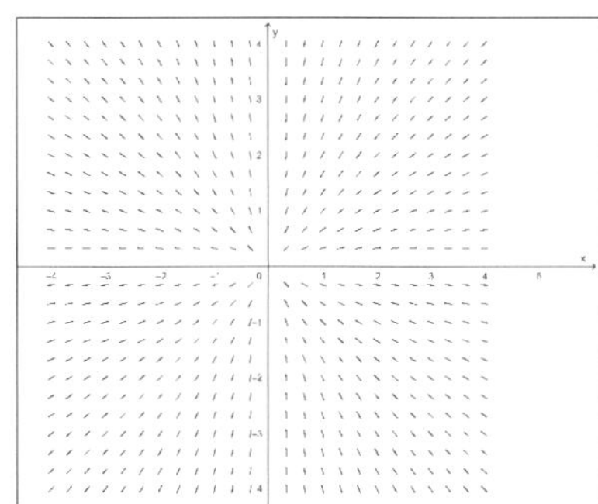

D

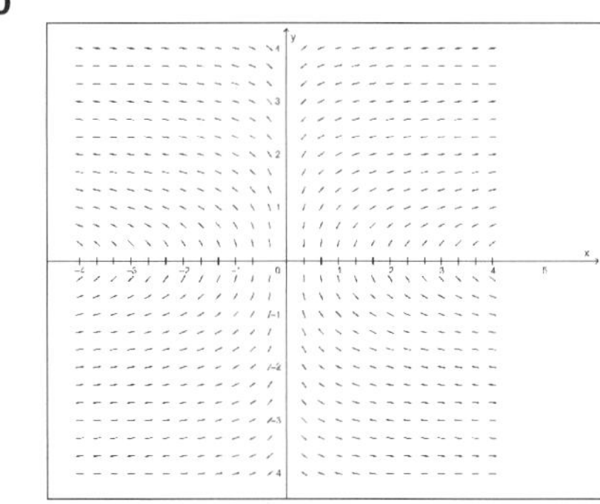

(1 mark)

Bonus question Easy

21 The population of a gold mining town is initially 8000 people. This population would increase at a rate of 1.5% per year, except that there is a steady flow of people leaving the town.

The population P after t years may be modelled by the differential equation $\frac{dP}{dt} = \frac{3P}{200} - k$, where k is the number of people leaving per year minus the number of people arriving per year.

i Show that, when $k = 600$, $P = 8000(5 - 4e^{0.015t})$ satisfies $\frac{dP}{dt} = \frac{3P}{200} - 600$. *(2 marks)* Hard

ii For $k = 600$, determine the number of years until the population is zero. *(2 marks)* Medium

Bonus question

22 For the differential equation

$\frac{dy}{dx} = (x-2)(y+1)$, given that $y = 0$ when $x = -4$, show that $y = -1 + e^{\frac{x^2}{2}-2x-16}$. *(3 marks)*

Bonus question Medium

23 Which of these diagrams best represents the direction field of the differential equation $\frac{dy}{dx} = \frac{x}{y-2}$?

A

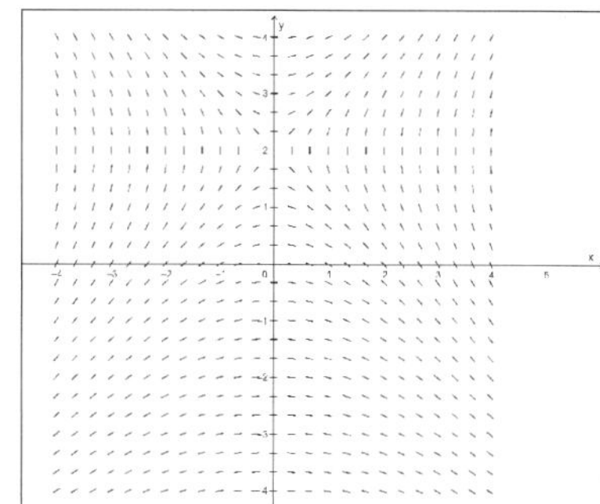

B

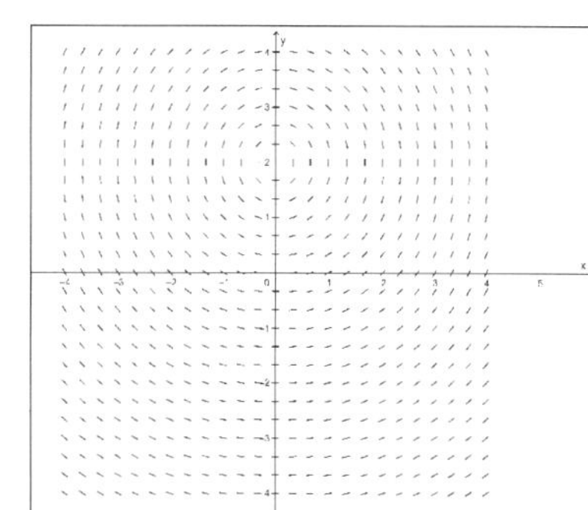

C

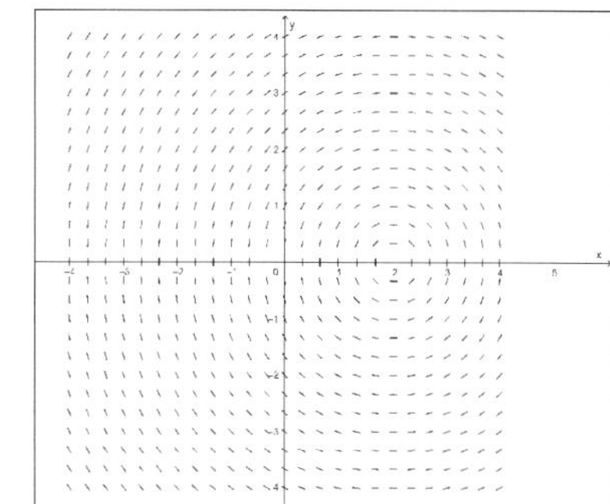

D

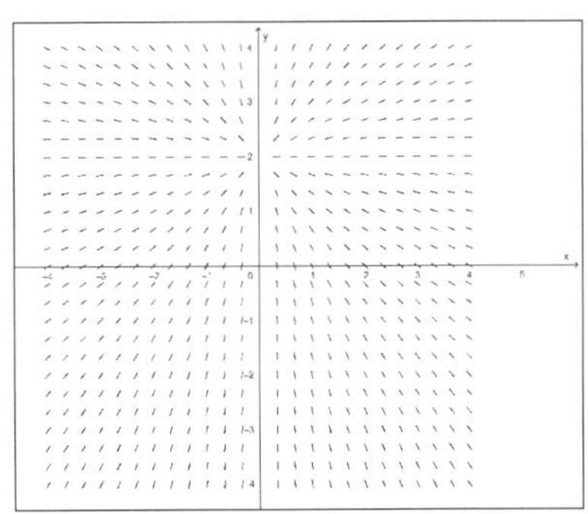

(1 mark)

Bonus question Easy

24 A small park in Canada is known to be capable of supporting no more than 100 grizzly bears. Ten bears are present in the park at present. The population growth of bears can be modelled by the logistic differential equation $\frac{dP}{dt} = 0.1P - 0.001P^2$, where t is measured in years.

i Show $\frac{1000}{100P - P^2} = \frac{10}{P} + \frac{10}{100 - P}$.

(1 mark) Easy

ii Using the result in part **i**, solve the differential equation $\frac{dP}{dt} = 0.1P - 0.001P^2$ to show $P = \frac{100e^{0.1t}}{9 + e^{0.1t}}$.

(4 marks) 

iii How long will it take for the population to reach 80? *(2 marks)* Medium

Bonus question

25 Solve the differential equation $\frac{dy}{dx} = \frac{x}{1 + y^2}$, given that $y = 3$ when $x = -2$.

(3 marks)

Bonus question Medium

26 Solve the differential equation $\frac{dy}{dx} = 1 + \frac{3y}{2}$, with the condition $x = 2, y = -\frac{1}{3}$. *(3 marks)*

Bonus question Medium

27 Solve the differential equation $\frac{dy}{dx} = \frac{xy - 2y}{y + 1}$, where $y \neq -1$, given that $y = 1$ when $x = 2$.

(3 marks)

Bonus question Medium

28 In a flock of 1000 chickens, the number P infected with a disease at time ***t* years** is given by

$$P = \frac{1000}{1 + ce^{-1000t}} \text{ where } c \text{ is a constant.}$$

i Show that, eventually, all the chickens will be infected. Easy

ii Suppose that when time $t = 0$, exactly one chicken was infected. After how many **days** will 500 chickens be infected?

Hard

iii Show that $\frac{dP}{dt} = P(1000 - P)$.

Hard

(Q5b, **1992 HSC**)

Year 12 Differential equations—Worked Answers

1 Considering the gradient function, $\frac{dy}{dx} = \sin y + 1$, the graph of the differential equation must have a maximum gradient at $y = \frac{\pi}{2}$.

Also $0 < \frac{dy}{dx} \le 2$. As y approaches $-\frac{\pi}{2}$ and $\frac{3\pi}{2}$, $\frac{dy}{dx}$ approaches 0, hence there are horizontal asymptotes at $y = -\frac{\pi}{2}$ and $y = \frac{3\pi}{2}$. This leaves B as the only option.
Alternatively, solving the differential equation gives:

$$\int \frac{dy}{1+\sin y} = \int dx$$

$$\begin{aligned}1 + \sin y &= 1+\cos\left(\frac{\pi}{2} - y\right)\\ &= 1+\cos\left(2\left(\frac{\pi}{4}-\frac{y}{2}\right)\right)\\ &= 1+2\cos^2\left(\frac{\pi}{4}-\frac{y}{2}\right)-1\\ &= 2\cos^2\left(\frac{\pi}{4}-\frac{y}{2}\right)\end{aligned}$$

$$\therefore \int \frac{dy}{2\cos^2\left(\frac{\pi}{4}-\frac{y}{2}\right)} = \int dx$$

$$\frac{1}{2}\int \sec^2\left(\frac{\pi}{4}-\frac{y}{2}\right)dy = \int dx$$

$$\frac{1}{2}\left[-2\tan\left(\frac{\pi}{4}-\frac{y}{2}\right)\right] = x + C$$

$$\tan\left(\frac{\pi}{4}-\frac{y}{2}\right) = -x + C$$

$$\frac{\pi}{4}-\frac{y}{2} = \tan^{-1}(-x + C)$$

$$\frac{\pi}{4} - \tan^{-1}(-x+C) = \frac{y}{2}$$

$$\therefore y = \frac{\pi}{2} + 2\tan^{-1}(x - C)$$

Answer B

2 $\frac{dy}{dx} = \frac{x-2y}{x^2+y^2}$

At $(-1, 1)$

$$\frac{dy}{dx} = \frac{(-1)-2(1)}{(-1)^2+(1)^2} = \frac{-3}{2}$$

At $(1, 1)$

$$\frac{dy}{dx} = \frac{(1)-2(1)}{(1)^2+(1)^2} = \frac{-1}{2}$$

At $(2, 1)$

$$\frac{dy}{dx} = \frac{(2)-2(1)}{(2)^2+(1)^2} = 0$$

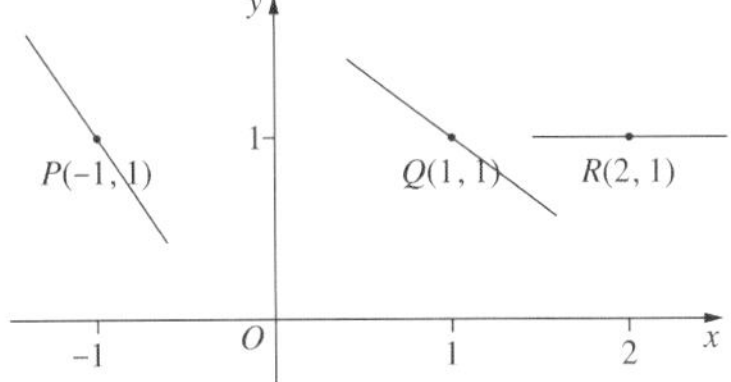

✓✓ *(2 marks)*

3 **i** Newton's law of cooling states that $\frac{dT}{dt} = k(T - T_1)$ where $T_1 = 12$

$$\frac{dT}{dt} = k(T - 12)$$

$$\frac{dT}{T-12} = k\,dt$$

$$\int \frac{1}{T-12}\,dT = \int k\,dt$$

$$\ln|T - 12| = kt + C$$ ✓

When $t = 0$, $T = 92$

$$\ln|92 - 12| = k(0) + C$$

$$\therefore C = \ln 80$$

$$\ln|T - 12| - \ln 80 = kt$$

$$\therefore \ln\left(\frac{T-12}{80}\right) = kt$$ ✓

When $t = 5$, $T = 76$

$$\ln\left(\frac{76-12}{80}\right) = 5k$$

$$\therefore k = \frac{1}{5}\ln\left(\frac{4}{5}\right)$$

$$\ln\left(\frac{T-12}{80}\right) = \frac{t}{5}\ln\left(\frac{4}{5}\right)$$

$$\frac{T-12}{80} = e^{\frac{t}{5}\ln\left(\frac{4}{5}\right)}$$

$$T - 12 = 80e^{\frac{t}{5}\ln\left(\frac{4}{5}\right)}$$

$$\therefore T = 12 + 80e^{\frac{t}{5}\ln\left(\frac{4}{5}\right)}$$ ✓

(3 marks)

ii Find when $T = 57$

$$T = 12 + 80e^{\frac{t}{5}\ln\left(\frac{4}{5}\right)}$$

$$57 = 12 + 80e^{\frac{t}{5}\ln\left(\frac{4}{5}\right)}$$

$$45 = 80e^{\frac{t}{5}\ln\left(\frac{4}{5}\right)}$$

$$\frac{9}{16} = e^{\frac{t}{5}\ln\left(\frac{4}{5}\right)}$$

$$\ln\left(\frac{9}{16}\right) = \frac{t}{5}\ln\left(\frac{4}{5}\right)$$

$$t = \frac{5\ln\left(\frac{9}{16}\right)}{\ln\left(\frac{4}{5}\right)}$$

$$t = 12.8922$$

$\therefore$ Coffee will reach 57 °C after 13 minutes. ✓

(1 mark)

4 $(x-2)\frac{dy}{dx} = xy$

$$\frac{dy}{y} = \frac{x}{x-2}dx$$

$$\int \frac{dy}{y} = \int \frac{x-2+2}{x-2}dx$$

$$\int \frac{dy}{y} = \int 1 + \frac{2}{x-2}dx$$ ✓

$$\ln y = x + 2\ln|x - 2| + C$$ ✓

When $x = 0$, $y = 1$

$$\ln 1 = 0 + 2\ln|0 - 2| + C$$

$$0 = 2\ln 2 + C$$

$$\therefore C = -2\ln 2$$

$\therefore \ln y = x + \ln(x-2)^2 - \ln 2^2$ ✓

$\ln y = x + \ln\left(\frac{x-2}{2}\right)^2$

$\ln y = \ln e^x + \ln\left(\frac{x-2}{2}\right)^2$

$\ln y = \ln\left[e^x\left(\frac{x-2}{2}\right)^2\right]$

$\therefore y = e^x\left(\frac{x-2}{2}\right)^2$ ✓

(4 marks)

5 $\frac{dy}{dx} = \frac{x}{y}$

$y\,dy = x\,dx$

$\int y\,dy = \int x\,dx$

$\frac{y^2}{2} = \frac{x^2}{2} + D$, for some constant D

$y^2 = x^2 + c$, for some constant c

Answer A

6 Two alternative solutions are drawn for the slope field.

✓

(1 mark)

7 $\frac{dP}{dt} = 0.1P\left(\frac{C-P}{C}\right)$

$\frac{C}{P(C-P)}dP = 0.1dt$

$\left(\frac{1}{P} + \frac{1}{C-P}\right)dP = \frac{1}{10}dt$

$\int \frac{1}{10}dt = \int\left(\frac{1}{P} + \frac{1}{C-P}\right)dP$ ✓

$\frac{t}{10} = \ln|P| - \ln|C-P| + K$ where K is a constant

$\frac{t}{10} = \ln\left|\frac{P}{C-P}\right| + K$

When $t = 0$, $P = 150000$:

$0 = \ln\left|\frac{150\,000}{C-150\,000}\right| + K$

$K = -\ln\left|\frac{150\,000}{C-150\,000}\right|$

$\frac{t}{10} = \ln\left|\frac{P}{C-P}\right| - \ln\left|\frac{150\,000}{C-150\,000}\right|$ ✓

When $t = 20$, $P = 600000$:

$2 = \ln\left|\frac{600\,000}{C-600\,000}\right| - \ln\left|\frac{150\,000}{C-150\,000}\right|$

$2 = \ln\left|\frac{4(C-150\,000)}{C-600\,000}\right|$

$e^2 = \frac{4(C-150\,000)}{C-600\,000}$ ✓

$e^2(C-600\,000) = 4(C-150\,000)$

$e^2C - 600\,000e^2 = 4C - 600\,000$

$C(e^2-4) = 600\,000(e^2-1)$

$C = \frac{600\,000(e^2-1)}{e^2-4}$

$C = 1\,131\,121$

$\therefore$ The carrying capacity is approximately 1 130 000. ✓

(4 marks)

8 $\frac{dy}{dx} = -\frac{x}{4y}$

$\frac{dy}{dx}$ is zero when $x = 0$ and undefined when $y = 0$.

At $(-0.5, 0.25)$, $\frac{dy}{dx} = 0.5$.

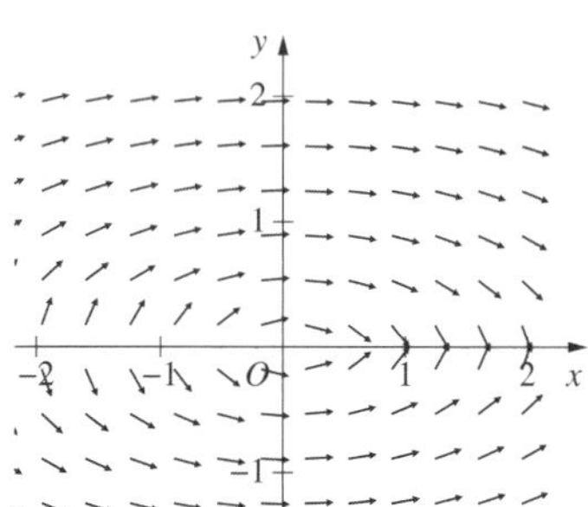

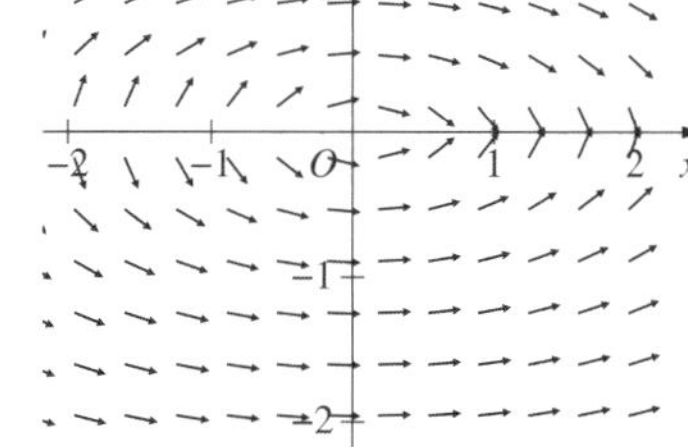

Answer A

9 $\frac{dy}{dx} = e^{2y}$

$\frac{dx}{dy} = \frac{1}{e^{2y}}$

$= e^{-2y}$

$x = \int e^{-2y}dy$ ✓

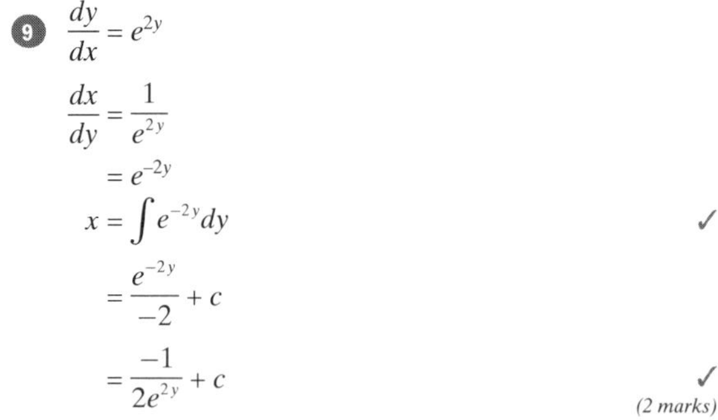

$= \frac{e^{-2y}}{-2} + c$

$= \frac{-1}{2e^{2y}} + c$ ✓

(2 marks)

10 $\frac{dy}{dx} = -\frac{x}{y}$

$y\,dy = -x\,dx$

$\frac{y^2}{2} = -\frac{x^2}{2} + c$ ✓

$y^2 = -x^2 + C$ (where $C = 2c$) ✓

When $x = 1$, $y = 0$

$0 = -1^2 + C$

$C = 1$

$y^2 = -x^2 + 1$

$\therefore x^2 + y^2 = 1$ ✓

(3 marks)

11 $2x^2\frac{dy}{dx} - 3x = 0$

$2x^2\frac{dy}{dx} = 3x$

$\frac{dy}{dx} = \frac{3}{2x}$

$$y = \frac{3}{2}\int \frac{1}{x}\,dx$$
$$y = \frac{3}{2}\log_e x + c$$

Answer D

12 i
$$\frac{dP}{dt} = 0.1P\left(1 - \frac{P}{2500}\right)$$
$$= \frac{P}{10}\left(\frac{2500 - P}{2500}\right)$$
$$= \frac{P(2500 - P)}{25000}$$
$$\frac{dt}{dP} = \frac{10}{P} + \frac{10}{2500 - P}$$
$$t = 10\int\left(\frac{1}{P} + \frac{10}{2500 - P}\right)dP$$
$$t = 10(\log_e|P| - \log_e|2500 - P|) + c \quad ✓$$
$$t = 10\log_e\frac{P}{2500 - P} + c \quad (\text{as } 0 < P < 2500)$$

When $t = 0$, $P = 200$.
$$\therefore 0 = 10\log_e\left(\frac{200}{2500 - 200}\right) + c \quad ✓$$
$$c = -10\log_e\frac{2}{23}$$
$$c = 10\log_e\frac{23}{2} \quad ✓$$
$$t = 10\log_e\frac{P}{2500 - P} + 10\log_e\frac{23}{2}$$
$$t = 10\log_e\frac{23P}{5000 - 2P}$$
$$\frac{t}{10} = \log_e\frac{23P}{5000 - 2P}$$
$$e^{0.1t} = \frac{23P}{5000 - 2P}$$
$$23P = 5000e^{0.1t} - 2Pe^{0.1t}$$
$$P(23 + 2e^{0.1t}) = 5000e^{0.1t}$$
$$P = \frac{5000e^{0.1t}}{23 + 2e^{0.1t}}$$
$$= \frac{5000}{\frac{23}{e^{0.1t}} + 2}$$
$$= \frac{5000}{2 + 23e^{-0.1t}} \quad ✓$$

(4 marks)

ii Substitute $t = 5$:
$$P = \frac{5000}{2 + 23e^{-0.1\times5}}$$
$= 313.4755914\ldots$
$= 313$ (nearest whole)
$\therefore$ there will be 313 wild dogs. ✓ *(1 mark)*

13
$$(x^2 + 5)\frac{dy}{dx} - xy = 0$$
$$(x^2 + 5)\frac{dy}{dx} = xy$$
$$\int \frac{1}{y}\,dy = \int \frac{x}{x^2 + 5}\,dx \quad ✓$$
$$\ln|y| = \frac{1}{2}\ln|x^2 + 5| + c$$
$$\ln|y| = \ln\sqrt{x^2 + 5} + c$$
$$c = \ln\left|\frac{y}{\sqrt{x^2 + 5}}\right| \quad ✓$$

When $x = 2$, $y = 3$.
$$c = \ln\left|\frac{3}{\sqrt{2^2 + 5}}\right|$$
$$c = \ln\frac{3}{3}$$
$$c = 0$$
$$\ln|y| = \ln\sqrt{x^2 + 5}$$
$$y = \sqrt{x^2 + 5} \quad ✓$$

When $x = 2\sqrt{5}$.
$$y = \sqrt{(2\sqrt{5})^2 + 5}$$
$$= \sqrt{20 + 5}$$
$$= \sqrt{25}$$
$$= 5 \quad ✓$$

(4 marks)

14
$$\frac{dy}{dx} = xy$$
$$\int \frac{1}{y}\,dy = \int x\,dx \quad ✓$$
$$\ln|y| = \frac{x^2}{2} + c$$
$$y = e^{\frac{x^2}{2} + c} \quad ✓$$

Substitute $y = e^2$, $x = \sqrt{2}$:
$$e^2 = e^{\frac{(\sqrt{2})^2}{2} + c}$$
$$e^2 = e^{1 + c}$$
$$c + 1 = 2$$
$$\therefore c = 1$$
$$\therefore y = e^{\frac{x^2}{2} + 1} \quad ✓$$

(3 marks)

15
$$\frac{dy}{dx} = \frac{x}{1 + y^2}$$
$$\int(1 + y^2)\,dy = \int x\,dx \quad ✓$$
$$y + \frac{y^3}{3} = \frac{x^2}{2} + c \quad ✓$$

Substitute $x = 4$, $y = 3$:
$$3 + \frac{3^3}{3} = \frac{4^2}{2} + c$$
$$3 + 9 = 8 + c$$
$$c = 4$$
$$y + \frac{y^3}{3} = \frac{x^2}{2} + 4 \quad ✓$$

Substitute $y = 2$:
$$2 + \frac{2^3}{3} = \frac{x^2}{2} + 4$$
$$\frac{x^2}{2} = \frac{2}{3}$$
$$x^2 = \frac{4}{3}$$
$$x = \pm\frac{2}{\sqrt{3}}$$

As $x > 0$, then $x = \frac{2}{\sqrt{3}}$. ✓

(4 marks)

16 At the point $(2, 4)$ gradient $= \frac{4 - 2(2)}{2(4) + 2} = 0$. So possible solutions are B and C.
At the point $(4, 0)$ gradient $= \frac{0 - 2(4)}{2(0) + 4} = -$ve. So the solution is C.
Hence,

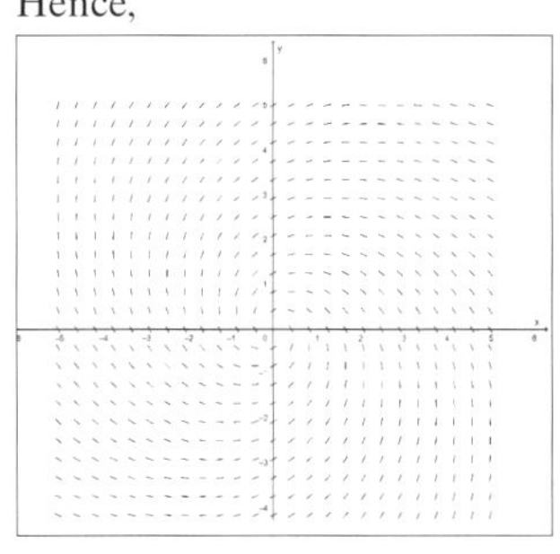

Answer C

17
$$\frac{dP}{dt} = 0.02P\left(1 - \frac{P}{1200}\right)$$
$$\frac{dP}{dt} = \frac{P}{50}\left(\frac{1200 - P}{1200}\right)$$
$$\frac{dP}{dt} = \frac{P(1200 - P)}{60000}$$
$$\frac{dt}{dP} = \frac{60000}{P(1200 - P)}$$
$$= \frac{50}{P} + \frac{50}{1200 - P}$$
$$t = 50\int\left(\frac{1}{P} + \frac{1}{1200 - P}\right)dP \quad ✓$$
$$t = 50(\log_e|P| - \log_e|1200 - P|) + c$$
$$t = 50\log_e\frac{P}{1200 - P} + c \quad ✓$$
$$(\text{as } 0 < P < 1200)$$

Substitute $t = 0$, $P = 800$:
$$0 = 50\log_e\frac{800}{1200 - 800} + c$$
$$= 50\log_e 2 + c$$
$$c = -50\log_e 2$$
$$t = 50\log_e\frac{P}{1200 - P} - 50\log_e 2 \quad ✓$$
$$t = 50\log_e\frac{P}{2400 - 2P}$$
$$\frac{t}{50} = \log_e\frac{P}{2400 - 2P}$$
$$e^{0.02t} = \frac{P}{2400 - 2P}$$
$$P = 2400e^{0.02t} - 2Pe^{0.02t}$$
$$P(1 + 2e^{0.02t}) = 2400e^{0.02t}$$
$$P = \frac{2400e^{0.02t}}{1 + 2e^{0.02t}}$$
$$P = \frac{2400}{2 + e^{-0.02t}} \quad ✓$$

(4 marks)

18 $(1 + x^2)\dfrac{dy}{dx} = 4xy$

$\displaystyle\int \frac{1}{y}\,dy = \int \frac{4x}{x^2+1}\,dx$

$\log_e y = 2\log_e(x^2 + 1) + c$ ✓

Substitute $x = 1, y = 2$:

$\log_e 2 = 2\log_e(1^2 + 1) + c$

$\log_e 2 = 2\log_e 2 + c$

$c = -\log_e 2$

$\log_e y = \log_e (x^2 + 1)^2 - \log_e 2$ ✓

$\log_e y = \log_e \dfrac{(x^2+1)^2}{2}$

$y = \dfrac{(x^2+1)^2}{2}$ ✓ *(3 marks)*

19 $\dfrac{dy}{dx} = y$

$\dfrac{dx}{dy} = \dfrac{1}{y}$

$x = \displaystyle\int \frac{1}{y}\,dy$

$= \ln|y| + c$

Substitute $y = 1$ when $x = 1$:

$1 = \ln|1| + c$

$c = 1$

$x = \ln|y| + 1$

$\ln|y| = x - 1$

$y = e^{x-1}$

Alternatively, consider each option:

For $y = e^x$ and $y = e^{x-1}$, $\dfrac{dy}{dx} = y$.

So possible answers are A and B.

Also, the point $(1, 1)$ lies on $y = e^{x-1}$. Hence, the answer is B.

Answer B

20 At the points $(1, 1), (2, 2)$ etc gradient = 1. So possible solutions are B and C.

At $(2, 0)$, the gradient is undefined. So the solution is B.

Hence,

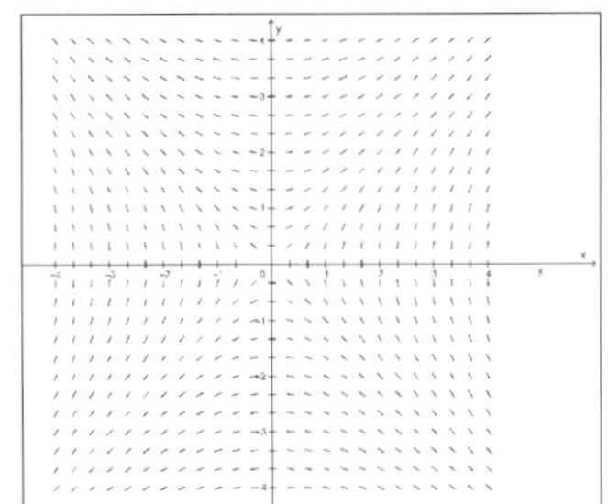

Answer B

21 **i** $P = 8000(5 - 4e^{0.015t})$

$= 40000 - 32000e^{0.015t}$ ... ①

$32000e^{0.015t} = 40\,000 - P$

$e^{0.015t} = \dfrac{40\,000 - P}{32\,000}$

Now, using ①:

$\dfrac{dP}{dt} = -480e^{0.015t}$ ✓

$= -480 \times \dfrac{40\,000 - P}{32\,000}$

$= -0.015(40000 - P)$

$= \dfrac{3P}{200} - 600$ ✓ *(2 marks)*

ii Let $P = 0$:

$0 = 8000(5 - 4e^{0.015t})$

$5 - 4e^{0.015t} = 0$

$4e^{0.015t} = 5$

$e^{0.015t} = \dfrac{5}{4}$ ✓

$0.015t = \log_e 1.25$

$t = \dfrac{\log_e 1.25}{0.015}$

$= 14.87623675\ldots$

$= 14.88$ (2 dec. pl.)

$\therefore$ It will take 14.88 years. ✓ *(2 marks)*

22 $\dfrac{dy}{dx} = (x - 2)(y + 1)$,

$\dfrac{1}{y+1}\,dy = (x - 2)\,dx$

$\displaystyle\int \frac{1}{y+1}\,dy = \int (x-2)\,dx$

$\ln|y + 1| = \dfrac{x^2}{2} - 2x + c$ ✓

When $x = -4, y = 0$.

$\ln|0 + 1| = \dfrac{(-4)^2}{2} - 2(-4) + c$

$0 = 16 + c$

$c = -16$

$\therefore \ln|y + 1| = \dfrac{x^2}{2} - 2x - 16$ ✓

$y + 1 = e^{\frac{x^2}{2} - 2x - 16}$

$\therefore y = -1 + e^{\frac{x^2}{2} - 2x - 16}$ ✓ *(3 marks)*

23 For $\dfrac{dy}{dx} = \dfrac{x}{y-2}$ the slope will be zero at the origin. So possible solutions are A and B.

At the point $(2, 4)$

gradient $= \dfrac{2}{4-2} = 1$.

Hence,

Answer A

24 **i** $\text{RHS} = \dfrac{10}{P} + \dfrac{10}{100-P}$

$= \dfrac{1000 - 10P + 10P}{P(100-P)}$

$= \dfrac{1000}{P(100-P)}$

$= \dfrac{1000}{100P - P^2}$

$= \text{LHS}$ ✓ *(1 mark)*

ii $\dfrac{dP}{dt} = 0.1P - 0.001P^2$

$= \dfrac{P}{10} - \dfrac{P^2}{1000}$

$= \dfrac{100P - P^2}{1000}$

$\dfrac{dt}{dP} = \dfrac{1000}{100P - P^2}$

$= \dfrac{10}{P} + \dfrac{10}{100-P}$ (from part **i**)

$t = \displaystyle\int\left(\frac{10}{P} + \frac{10}{100-P}\right)dP$ ✓

$= 10\log_e|P| - 10\log_e|100 - P| + c$

$= 10\log_e \dfrac{P}{100-P} + c$ ✓ (as $0 < P < 100$)

Substitute $t = 0, P = 10$:

$0 = 10\log_e \dfrac{10}{100-10} + c$

$c = -10\log_e \dfrac{1}{9}$

$c = 10\log_e 9$

$t = 10\log_e \dfrac{P}{100-P} + 10\log_e 9$

$t = 10\log_e \dfrac{9P}{100-P}$ ✓

$\dfrac{t}{10} = \log_e \dfrac{9P}{100-P}$

$0.1t = \log_e \dfrac{9P}{100-P}$

$e^{0.1t} = \dfrac{9P}{100-P}$

$9P = 100e^{0.1t} - Pe^{0.1t}$

$P(9 + e^{0.1t}) = 100e^{0.1t}$

$P = \dfrac{100e^{0.1t}}{9 + e^{0.1t}}$ ✓ *(4 marks)*

iii Substitute $P = 80$:

$t = 10\log_e \dfrac{9(80)}{100-80}$

$= 10\log_e 36$ ✓

$= 35.83518938\ldots$

$= 36$ (nearest whole)

$\therefore$ 36 years. ✓ *(2 marks)*

25

$$\frac{dy}{dx} = \frac{x}{1+y^2}$$

$$(1+y^2)\,dy = x\,dx$$

$$\int (1+y^2)\,dy = \int x\,dx \quad ✓$$

$$y + \frac{y^3}{3} = \frac{x^2}{2} + c \quad ✓$$

Substitute $x = -2$ and $y = 3$.

$$3 + \frac{3^3}{3} = \frac{(-2)^2}{2} + c$$

$$12 = 2 + c$$

$$c = 10$$

$$\therefore y + \frac{y^3}{3} = \frac{x^2}{2} + 10$$

$$\therefore 6y + 2y^3 = 3x^2 + 60$$

$$\therefore 3x^2 - 2y^3 - 6y + 60 = 0 \quad ✓$$

(3 marks)

26

$$\frac{dy}{dx} = 1 + \frac{3y}{2}$$

$$\frac{dy}{dx} = \frac{2+3y}{2}$$

$$\frac{dx}{dy} = \frac{2}{2+3y}$$

$$x = 2\int \frac{1}{2+3y}\,dy$$

$$x = \frac{2}{3}\ln|2+3y| + c \quad ✓$$

$$\ln|2+3y| = \frac{3x}{2} + C$$

Substitute $x = 2, y = -\frac{1}{3}$:

$$\ln\left|2 + 3\left(-\frac{1}{3}\right)\right| = \frac{3(2)}{2} + C$$

$$0 = 3 + C$$

$$C = -3$$

$$\ln|2+3y| = \frac{3x}{2} - 3 \quad ✓$$

$$\ln|2+3y| = \frac{3x-6}{2}$$

$$2 + 3y = e^{\frac{3x-6}{2}}$$

$$3y = e^{\frac{3x-6}{2}} - 2$$

$$y = \frac{1}{3}\left(e^{\frac{3x-6}{2}} - 2\right) \quad ✓$$

(3 marks)

27

$$\frac{dy}{dx} = \frac{xy-2y}{y+1}$$

$$\frac{dy}{dx} = \frac{y(x-2)}{y+1}$$

$$\int \frac{y+1}{y}\,dy = \int (x-2)\,dx \quad ✓$$

$$\int \left(1 + \frac{1}{y}\right) dy = \int (x-2)\,dx$$

$$y + \log_e|y| = \frac{x^2}{2} - 2x + c \quad ✓$$

Substitute $x = 2, y \neq -1$:

$$1 + \log_e 1 = \frac{2^2}{2} - 2(2) + c$$

$$1 = 2 - 4 + c$$

$$c = 3$$

$$\therefore y + \log_e|y| = \frac{x^2}{2} - 2x + 3 \quad ✓$$

(3 marks)

28 **i**

$$\lim_{t\to\infty} P = \lim_{t\to\infty} \frac{1000}{1 + ce^{-1000t}}$$

$$= \frac{1000}{1+0} \quad \left(\text{since } \lim_{t\to\infty} e^{-1000t} = 0\right)$$

$$= 1000$$

$\therefore$ Eventually all 1000 chickens will be infected.

ii At $t = 0$, $P = 1$

$$\therefore 1 = \frac{1000}{1 + ce^0}$$

$$\therefore 1 + c = 1000$$

$$c = 999$$

$$\therefore P = \frac{1000}{1 + 999e^{-1000t}}$$

When $P = 500$,

$$500 = \frac{1000}{1 + 999e^{-1000t}}$$

$$1 + 999e^{-1000t} = 2$$

$$999e^{-1000t} = 1$$

$$\therefore e^{1000t} = 999$$

$$1000t = \ln 999$$

$$t = \frac{\ln 999}{1000} \text{ (years)}$$

$$\therefore \text{Number of days} = \frac{\ln 999}{1000} \times 365$$

$$= 2.5209\ldots$$

$= 2.5$ days to 1 decimal place.

iii

$$P = 1000\left(1 + ce^{-1000t}\right)^{-1}$$

$$\frac{dP}{dt} = -1000\left(1 + ce^{-1000t}\right)^{-2}\left(-1000ce^{-1000t}\right)$$

$$= \frac{1000^2 ce^{-1000t}}{\left(1 + ce^{-1000t}\right)^2}$$

$$= \frac{1000}{1 + ce^{-1000t}}\left[\frac{1000ce^{-1000t}}{1 + ce^{-1000t}}\right]$$

$$= \frac{1000}{1 + ce^{-1000t}}\left[\frac{1000 + 1000ce^{-1000t} - 1000}{1 + ce^{-1000t}}\right]$$

$$= \frac{1000}{1 + ce^{-1000t}}\left[\frac{1000\left(1 + ce^{-1000t}\right)}{1 + ce^{-1000t}} - \frac{1000}{1 + ce^{-1000t}}\right]$$

$$= P[1000 - P].$$

(No mark allocation in exam)

1 A game consists of randomly selecting 4 balls from a bag. After each ball is selected it is replaced in the bag. The bag contains 3 red balls and 7 green balls.

For each red ball selected, 10 points are earned and for each green ball selected, 5 points are deducted. For instance, if a player picks 3 red balls and 1 green ball, the score will be $3 \times 10 - 1 \times 5 = 25$ points.

What is the expected score in the game?

(2 marks)

(Q12e, **2022 HSC**) Medium

2 The random variable X represents the number of successes in 10 independent Bernoulli trials. The probability of success is $p = 0.9$ in each trial.
Let $r = P(X \geq 1)$.
Which of the following describes the value of r?

A $r > 0.9$ **B** $r = 0.9$
C $0.1 < r < 0.9$ **D** $r \leq 0.1$ *(1 mark)*

(Q6, **2021 HSC**) Medium

3 When a particular biased coin is tossed, the probability of obtaining a head is $\frac{3}{5}$.

This coin is tossed 100 times.

Let X be the random variable representing the number of heads obtained. This random variable will have a binomial distribution.

i Find the expected value, $E(X)$. *(1 mark)* Easy

ii By finding the variance, $\text{Var}(X)$, show that the standard deviation of X is approximately 5. *(1 mark)* Easy

iii By using a normal approximation, find the approximate probability that X is between 55 and 65. *(1 mark)* Easy

(Q12b, **2020 HSC**)

4 In a binomial distribution, X is the random variable representing the number of successes in n trials and p is the probability of success.

If the mean is 30 and the variance is 22.5, find the values of n and p. *(2 marks)*

Bonus question (see page iv) Easy

5 A binomial random variable X has a probability of success of p after 5 trials.

Use the table for the binomial distribution to find the value of p, correct to 2 decimal places.

x	0	1	2	3	4	5
$P(X = x)$	0.00525	0.04877	0.18115	0.33642	0.31239	0.11603

A $p = 0.56$ **B** $p = 0.61$
C $p = 0.63$ **D** $p = 0.65$ *(1 mark)*

Bonus question Easy

6 An examination consists of 50 multiple choice questions, each question having four possible answers.

A student guesses the answer to every question.

Let X be the number of correct answers.

Find the value of $\text{Var}(X)$. *(2 marks)*

Bonus question Medium

7 When a standard 6-sided dice is thrown, the probability that it shows a multiple of 3 is $\frac{1}{3}$.
If 30 standard dice are thrown, the number of times, X, a multiple of 3 is showing has a binomial distribution.

What is the standard deviation of X, correct to 2 decimal places?

A 2.58 **B** 3.26
C 3.81 **D** 6.67 *(1 mark)*

Bonus question Easy

8 A binomial random variable X has a probability of success of p after 4 trials.

Use the table for the binomial distribution to find the value of p.

x	0	1	2	3	4
$P(X = x)$	0.1296	0.3456	0.3456	0.1536	0.0256

(1 mark)

Bonus question Easy

9 A binomial random variable X has a probability of success of p after 3 trials.

i Use the table for the binomial distribution to find the value of p. *(1 mark)* Easy

x	0	1	2	3
$P(X = x)$	0.064	a	b	0.216

Two values are missing from the table.

ii Show that $b = 0.432$. *(1 mark)* Easy

iii Find $P(X \geq 1)$ *(1 mark)* Easy

iv Find the variance, $\text{Var}(X)$. *(1 mark)* Easy

Bonus question

10 A binomial random variable X has a mean of 15 and standard deviation of $\sqrt{6}$.

Find the parameters n and p and hence find $P(X = 10)$, correct to 4 decimal places. *(3 marks)*

Bonus question Medium

11 Suppose $X \sim \text{Bin}(n, p)$, where the mean is 20 and the standard deviation is 3.

Which of these is the probability of success p?

A 0.35 **B** 0.45

C 0.55 **D** 0.65 *(1 mark)*

Bonus question Easy

12 A binomial random variable X has a probability of success of p after 4 trials.

Use the table for the binomial distribution to find the value of p, correct to 2 decimal places.

x	0	1	2	3	4
$P(X = x)$	0.03419	0.18128	0.36044	0.31853	0.10556

A $p = 0.51$ **B** $p = 0.53$

C $p = 0.57$ **D** $p = 0.61$ *(1 mark)*

Bonus question Easy

13 An examination consists of 20 multiple-choice questions, each question having four possible answers.

Marie guesses the answer to each question.

Let X be the number of correct answers.

Which of these is the value of $\text{Var}(X)$?

A $\frac{5}{4}$ **B** $\frac{15}{4}$

C 5 **D** $\frac{25}{4}$ *(1 mark)*

Bonus question Easy

14 In a binomial distribution, X is the random variable representing the number of successes in n trials and p is the probability of success.

If the mean is 60 and the standard deviation is $2\sqrt{5}$, find the values of n and p. *(2 marks)*

Bonus question Easy

15 From past experience, Julio knows that only 70% of his daffodil bulbs will flower.

This year he has planted 30 bulbs.

Find, correct to 4 decimal places, the probability that:

i exactly half of his bulbs will flower *(1 mark)* Easy

ii more than 95% of his bulbs will flower. *(1 mark)* Easy

Bonus question

16 The probability of winning a prize with a lucky ticket on a Spin-the-Wheel is 0.15.

What is the smallest number of tickets that should be purchased to ensure a probability of more than 80% of winning at least once? *(3 marks)*

Bonus question Medium

17 A machine produces a computer component. It has been found that 95% of the components are satisfactory, but the remainder defective. From a sample of 20 components, what is the probability that at least 3 will be defective? Give your answer correct to 4 decimal places. *(2 marks)*

Bonus question Medium

18 Prize-winning symbols are printed on 5% of ice-cream sticks. The ice-creams are randomly packed into boxes of 8.

i What is the probability that a box contains no prize-winning symbols? *(1 mark)* Easy

ii What is the probability that a box contains at least 2 prize-winning symbols? *(2 marks)* Easy

(Q11f, **2019 HSC**)

19 A group of 12 people sets off on a trek. The probability that a person finishes the trek within 8 hours is 0.75.

Find an expression for the probability that at least 10 people from the group complete the trek within 8 hours. *(2 marks)*

(Q12d, **2018 HSC**) Medium

20 The probability that a particular type of seedling produces red flowers is $\frac{1}{5}$.

Eight of these seedlings are planted.

i Write an expression for the probability that exactly three of the eight seedlings produce red flowers. *(1 mark)* Easy

ii Write an expression for the probability that none of the eight seedlings produces red flowers. *(1 mark)* Easy

iii Write an expression for the probability that at least one of the eight seedlings produces red flowers. *(1 mark)* Easy

(Q11g, **2017 HSC**)

21 A darts player calculates that when she aims for the bullseye the probability of her hitting the bullseye is $\frac{3}{5}$ with each throw.

- **i** Find the probability that she hits the bullseye with exactly one of her first three throws. *(1 mark)* Easy
- **ii** Find the probability that she hits the bullseye with at least two of her first six throws. *(2 marks)* Medium

(Q11f, **2016 HSC**)

22 The probability that it rains on any particular day during the 30 days of November is 0.1.

Write an expression for the probability that it rains on fewer than 3 days in November. *(2 marks)*

(Q11b, **2014 HSC**) Medium

23 An examination has 10 multiple-choice questions, each with 4 options. In each question, only one option is correct. For each question a student chooses one option at random.

Write an expression for the probability that the student chooses the correct option for exactly 7 questions. *(2 marks)*

(Q11c, **2013 HSC**) Easy

24 Kim and Mel play a simple game using a spinner marked with the numbers 1, 2, 3, 4 and 5.

The game consists of each player spinning the spinner once. Each of the five numbers is equally likely to occur.

The player who obtains the higher number wins the game.

If both players obtain the same number, the result is a draw.

- **i** Kim and Mel play one game. What is the probability that Kim wins the game? *(1 mark)* Hard
- **ii** Kim and Mel play six games. What is the probability that Kim wins exactly three games? *(2 marks)* Medium

(Q12c, **2012 HSC**)

25 A game is played by throwing darts at a target. A player can choose to throw two or three darts.

Darcy plays two games. In Game 1, he chooses to throw two darts, and wins if he hits the target at least once. In Game 2, he chooses to throw three darts, and wins if he hits the target at least twice.

The probability that Darcy hits the target on any throw is p, where $0 < p < 1$.

- **i** Show that the probability that Darcy wins Game 1 is $2p - p^2$. *(1 mark)* Medium
- **ii** Show that the probability that Darcy wins Game 2 is $3p^2 - 2p^3$. *(1 mark)* Medium
- **iii** Prove that Darcy is more likely to win Game 1 than Game 2. *(2 marks)* Hard
- **iv** Find the value of p for which Darcy is twice as likely to win Game 1 as he is to win Game 2. *(2 marks)* Hard

(Q6c, **2011 HSC**)

26 Five ordinary six-sided dice are thrown.

What is the probability that exactly two of the dice land showing a four? Leave your answer in unsimplified form. *(1 mark)*

(Q1f, **2010 HSC**) Medium

27 A test consists of five multiple-choice questions. Each question has four alternative answers. For each question only one of the alternative answers is correct.

Huong randomly selects an answer to each of the five questions.

- **i** What is the probability that Huong selects three correct and two incorrect answers? *(2 marks)* Medium
- **ii** What is the probability that Huong selects three or more correct answers? *(2 marks)* Medium
- **iii** What is the probability that Huong selects at least one incorrect answer? *(1 mark)* Easy

(Q4a, **2009 HSC**)

28 In a large city, 10% of the population has green eyes.

- **i** What is the probability that two randomly chosen people both have green eyes? *(1 mark)* Easy
- **ii** What is the probability that exactly two of a group of 20 randomly chosen people have green eyes? Give your answer correct to three decimal places. *(1 mark)* Easy

iii What is the probability that more than two of a group of 20 randomly chosen people have green eyes? Give your answer correct to two decimal places. *(2 marks)* **Medium**

(Q4a, **2007 HSC**)

29 In an endurance event, the probability that a competitor will complete the course is p and the probability that a competitor will not complete the course is $q = 1 - p$. Teams consist of either two or four competitors. A team scores points if at least half its members complete the course.

i Show that the probability that a four-member team will have at least three of its members not complete the course is $4pq^3 + q^4$. *(1 mark)* **Medium**

ii Hence, or otherwise, find an expression in terms of q only for the probability that a four-member team will score points. *(2 marks)* **Hard**

iii Find an expression in terms of q only for the probability that a two-member team will score points. *(1 mark)* **Hard**

iv Hence, or otherwise, find the range of values of q for which a two-member team is more likely than a four-member team to score points. *(2 marks)* **Hard**

(Q6b, **2006 HSC**)

30 There are five matches on each weekend of a football season. Megan takes part in a competition in which she earns one point if she picks more than half of the winning teams for a weekend, and zero points otherwise. The probability that Megan correctly picks the team that wins any given match is $\frac{2}{3}$.

i Show that the probability that Megan earns one point for a given weekend is 0.7901, correct to four decimal places. *(2 marks)* **Medium**

ii Hence find the probability that Megan earns one point every week of the eighteen-week season. Give your answer correct to two decimal places. *(1 mark)* **Easy**

iii Find the probability that Megan earns at most 16 points during the eighteen-week season. Give your answer correct to two decimal places. *(2 marks)* **Medium**

(Q6a, **2005 HSC**)

31 Katie is one of ten members of a social club. Each week one member is selected at random to win a prize.

i What is the probability that in the first 7 weeks Katie will win at least 1 prize? *(1 mark)* **Easy**

ii Show that in the first 20 weeks Katie has a greater chance of winning exactly 2 prizes than of winning exactly 1 prize. *(2 marks)* **Medium**

iii For how many weeks must Katie participate in the prize drawing so that she has a greater chance of winning exactly 3 prizes than of winning exactly 2 prizes? *(2 marks)* **Hard**

(Q4c, **2004 HSC**)

32 i Explain why the probability of getting a sum of 5 when one pair of fair dice is tossed is $\frac{1}{9}$. *(1 mark)* **Easy**

ii Find the probability of getting a sum of 5 at least twice when a pair of dice is tossed 7 times. *(2 marks)* **Medium**

(Q3c, **2003 HSC**)

33 Lyndal hits the target on average 2 out of every 3 shots in archery competitions. During a competition she has 10 shots at the target.

i What is the probability that Lyndal hits the target exactly 9 times? Leave your answer in unsimplified form. *(1 mark)* **Easy**

ii What is the probability that Lyndal hits the target fewer than 9 times? Leave your answer in unsimplified form. *(2 marks)* **Easy**

(Q4a, **2002 HSC**)

34 A fair six-sided die is randomly tossed n times.

i Suppose $0 \le r \le n$. What is the probability that exactly r 'sixes' appear in the uppermost position? *(2 marks)* **Medium**

ii By using the result of part (b)*, or otherwise, show that the probability that an odd number of 'sixes' appears is

$$\frac{1}{2}\left\{1 - \left(\frac{2}{3}\right)^n\right\}.$$ *(2 marks)* **Medium**

$$*[(q+p)^n - (q-p)^n = 2\binom{n}{1}(q^{n-1}p + 2\binom{n}{3}q^{n-3}p^3 + \ldots]$$

(Q5c, **2001 HSC**)

35 A fair, six-sided die is thrown seven times. What is the probability that a '6' occurs on exactly 2 of the 7 throws? *(2 marks)*

(Q3b, **1999 HSC**) **Easy**

36 A tank contains 10 tagged fish and 50 untagged fish. On each day, 4 fish are selected at random from the tank and placed together in a separate tank for observation.

Later the same day the 4 fish are returned to the original tank.

- **i** What is the probability of selecting no tagged fish on a given day? Easy
- **ii** What is the probability of selecting at least one tagged fish on a given day? Easy
- **iii** Calculate the probability of selecting no tagged fish on every day for 7 given days. Easy
- **iv** What is the probability of selecting no tagged fish on exactly 2 of the 7 days? Medium

(5 marks) (Q5c, **1998 HSC**)

37 In each game of Sic Bo, three regular, six-sided dice are thrown once.

- **i** In a single game, what is the probability that all three dice show 2? Easy
- **ii** What is the probability that exactly two of the dice show 2? Easy
- **iii** What is the probability that exactly two of the dice show the same number? Easy
- **iv** A player claims that you expect to see three different numbers on the dice in at least half of the games. Is the player correct? Justify your answer. Medium

(5 marks) (Q3c, **1997 HSC**)

38 In a Jackpot Lottery, 1500 numbers are drawn from a barrel containing the 100 000 ticket numbers available.

After all the 1500 prize-winning numbers are drawn, they are returned to the barrel and a jackpot number is drawn. If the jackpot number is the same as one of the 1500 numbers that have already been selected, then the additional jackpot prize is won.

The probability that the jackpot prize is won in a given game is thus

$$p = \frac{1500}{100\,000} = 0{\cdot}015.$$

- **i** Calculate the probability that the jackpot prize will be won *exactly* once in 10 independent lottery games. Easy
- **ii** Calculate the probability that the jackpot prize will be won *at least* once in 10 independent lottery games. Easy
- **iii** The jackpot prize is initially \$8000, and it increases by \$8000 each time the prize is NOT won.
  Calculate the probability that the jackpot prize will exceed \$200 000 when it is finally won. Medium

(6 marks) (Q5b, **1995 HSC**)

39 New cars are subjected to a quality check, which 75% pass. Calculate the probability that of the next ten cars checked, more than seven will pass. Leave your answer in unsimplified form. Easy

(Q3a, **1994 HSC**)

40 The probability that any one of the thirty-one days in December is rainy is 0·2. What is the probability that December has exactly ten rainy days? Leave your answer in index form. Easy

(Q1d, **1992 HSC**)

Year 12 Bernoulli and binomial distributions—Worked Answers

1 Expected score from one draw:

$E(X_1) = \frac{3}{10} \times 10 + \frac{7}{10} \times -5 = -\frac{1}{2}$ ✓

Since events are independent, then the expected score from four draws:

$E(X) = -\frac{1}{2} \times 4$

$= -2$ ✓

(2 marks)

2 If $p = 0.9$, then $q = 0.1$.

$r = P(X \geq 1)$

$= 1 - P(X = 0)$

$= 1 - {}^{10}C_0 \times 0.1^{10}$

$= 0.999...$

$\therefore r > 0.9$

Answer A

3 **i** $P(\text{head}) = \frac{3}{5}$

So $p = \frac{3}{5}$

The coin is tossed 100 times so $n = 100$.

Binomial distribution.

$E(X) = np$

$= 100 \times \frac{3}{5}$

$= 60$ ✓

(1 mark)

ii $\text{Var}(X) = np(1-p)$

$= 60 \times \left(1 - \frac{3}{5}\right)$

$= 24$

Standard deviation

$= \sqrt{24}$

$= 4.898\,979\,48...$

So the standard deviation is approximately 5. ✓

(1 mark)

iii A score of 55 is approximately one standard deviation below the mean and a score of 65 is approximately one standard deviation above the mean.

So the approximate probability that X is between 55 and 65 is 68%. ✓

(1 mark)

4

$\mu = np$

$30 = np$

$\sigma^2 = np(1-p)$

$22.5 = 30(1-p)$

$1 - p = \frac{22.5}{30}$

$1 - p = 0.75$

$p = 0.25$ ✓

Hence, $30 = n(0.25)$

$n = 120$ ✓

$\therefore n = 120, p = 0.25$

(2 marks)

5 $P(X = 5) = \binom{5}{5} p^5(1-p)^0$

$p^5 = 0.116\,03$

$p = 0.650\,001\,05...$

$= 0.65$ (2 dec. pl.)

Answer D

6 The number of correct answers, X, is a binomial variable with $n = 50$ and $p = 0.25$.

$\mu = np$

$= 50(0.25)$

$= 12.5$ ✓

$\text{Var}(X) = np(1-p)$

$= 12.5(1 - 0.25)$

$= 9.375$ ✓

(2 marks)

7 $E(X) = np$

$= 30 \times \frac{1}{3}$

$= 10$

$\text{Var}(X) = np(1-p)$

$= 10 \times \frac{2}{3}$

$= \frac{20}{3}$

Stand. dev. $= \sqrt{\frac{20}{3}}$

$= 2.581\,988\,97...$

$= 2.58$ (2 dec. pl.)

Answer A

8 $\binom{4}{4} p^4(1-p)^0 = 0.0256$

$p^4 = 0.0256$

$p = 0.4$ ✓

(1 mark)

9 **i** $P(X = 3) =$

$\binom{3}{3} p^3(1-p)^0 = 0.216$

$p^3 = 0.216$

$p = 0.6$ ✓

(1 mark)

ii $P(X = 2) = \binom{3}{2} 0.6^2(1 - 0.6)$

$= 0.432$ ✓

(1 mark)

iii $P(X > 1) =$

$P(X = 2) + P(X = 3)$

$= 0.432 + 0.216$

$= 0.648$ ✓ *(1 mark)*

iv $E(X) = \mu = np$

$= 3 \times 0.6$

$= 1.8$

$\text{Var}(X) = np(1-p)$

$= 1.8(1 - 0.6)$

$= 0.72$ ✓ *(1 mark)*

10

$\mu = np$

$15 = np$

$\sigma^2 = np(1-p)$

$6 = 15(1-p)$

$1 - p = 0.4$

$p = 0.6$ ✓

Hence, $15 = n(0.6)$

$n = 25$ ✓

$\therefore n = 25, p = 0.6$

$\therefore P(X = 10) = \binom{25}{10}(0.6)^{10}(1 - 0.6)^{15}$

$= 0.021\,222\,444...$

$= 0.0212$ (4 dec. pl.) ✓

(3 marks)

11 $E(X) = np = 20$

$\sigma = \sqrt{\text{Var}(X)}$

$\therefore \sqrt{np(1-p)} = 3$

$\sqrt{20(1-p)} = 3$

$20(1-p) = 9$

$1 - p = 0.45$

$p = 0.55$

Answer C

12 $P(X = 4) = \binom{4}{4} p^4(1-p)^0$

$p^4 = 0.105\,56$

$p = 0.569\,999\,986...$

$= 0.57$ (2 dec. pl.)

Answer C

13 $E(X) = \mu = np$

$= 20 \times \frac{1}{4}$

$= 5$

$\text{Var}(X) = np(1-p)$

$= 5\left(1 - \frac{1}{4}\right)$

$= 5\left(\frac{3}{4}\right)$

$= \frac{15}{4}$

Answer B

14

$E(x) = np$

$60 = np$

$\text{Var}(x) = np(1-p)$

$(2\sqrt{5})^2 = 60(1-p)$

$20 = 60(1-p)$

$1 - p = \frac{1}{3}$

$p = \frac{2}{3}$ ✓

Hence, $60 = n\left(\frac{2}{3}\right)$

$n = 90$ ✓

$\therefore n = 90, p = \frac{2}{3}$

(2 marks)

15 Consider X~Bin(30, 0.7).

i As half of 30 is 15,

$P(X = 15)$

$= {}^{30}C_{15}\, 0.7^{15}(1 - 0.7)^{15}$

$= 0.010\,566\,965...$

$= 0.0106$ (4 dec. pl.) ✓

(1 mark)

ii As 95% of 30 is 28.5,
$P(X = 29) + P(X = 30)$
$= {}^{30}C_{29}\, 0.7^{29}(1 - 0.7) + {}^{30}C_{30}\, 0.7^{30}$
$= 0.000\,312\,33\ldots$
$= 0.0003$ (4 dec. pl.) ✓
(1 mark)

16 Let $p = 0.15$, n = number of tickets.

$$\begin{aligned} P(X \geq 1) &> 0.8 \\ 1 - P(X = 0) &> 0.8 \\ P(X = 0) &< 0.2 \\ {}^{n}C_0(0.15)^0(0.85)^n &< 0.2 \quad ✓ \\ 0.85^n &< 0.2 \\ \log_e 0.85^n &< \log_e 0.2 \\ n\log_e 0.85 &< \log_e 0.2 \\ n &> \frac{\log_e 0.2}{\log_e 0.85} \quad ✓ \end{aligned}$$

$n > 9.903\,079\,705\ldots$
(as $\log_e 0.85 < 0$)
$\therefore$ there needs to be at least 10 tickets purchased. ✓
(3 marks)

17 Using $X \sim \text{Bin}(20, 0.05)$:
$P(X \geq 3)$
$= 1 - [P(X = 0) + P(X = 1) + P(X = 2)]$
$= 1 - [{}^{20}C_0(0.05)^0(1 - 0.05)^{20} + {}^{20}C_1(0.05)^1(1 - 0.05)^{19} + {}^{20}C_2(0.05)^2(1 - 0.05)^{18}]$ ✓
$= 0.075\,483\,673\ldots$
$= 0.0755$ (4 dec. pl.) ✓
(2 marks)

18 **i** P(prize on stick) $= 0.05$
$\therefore P$(no prize on stick) $= 0.95$
P(no prize on 8 sticks)
$= 0.95^8$
$= 0.663\,420\,431\ldots$
So the probability that a box has no prize-winning sticks is about 66%. ✓
(1 mark)

ii P(1 prize stick in box)
$= 8 \times 0.95^7 \times 0.05$
P(at least 2 prize-winning sticks in a box)
$= 1 - P$(0 prize or 1 prize)
$= 1 - (0.95^8 + 0.4(0.95)^7)$ ✓
$= 0.057\,244\,6502\ldots$
$= 5.7\%$ (1 d.p.)
The probability that a box holds at least 2 prize-winning symbols is about 5.7%. ✓
(2 marks)

19 P(finishes in 8 h) $= 0.75$
$\therefore P$(does not finish in 8 h) $= 0.25$
P(at least 10 finish in 8 h)
$= P$(10 finish) $+ P$(11 finish) $+ P$(12 finish)

$$= \binom{12}{10}(0.75)^{10}(0.25)^2 + \binom{12}{11}(0.75)^{11}(0.25) + \binom{12}{12}(0.75)^{12} \quad ✓$$

$$= 66\left(\frac{3}{4}\right)^{10}\left(\frac{1}{4}\right)^2 + 12\left(\frac{3}{4}\right)^{11}\left(\frac{1}{4}\right) + \left(\frac{3}{4}\right)^{12}$$

$$= \frac{66 \times 3^{10} + 12 \times 3^{11} + 3^{12}}{4^{12}}$$

$$= \frac{3^{11}(22 + 12 + 3)}{4^{12}}$$

$$= \frac{37 \times 3^{11}}{4^{12}} \quad ✓$$

(2 marks)

20 $P(\text{red}) = \dfrac{1}{5}$, $\quad P(\text{not red}) = \dfrac{4}{5}$

i P(exactly 3 red in 8)
$$= \binom{8}{3}\left(\frac{1}{5}\right)^3\left(\frac{4}{5}\right)^5 \quad ✓$$
(1 mark)

ii $P(\text{no red flowers in } 8) = \left(\dfrac{4}{5}\right)^8$ ✓
(1 mark)

iii P(at least 1 red in 8)
$$= 1 - \left(\frac{4}{5}\right)^8 \quad ✓$$
(1 mark)

21 Assuming she aims for the bullseye

i $P(\text{bullseye}) = \dfrac{3}{5}$,
$P(\text{no bullseye}) = \dfrac{2}{5}$

P(1 bullseye in 3 throws)
$$= 3 \times \frac{3}{5} \times \left(\frac{2}{5}\right)^2$$
$$= \frac{36}{125} \quad ✓$$
(1 mark)

ii P(at least 2 bullseyes in 6 throws)
$= 1 - P$(0 bullseyes) $- P$(1 bullseye)
$$= 1 - \left(\frac{2}{5}\right)^6 - 6\left(\frac{2}{5}\right)^5\left(\frac{3}{5}\right) \quad ✓$$
$$= \frac{2997}{3125} \quad ✓$$
(2 marks)

22 P(rain on any day) $= 0.1$
$\therefore P$(no rain on any day) $= 0.9$
P(rain on fewer than 3 days)
$= P$(rain 0 days) $+ P$(rain 1 day) $+ P$(rain 2 days) ✓
$$= \binom{30}{0}(0.9)^{30} + \binom{30}{1}(0.1)(0.9)^{29} + \binom{30}{2}(0.1)^2(0.9)^{28} \quad ✓$$
(2 marks)

23 $p = P(\text{correct answer}) = \dfrac{1}{4}$

$q = P(\text{incorrect answer}) = \dfrac{3}{4}$ ✓

P(exactly 7 correct in 10)
$$= \binom{10}{7}\left(\frac{1}{4}\right)^7\left(\frac{3}{4}\right)^3 \quad ✓$$
(2 marks)

24 **i** $P(\text{draw}) = \dfrac{1}{5}$

[P(Second player spins the same number as the first player did)]

$P(\text{not a draw}) = \dfrac{4}{5}$

$$P(\text{Kim wins}) = \frac{1}{2} \times \frac{4}{5} = \frac{2}{5} \quad ✓$$
(1 mark)

ii $P(\text{Kim does not win}) = \dfrac{3}{5}$

P(Kim wins exactly 3 games out of 6)
$$= {}^{6}C_3\left(\frac{2}{5}\right)^3\left(\frac{3}{5}\right)^3 \quad ✓$$
$$= \frac{864}{3125} \quad ✓$$
(2 marks)

25 **i** $P(\text{hit}) = p$
$P(\text{miss}) = 1 - p$
In Game 1,
P(at least 1 hit)
$= 1 - P$(miss both times)
$= 1 - (1 - p)^2$
$= 1 - (1 - 2p + p^2)$
$= 1 - 1 + 2p - p^2$
$= 2p - p^2$ ✓
(1 mark)

ii In Game 2,
P(at least 2 hits)
$= P$(2 hits) $+ P$(3 hits)
$= 3p^2(1 - p) + p^3$
$= 3p^2 - 3p^3 + p^3$
$= 3p^2 - 2p^3$ ✓
(1 mark)

iii Assume Darcy is not more likely to win Game 1 than Game 2
i.e. $\quad 3p^2 - 2p^3 \geq 2p - p^2$
$4p^2 - 2p^3 - 2p \geq 0$
$2p(2p - p^2 - 1) \geq 0$
$2p - p^2 - 1 \geq 0 \quad (p > 0)$
$p^2 - 2p + 1 \leq 0$ ✓
$(p - 1)^2 \leq 0$
But $(p - 1)^2 > 0 \quad (0 < p < 1)$

$\therefore$ Darcy is more likely to win Game 1 than Game 2. ✓ *(2 marks)*

iv If Darcy is twice as likely to win Game 1:

$$2p - p^2 = 2(3p^2 - 2p^3)$$
$$= 6p^2 - 4p^3$$
$$4p^3 - 7p^2 + 2p = 0$$
$$4p^2 - 7p + 2 = 0 \quad (p \neq 0) ✓$$
$$p = \frac{-(-7) \pm \sqrt{(-7)^2 - 4 \times 4 \times 2}}{2 \times 4}$$
$$= \frac{7 \pm \sqrt{17}}{8}$$

But $0 < p < 1$

$$\therefore p = \frac{7 - \sqrt{17}}{8}$$
$$= 0.359\,611\,796...$$
$$= 0.36 \quad [2 \text{ d.p.}] ✓$$ *(2 marks)*

26 $P(\text{four}) = \frac{1}{6}$

$P(\text{not a four}) = \frac{5}{6}$

$P(\text{exactly 2 fours in 5 throws})$

$= {}^5C_2\left(\frac{1}{6}\right)^2\left(\frac{5}{6}\right)^3$ ✓

(1 mark)

27 i $P(\text{Correct answer}) = p = \frac{1}{4}$

$P(\text{Incorrect answer}) = q = \frac{3}{4}$

P(3 correct, 2 incorrect)

$= {}^5C_3\left(\frac{3}{4}\right)^2\left(\frac{1}{4}\right)^3$ ✓

$= 10 \times \frac{9}{16} \times \frac{1}{64}$

$= \frac{45}{512}$. ✓

(2 marks)

ii P(3 or more correct)

= P(3 correct) + P(4 correct) + P(5 correct)

$= \frac{45}{512} + {}^5C_4\left(\frac{3}{4}\right)\left(\frac{1}{4}\right)^4 + {}^5C_5\left(\frac{1}{4}\right)^5$ ✓

$= \frac{45}{512} + \frac{15}{1024} + \frac{1}{1024}$

$= \frac{106}{1024}$

$= \frac{53}{512}$. ✓

(2 marks)

iii P(At least one incorrect)

= 1 – P(all correct)

$= 1 - {}^5C_5\left(\frac{3}{4}\right)^0\left(\frac{1}{4}\right)^5$

$= 1 - \frac{1}{1024}$

$= \frac{1023}{1024}$. ✓

(1 mark)

28 i P(green, green) $= 0.1 \times 0.1$

$= 0.01$. ✓

(1 mark)

ii $P = {}^{20}C_2(0.9)^{18}(0.1)^2$

$= 190 \times 0.9^{18} \times 0.1^2$

$= 0.2851 \ldots$

$= 0.285$ to 3 decimal places. ✓

(1 mark)

iii P(more than 2 out of 20)

= 1 – P(0 of 20) – P(1 of 20) – P(2 of 20)

$= 1 - {}^{20}C_0(0.9)^{20}(0.1)^0 - {}^{20}C_1(0.9)^{19}(0.1) - {}^{20}C_2(0.9)^{18}(0.8)^2$ ✓

$= 1 - 0.1215\ldots - 0.2701\ldots - 0.2851\ldots$

$= 0.3230\ldots$

$= 0.32$ to 2 decimal places. ✓

(2 marks)

29 i $P(\text{at least 3 do not complete})$

$= P(\text{3 not complete}) + P(\text{4 not complete})$

$= {}^4C_1q^3p + q^4$

$= 4pq^3 + q^4$. ✓

(1 mark)

ii $P(\text{at least 2 complete the course})$

$= 1 - (4pq^3 + q^4)$ ✓

$= 1 - 4(1 - q)q^3 - q^4$

$= 1 - 4q^3 + 4q^4 - q^4$

$= 1 - 4q^3 + 3q^4$. ✓

(2 marks)

iii $P(\text{2-member team scores})$

$= P(\text{1 completes}) + P(\text{2 completes})$

$= {}^2C_1qp + p^2$

$= 2qp + p^2$

$= 2q(1 - q) + (1 - q)^2$

$= 2q - 2q^2 + 1 - 2q + q^2$

$= 1 - q^2$. ✓

(1 mark)

iv For a 2-member team to be more likely to score points:

$$1 - q^2 > 1 - 4q^3 + 3q^4$$
$$3q^4 - 4q^3 + q^2 < 0 \quad ✓$$
$$q^2(3q^2 - 4q + 1) < 0$$
$$q^2(3q - 1)(q - 1) < 0$$

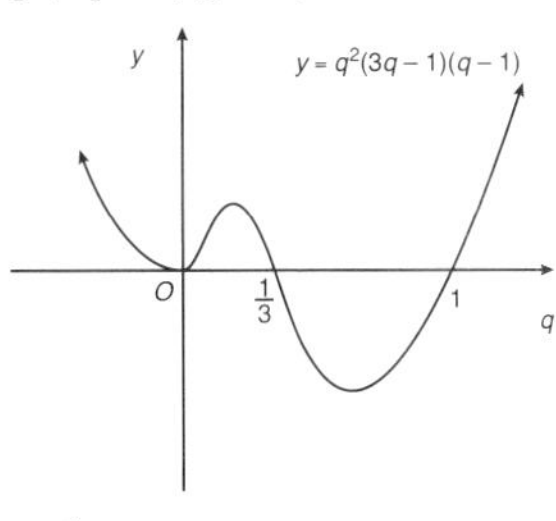

$\therefore \frac{1}{3} < q < 1$. ✓

(2 marks)

30 i $P(\text{correct pick}) = \frac{2}{3}$

$P(\text{incorrect pick}) = \frac{1}{3}$

One point will be earned if 3, 4 or 5 winning teams are picked.

$P(\text{3 correct picks})$

$= {}^5C_3\left(\frac{1}{3}\right)^2\left(\frac{2}{3}\right)^3$

$P(\text{4 correct picks})$

$= {}^5C_4\left(\frac{1}{3}\right)^1\left(\frac{2}{3}\right)^4$

$P(\text{5 correct picks})$

$= {}^5C_5\left(\frac{1}{3}\right)^0\left(\frac{2}{3}\right)^5$ ✓

$\therefore P(\text{Earning 1 point})$

$= {}^5C_3\left(\frac{1}{3}\right)^2\left(\frac{2}{3}\right)^3 + {}^5C_4\left(\frac{1}{3}\right)^1\left(\frac{2}{3}\right)^4 + {}^5C_5\left(\frac{1}{3}\right)^0\left(\frac{2}{3}\right)^5$

$= 0.790123\ldots$

$= 0.7901$ to 4 decimal places. ✓

(2 marks)

ii $P(\text{1 point for 18 weeks})$

$= (0.7901)^{18}$

$= 0.01439\ldots$

$= 0.01$ to 2 decimal places. ✓

(1 mark)

iii $P(\text{At most 16 points})$

$= 1 - P(\text{18 points}) - P(\text{17 points})$

$P(\text{18 points})$

$= (0.7901)^{18}$ (from **ii**)

$P(\text{17 points})$

$= {}^{18}C_{17}(1 - 0.7901)(0.7901)^{17}$

where $(1 - 0.7901)$ is the probability of NOT scoring 1 point over 18 weeks. ✓

$\therefore P(\text{At most 16 points})$

$= 1 - (0.7901)^{18} - {}^{18}C_{17}(1 - 0.7901)(0.7901)^{17}$

$= 1 - 0.01439\ldots - 0.06884\ldots$

$= 0.9167\ldots$

$= 0.92$ to 2 decimal places. ✓

(2 marks)

31 i $P(\text{Katie wins}) = \frac{1}{10}$

$P(\text{Katie loses}) = \frac{9}{10}$

$P(\text{Katie wins at least 1 prize})$

$= 1 - P(\text{no prize})$

$= 1 - \left(\frac{9}{10}\right)^7$

$= 0.5217\ldots$

$\doteqdot 0.52$ ✓

(1 mark)

ii $P(\text{Katie wins exactly 2 prizes})$

$= {}^{20}C_2\left(\frac{9}{10}\right)^{18}\left(\frac{1}{10}\right)^2$

$= \frac{190 \times 9^{18}}{10^{20}}$

$= 0.2851\ldots$

$\doteqdot 0.285$ ✓

$P(\text{Katie wins exactly 1 prize})$

$= {}^{20}C_1\left(\frac{9}{10}\right)^{19}\left(\frac{1}{10}\right)$

$= \frac{20 \times 9^{19}}{10^{20}}$

$= 0.2701\ldots$

$\doteqdot 0.27$

$\therefore$ Katie has a greater chance of winning exactly 2 prizes. ✓

(2 marks)

iii $P(\text{exactly 3 prizes}) > P(\text{exactly 2 prizes})$

$$\therefore {}^nC_3\left(\frac{9}{10}\right)^{n-3}\left(\frac{1}{10}\right)^3 > {}^nC_2\left(\frac{9}{10}\right)^{n-2}\left(\frac{1}{10}\right)^2$$

$$\therefore \frac{n!}{(n-3)!3!}\left(\frac{9}{10}\right)^{n-3}\left(\frac{1}{10}\right)^3 > \frac{n!}{(n-2)!2!}\left(\frac{9}{10}\right)^{n-2}\left(\frac{1}{10}\right)^2 \checkmark$$

$$\therefore \frac{(n-2)!2!}{(n-3)!3!} > \left(\frac{10^3}{10^2}\right)\left(\frac{10^{n-3}}{10^{n-2}}\right)\left(\frac{9^{n-2}}{9^{n-3}}\right)$$

$$\therefore \frac{(n-2)}{3} > 9$$

$$\therefore n-2 > 27$$

$$\therefore n > 29$$

$\therefore$ Katie must participate for 30 weeks. ✓

(2 marks)

32 i There are $6^2 = 36$ possible results of tossing a pair of dice.
A sum of 5 can be obtained in 4 ways, i.e. (1, 4), (2, 3), (3, 2) and (4, 1).
Hence the probability of a sum of 5 is $\frac{4}{36} = \frac{1}{9}$. ✓ *(1 mark)*

ii The probability of not getting a sum of 5 in a toss

$= 1 - \frac{1}{9}$

$= \frac{8}{9}$

The probability of getting a sum of 5 exactly 0 times in 7 tosses is

${}^7C_0\left(\frac{8}{9}\right)^7\left(\frac{1}{9}\right)^0 = \left(\frac{8}{9}\right)^7 = \frac{8^7}{9^7}$

The probability of getting a sum of 5 exactly 1 time in 7 tosses is

${}^7C_1\left(\frac{8}{9}\right)^6\left(\frac{1}{9}\right)^1 = 7 \times \frac{8^6}{9^6} \times \frac{1}{9}$

$= 7 \times \frac{8^6}{9^7}$ ✓

$\therefore$ The probability of getting a sum of 5 at most 1 time is

$\frac{8^7}{9^7} + 7 \times \frac{8^6}{9^7} = \frac{8^6}{9^7}(8+7)$

$= \frac{8^6}{9^7} \times 15$

$\therefore$ The probability of getting a sum of 5 at least twice is

$1 - \left(\frac{8^6}{9^7} \times 15\right) \doteqdot 0.178.$ ✓

(2 marks)

33 i $p = P(\text{hit}) = \frac{2}{3}$.

$q = P(\text{miss}) = \frac{1}{3}$.

The probability of 9 hits in 10 trials

$= {}^{10}C_9\left(\frac{1}{3}\right)^1\left(\frac{2}{3}\right)^9$

$= 10 \times \frac{2^9}{3^{10}}$. ✓

(1 mark)

ii The probability of 10 hits in 10 trials

$= {}^{10}C_{10}\left(\frac{2}{3}\right)^{10} = \left(\frac{2}{3}\right)^{10}$

The probability of 9 or 10 hits

$= 10 \times \frac{2^9}{3^{10}} + \left(\frac{2}{3}\right)^{10}$ ✓

$= \frac{2^9}{3^{10}}(10+2)$

$\therefore$ The probability of fewer than 9 hits

$= 1 - \left(12 \times \frac{2^9}{3^{10}}\right)$ ✓

(2 marks)

34 i $p = P(\text{1 six in 1 toss}) = \frac{1}{6}$

$q = P(\text{no six in 1 toss}) = \frac{5}{6}$

The probability of exactly r 'sixes' in n trials is

$\binom{n}{r} q^{n-r}p^r = \binom{n}{r}\left(\frac{5}{6}\right)^{n-r}\left(\frac{1}{6}\right)^r$ ✓

$= \binom{n}{r}\frac{5^{n-r}}{6^n}$ ✓

(2 marks)

ii The probability of an odd number of 'sixes' is therefore:

$P = \binom{n}{1}q^{n-1}p + \binom{n}{3}q^{n-3}p^3 + \ldots$

From (b):

$2\binom{n}{1}q^{n-1}p + 2\binom{n}{3}q^{n-3}p^3 + \ldots$

$= (q+p)^n - (q-p)^n$

$\therefore \binom{n}{1}q^{n-1}p + \binom{n}{3}q^{n-3}p^3 + \ldots$

$= \frac{1}{2}[(q+p)^n - (q-p)^n]$ ✓

$\therefore P = \frac{1}{2}[(q+p)^n - (q-p)^n]$

$= \frac{1}{2}\left[\left(\frac{5}{6}+\frac{1}{6}\right)^n - \left(\frac{5}{6}-\frac{1}{6}\right)^n\right]$

$= \frac{1}{2}\left(1 - \left(\frac{2}{3}\right)^n\right).$ ✓

(2 marks)

35 On each throw

$P(6) = \frac{1}{6}$

$P(\text{Not } 6) = \frac{5}{6}$

$\therefore$ P('6' on 2 of 7 throws)

$= {}^7C_2\left(\frac{5}{6}\right)^5\left(\frac{1}{6}\right)^2$ ✓

$= 21 \times \frac{5^5}{6^5} \times \frac{1}{6^2}$

$= 21 \times \frac{3125}{279\,936}$

$= 0.234$ to 3 sig. figures. ✓ *(2 marks)*

36 **i** $P(4 \text{ untagged})$

$= \frac{50}{60} \times \frac{49}{59} \times \frac{48}{58} \times \frac{47}{57}$

$= \frac{46\,060}{97\,527}$

≈ 0.47 ✓

ii $P(\text{At least 1 tagged})$

$= 1 - \frac{46\,060}{97\,527}$

$= \frac{51\,467}{97\,527}$

≈ 0.53 ✓

iii $P(\text{all untagged for 7 days})$

$= \left(\frac{46\,060}{97\,527}\right)^7$

$\approx 0.47^7$

$\approx 0.005.$ ✓

iv $P(\text{Untagged on 2 days})$

$= {}^7C_2\left(\frac{51\,467}{97\,527}\right)^5\left(\frac{46\,060}{97\,527}\right)^2$ ✓

$\approx {}^7C_2\, 0.53^5 \times 0.47^2$

$\approx 0.19.$ ✓ *(5 marks)*

(Total mark allocation only in exam)

37 **i** $P(2, 2, 2) = \left(\frac{1}{6}\right)^3$

$= \frac{1}{6^3}\ \left(= \frac{1}{216}\right)$ ✓

ii $P(\text{Exactly two 2's})$

$= {}^3C_2\left(\frac{5}{6}\right)\left(\frac{1}{6}\right)^2$

$= 3 \times \frac{5}{6^3}$

$= \frac{15}{6^3}$

$= \frac{15}{216}$

$= \frac{5}{72}.$ ✓

iii $P(\text{Exactly 2 same}) = 6 \times \frac{5}{72}$

$= \frac{5}{12}$ ✓

(This is the probability of exactly two 2s or two 1s or two 3s etc.)

iv $P(3 \text{ different nos.})$

$= 1 - [P(\text{All nos. same}) + P(\text{Exactly 2 same})]$

$= 1 - \left(6 \times \frac{1}{6^3} + \frac{5}{12}\right)$ ✓

$= 1 - \left(\frac{1}{36} + \frac{5}{12}\right)$

$= 1 - \left(\frac{1}{36} + \frac{15}{36}\right)$

$= 1 - \frac{16}{36}$

$= \frac{20}{36}$

$= \frac{5}{9} > \frac{1}{2}$

$\therefore$ The player is correct. ✓

(5 marks)

(Total mark allocation only in exam)

38 **i** $p = 0.015$

$q = 1 - 0.015$

$= 0.985$ ✓

Probability of 1 jackpot prize

$= {}^{10}C_1 q^9 p^1$

$= 10 \times (0.985)^9 \times 0.015$

$\doteqdot 0.1309.$ ✓

ii $P(\text{At least 1 jackpot})$

$= 1 - P(\text{No jackpots})$

$= 1 - q^{10}$ ✓

$= 1 - 0.985^{10}$

$\doteqdot 0.1403.$ ✓

iii $\frac{200\,000}{8000} = 25$

The jackpot is \$8000 if won at the first game.

It is \$200 000 if won at the 25th game for the first time.

The jackpot is more than \$200 000 if it is not won in any of the first 25 games. ✓

$\therefore$ Required probability

$= q^{25}$

$= 0.985^{25}$

$\doteqdot 0.6853.$ ✓

(6 marks)

(Total mark allocation only in exam)

39 $p = \frac{75}{100} = \frac{3}{4}$

$q = \frac{25}{100} = \frac{1}{4}$

Required probability

$= {}^{10}C_8 q^2 p^8 + {}^{10}C_9 q^1 p^9 + {}^{10}C_{10} p^{10}$

$= \binom{10}{8}\left(\frac{1}{4}\right)^2\left(\frac{3}{4}\right)^8 + \binom{10}{9}\left(\frac{1}{4}\right)\left(\frac{3}{4}\right)^9 + \left(\frac{3}{4}\right)^{10}$

$= \frac{3^8}{4^{10}}\left[\binom{10}{8} + 3\binom{10}{9} + 3^2\right]$

$= \frac{3^8}{4^{10}}[45 + 3(10) + 9]$

$= \frac{3^8}{4^{10}} \times 84$

$= \frac{3^8}{4^9} \times 21$

$= \frac{3^9}{4^9} \times 7$

$= 7 \times \left(\frac{3}{4}\right)^9.$

(No mark allocation in exam)

40 $p = 0.2$

$q = 1 - p = 0.8$

$P(\text{exactly 10 rainy days})$

$= {}^{31}C_{10}\, q^{21} p^{10}$

$= {}^{31}C_{10}(0.8)^{21}(0.2)^{10}.$

(No mark allocation in exam)

1 *You may use the information on page 18 to answer this question.* [provided after the questions in this chapter]

A chocolate factory sells 150-gram chocolate bars. There has been a complaint that the bars actually weigh less than 150 grams, so a team of inspectors was sent to the factory to check. They randomly selected 16 bars, weighed them and noted that 8 bars weighed less than 150 grams.

The factory manager claims 80% of the chocolate bars produced by the factory weigh 150 grams or more.

i The inspectors used the normal approximation to the binomial distribution to calculate the probability, $\mathcal{P}$, of having at least 8 bars weighing less than 150 grams in a random sample of 16, assuming the factory manager's claim is correct. Calculate the value of $\mathcal{P}$. *(2 marks)* **Hard**

ii The factory manager disagrees with the method used by the inspectors as described in part **i**.

Explain why the method used by the inspectors might not be valid.

(1 mark) **Easy**

(Q13e, **2022 HSC**)

2 *You may use the information on page 18 to answer this question.* [provided after the questions in this chapter]

An airline company that has empty seats on a flight is not maximising its profit.

An airline company has found that there is a probability of 5% that a passenger books a flight but misses it. The management of the airline company decides to allow for overbooking, which means selling more tickets than the number of seats available on each flight.

To protect their reputation, management makes the decision that no more than 1% of their flights should have more passengers showing up for the flight than available seats.

Given management's decision and using a suitable approximation, find the maximum number of tickets that can be sold for a flight which has 350 seats. *(4 marks)*

(Q14d, **2022 HSC**) **Hard**

3 At a certain factory, the proportion of faulty items produced by a machine is $p = \frac{3}{500}$, which is considered to be acceptable. To confirm that the machine is working to this standard, a sample of size n is taken and the sample proportion $\hat{p}$ is calculated.

It is assumed that $\hat{p}$ is approximately normally distributed with $\mu = p$ and $\sigma^2 = \frac{p(1-p)}{n}$.

Production by this machine will be shut down if $\hat{p} \geq \frac{4}{500}$. The sample size is to be chosen so that the chance of shutting down the machine unnecessarily is less than 2.5%.

Find the approximate sample size required, giving your answer to the nearest thousand.

(3 marks)

(Q14d, **2021 HSC**) **Hard**

4 Using airline passenger cards to determine reasons for international travel in 2012, the ABC determined that 23% of Australian residents travelled overseas to visit friends and family.

100 samples of 50 people were taken and the proportion of people travelling to visit friends and family were recorded.

a Describe the type of distribution that $\hat{p}$ will follow. *(1 mark)* **Easy**

b Find the expected value and the standard deviation of this distribution.

(2 marks) **Easy**

Bonus question (see page iv)

5 A company does quality control checks on its products. From a random sample it found that 12 were defective, with a sample proportion of 0.015.

a Find the sample size. *(1 mark)* **Easy**

b Hence find the variance of the sample distribution (answer to three significant figures). *(1 mark)* **Easy**

Bonus question

6 A research organisation suggest men have a 40.1% chance of getting cancer. A sample size of 132 men was taken.

a How many men would you expect to have/had cancer in the sample taken.

(1 mark) **Easy**

b Find the sample proportion, $\hat{p}$.

(1 mark) Easy

c Assuming the sample proportion is approximately normally distributed, what is the probability that the percentage of men in this sample who had cancer was:

i less than 30%? *(3 marks)* Medium

ii more than 45%?

(You may use the table provided after these questions.)

(2 marks) Medium

Bonus question

---

7 In a certain town, the proportion of people who support the local team is given by p, where $p > 50\%$. A random sample of 12 people were asked if they supported the local team and the standard deviation of their responses was 0.125. Find the value of p.

(3 marks)

Bonus question Medium

---

8 In 2015, the proportion of Australians aged 14 and over eating chocolate at least once in an average four weeks was 69%.*
42 people were sampled.

a How many people would you expect aged 14 and over to have eaten chocolate in an average four weeks in the sample taken.

(1 mark) 

b Find the sample proportion, $\hat{p}$.

(1 mark) 

c Assuming the sample proportion is approximately normally distributed, what is the probability that the percentage of people in this sample who have eaten chocolate in an average four weeks was:

i less than 50%? *(3 marks)* Medium

ii more than 90%?

(You may use the table provided after these questions.)

(3 marks) 

Bonus question

(*Source: www.roymorgan.com/findings/more-aussies-chomping-on-chocolate-20160321233)

---

9 It is known that 84% of houses in a gated community pay their fees on time. A sample of 20 homes was taken and 18 say they have paid on time.

What is the probability of finding a sample proportion, $\hat{p}$, less than the one found by this sample? Answer to 3 decimal places.

You can assume the sample is approximately normally distributed.
(You may use the table provided after these questions.) *(4 marks)* Hard

Bonus question

---

10 The population in a particular city is known to be 50% female.

What is the probability that a random sample of 100 people will contain less than 45% female? *(2 marks)*

Bonus question Medium

---

11 In a large population the probability that a person is left-handed is 0.1.

Let $\hat{p}$ be the random variable that represents the sample proportion of left-handed people for sample of size n drawn from the population.

What is the smallest value of n such that the standard deviation of $\hat{p}$ is less than or equal to 0.04? Select the correct option.

A $n = 50$ **B** $n = 52$

C $n = 54$ **D** $n = 57$ *(1 mark)*

Bonus question Medium

---

12 It is known that 60% of gym members continue their membership after 6 months if they attend on more than two times each week.

What is the probability that in a random sample of 600 gym members more than 64% of those surveyed will continue their membership after 6 months? *(2 marks)*

Bonus question Medium

Table of values $P(Z \leq z)$ for the normal distribution $N(0, 1)$

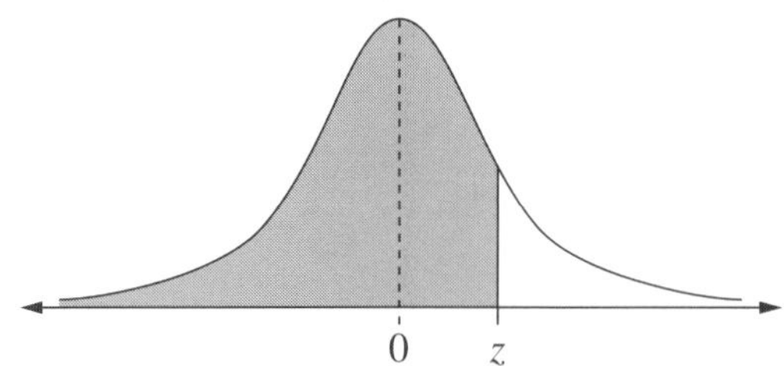

Z	0.00	0.01	0.02	0.03	0.04	0.05	0.06	0.07	0.08	0.09
0.0	0.5000	0.5040	0.5080	0.5120	0.5160	0.5199	0.5239	0.5279	0.5319	0.5359
0.1	0.5398	0.5438	0.5478	0.5517	0.5557	0.5596	0.5636	0.5675	0.5714	0.5753
0.2	0.5793	0.5832	0.5871	0.5910	0.5948	0.5987	0.6026	0.6064	0.6103	0.6141
0.3	0.6179	0.6217	0.6255	0.6293	0.6331	0.6368	0.6406	0.6443	0.6480	0.6517
0.4	0.6554	0.6591	0.6628	0.6664	0.6700	0.6736	0.6772	0.6808	0.6844	0.6879
0.5	0.6915	0.6950	0.6985	0.7019	0.7054	0.7088	0.7123	0.7157	0.7190	0.7224
0.6	0.7257	0.7291	0.7324	0.7357	0.7389	0.7422	0.7454	0.7486	0.7517	0.7549
0.7	0.7580	0.7611	0.7642	0.7673	0.7704	0.7734	0.7764	0.7794	0.7823	0.7852
0.8	0.7881	0.7910	0.7939	0.7967	0.7995	0.8023	0.8051	0.8078	0.8106	0.8133
0.9	0.8159	0.8186	0.8212	0.8238	0.8264	0.8289	0.8315	0.8340	0.8365	0.8389
1.0	0.8413	0.8438	0.8461	0.8485	0.8508	0.8531	0.8554	0.8577	0.8599	0.8621
1.1	0.8643	0.8665	0.8686	0.8708	0.8729	0.8749	0.8770	0.8790	0.8810	0.8830
1.2	0.8849	0.8869	0.8888	0.8907	0.8925	0.8944	0.8962	0.8980	0.8997	0.9015
1.3	0.9032	0.9049	0.9066	0.9082	0.9099	0.9115	0.9131	0.9147	0.9162	0.9177
1.4	0.9192	0.9207	0.9222	0.9236	0.9251	0.9265	0.9279	0.9292	0.9306	0.9319
1.5	0.9332	0.9345	0.9357	0.9370	0.9382	0.9394	0.9406	0.9418	0.9429	0.9441
1.6	0.9452	0.9463	0.9474	0.9484	0.9495	0.9505	0.9515	0.9525	0.9535	0.9545
1.7	0.9554	0.9564	0.9573	0.9582	0.9591	0.9599	0.9608	0.9616	0.9625	0.9633
1.8	0.9641	0.9649	0.9656	0.9664	0.9671	0.9678	0.9686	0.9693	0.9699	0.9706
1.9	0.9713	0.9719	0.9726	0.9732	0.9738	0.9744	0.9750	0.9756	0.9761	0.9767
2.0	0.9772	0.9778	0.9783	0.9788	0.9793	0.9798	0.9803	0.9808	0.9812	0.9817
2.1	0.9821	0.9826	0.9830	0.9834	0.9838	0.9842	0.9846	0.9850	0.9854	0.9857
2.2	0.9861	0.9864	0.9868	0.9871	0.9875	0.9878	0.9881	0.9884	0.9887	0.9890
2.3	0.9893	0.9896	0.9898	0.9901	0.9904	0.9906	0.9909	0.9911	0.9913	0.9916
2.4	0.9918	0.9920	0.9922	0.9925	0.9927	0.9929	0.9931	0.9932	0.9934	0.9936
2.5	0.9938	0.9940	0.9941	0.9943	0.9945	0.9946	0.9948	0.9949	0.9951	0.9952
2.6	0.9953	0.9955	0.9956	0.9957	0.9959	0.9960	0.9961	0.9962	0.9963	0.9964
2.7	0.9965	0.9966	0.9967	0.9968	0.9969	0.9970	0.9971	0.9972	0.9973	0.9974
2.8	0.9974	0.9975	0.9976	0.9977	0.9977	0.9978	0.9979	0.9979	0.9980	0.9981
2.9	0.9981	0.9982	0.9982	0.9983	0.9984	0.9984	0.9985	0.9985	0.9986	0.9986
3.0	0.9987	0.9987	0.9987	0.9988	0.9988	0.9989	0.9989	0.9989	0.9990	0.9990
3.1	0.9990	0.9991	0.9991	0.9991	0.9992	0.9992	0.9992	0.9992	0.9993	0.9993
3.2	0.9993	0.9993	0.9994	0.9994	0.9994	0.9994	0.9994	0.9995	0.9995	0.9995

Year 12 Normal approximation for the sample proportion—Worked Answers

1 **i** $n = 16, p = 0.2$

$$\mu = np = 16 \times 0.2 = 3.2$$

$$\sigma = \sqrt{np(1-p)} = \sqrt{3.2(1-0.2)} = 1.6$$

$$P(x \geq 8) = P\left(Z \geq \frac{8-3.2}{1.6}\right) \checkmark$$

$$= P(Z \geq 3) = 1 - P(Z \leq 3) = 1 - 0.9987 = 0.0013 \checkmark$$

(2 marks)

ii The method may not be valid as the sample size is insufficient. For approximation to be accurate $np \geq 10$.

$np = 3.2$
< 10

$\therefore$ Approximation is not considered to be valid. ✓

(1 mark)

2 $\mu = np$
$= 0.95n$ where p is Pr(not missing a flight)

$\sigma^2 = npq$
$= 0.05 \times 0.95n$
$= 0.0475n$ ✓

$$\Pr(X > 350) = \Pr\left(z > \frac{350 - 0.95n}{\sqrt{0.0475n}}\right) \leq 0.01$$

$$\Pr\left(z \leq \frac{350 - 0.95n}{\sqrt{0.0475n}}\right) \geq 0.99$$

From the z-score table provided:

$\Pr(z \leq 2.33) \geq 0.99$ ✓

$$\therefore \frac{350 - 0.95n}{\sqrt{0.0475n}} = 2.33$$

$$350 - 0.95n = 2.33\sqrt{0.0475n}$$

$$(350 - 0.95n)^2 = 2.33^2 \times 0.0475n$$

$$122\,500 - 665n + 0.9025n^2 = 0.25\,787\ldots n$$

$$0.9025n^2 - 665.258n + 122\,500 = 0 \checkmark$$

$$n = \frac{665.258 \pm \sqrt{(-665.258)^2 - 4 \times 0.9025 \times 122\,500}}{2 \times 0.9025}$$

$n = 358$ or 379

But since $\frac{350 - 0.95n}{\sqrt{0.0475n}}$, only $n = 358$ can be a possible solution.

$\therefore$ The maximum number of tickets that can be sold is 358. ✓

(4 marks)

3 $p = \frac{3}{500}$

$$\sigma^2 = \frac{p(1-p)}{n}$$

$$\sigma^2 = \frac{\frac{3}{500}\left(1 - \frac{3}{500}\right)}{n} = \frac{1491}{500^2 n} \checkmark$$

$$P\left(\hat{p} \geq \frac{4}{500}\right) = P\left(Z \geq \frac{\frac{4}{500} - \mu}{\sigma}\right)$$

$$= P\left(Z \geq \frac{\frac{4}{500} - \frac{3}{500}}{\frac{1}{500}\sqrt{\frac{1491}{n}}}\right)$$

$$= P\left(Z \geq \frac{\sqrt{n}}{\sqrt{1491}}\right) \checkmark$$

Hence we require $\frac{\sqrt{n}}{\sqrt{1491}} \geq 2$ (using the Empirical rule).

$$\sqrt{n} \geq 2\sqrt{1491}$$

$$n \geq 4 \times 1491$$

$$n \geq 5964$$

$\therefore$ The sample size required is 6000, to the nearest thousand. ✓

(3 marks)

4 **a** $n = 50$ and $p = 0.23$

$np = 50 \times 0.23 = 11.5$

$\therefore np \geq 10$

$n(1-p) = 50 \times 0.77 = 38.5$

$\therefore n(1-p) \geq 10$

Therefore, we can assume that $\hat{p}$ follows a normal distribution. ✓

(1 mark)

b $E(\hat{p}) = p = 0.23$

$$\sigma = \sqrt{Var(\hat{p})} = \sqrt{\frac{p(1-p)}{n}} \checkmark$$

$$= \sqrt{\frac{0.23(1-0.23)}{50}} = \sqrt{0.003542} = 0.060 \checkmark$$

(3 decimal places)

Note: some HSC resources refer to standard deviation as standard error.

(2 marks)

5 **a** Using $\hat{p} = \frac{x}{n}$, $\hat{p} = 0.015$ and $x = 12$

$$0.015 = \frac{12}{n}$$

$\therefore n = 800$

So the sample size was 800 products. ✓

(1 mark)

b $$\text{Var}(\hat{p}) = \frac{0.015(1-0.015)}{800} = 0.000\,0185$$

or 1.85×10^{-5} ✓

(1 mark)

6 **a** $40.1\% \times 132 = 52.932$

$\therefore$ 53 men would be expected to have/had cancer in the sample taken. ✓

(1 mark)

b $\hat{p} = \dfrac{53}{132}$ ✓ *(1 mark)*

c **i** $\mu = p$
$= 0.401$
$\sigma = \sqrt{Var(\hat{p})}$
$= \sqrt{\dfrac{0.401(1-0.401)}{132}}$
$\cong 0.04266$ ✓

Using $x = 30\%$ *or* 0.3

$z-\text{score} = \dfrac{x-\mu}{\sigma}$
$= \dfrac{0.3-0.401}{0.04266}$
$\cong -2.37$ ✓

$\therefore P(z-\text{score} < -2.37)$
$= 0.0089$

So the probability that less than 30% of the men had cancer 0.9%. ✓ *(3 marks)*

ii Using $x = 45\%$ *or* 0.45

$z-\text{score} = \dfrac{0.45-0.401}{0.04266}$
$\cong 1.15$ ✓

$\therefore P(z-\text{score} > 1.15)$
$= 1 - P(z-\text{score} < 1.15)$
$= 1 - 0.8749$
$= 0.1251$

So the probability that more than 45% of the men had cancer is 12.5%. ✓ *(2 marks)*

7 $n = 12$ and $\sigma = 0.125$

$\dfrac{1}{8} = \sqrt{\dfrac{p(1-p)}{12}}$ ✓

$\dfrac{1}{64} = \dfrac{p-p^2}{12}$

$16p - 16p^2 = 3$

$16p^2 - 16p + 3 = 0$ ✓

$(4p-1)(4p-3) = 0$

$\therefore p = \dfrac{1}{4}$ or $\dfrac{3}{4}$

Since $p > 50\%$, then $p = \dfrac{3}{4}$. ✓ *(3 marks)*

8 **a** $69\% \times 42 = 28.98$

$\therefore$ 29 Australians aged 14 and over would be expected to have eaten chocolate in an average four weeks in the sample taken. ✓ *(1 mark)*

b $\hat{p} = \dfrac{29}{42}$ *(1 mark)*

c **i** $\mu = p$
$= 0.69$
$\sigma = \sqrt{Var(\hat{p})}$
$= \sqrt{\dfrac{0.69(1-0.69)}{42}}$
$\cong 0.7136$ ✓

Using $x = 50\%$ or 0.5

$z-\text{score} = \dfrac{x-\mu}{\sigma}$
$= \dfrac{0.5-0.69}{0.07136}$
$\cong -2.66$ ✓

$\therefore P(z-\text{score} < -2.66)$
$= 0.0039$

So the probability that less than 50% of the sample had eaten chocolate in an average four weeks is 0.39%. ✓ *(3 marks)*

ii Using $x = 90\%$ *or* 0.9

$z-\text{score} = \dfrac{0.9-0.69}{0.07136}$
$\cong 2.94$ ✓

$\therefore P(z-\text{score} > 2.94)$
$= 1 - P(z-\text{score} < 2.94)$ ✓
$= 1 - 0.9984$
$= 0.0016$

So the probability that more than 90% of the sample had eaten chocolate in an average four weeks is 0.16%. ✓ *(3 marks)*

9 Using $p = 0.84$, $n = 20$ and $x = 18$.

$\hat{p} = \dfrac{18}{20}$
$= 0.9$ ✓

$\sigma = \sqrt{\dfrac{0.84(1-0.84)}{20}}$
$\cong 0.082$ ✓

Using $x = \hat{p}$ as comparison value,

$z-\text{score} = \dfrac{x-\mu}{\sigma}$
$= \dfrac{0.9-0.84}{0.082}$
$\cong 0.73$ ✓

$\therefore P(z-\text{score} < 0.73) = 0.7673$

So the probability of finding a sample proportion less than the one found using this sample is 76.73%. ✓ *(4 marks)*

10 Let X denote the number of females in the sample.

X has a binomial distribution with $n = 100$ and $p = 0.5$

The distribution of the sample proportion $\hat{p}$ is approximately normal with $E(\hat{p}) = 0.5$

$Var(\hat{p}) = \dfrac{p(1-p)}{n}$
$= \dfrac{0.5(1-0.5)}{100}$
$= 0.0025$
$\sigma = \sqrt{0.025}$
$= 0.05$ ✓

$P(\hat{p} < 0.45) = P\left(z < \dfrac{0.45-0.5}{0.05}\right)$
$= P(z < -1)$
$= \dfrac{100\% - 68\%}{2}$
$= 16\%$ ✓ *(2 marks)*

11 Standard deviation $= \sqrt{\dfrac{p(1-p)}{n}}$

$\therefore \sqrt{\dfrac{0.1(1-0.1)}{n}} \le 0.04$

$\dfrac{0.1(1-0.1)}{n} \le 0.0016$

$0.0016n \ge 0.09 \quad (n > 0)$

$n \ge \dfrac{0.09}{0.0016}$

$n \ge 56.25$

$\therefore$ the smallest value is 57.

Answer D

12 Let X denote the number of members who continue their membership in the sample.

X has a binomial distribution with $n = 600$ and $p = 0.6$

The distribution of the sample proportion $\hat{p}$ is approximately normal with $E(\hat{p}) = 0.6$

$Var(\hat{p}) = \dfrac{p(1-p)}{n}$
$= \dfrac{0.6(1-0.6)}{600}$
$= 0.0004$
$\sigma = \sqrt{0.0004}$
$= 0.02$ ✓

$P(\hat{p} > 0.64) = P\left(z > \dfrac{0.64-0.6}{0.02}\right)$
$= P(z > 2)$
$= \dfrac{100\% - 95\%}{2}$
$= 2.5\%$ ✓ *(2 marks)*

NSW Education Standards Authority

2022 HIGHER SCHOOL CERTIFICATE EXAMINATION

Mathematics Advanced
Mathematics Extension 1
Mathematics Extension 2

REFERENCE SHEET

Measurement

Length

$$l = \frac{\theta}{360} \times 2\pi r$$

Area

$$A = \frac{\theta}{360} \times \pi r^2$$

$$A = \frac{h}{2}(a+b)$$

Surface area

$$A = 2\pi r^2 + 2\pi rh$$

$$A = 4\pi r^2$$

Volume

$$V = \frac{1}{3}Ah$$

$$V = \frac{4}{3}\pi r^3$$

Functions

$$x = \frac{-b \pm \sqrt{b^2 - 4ac}}{2a}$$

For $ax^3 + bx^2 + cx + d = 0$:

$$\alpha + \beta + \gamma = -\frac{b}{a}$$

$$\alpha\beta + \alpha\gamma + \beta\gamma = \frac{c}{a}$$

and $\alpha\beta\gamma = -\frac{d}{a}$

Relations

$$(x-h)^2 + (y-k)^2 = r^2$$

Financial Mathematics

$$A = P(1+r)^n$$

Sequences and series

$$T_n = a + (n-1)d$$

$$S_n = \frac{n}{2}[2a + (n-1)d] = \frac{n}{2}(a+l)$$

$$T_n = ar^{n-1}$$

$$S_n = \frac{a(1-r^n)}{1-r} = \frac{a(r^n-1)}{r-1}, \; r \neq 1$$

$$S = \frac{a}{1-r}, \; |r| < 1$$

Logarithmic and Exponential Functions

$$\log_a a^x = x = a^{\log_a x}$$

$$\log_a x = \frac{\log_b x}{\log_b a}$$

$$a^x = e^{x \ln a}$$

Trigonometric Functions

$$\sin A = \frac{\text{opp}}{\text{hyp}}, \quad \cos A = \frac{\text{adj}}{\text{hyp}}, \quad \tan A = \frac{\text{opp}}{\text{adj}}$$

$$A = \frac{1}{2}ab\sin C$$

$$\frac{a}{\sin A} = \frac{b}{\sin B} = \frac{c}{\sin C}$$

$$c^2 = a^2 + b^2 - 2ab\cos C$$

$$\cos C = \frac{a^2 + b^2 - c^2}{2ab}$$

$$l = r\theta$$

$$A = \frac{1}{2}r^2\theta$$

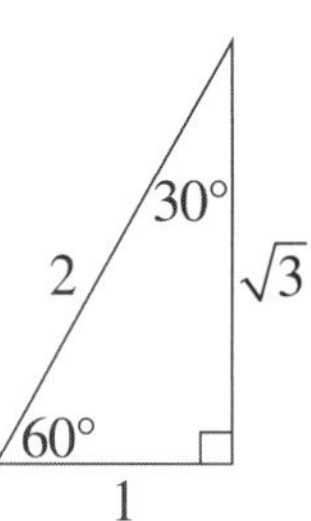

Trigonometric identities

$$\sec A = \frac{1}{\cos A}, \cos A \neq 0$$

$$\operatorname{cosec} A = \frac{1}{\sin A}, \sin A \neq 0$$

$$\cot A = \frac{\cos A}{\sin A}, \sin A \neq 0$$

$$\cos^2 x + \sin^2 x = 1$$

Compound angles

$$\sin(A + B) = \sin A\cos B + \cos A\sin B$$

$$\cos(A + B) = \cos A\cos B - \sin A\sin B$$

$$\tan(A + B) = \frac{\tan A + \tan B}{1 - \tan A\tan B}$$

If $t = \tan\frac{A}{2}$ then $\sin A = \frac{2t}{1+t^2}$

$$\cos A = \frac{1-t^2}{1+t^2}$$

$$\tan A = \frac{2t}{1-t^2}$$

$$\cos A\cos B = \frac{1}{2}\left[\cos(A - B) + \cos(A + B)\right]$$

$$\sin A\sin B = \frac{1}{2}\left[\cos(A - B) - \cos(A + B)\right]$$

$$\sin A\cos B = \frac{1}{2}\left[\sin(A + B) + \sin(A - B)\right]$$

$$\cos A\sin B = \frac{1}{2}\left[\sin(A + B) - \sin(A - B)\right]$$

$$\sin^2 nx = \frac{1}{2}(1 - \cos 2nx)$$

$$\cos^2 nx = \frac{1}{2}(1 + \cos 2nx)$$

Statistical Analysis

$$z = \frac{x - \mu}{\sigma}$$

An outlier is a score
less than $Q_1 - 1.5 \times IQR$
or
more than $Q_3 + 1.5 \times IQR$

Normal distribution

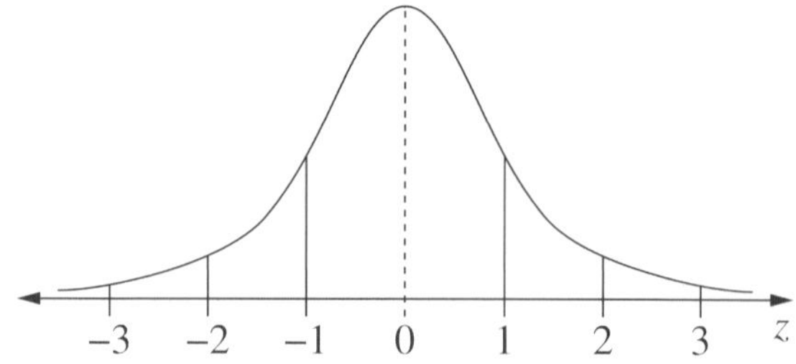

- approximately 68% of scores have z-scores between –1 and 1
- approximately 95% of scores have z-scores between –2 and 2
- approximately 99.7% of scores have z-scores between –3 and 3

$$E(X) = \mu$$

$$\text{Var}(X) = E\left[(X - \mu)^2\right] = E\left(X^2\right) - \mu^2$$

Probability

$$P(A \cap B) = P(A)P(B)$$

$$P(A \cup B) = P(A) + P(B) - P(A \cap B)$$

$$P(A|B) = \frac{P(A \cap B)}{P(B)}, P(B) \neq 0$$

Continuous random variables

$$P(X \le r) = \int_a^r f(x)\,dx$$

$$P(a < X < b) = \int_a^b f(x)\,dx$$

Binomial distribution

$$P(X = r) = {}^nC_r p^r(1 - p)^{n-r}$$

$$X \sim \text{Bin}(n, p)$$

$$\Rightarrow \quad P(X = x)$$

$$= \binom{n}{x}p^x(1 - p)^{n-x}, x = 0, 1, \ldots, n$$

$$E(X) = np$$

$$\text{Var}(X) = np(1 - p)$$

Differential Calculus

Function	Derivative
$y = f(x)^n$	$\frac{dy}{dx} = nf'(x)[f(x)]^{n-1}$
$y = uv$	$\frac{dy}{dx} = u\frac{dv}{dx} + v\frac{du}{dx}$
$y = g(u)$ where $u = f(x)$	$\frac{dy}{dx} = \frac{dy}{du} \times \frac{du}{dx}$
$y = \frac{u}{v}$	$\frac{dy}{dx} = \frac{v\frac{du}{dx} - u\frac{dv}{dx}}{v^2}$
$y = \sin f(x)$	$\frac{dy}{dx} = f'(x)\cos f(x)$
$y = \cos f(x)$	$\frac{dy}{dx} = -f'(x)\sin f(x)$
$y = \tan f(x)$	$\frac{dy}{dx} = f'(x)\sec^2 f(x)$
$y = e^{f(x)}$	$\frac{dy}{dx} = f'(x)e^{f(x)}$
$y = \ln f(x)$	$\frac{dy}{dx} = \frac{f'(x)}{f(x)}$
$y = a^{f(x)}$	$\frac{dy}{dx} = (\ln a)f'(x)a^{f(x)}$
$y = \log_a f(x)$	$\frac{dy}{dx} = \frac{f'(x)}{(\ln a)f(x)}$
$y = \sin^{-1} f(x)$	$\frac{dy}{dx} = \frac{f'(x)}{\sqrt{1-[f(x)]^2}}$
$y = \cos^{-1} f(x)$	$\frac{dy}{dx} = -\frac{f'(x)}{\sqrt{1-[f(x)]^2}}$
$y = \tan^{-1} f(x)$	$\frac{dy}{dx} = \frac{f'(x)}{1+[f(x)]^2}$

Integral Calculus

$$\int f'(x)[f(x)]^n dx = \frac{1}{n+1}[f(x)]^{n+1} + c$$

where $n \neq -1$

$$\int f'(x)\sin f(x)dx = -\cos f(x) + c$$

$$\int f'(x)\cos f(x)dx = \sin f(x) + c$$

$$\int f'(x)\sec^2 f(x)dx = \tan f(x) + c$$

$$\int f'(x)e^{f(x)}dx = e^{f(x)} + c$$

$$\int \frac{f'(x)}{f(x)}dx = \ln|f(x)| + c$$

$$\int f'(x)a^{f(x)}dx = \frac{a^{f(x)}}{\ln a} + c$$

$$\int \frac{f'(x)}{\sqrt{a^2-[f(x)]^2}}dx = \sin^{-1}\frac{f(x)}{a} + c$$

$$\int \frac{f'(x)}{a^2+[f(x)]^2}dx = \frac{1}{a}\tan^{-1}\frac{f(x)}{a} + c$$

$$\int u\frac{dv}{dx}dx = uv - \int v\frac{du}{dx}dx$$

$$\int_a^b f(x)dx$$

$$\approx \frac{b-a}{2n}\left\{f(a) + f(b) + 2\left[f(x_1) + \cdots + f(x_{n-1})\right]\right\}$$

where $a = x_0$ and $b = x_n$

Combinatorics

$$^nP_r = \frac{n!}{(n-r)!}$$

$$\binom{n}{r} = {}^nC_r = \frac{n!}{r!(n-r)!}$$

$$(x+a)^n = x^n + \binom{n}{1}x^{n-1}a + \cdots + \binom{n}{r}x^{n-r}a^r + \cdots + a^n$$

Vectors

$$|\underset{\sim}{u}| = |x\underset{\sim}{i} + y\underset{\sim}{j}| = \sqrt{x^2 + y^2}$$

$$\underset{\sim}{u} \cdot \underset{\sim}{v} = |\underset{\sim}{u}||\underset{\sim}{v}|\cos\theta = x_1x_2 + y_1y_2,$$

where $\underset{\sim}{u} = x_1\underset{\sim}{i} + y_1\underset{\sim}{j}$

and $\underset{\sim}{v} = x_2\underset{\sim}{i} + y_2\underset{\sim}{j}$

$$\underset{\sim}{r} = \underset{\sim}{a} + \lambda\underset{\sim}{b}$$

Complex Numbers

$$\begin{aligned} z = a + ib &= r(\cos\theta + i\sin\theta) \\ &= re^{i\theta} \end{aligned}$$

$$\begin{aligned} [r(\cos\theta + i\sin\theta)]^n &= r^n(\cos n\theta + i\sin n\theta) \\ &= r^n e^{in\theta} \end{aligned}$$

Mechanics

$$\frac{d^2x}{dt^2} = \frac{dv}{dt} = v\frac{dv}{dx} = \frac{d}{dx}\left(\frac{1}{2}v^2\right)$$

$$x = a\cos(nt + \alpha) + c$$

$$x = a\sin(nt + \alpha) + c$$

$$\ddot{x} = -n^2(x - c)$$

Notes

Notes